Marine Viruses 2016

Special Issue Editors

Mathias Middelboe
Corina P.D. Brussaard

MDPI • Basel • Beijing • Wuhan • Barcelona • Belgrade

MDPI

Special Issue Editors
Mathias Middelboe   Corina P.D. Brussaard
University of Copenhagen   University of Amsterdam
Denmark   The Netherlands

Editorial Office
MDPI AG
St. Alban-Anlage 66
Basel, Switzerland

This edition is a reprint of the Special Issue published online in the open access journal *Viruses* (ISSN 1999-4915) in 2017 (available at: http://www.mdpi.com/journal/viruses/special_issues/marine_viruses_2016).

For citation purposes, cite each article independently as indicated on the article page online and as indicated below:

Author 1; Author 2. Article title. *Journal Name* **Year**, *Article number*, page range.

First Edition 2017

ISBN 978-3-03842-620-2 (Pbk)
ISBN 978-3-03842-621-9 (PDF)

Table of Contents

About the Special Issue Editors

Mathias Middelboe is a professor in marine viral ecology at the Department of Biology, University of Copenhagen. He earned his PhD in aquatic microbial ecology from the University of Copenhagen in 1994 and established in 1997 a research group with a focus on the role of bacteriophages in marine environments (water column and sediments). He is especially interested in exploring how interactions between bacteriophages and bacteria drive phage and host diversity and evolution, and in resolving the role of bacteriophages in marine biogeochemical cycling. More recently, his research has included more applied aspects of phage–bacteria interactions, exploring the potential and challenges of phages to control pathogenic bacteria in aquaculture.

Corina P.D. Brussaard works in the Department of Marine Microbiology and Biogeochemistry at the NIOZ Royal Netherlands Institute for Sea Research, and is Professor of Viral Ecology at the Institute for Biodiversity and Ecosystem Dynamics, University of Amsterdam. Her research in microbial oceanography focuses largely on the ecological role of viruses in marine ecosystems. Her research has led to crucial developments in the isolation of novel algal viruses, virus enumeration and virus-induced mortality rate measurements. These observations helped to expand our understanding of the global importance of viruses as drivers of biodiversity and ecosystem productivity. Her latest research has revolved around the influence of environmental factors on microbial host–virus interactions, with particular emphasis on the effects of global climate change. Dr. Brussaard has taken many leadership roles in the community and is currently president-elect of the International Society for Viruses of Microorganisms (ISVM), and secretary of the Scientific Committee on Oceanic Research (SCOR).

Preface to "Marine Viruses 2016"

The exploration of marine viruses is a rapidly expanding research field, revealing increasingly complex interplays between viruses and their prokaryotic and eukaryotic hosts. The aim of this Special Issue is to highlight the progress in our understanding of the role of viruses in the marine environment by presenting novel research on the ecology, distribution and diversity of marine viruses and the influence of viruses on the mortality of fish, phytoplankton and bacteria, and the role of viruses for marine element cycling and evolution of microbial communities.

The compilation of papers included in this Special Issue is a significant contribution to the advancement of the field, emphasizing the importance of viruses in virtually every drop of water in the marine environment, covering the range from bacteriophages and prophages to large eukaryotic viruses and pathogenic viruses infecting multicellular animals.

Mathias Middelboe and Corina P.D. Brussaard

Special Issue Editors

viruses

MDPI

Editorial

Marine Viruses: Key Players in Marine Ecosystems

Mathias Middelboe [1,*] and Corina P. D. Brussaard [2]

[1]   Marine Biological Section, University of Copenhagen, DK-3000 Helsingør, Denmark
[2]   Department of Marine Microbiology and Biogeochemistry, NIOZ Royal Netherlands Institute of Sea
      Research, and University of Utrecht, P.O. Box 59, 1790 AB Den Burg, Texel, The Netherlands;
      corina.brussaard@nioz.nl
*   Correspondence: mmiddelboe@bio.ku.dk; Tel.: +45-3532-1991

Received: 12 October 2017; Accepted: 16 October 2017; Published: 18 October 2017

Viruses were recognized as the causative agents of fish diseases, such as infectious pancreatic necrosis and Oregon sockeye disease, in the early 1960s [1], and have since been shown to be responsible for diseases in all marine life from bacteria to protists, mollusks, crustaceans, fish and mammals [2]. However, it was not until the early 1990s that viral infections were discovered to affect marine systems beyond their role as pathogens of plants and animals, and viruses infecting unicellular organisms such as bacteria (i.e., the bacteriophages) and phytoplankton were shown to have a large influence on ecosystem processes. Since Karl-Heinz Moebus' pioneering work on bacteriophage isolation and infection patterns obtained during a transect across the North Atlantic [3,4], research in marine viruses has developed into a significant and independent research field in marine biology, prompted by the increasing realization of the important and diverse roles of viruses in the marine ecosystem (e.g., [5,6]). The discovery that viruses were the most abundant biological entities in oceanic marine environments [7], reaching up to 10^8 viruses mL^{-1}, further stimulated marine virus research. Technical improvements in detection and enumeration of marine viruses (e.g., [8]) promoted advances in more detailed studies of viral abundance and diversity at high spatial and temporal resolution. Later, the expansion of virus research to coral reefs [9], sediments [10,11], the deep biosphere [12], and freshwater environments [13] emphasized that viruses are integrated inhabitants of all aquatic environments. Consequently, the past decades' research has revealed viruses as key players in the marine ecosystem, from driving bacterial and algal mortality and evolution at the nanoscale, to influencing global-scale biogeochemical cycles and ocean productivity. The research has fundamentally changed our conceptual understanding of the function and regulation of aquatic ecosystems, and the development of molecular tools and DNA sequencing techniques has opened up for the exploration of viral diversity and the genetic mechanisms of virus-host interactions.

The present special issue aims at highlighting the progress in our understanding of the role of viruses and virus-host interactions in the marine environment by presenting novel research on the ecology, pathogenicity, distribution and diversity of marine viruses and the influence of virus-host interactions on mortality and evolution of marine microbial communities (Figure 1).

With the global increase in aquaculture, many of the viral pathogens have become severe causes of mortality in farmed organisms. Piscine orthoreovirus (PRV) is an example of a ubiquitous virus in sea water, which causes muscle inflammation in Atlantic salmon. In this special issue, Haatveit et al. [14] provide new insight on the infection kinetics of PRV, showing that the acute infection phase with high virus production is followed by reduced transcription of viral RNA, and the virus is maintained in the fish at a low persistent level. Pathogens of marine animals, however, constitute a very small fraction of marine viruses, as the majority of viruses infect bacteria and protists. Recent studies have shown that, even though viruses are typically 10-fold more abundant than bacteria in marine surface water, there are large variations in the virus-bacteria ratio across marine environments [15]. By examining the influence of environmental conditions on the relationship between viruses and bacteria using multivariate models, Finke et al. [16], here, demonstrate that environmental factors, such as inorganic nutrient concentrations, are important predictors of host and, consequently, viral abundance—and thus virus-host ratios—across a broad range of temporal and spatial scales. Similarly, trophic interactions in the microbial food web, such as predation and the availability of limiting nutrients, were shown to affect the structure and function of viral and prokaryote communities [17].

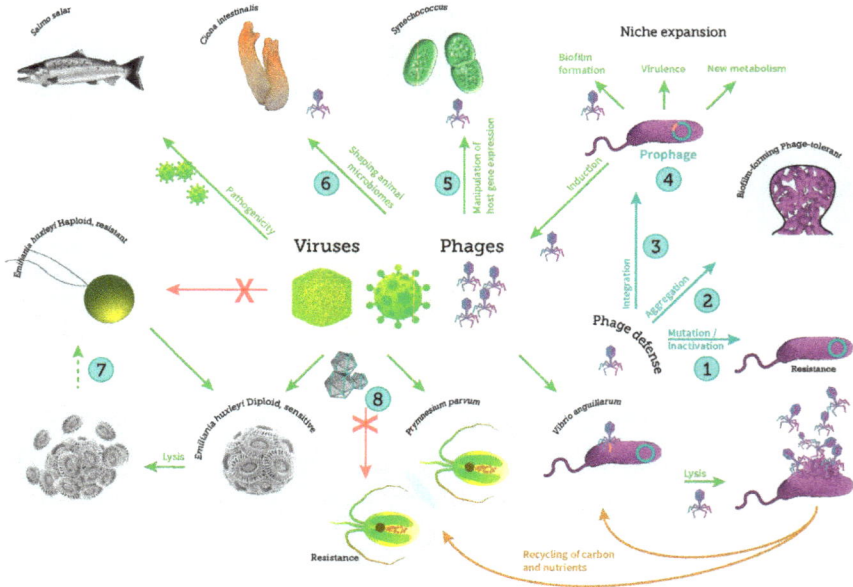

Figure 1. Schematic overview of important virus-host interactions in the marine ecosystem covered in this special issue, including viral infection of bacteria, phytoplankton, and fish. Specific explanations of key interactions: (1) Bacteria can prevent phage infection by mutational modification of surface receptors or by enzymatic degradation of the incoming phage DNA; (2) Alternatively, protection of cells in aggregates or biofilms can be a defense strategy against phage infection; (3) Infection by temperate phages can result in the integration of the phage DNA in the host genome as prophage; The integrated prophage can prevent infection by similar phages (Superinfection exclusion mechanism) and (4) contribute with important genetic information to the host that may expand its metabolic or virulence properties. Prophage induction leads to the release of new phages and may also stimulate biofilm formation; (5) Phages can manipulate host gene expression in cyanobacteria for improved infection efficiency, either by exploiting the host genes or by encoding host photosynthesis genes which are then expressed during infection; (6) Phage interaction with their bacterial hosts contributes to shaping the gut microbiome of invertebrates (e.g., tunicates), thus affecting the symbiotic relationship between gut microbes and their hosts; (7) In the coccolithophorid phytoplankton *Emiliania huxleyi* the diploid virally infected cells may undergo viral induced lysis or re-emerge (dotted arrow) as haploid cells containing viral RNA and lipids. These haploid cells are thought to resist virus infection (as indicated with the X) and develop into the diploid cells by karyogamy; (8) The large and diverse group of nucleocytoplasmic large DNA viruses (NCLDV) infects a range of photosynthetic protists such as the prasinophytes *Micromonas pusilla* and *Ostreococcus tauri* and, as exemplified in the figure, the toxin-producing haptophyte *Prymnesium parvum*, thus affecting mortality, diversity and production of phytoplankton. These interactions are strongly controlled by environmental factors such as temperature, nutrient availability and light. As for bacteria, several mechanisms of resistance (indicated with an X) to viruses have been described in the photosynthetic protists (see text for further details).

A three-year study on the virus-host dynamics of haptophyte phytoplankton and their dsDNA viruses showed seasonal fluctuations in specific virus populations indicating shifts in viral communities in response to seasonal variations in host diversity [18]. At the same time, the presence of persistent viral genotypes throughout the study period suggested co-existence between specific viruses and their hosts. Understanding of the environmental drivers of virus-host interaction is essential for predicting how environmental changes will affect virus-driven processes in a future ocean [19]. Highfield et al. [20] showed that elevated pCO_2 levels can affect the composition and diversity of *Emiliania huxleyi* viruses (EhV). Maat et al. [21] found highly specific temperature sensitivity in virus infectivity and production for four newly isolated viruses infecting the Arctic picophytoplankter *Micromonas polaris*. As the predicted warming of the Arctic regions will stimulate *Micromonas* growth rates and promote growth earlier in the season, the authors suggest that viral production will likely do the same.

The increased accumulation of viral metagenomic data the past decade has revealed a huge viral diversity (e.g., [22,23]), and the marine virome is considered the largest pool of unexplored genetic diversity on the globe, with 63–93% of the sequences not represented in the public databases [24]. A recent analysis of viral metagenomic sequence data from 43 surface ocean sites identified ~5500 populations of dsDNA viruses, of which only 39 could be affiliated to cultured viruses [22]. Even at very small scales, viral diversity may be high, as demonstrated by Flaviani et al. [25], who found 254 unique virus phylotypes in a 250 mL oceanic water sample, supporting previous suggestions that local viral diversity is relatively similar to global diversity [22].

The large number of unknown viral populations in the marine metagenome emphasizes the need for further isolation, characterization and sequencing of specific marine viruses. This special issue presents several new marine viruses of eukaryotes (*Prymnesium parvum*, [26]) and bacteria (*Shewanella*, [27], *Vibrio anguillarum* [28] and *Dinoroseobacter shibae* [29]), adding to the rapidly growing database of genome-sequenced and characterized marine viruses. Several auxiliary metabolic genes and other functional genes were identified in the phage genomes, suggesting a mutual benefit for both phage and host that could potentially be disseminated to other hosts by horizontal gene transfer (Figure 1). Prophage-encode genes can thus contribute to host functional properties, including virulence, by so-called lysogenic conversion, potentially expanding the niches occupied by the lysogenized hosts (Figure 1). Further, prophage induction can stimulate biofilm formation by promoting the release of extracellular DNA, which becomes a component in the biofilm matrix [30]. The paper by Leigh et al. [27] shows that lytic phage infections also enhance biofilm formation in *Shewanella*, which forms biofilms in the gut of the tunicate *Ciona intestinalis*. *Shewanella* is part of a complex relationship between the *C. intestinalis* and its associated microbiome, and the study demonstrates that phage interaction with its *Shewanella* host contributes to the symbiotic relationships between the gut microbiome and the tunicate host (Figure 1).

Viruses may also acquire accessory genes from their eukaryotic or prokaryotic hosts [31] (Figure 1). By expressing these genes during infection, the viruses may augment key steps in cellular metabolism and ultimately increase virus production [32]. In addition to using viral genes acquired from the host, viruses may also control the expression of host genes during infection to promote viral production or inhibit host defense systems. This was demonstrated by Fedida & Lindell [33], where expression patterns of specific host genes in the cyanobacterium *Synechococcus* sp strain WH8102 during cyanophage infection suggested that the phage exploited the host genes for improved infection efficiency.

The high local viral diversity obtained from oceanic metagenomic data [22] suggests a high dispersal of viral genes across the sampled ocean viral communities. Moebus [4] had already demonstrated that bacterial viruses with specific infectious properties were distributed across large spatial scales in the North Atlantic. Later, a worldwide distribution of a virus infecting the picophytoplankton *Micromonas pusilla* was reported by Cottrell and Suttle [34], suggesting that viruses are efficiently spread in the marine environment. This is supported by the study by Kalatzis et al. [28], which demonstrates that H20-like vibriophages infecting the fish pathogen *Vibrio anguillarum* are globally distributed either as free phages or as prophages inside bacterial genomes. The authors argue that selection for co-existence, rather than arms race dynamics, might explain the global distribution of near-identical H20-like bacteriophages and their prevalence as prophages in *Vibrio* genomes.

Viral host cells have developed multiple defense strategies against lytic viral infections (Figure 1). These include both mutational changes in the cell surface receptors providing resistance to phage adsorption, and various mechanisms for destroying the viral DNA upon infection (e.g., restriction modification and CRISPR-Cas defense) [35]. Mordecai et al. [36] propose a different life cycle strategy of the dsDNA EhV viruses infecting the coccolithophorid *Emiliania huxleyi*, where the detection of viral RNA in the virus-resistant haploid cell of *E. huxelyi* suggested a new mechanism of infection, and the co-existence of viruses and host. Defense strategies are often associated with a fitness cost, as surface modification mutations may have an influence on, e.g., substrate uptake or enzyme secretion [37], and because virus inactivation mechanisms may be expensive to maintain [38]. Such trade-offs between resistance and fitness costs were explored in the two groups of eukaryotic phytoplankton, *Ostreococcus tauri* [39], and *E. huxleyi* [40]. Surprisingly, no direct cost of resistance was detected in these systems, emphasizing the complexity of interplay between virus-host co-evolution and the environmental conditions.

The large and diverse group of nucleocytoplasmic large DNA viruses (NCLDV) includes a number of viral families infecting small photosynthetic protists, thus affecting mortality, evolution and production of these phytoplankton. In the current special issue, the research on NCLDV infecting phytoplankton is represented by the Prasinoviruses infecting the ubiquitous group of pico-sized Prasinophycea such as *Micromonas* and *Ostreococcus* [21,39,41], and viruses infecting bloom-forming haptophytes such as *Prymnesium parvum* [18,26] and *Emiliania huxleyi* [20,31,36,40]. These studies highlight the progression in our understanding of the role of viruses infecting eukaryotic algae, provide a synthesis of the current knowledge in the field, and identify gaps in our knowledge surrounding viral life history and interactions with their hosts.

The discovery of the giant *Acanthamoeba polyphaga* mimivirus stimulated a new line of research, exploring the ecology and evolution of the group of large DNA viruses infecting eukaryotic protists, including the haptophyte *Phaeocystis globosa* [42]. Here, Wilhelm et al. [43] synthesize the current knowledge and common characteristics of this group of novel viruses and their interactions with their hosts, as well as their virophage parasites.

The compilation of papers included in the current special issue highlights the exploration of eukaryotic and prokaryotic viruses, from discovery to complex interplays between virus and host and the interactions with ecologically relevant environmental variables. The discovery of novel viruses and new mechanisms underlying virus distribution and diversity exemplify the fascinating world of marine viruses. The oceans greatly shape Earth's climate, hold 1.37 billion km^3 of seawater, produce half the half of the oxygen in the atmosphere, and are integral to all known life. In a time where life in the oceans is under increasing threat (global warming, acidification, pollution, economic use), it is pressing to understand how viruses affect host population dynamics, biodiversity, biogeochemical cycling and ecosystem efficiency.

Acknowledgments: Mathias Middelboe was supported by The Independent Research Fund Denmark (DFF—7014-00080) and the BONUS BLUEPRINT project, supported by BONUS (Art 185), funded jointly by the EU and Danish Agency for Science, Technology and Innovation.

References

1. Crane, M.; Hyatt, A. Viruses of fish: An overview of significant pathogens. *Viruses* **2011**, *3*, 2025–2046. [CrossRef] [PubMed]
2. Munn, C.B. Viruses as pathogens of marine organisms—From bacteria to whales. *J. Mar. Biol. Assoc. UK* **2006**, *86*, 453. [CrossRef]
3. Moebus, K. A method for the detection of bacteriophages from ocean water. *Helgol. Meeresunters.* **1980**, *34*, 1–14. [CrossRef]

4. Moebus, K.; Nattkemper, H. Bacteriophage sensitivity patterns among bacteria isolated from marine waters. *Helgol. Meeresunters.* **1981**, *34*, 375–385. [CrossRef]
5. Brussaard, C.P.D.; Wilhelm, S.W.; Thingstad, F.; Weinbauer, M.G.; Bratbak, G.; Heldal, M.; Kimmance, S.; Middelboe, M.; Nagasaki, K.; Paul, J.H.; et al. Global-scale processes with a nanoscale drive: The role of marine viruses. *ISME J.* **2008**, *2*, 575–578. [CrossRef] [PubMed]
6. Rohwer, F.; Thurber, R.V. Viruses manipulate the marine environment. *Nature* **2009**, *459*, 207–212. [CrossRef] [PubMed]
7. Bergh, Ø.; Børsheim, K.; Bratbak, G.; Heldal, M. High abundance of virues found in aquatic environments. *Nature* **1989**, *340*, 467–468. [CrossRef] [PubMed]
8. Brussaard, C.P.; Marie, D.; Bratbak, G. Flow cytometric detection of viruses. *J. Virol. Methods* **2000**, *85*, 175–182. [CrossRef]
9. Thurber, R.V.; Payet, J.P.; Thurber, A.R.; Correa, A.M.S. Virus–host interactions and their roles in coral reef health and disease. *Nat. Rev. Microbiol.* **2017**, *15*, 205–216. [CrossRef] [PubMed]
10. Danovaro, R.; Dell'Anno, A.; Corinaldesi, C.; Magagnini, M.; Noble, R.; Tamburini, C.; Weinbauer, M. Major viral impact on the functioning of benthic deep-sea ecosystems. *Nature* **2008**, *454*, 1084–1087. [CrossRef] [PubMed]
11. Middelboe, M.; Glud, R.N. Viral activity along a trophic gradient in continental margin sediments off central Chile. *Mar. Biol. Res.* **2006**, *2*, 41–51. [CrossRef]
12. Engelhardt, T.; Kallmeyer, J.; Cypionka, H.; Engelen, B. High virus-to-cell ratios indicate ongoing production of viruses in deep subsurface sediments. *ISME J.* **2014**, *8*, 1–7. [CrossRef] [PubMed]
13. Middelboe, M.; Jacquet, S.; Weinbauer, M. Viruses in freshwater ecosystems: An introduction to the exploration of viruses in new aquatic habitats. *Freshw. Biol.* **2008**, *53*, 1069–1075. [CrossRef]
14. Haatveit, H.M.; Wessel, Ø.; Markussen, T.; Lund, M.; Thiede, B.; Nyman, I.B.; Braaen, S.; Dahle, M.K.; Rimstad, E. Viral protein kinetics of piscine orthoreovirus infection in atlantic salmon blood cells. *Viruses* **2017**, *9*. [CrossRef] [PubMed]
15. Wigington, C.H.; Sonderegger, D.L.; Brussaard, C.P.D.; Buchan, A.; Finke, J.F.; Fuhrman, J.; Lennon, J.T.; Middelboe, M.; Suttle, C.A.; Stock, C.; et al. Re-examining the relationship between virus and microbial cell abundances in the global oceans. *Nat. Microbiol.* **2016**. [CrossRef] [PubMed]
16. Finke, J.F.; Hunt, B.P.V.; Winter, C.; Carmack, E.C.; Suttle, C.A. Nutrients and other environmental factors influence virus abundances across oxic and hypoxic marine environments. *Viruses* **2017**, *9*, 1–15. [CrossRef] [PubMed]
17. Sandaa, R.A.; Pree, B.; Larsen, A.; Våge, S.; Töpper, B.; Töpper, J.P.; Thyrhaug, R.; Thingstad, T.F. The response of heterotrophic prokaryote and viral communities to labile organic carbon inputs is controlled by the predator food chain structure. *Viruses* **2017**, *9*. [CrossRef] [PubMed]
18. Johannessen, T.V.; Larsen, A.; Bratbak, G.; Pagarete, A.; Edvardsen, B.; Egge, E.D.; Sandaa, R.A. Seasonal dynamics of haptophytes and dsDNA algal viruses suggest complex virus-host relationship. *Viruses* **2017**, *9*. [CrossRef] [PubMed]
19. Mojica, K.D.A.; Brussaard, C.P.D. Factors affecting virus dynamics and microbial host-virus interactions in marine environments. *FEMS Microbiol. Ecol.* **2014**, *89*, 495–515. [CrossRef] [PubMed]
20. Highfield, A.; Joint, I.; Gilbert, J.A.; Crawfurd, K.J.; Schroeder, D.C. Change in Emiliania huxleyi virus assemblage diversity but not in host genetic composition during an ocean acidification mesocosm experiment. *Viruses* **2017**, *9*. [CrossRef] [PubMed]
21. Maat, D.S.; Biggs, T.; Evans, C.; van Bleijswijk, J.D.L.; van Der Wel, N.N.; Dutilh, B.E.; Brussaard, C.P.D. Characterization and temperature dependence of arctic micromonas polaris viruses. *Viruses* **2017**, *9*, 6–9. [CrossRef] [PubMed]
22. Brum, J.R.; Sullivan, M.B. Rising to the challenge: Accelerated pace of discovery transforms marine virology. *Nat. Rev. Microbiol.* **2015**, *13*, 147–159. [CrossRef] [PubMed]
23. Paez-Espino, D.; Eloe-Fadrosh, E.A.; Pavlopoulos, G.A.; Thomas, A.D.; Huntemann, M.; Mikhailova, N.; Rubin, E.; Ivanova, N.N.; Kyrpides, N.C. Uncovering Earth's virome. *Nature* **2016**, *536*, 425–430. [CrossRef] [PubMed]
24. Hurwitz, B.L.; Sullivan, M.B. The Pacific Ocean Virome (POV): A Marine Viral Metagenomic Dataset and Associated Protein Clusters for Quantitative Viral Ecology. *PLoS ONE* **2013**, *8*. [CrossRef] [PubMed]

25. Flaviani, F.; Schroeder, D.C.; Balestreri, C.; Schroeder, J.L.; Moore, K.; Paszkiewicz, K.; Pfaff, M.C.; Rybicki, E.P. A pelagic microbiome (Viruses to protists) from a small cup of seawater. *Viruses* **2017**, *9*. [CrossRef] [PubMed]

26. Wagstaff, B.A.; Vladu, I.C.; Barclay, J.E.; Schroeder, D.C.; Malin, G.; Field, R.A. Isolation and characterization of a double stranded DNA megavirus infecting the toxin-producing haptophyte *Prymnesium parvum*. *Viruses* **2017**, *9*. [CrossRef] [PubMed]

27. Leigh, B.; Karrer, C.; Cannon, J.P.; Breitbart, M.; Dishaw, L.J. Isolation and characterization of a *Shewanella* phage–host system from the gut of the tunicate, *Ciona intestinalis*. *Viruses* **2017**, *9*. [CrossRef] [PubMed]

28. Kalatzis, P.G.; Rørbo, N.; Castillo, D.; Mauritzen, J.J.; Jørgensen, J.; Kokkari, K.; Zhang, F.; Katharios, P.; Middelboe, M. Stumbling across the Same Phage: Comparative genomics of widespread temperate phages infecting the fish pathogen *Vibrio anguillarum*. *Viruses* **2017**, *9*. [CrossRef] [PubMed]

29. Yang, Y.; Cai, L.; Ma, R.; Xu, Y.; Tong, Y.; Huang, Y.; Jiao, N.; Zhang, R. A novel roseosiphophage isolated from the oligotrophic South China Sea. *Viruses* **2017**, *9*, 1–16. [CrossRef] [PubMed]

30. Gödeke, J.; Paul, K.; Lassak, J.; Thormann, K.M. Phage-induced lysis enhances biofilm formation in *Shewanella oneidensis* MR-1. *ISME J.* **2011**, *5*, 613–626. [CrossRef] [PubMed]

31. Nissimov, J.I.; Pagarete, A.; Ma, F.; Cody, S.; Dunigan, D.D.; Kimmance, S.A.; Allen, M.J. Coccolithoviruses: A review of cross-kingdom genomic thievery and metabolic thuggery. *Viruses* **2017**, *9*. [CrossRef] [PubMed]

32. Lindell, D.; Jaffe, J.D.; Johnson, Z.I.; Church, G.M.; Chisholm, S.W. Photosynthesis genes in marine viruses yield proteins during host infection. *Nature* **2005**, *438*, 86–89. [CrossRef] [PubMed]

33. Fedida, A.; Lindell, D. Two synechococcus genes, two different effects on cyanophage infection. *Viruses* **2017**, *9*. [CrossRef] [PubMed]

34. Cottrell, M.T.; Suttle, C.A. Wide-spread occurrence and clonal variation in viruses which cause lysis of a cosmopolitan, eukaryotic marine phytoplankter, *Micromonas pusilla*. *Mar. Ecol. Prog. Ser.* **1991**, *78*, 1–9. [CrossRef]

35. Labrie, S.J.; Samson, J.E.; Moineau, S. Bacteriophage resistance mechanisms. *Nat. Rev. Microbiol.* **2010**, *8*, 317–327. [CrossRef] [PubMed]

36. Mordecai, G.J.; Verret, F.; Highfield, A.; Schroeder, D.C. Schrödinger's cheshire cat: Are Haploid *Emiliania huxleyi* cells resistant to viral infection or not? *Viruses* **2017**, *9*. [CrossRef] [PubMed]

37. Castillo, D.; Christiansen, R.H.; Dalsgaard, I.; Madsen, L.; Middelboe, M. Bacteriophage resistance mechanisms in the fish pathogen *Flavobacterium psychrophilum*: Linking genomic mutations to changes in bacterial virulence factors. *Appl. Environ. Microbiol.* **2015**, *81*, 1157–1167. [CrossRef] [PubMed]

38. Westra, E.R.; van houte, S.; Oyesiku-Blakemore, S.; Makin, B.; Broniewski, J.M.; Best, A.; Bondy-Denomy, J.; Davidson, A.; Boots, M.; Buckling, A. Parasite exposure drives selective evolution of constitutive versus inducible defense. *Curr. Biol.* **2015**, *25*, 1043–1049. [CrossRef] [PubMed]

39. Heath, S.E.; Knox, K.; Vale, P.F.; Collins, S. Virus resistance is not costly in a marine alga evolving under multiple environmental stressors. *Viruses* **2017**, *9*. [CrossRef] [PubMed]

40. Ruiz, E.; Oosterhof, M.; Sandaa, R.A.; Larsen, A.; Pagarete, A. Emerging interaction patterns in the *Emiliania huxleyi*-EhV system. *Viruses* **2017**, *9*. [CrossRef] [PubMed]

41. Weynberg, K.D.; Allen, M.J.; Wilson, W.H. Marine prasinoviruses and their tiny plankton hosts: A review. *Viruses* **2017**, *9*, 1–20. [CrossRef] [PubMed]

42. Fischer, M.G. Giant viruses come of age. *Curr. Opin. Microbiol.* **2016**, *31*, 50–57. [CrossRef] [PubMed]

43. Wilhelm, S.W.; Bird, J.T.; Bonifer, K.S.; Calfee, B.C.; Chen, T.; Coy, S.R.; Jackson Gainer, P.; Gann, E.R.; Heatherly, H.T.; Lee, J.; et al. A student's guide to giant viruses infecting small eukaryotes: From *Acanthamoeba* to *Zooxanthellae*. *Viruses* **2017**, *9*. [CrossRef] [PubMed]

viruses

MDPI

Article

Viral Protein Kinetics of Piscine Orthoreovirus Infection in Atlantic Salmon Blood Cells

Hanne Merethe Haatveit [1], Øystein Wessel [1], Turhan Markussen [1], Morten Lund [2], Bernd Thiede [3], Ingvild Berg Nyman [1], Stine Braaen [1], Maria Krudtaa Dahle [2] and Espen Rimstad [1,*]

[1]   Department of Food Safety and Infectious Biology, Faculty of Veterinary Medicine, Norwegian University of Life Sciences, 0454 Oslo, Norway; hanne.merethe.haatveit@nmbu.no (H.M.H.); oystein.wessel@nmbu.no (Ø.W.); turhan.markussen@nmbu.no (T.M.); ingvild.nyman@nmbu.no (I.B.N.); stine.braaen@nmbu.no (S.B.)
[2]   Department of Immunology, Norwegian Veterinary Institute, 0454 Oslo, Norway; morten.lund@vetinst.no (M.L.); maria.dahle@vetinst.no (M.K.D.)
[3]   Department of Biosciences, University of Oslo, 0316 Oslo, Norway; bernd.thiede@ibv.uio.no
*   Correspondence: espen.rimstad@nmbu.no; Tel.: +47-672-32-227

Academic Editors: Corina P.D. Brussaard and Mathias Middelboe
Received: 15 December 2016; Accepted: 10 March 2017; Published: 18 March 2017

Abstract: *Piscine orthoreovirus* (PRV) is ubiquitous in farmed Atlantic salmon (*Salmo salar*) and the cause of heart and skeletal muscle inflammation. Erythrocytes are important target cells for PRV. We have investigated the kinetics of PRV infection in salmon blood cells. The findings indicate that PRV causes an acute infection of blood cells lasting 1–2 weeks, before it subsides into persistence. A high production of viral proteins occurred initially in the acute phase which significantly correlated with antiviral gene transcription. Globular viral factories organized by the non-structural protein µNS were also observed initially, but were not evident at later stages. Interactions between µNS and the PRV structural proteins λ1, µ1, σ1 and σ3 were demonstrated. Different size variants of µNS and the outer capsid protein µ1 appeared at specific time points during infection. Maximal viral protein load was observed five weeks post cohabitant challenge and was undetectable from seven weeks post challenge. In contrast, viral RNA at a high level could be detected throughout the eight-week trial. A proteolytic cleavage fragment of the µ1 protein was the only viral protein detectable after seven weeks post challenge, indicating that this µ1 fragment may be involved in the mechanisms of persistent infection.

Keywords: *Piscine orthoreovirus*; PRV; non-structural protein; µNS; µ1; expression kinetics; proteolytic cleavage; pathogenesis; blood cells; Atlantic Salmon

1. Introduction

Piscine orthoreovirus (PRV) belongs to the genus *Orthoreovirus* in the family *Reoviridae* [1,2]. The orthoreoviruses are ubiquitous in various animal species, but only found to be of pathogenic significance in poultry and recently in fish [3–7]. PRV is abundant in farmed Atlantic salmon (*Salmo salar*), detected both in apparently healthy and diseased fish [8–11]. The infection causes heart and skeletal muscle inflammation (HSMI) and is associated with melanised foci in white muscle in Atlantic salmon [1,7,12]. HSMI is a prevalent disease and melanised foci is a quality problem; both conditions are of major economic importance to salmon aquaculture. The pathogenesis of HSMI is not completely elucidated. Outbreaks of the disease are primarily observed in the seawater phase and last for several weeks in the population [13], after which the PRV infection becomes persistent [9,11,14]. In experimental cohabitant infection trials, disease onset occurs after 8–10 weeks [15].

The study of molecular mechanisms linked to PRV infection has been limited by the lack of susceptible cell lines. Studies of the viral infection have therefore been performed in vivo or by infecting erythrocytes ex vivo [16]. Piscine erythrocytes are nucleated and contain the transcriptional and translational machinery necessary for expression of mRNA and proteins [17]. Erythrocytes are important target cells for PRV and the infection activates an innate antiviral immune response typical for RNA viruses in these cells [18]. During the peak phase of infection, more than 50% of all erythrocytes may be infected [19]. Interestingly, severe anemia has not been reported from HSMI outbreaks in the seawater phase, indicating low or no virus-induced lysis of infected erythrocytes [20]. Recently, a variant of PRV was demonstrated to be the etiologic agent of erythrocytic inclusion body syndrome (EIBS), a condition associated with anemia and mass mortality in juvenile Coho salmon (*Onchorhynchus kisutchi*). The level of anemia in EIBS affected fish corresponded with the level of viral replication in blood [7]. In addition, infection of rainbow trout in fresh water by yet another PRV variant is also associated with anemia and an HSMI-like disease [5].

The *Orthoreovirus* genome consists of ten double-stranded RNA (dsRNA) segments enclosed in a double protein capsid. The genomic segments are classified according to size with three large (L), three medium (M) and four small (S) segments encoding the λ, μ and σ class proteins, respectively [3,21]. In mammalian orthoreovirus (MRV), the species type of genus *Orthoreovirus*, the viral transcription machinery is located in the inner core and consists of $\lambda 1$, $\lambda 2$, $\lambda 3$, $\mu 2$ and $\sigma 2$ [22]. The outer capsid proteins $\mu 1$, $\sigma 1$ and $\sigma 3$ are involved in cell attachment and membrane penetration during the initial stages of infection [23–25]. The two non-structural proteins μNS and σNS participate in the formation of viral factories where viral genome replication and particle assembly occur [21,26,27]. Although some important amino acid motifs are conserved between MRV and PRV, sequence identities between homologous proteins are generally low [2]. MRV enters the cell by receptor-mediated endocytosis. The outer capsid is largely removed and $\mu 1$ is cleaved at two positions that generate, in addition to the full-length protein, five different fragments [24,28]. The N-terminal autolytic cleavage site, which produces $\mu 1$N and $\mu 1$C, seems conserved across orthoreoviruses, including PRV [2,29,30]. Further cleavage of $\mu 1$C, mediated by exogenous proteases, generate fragments δ and ϕ [24].

Structures resembling viral factories have also been observed in PRV-infected erythrocytes, and recombinant expression of the protein in fish cell lines indicate that PRV μNS has an analogous role in factory formation [16,19,31]. The majority of virus-encoded proteins localize completely or partially within these viral factories [2,3,21]. The viral factories in PRV-infected cells resemble the globular structures observed for the MRV type 3 Dearing (T3D) strain, in contrast to the filamentous-like viral factories generated by MRV Type 1 Lang (T1L) [19]. The latter is considered the most common morphology type of orthoreoviral factories [21,32]. Gene segment M3 in MRV and avian orthoreovirus (ARV) are reported to produce two isoforms of the factory forming μNS protein in infected cells [33–35]. The second isoform is produced by different mechanisms in the two viruses; in MRV, μNSC is expressed by a second in-frame AUG (Met_{41}) while in ARV, post-translational cleavage in the N-terminal region releases μNSN [33,35,36]. In ARV, only full-length μNS interacts with σNS in infected cells, suggesting that the two isoforms play different roles during ARV infection [34].

Considering the emerging occurrence of HSMI, PRV exhibits a considerable risk for the aquaculture industry and proper disease control is highly desired. To understand the association between PRV infection and disease outcome, and also to limit further disease outbreaks, more information regarding PRV protein kinetics is essential. In the present study, the kinetics of viral RNA, viral protein and antiviral immune response in blood cells from experimentally PRV-infected Atlantic salmon were investigated. We hypothesized that PRV causes an acute infection in blood cells correlating with innate antiviral gene expression, before the infection subsides to a low persistent level.

2. Materials and Methods

2.1. Construction and Expression of Recombinant Piscine orthoreovirus (PRV) µNS

Following the supplier's protocol, the BaculoDirect™ Baculovirus Expression System (Invitrogen, Carlsbad, CA, USA) was used to generate recombinant µNS. The µNS open reading frame (ORF) (acc. no. KR337478) was obtained by polymerase chain reaction (PCR; primers listed in Table S1) of the plasmid construct pcDNA3.1 µNS N-FLAG [31] and cloned into the pENTR™ TOPO® vector (Invitrogen). The pENTR µNS construct was used in a recombination reaction to generate the recombinant baculovirus DNA. Sanger sequencing (GATC Biotech AG, Konstanz, Germany) confirmed the sequence of the construct. Spodoptera frugiperda (Sf9) insect cells (BD Bioscience, Erembodegem, Belgium) cultured in Grace Insect Medium (Invitrogen) supplemented with 10% heat inactivated fetal bovine serum (FBS, Life Technologies, Paisley, Scotland, UK), 100 U/mL Penicillin, 100 µg/mL Streptomycin and 0.25 µg/mL Fungizone (Life Technologies), were transfected with recombinant baculovirus DNA. Passage 1 (P1) viral stock was harvested 11 days post transfection and used to produce high titer viral stocks according to the supplier's protocol. The BacPAK quantitative PCR (qPCR) Titration kit (Clontech, Mountain View, CA, USA) was used to determine the viral titer. Finally, Sf9 insect cells were infected with Passage 2 (P2) or higher passage of recombinant baculovirus stock (>1 × 10^8 copies/mL) and incubated at 27 °C for 96 h for expression of the recombinant µNS protein containing a C-terminal 6xHis-tag.

2.2. Construction and Expression of Recombinant PRV λ1

The ORF of PRV structural protein λ1 (acc. no. KR337475) encoded by gene segment L3 was amplified (primers listed in Table S1) using cDNA originating from a HSMI outbreak [31] as template. The PCR product was cloned into pET100/D-TOPO (Invitrogen) and the sequence verified by Sanger sequencing (GATC Biotech AG). The pET100-λ1 plasmid was transfected into *E. coli* (BL21 DE3 strain, Invitrogen) and expressed with a N-terminal 6xHis-tag, following the manufacturer's instructions. Protein expression was monitored by sodium dodecyl sulfate polyacrylamide gel electrophoresis (SDS-PAGE).

2.3. Protein Purification

The Sf9 insect cells and the *E. coli* cells expressing recombinant PRV µNS and λ1 proteins, respectively, were pelleted by centrifugation at 5000× *g* for 10 min, then dissolved and washed in phosphate-buffered solution (PBS). Purification of recombinant proteins was carried out using ProBond Purification System (Life Technologies) following the manufacturer's instructions. The recombinant µNS protein was eluted with an elution buffer containing 8 M Urea, 20 mM $Na_2H_2PO_4$ (pH 4.0), and 500 mM NaCl. The purity of the recombinant protein was monitored by SDS-PAGE using a 4%–12% Bis–Tris Criterion XT gel (Bio-Rad, Hercules, CA, USA). To purify λ1, the Ni-NTA agarose was run on a SDS-PAGE where a band matching the size of λ1 was excised. The gel sample containing λ1 protein was solubilized in 250 mM Tris-HCl with 0.1% SDS, pH 6.8, sonicated 3 × 5 s and incubated at 4 °C with shaking overnight. The sample was centrifuged at 10,000× *g* for 10 min and the supernatant was dialyzed using the Slide-A-Lyser® Dialysis cassette with 20,000 molecular weight cut-off (MWCO) and 0.5–3.0 mL capacity (Thermo Scientific, Waltham, MA, USA) following the manufacturer's protocol. SDS-PAGE confirmed the purity of the recombinant λ1 protein. Protein concentrations for both µNS and λ1 were determined using the DC Protein Assay Reagent Package (Bio-Rad), with bovine serum albumin (BSA; Sigma-Aldrich, St. Louis, MO, USA) as protein standard.

2.4. Immunization of Rabbits

The purified recombinant proteins were used for immunization of rabbits and generation of antisera named anti-µNS #R320684 and anti-λ1 #K273. In the first injection, Freund's complete adjuvant was added, thereafter the rabbits were boosted three times with Freund's incomplete adjuvant weekly. The amount of µNS and λ1 antigen used per immunization was in the range of 45–500 µg. The rabbit sera produced were tested by Western blotting (WB) and fluorescent microscopy after transfection of epithelioma papulosum cyprini (EPC; ATCC CRL-2872) cells with pcDNA3.1 µNS N-FLAG [31] or pcDNA3.1 λ1 N-HA [31] (see description below). Antisera controls were collected prior to immunization. WB and immunofluorescent microscopy confirmed that the rabbit µNS and λ1 antisera recognized the µNS and λ1 proteins in transfected EPC cells (Figure S1). No staining was detected using the pre-immunization sera (data not shown).

2.5. Specificity of Antisera

EPC cells were cultivated in Leibovitz-15 medium (L15; Life Technologies) supplemented with 10% heat inactivated FBS, 2 mM L-glutamine, 0.04 mM mercaptoethanol and 0.05 mg/mL gentamycin-sulphate (Life Technologies), and seeded at a density of 1.5×10^4 cells/well in a 24-well plate 24 h prior to transfection. Plasmids pcDNA3.1-µNS N-FLAG and pcDNA3.1-λ1 N-HA were transfected using Lipofectamine LTX reagent (Life Technologies) according to the manufacturer's instructions. The cells were fixed and stained 48 h post-transfection with an Intracellular Fixation and Permeabilization Buffer Set (eBioscience, San Diego, CA, USA) following the manufacturer's protocol. Antisera against µNS (1:1000) and λ1 (1:500); secondary antibody against rabbit IgG conjugated with Alexa Fluor 488 (Life Technologies) and Hoechst trihydrochloride trihydrate (Life Technologies) were used for staining. Images were captured on an inverted fluorescence microscope (Olympus IX81). Transfected EPC cells were also used to further verify anti-µNS and anti-λ1 in WB. A total of 3×5 million EPC cells were pelleted by centrifugation, resuspended in 100 µL Ingenio Electroporation Solution (Mirus, Madison, WI, USA) and transfected with 4 µg pcDNA3.1 µNS N-FLAG or pcDNA3.1 λ1 N-HA. The transfected cells were transferred to 75 cm^2 culture flasks containing 20 mL pre-equilibrated L-15 growth medium (described above) and collected 72 h post-transfection. The cell pellets were lysed in Nonidet-P40 lysis buffer (1% NP-40, 50 mM Tris–HCl pH 8.0, 150 mM NaCl, 2 mM EDTA) containing Complete ultra mini protease inhibitor cocktail (Roche, Mannheim, Germany). The mix was incubated on ice for 30 min, and then centrifuged at $5000\times g$ for 5 min at 4 °C. The supernatant was mixed with Sample Buffer (Bio-Rad) and Reducing Agent (Bio-Rad), denatured for 5 min at 95 °C and run in SDS-PAGE, using 4%–12% Bis–Tris Criterion XT gel (Bio-Rad). Magic Mark^{TM} XP Standard (Invitrogen) was used as a molecular size marker. Following SDS-PAGE, the proteins were blotted onto a polyvinylidene fluoride (PVDF) membrane (Bio-Rad) and anti-µNS and anti-λ1 were used as primary antibodies and anti-Rabbit IgG-HRP (GE Healthcare, Buchinghamshire, UK) as secondary antibody. Protein bands were detected by chemiluminescence (Amersham ECL Plus, GE Healthcare).

2.6. Experimental Challenge of Salmon

A cohabitation challenge experiment was performed at VESO Vikan aquatic research facility, (Vikan, Norway). The fish had an average weight of 30 grams at the onset of the experiment with a maximum stocking density of 80 kg/m^3, and were kept in 0.4 m^3 tanks supplied with filtered and UV-radiated fresh water, 12 °C $\pm$ 1 °C with a 12 h light/12 h dark regime. Water discharge of the tanks was provided by a tube overflow system with 7.2 L/min flow rate. The fish were acclimatized for two weeks prior to challenge, fed according to standard procedures and anesthetized by bath immersion (2–5 min) in benzocaine chloride (0.5 g/10 L water, Apotekproduksjon AS, Oslo, Norway) before handling. Briefly, the experimental study included one group of shedder fish (50%) marked at the time of PRV-injection by cutting off the adipose fin and one naïve cohabitant group (50%). The PRV inoculum was prepared from a batch of pooled heparinized blood samples from a previous PRV challenge experiment [19].

On day 0 of the challenge, the heparinized blood was diluted 1:2 in PBS and 0.1 mL of the inoculum was intraperitoneal (i.p.) injected into the shedders. The inoculum was confirmed negative for salmon viruses such as infectious pancreatic necrosis virus (IPNV), infectious salmon anemia virus

(ISAV), salmonid alphavirus (SAV) and piscine myocarditis virus (PMCV) by reverse transcription quantitative PCR (RT-qPCR). Samples from six fish were collected before initiation of the experiment to provide time-0 uninfected control material for protein assays. Heparinized blood was collected from six cohabitant fish at each sampling point; 3, 4, 5, 6, 7 and 8 weeks post challenge (wpc). In addition, a second cohabitation challenge experiment lasting 10 weeks was performed at the same facility following a similar experimental design. In this study, six fish sampled prior to PRV challenge were used to provide uninfected control material for protein and RT-qPCR assays, and heparinized blood was collected from six cohabitant fish at 4, 6, 8 and 10 wpc. The second challenge experiment was otherwise performed under the same conditions as the first experiment. Both experiments were approved by the Norwegian Animal Research Authority and followed the European Union Directive 2010/63/EU for animal experiments.

2.7. RNA Isolation and Reverse Transcription Quantiative Polymerase Chain Reaction (RT-qPCR)

Total RNA was isolated from 20 µL heparinized blood homogenized in 650 µL QIAzol Lysis Reagent (Qiagen, Hilden, Germany) using 5 mm steel beads, TissueLyser II (Qiagen) and RNeasy Mini spin column (Qiagen) as recommended by the manufacturer. RNA was quantified using a NanoDrop, ND-1000 spectrophotometer (Thermo Fisher Scientific, Wilmington, DE, USA). The Qiagen OneStep kit (Qiagen) was used for RT-qPCR with a standard input of 100 ng (5 µL of 20 ng/µL) of the isolated total RNA per reaction in a total reaction volume of 12.5 µL. The template RNA was denaturated at 95 °C for 5 min prior to RT-qPCR targeting PRV gene segments S1, M2 and M3. The following conditions were used for S1: 400 nM primer, 300 nM probe, 400 nM dNTPs, 1.26 mM $MgCl_2$, 1:100 RNase Out (Invitrogen) and $1 \times$ ROX reference dye with the following cycle parameters: 30 min at 50 °C, 15 min at 94 °C, 40 cycles of 94 °C/15 s, 54 °C/30 s and 72 °C/15 s in an AriaMx (Agilent, Santa Clara, CA, USA). Similar conditions and cycle parameters were also used targeting M2 and M3, although primer concentration was adjusted to 600 nM and annealing temperature to 58 °C. All samples were run in duplicates, and a sample was defined as positive if both parallel samples had a Ct <35. The fluorescence threshold for S1, M2 and M3 was set at ΔRn 0.261, 0.028 and 0.021, respectively. The primers and probes are listed in Table S1. For analysis of antiviral gene expression, cDNA was prepared from 500 ng RNA using the QuantiTect reverse transcription kit with gDNA elimination (Qiagen) following the instructions from the manufacturer. Quantitative PCR was performed in triplets on 384-well plates using cDNA corresponding to 5 ng RNA in a total volume of 10 µL per parallel, SsoAdvanced™ Universal SYBR® Green Supermix, and 500 nM forward and reverse primers (Table S2). The qPCRs were run for 40 cycles of 94 °C/15 s and 60 °C/30 s. All samples in the sample set were analyzed on the same plate using the same fluorescence threshold, and the cut-off value was set to Ct 37. The specificity of the SYBR green assays was confirmed by melting point analysis. Levels of Elongation factor (EF1α) mRNA were used for normalization of all assays by the $\Delta\Delta$Ct method.

2.8. Flow Cytometry

Samples consisting of 1.25 µL heparinized blood (diluted 1:20 in PBS) from each of the cohabitant fish in the first challenge experiment were plated into 96-well plates for intracellular staining as previously described [19] using anti-µNS and anti-σ1 [4]. The corresponding zero serum, anti-µNS Zero and anti-σ1 Zero [4] were used as negative controls for background staining. Samples originating from 5 and 8 wpc were fixed, stained and analyzed immediately, while samples from 4 and 7 wpc were fixed and stored for one week and samples from 0, 3 and 6 wpc were fixed and stored for two weeks in flowbuffer (PBS, 1% BSA, 0.05% azide) before analysis. The cells were analyzed on a Gallios Flow Cytometer (Beckman Coulter, Miami, FL, USA), counting 50,000 cells per sample, and the data were analyzed using the Kaluza software (Becton Dickinson). Cells were gated according to size and granularity to include only intact cells and samples from 0 wpc were used as negative controls.

Due to slight variation in background staining, the flow charts were gated individually to discriminate between negative and positive peaks.

2.9. Immunofluorescence Microscopy

Following flow cytometry analysis, the cells were prepared for immunofluorescence microscopy. The nuclei were stained with Hoechst trihydrochloride trihydrate (Life Technologies) and the cells were mounted to glass slides using Fluoroshield (Sigma-Aldrich, St. Louis, MO, USA) and cover slips. Images were captured on an inverted fluorescence microscope (Olympus IX81).

2.10. Transmission Electron Microscopy (TEM)

Samples consisting of 20 µL heparinized blood from each cohabitant fish in the first experimental challenge were diluted in 1 mL PBS, centrifuged at $1000 \times g$ for 5 min at 4 °C, washed twice in PBS and fixed in 3% glutaraldehyde overnight at 4 °C. All samples were further washed twice in PBS and prepared for transmission electron microscopy (TEM) as described earlier [19]. The sections were examined in a FEI MORGAGNI 268, and photographs were recorded using a VELETA camera.

2.11. Western Blotting (WB)

Heparinized blood from each cohabitant fish in the first challenge experiment was analyzed separately and as pooled samples from the different time-points. The samples were centrifuged at $5000 \times g$ and the blood pellets was lysed in Nonidet-P40 lysis buffer containing Complete ultra mini protease inhibitor cocktail and prepared for WB as described above. Anti-µNS (1:1000), anti-µ1C (1:500) [4], anti-σ1 (1:1000) [4], anti-σ3 (1:500) [2] and anti-λ1 (1:500) were used as primary antisera, Rabbit Anti-Actin (Sigma-Aldrich, St. Louis, MO, USA) was used to standardize the blots and Anti-Rabbit IgG-HRP (GE Healthcare) was used as secondary antibody. Blood collected at 0 wpc was used as negative control. In addition, heparinized blood from six of the cohabitant fish sampled at 0, 4, 6, 8 and 10 wpc in the second challenge experiment were prepared and analyzed in the same manner.

2.12. Immunoprecipitation (IP)

Blood from six cohabitants in the first challenge experiment sampled at 4, 5 and 8 wpc were pooled and lysed in Nonidet-P40 lysis buffer containing Complete ultra mini protease inhibitor cocktail as described above. The supernatants were transferred to new tubes and added anti-µNS or anti-µ1C (1:50) and incubated at 4 °C overnight with rotation. The Immunoprecipitation Kit Dynabeads Protein G (Novex, Life Technologies) was used for protein extraction and the beads were prepared according to the manufacturer's protocol. The cell–lysate–antibody mixtures were mixed with the protein G-coated beads and incubated 2 h at 4 °C. The beads–antibody–protein complexes were washed according to the manufacturer's protocol and run in SDS-PAGE. The SDS-gel was blotted onto PVDF membranes (Bio-Rad) and the proteins were detected using anti-µNS, anti-µ1C [4], anti-σ1 [4], anti-σ3 [2] and anti-λ1.

2.13. Liquid Chromatography–Mass Spectrometry (LC–MS)

Five and three fragments immunoprecipitated with anti-µNS (4 and 5 wpc) and anti-µ1C (5 wpc), respectively, that were not observed at 0 wpc, were excised and in-gel digested with 0.1 µg of trypsin in 20 µL of 50 mM ammonium bicarbonate, pH 7.8 for 16 h at 37 °C (Promega, Madison, WI, USA). The peptides were purified with µ-C18 ZipTips (Millipore, Billerica, MA, USA), and analyzed using an Ultimate 3000 nano-UHPLC system (Dionex, Sunnyvale, CA, USA) connected to a Q Exactive mass spectrometer (ThermoElectron, Bremen, Germany). Liquid chromatography and mass spectrometry was performed as previously described [37]. Data were acquired using Xcalibur v2.5.5 and raw files were processed to generate peak list in Mascot generic format (*.mgf) using ProteoWizard release (Version 3.0.331). Database searches were performed using Mascot (Version 2.4.0) against the protein sequences of λ1, λ2, λ3, μNS, μ1, μ2, σNS, σ1, σ2 and σ3 assuming the digestion enzyme trypsin and semi-trypsin, at a maximum of one missed cleavage site, fragment ion mass tolerance of 0.05 Da, parent ion tolerance of 10 ppm and oxidation of methionines, propionamidylation of cysteines, acetylation of the protein N-terminus as variable modifications. Scaffold 4.4.8 (Proteome Software Inc., Portland, OR, USA) was used to validate MS/MS based peptide and protein identifications.

2.14. Computational Analysis

Theoretical molecular weights for proteins were calculated using the Compute pI/Mw tool [38]. PSI-blast based secondary structure PREDiction (PSIPRED; Version 3.3) was used to predict protein secondary structure [39].

2.15. Statistical Analysis

Differences in gene expression levels of innate antiviral genes was analyzed using one-way Anova with Tukey's multiple comparison test. Correlation analysis between PRV S1/M3 RNA levels and antiviral and immune gene expression were performed using nonparametric Spearman correlation.

3. Results

3.1. Viral RNA Load in Blood Cells

RT-qPCR targeting PRV genomic segments S1, M2 and M3 revealed high viral RNA loads in blood cells from 3 to 8 wpc (Figure 1). RNA from segments S1, M2 and M3 were first detected at 3 wpc and peaked at 5 wpc with mean Ct-values of 17.2 (±0.4), 14.5 (±0.3) and 14.6 (±0.4). From 5 wpc, the S1 RNA load decreased, and by 8 wpc the mean Ct-value was 26.4 (±0.6). However, a similar decrease was not observed for the M2 and M3 RNAs, and by 8 wpc mean Ct-values for these genomic segments were 17.4 (±0.5) and 17.7 (±0.4), respectively. RT-qPCR targeting genomic segment S1 in blood from six fish sampled at 0, 4, 6, 8 and 10 wpc in the second challenge experiment was also performed and gave similar results (Figure S2).

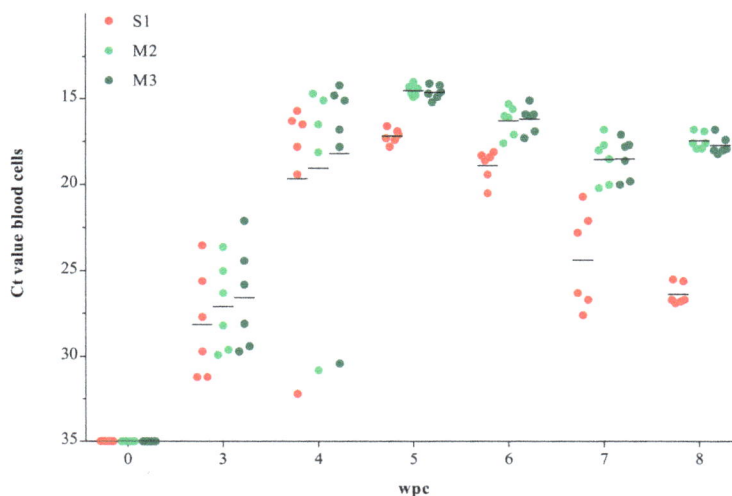

Figure 1. *Piscine orthoreovirus* (PRV) RNA load in blood cells. Reverse transcription quantitative polymerase chain reaction (RT-qPCR) of PRV gene segments S1, M2 and M3 in blood cells from cohabitant fish. Individual (**dots**) and mean (**line**) Ct-values, *n* = 6 per time-point. wpc = weeks post challenge.

3.2. Expression of Innate Antiviral Genes in PRV Infected Blood Cells

The innate antiviral immune response in blood following PRV infection was studied by RT-qPCR targeting Atlantic salmon type I interferon (IFNab), viperin, interferon-stimulated gene 15 (ISG15), dsRNA-activated protein kinase (PKR) and IFNγ. All innate antiviral genes analyzed were statistically significantly upregulated during the peak phase of PRV infection from 4 to 6 wpc, increasing 5- to 20-fold compared to the level at 3 wpc (Figure 2a, Figure S3). The Ct values for S1 and M3 RNA correlated with the relative levels of gene expression for all innate antiviral genes, but not for the T-cell marker genes CD4 and CD8 (Figure 2b). When comparing the early phase up to the peak of infection (3–5 wpc) with the later phase (6–8 wpc), S1 RNA was correlated with the innate antiviral response in both phases, whereas M3 only showed significant correlation in the early phase (Figure 2b). EF1α were stably expressed during PRV infection and were used for normalization of all other assays by the ΔΔCt method (Figure S4) [15,40].

(a)

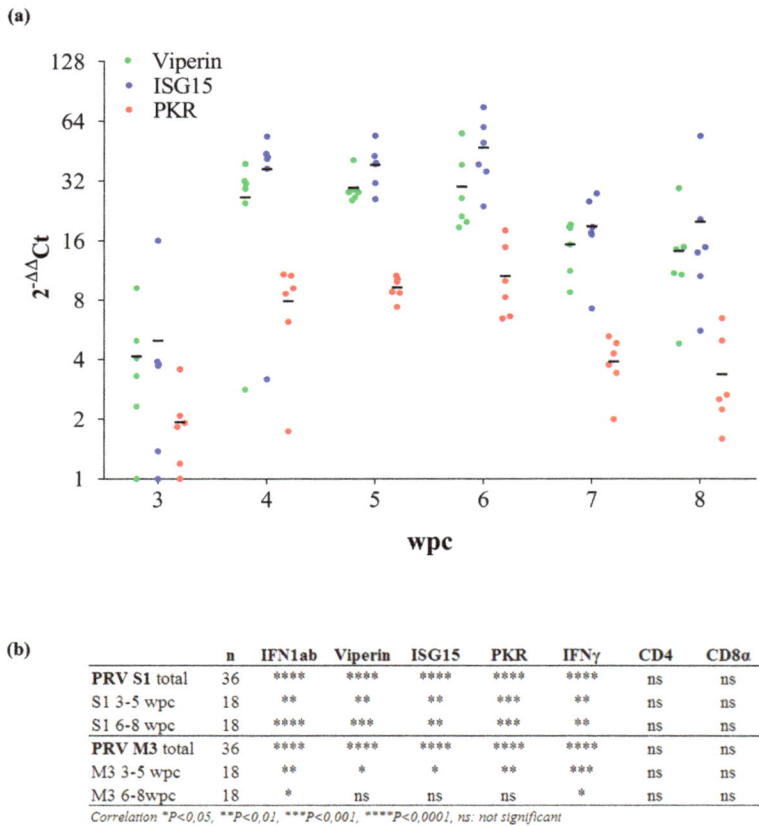

(b)

	n	IFN1ab	Viperin	ISG15	PKR	IFNγ	CD4	CD8α
PRV S1 total	36	****	****	****	****	****	ns	ns
S1 3-5 wpc	18	**	**	**	***	**	ns	ns
S1 6-8 wpc	18	****	***	**	***	**	ns	ns
PRV M3 total	36	****	****	****	****	****	ns	ns
M3 3-5 wpc	18	**	*	*	**	***	ns	ns
M3 6-8wpc	18	*	ns	ns	ns	*	ns	ns

*Correlation *P<0,05, **P<0,01, ***P<0,001, ****P<0,0001, ns: not significant*

Figure 2. Expression of immune genes in blood cells. (**a**) Immune genes were assayed at 3–8 wpc by RT-qPCR in blood cells from cohabitant fish (*n* = 6 per time point). Data are normalized against EF1α and the lowest ΔCt level at 3 wpc (*n* = 6), and 2-ΔΔCt values are calculated. Mean relative expression is indicated. ISG = interferon-stimulated gene, PKR = double-stranded RNA (dsRNA)-activated protein kinase; (**b**) Correlation between Ct values for S1/M3 RNA and relative levels of antiviral gene expression for a set of immune genes.

3.3. Flow Cytometry Indicates a Transient Peak in Blood Cells

Blood cells stained intracellularly with anti-μNS and anti-σ1 were analyzed by flow cytometry (Figure 3a, Figure S5). A PRV positive population of blood cells was observed from 4 wpc as a marked shift in the histograms compared to negative samples. Five out of six fish were positive for μNS by flow cytometry at 4 wpc, consistent with the RT-qPCR data where the positive fish had lower Ct-values (18.2 ± 5.6) compared to the negative fish (30.4). At 5 wpc, the PRV positive blood cell population decreased, but was still visible for all individuals. From 6 wpc and onwards, no PRV-positive cell populations were observed. The pattern for σ1 positive cells was similar to that described for μNS.

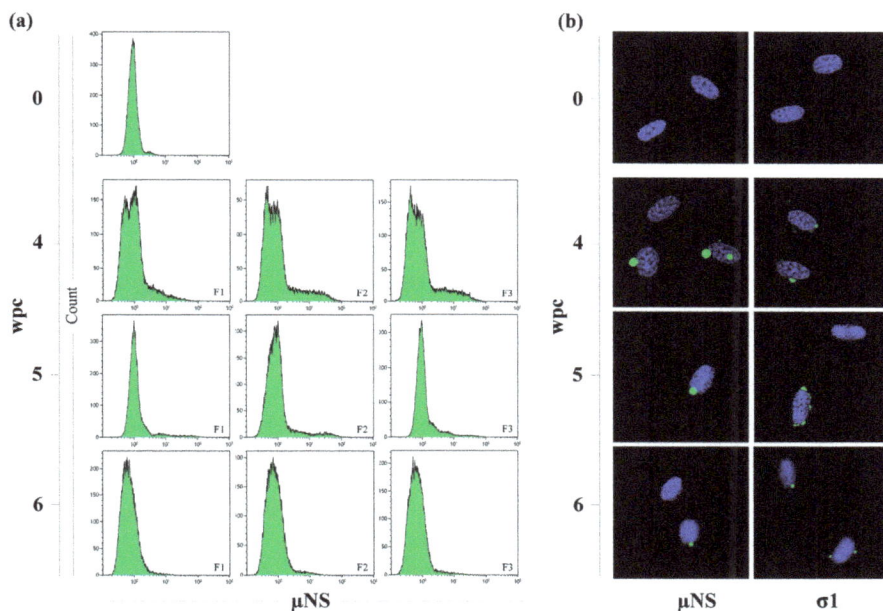

Figure 3. Presence of PRV μNS and σ1 in blood cells. (**a**) Intracellular staining of μNS in blood cells analyzed by flow cytometry from three cohabitant fish sampled at 4, 5 and 6 wpc. The negative control staining is one fish sampled at 0 wpc. A total of 50,000 cells were counted per sample and 30,000 were gated for analysis; (**b**) Fluorescent labeling of μNS (left) and σ1 (right) displaying viral factory-like inclusions (green) in infected red blood cells sampled 0 (negative control), 4, 5 and 6 wpc. The nuclei were stained with Hoechst (blue).

3.4. Viral Factories Observed in Blood Cells

Both μNS and σ1 were detected by immunofluorescence as cytoplasmic globular inclusions in erythrocytes at 4, 5 and 6 wpc (Figure 3b). The inclusions varied in both size and number. At 4 and 5 wpc, they were predominantly large and perinuclear. Inclusions stained with anti-σ1 were generally smaller and more variable in size than those stained with anti-μNS. At 6 wpc, the number and size of the inclusions were considerably reduced and at 7 wpc and onward no inclusions were detected. These findings correlated with the results obtained from flow cytometry.

3.5. TEM of PRV Infected Blood Cells

TEM of PRV infected blood cells sampled at 0, 4, 5 and 6 wpc are shown in Figure 4. The control cells (0 wpc) contained circular cytoplasmic vesicles (200–500 nm) that were apparently devoid of specific content. In addition, a few control cells contained lamellar structures up to 300 nm in size. At 4 wpc, lamellar structures were frequent and a few large cytoplasmic inclusions (~800 nm) containing particles with reovirus-like morphology were observed. The viral particles were naked with an electron dense core that resembled previous TEM descriptions of PRV [19]. At 5 wpc, several small (200–500 nm) and large (~800 nm) cytoplasmic inclusions containing reovirus-like particles were detected. The larger inclusions contained a mixture of reovirus-like particles and lamellar structures, some enclosed within membrane-like structures. At 6 wpc, large inclusions were frequent, but only a few contained viral particles.

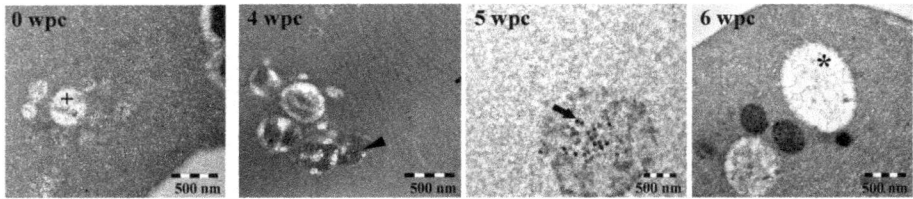

Figure 4. Transmission electron microscopy (TEM) of blood cells. PRV-infected red blood cells sampled at 0 (negative control), 4, 5 and 6 wpc show small empty vesicles (cross), lamellar structures (arrowhead), reovirus-like particles (arrow) and large empty inclusions (star).

3.6. µNS Protein Expression in Individual Fish Correlate with viral RNA only during the Acute Phase of Infection

Blood cells from six fish sampled at 3, 4, 5 and 6 wpc were analyzed by WB using anti-µNS and compared to Ct-values targeting the corresponding genomic segment M3 of the same samples (Figure 5). No fish were positive by WB at 3 wpc, while five samples at 4 wpc demonstrated bands at molecular weight (MW) 83.5 (putative full-length µNS) and 70 kDa. The Ct-values from the same samples corresponded to the positive staining of the putative full-length µNS bands. Fish 6 at 4 wpc, was negative for µNS by WB; this individual also displayed a higher Ct-value (30.4) than the other cohabitants. The amount of µNS decreased markedly from 4 to 5 wpc, and the 70 kDa band was barely detectable at 5 wpc. At 6 wpc, the µNS protein was non-detectable by WB in fish 1, 5 and 6, and only barely detectable in the remaining fish. Although the µNS protein level decreased below the detection limit for WB, the corresponding viral RNA levels (genomic segment M3) remained high throughout the challenge. Thus, µNS protein and M3 RNA levels only correlated at 4 wpc.

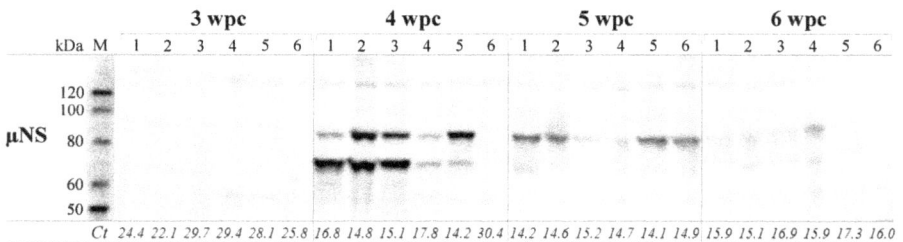

Figure 5. Detection of PRV uNS protein in blood cells compared to viral RNA load. Blood cells from 3, 4, 5 and 6 wpc (*n* = 6) analyzed for µNS by Western blotting. Ct-values for gene segment M3 (µNS) from the same samples are shown below each lane. M = molecular weight standard; Lane 1–6 refers to individual fish (1–6) per time point.

3.7. PRV Protein Levels Display a Transient Peak in Blood Cells

The load of structural proteins λ1, μ1, σ1 and σ3, and the non-structural protein μNS, displayed a similar transient peak at 4–6 wpc in blood cells (Figure 6). All five proteins appeared at 4 wpc and were non-detectable at 7 wpc. In addition to the putative full-length μNS, a band with the MW of about 70 kDa was observed at 4 wpc, consistent with findings from individual fish (Figure 5). The putative full-length μ1 protein (74.2 kDa) was detected at 4 wpc. However, at 5 wpc, this band was not present but replaced by three bands of approximately 70 kDa, 37 kDa and 32 kDa in size. At 7 and 8 wpc, only one band of approximately 35 kDa was detected. The same staining patterns for the λ1, μNS, μ1, σ1 and σ3 proteins were observed when blood from the second challenge experiment was analyzed (Figure S6).

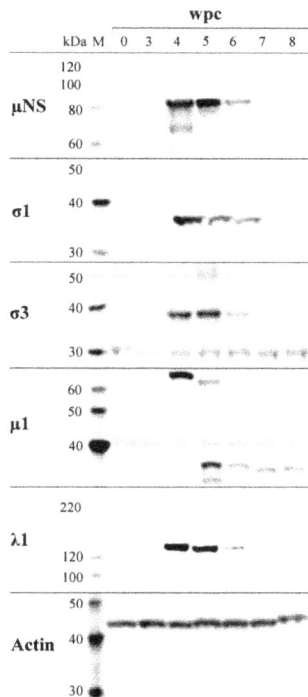

Figure 6. Presence of PRV proteins in blood cells. Pooled blood cell samples (*n* = 6) from each week were analyzed by Western blotting, targeting μNS, σ1, σ3, μ1 and λ1. M = molecular weight standard. Actin was used as control for protein load.

3.8. PRV Proteins Interact with μNS

Interaction between μNS and other PRV proteins was studied by IP and WB (Figure 7). At 4 wpc, μNS was detected as a 70 kDa protein and at the same time-point the structural proteins λ1, μ1, σ1 and σ3 were co-immunoprecipitated. At 5 wpc, μNS was detected in three different sizes ranging from 70 kDa to 83.5 kDa (putative full-length μNS). However, the only structural proteins co-immunoprecipitating with μNS at 5 wpc were σ3 and the 35 kDa fragment of μ1 (see above). Interactions between μNS and other viral proteins were also investigated by liquid chromatography–mass spectrometry (LC–MS; Table 1) and peptides corresponding to λ1, λ2, λ3, μNS, μ1, σNS and σ1 were identified.

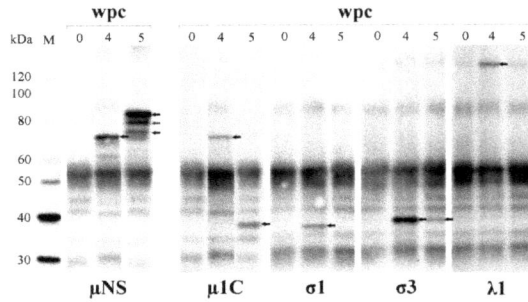

Figure 7. μNS interacts with multiple PRV proteins. Pooled blood cell lysate (*n* = 6) immunoprecipitated with μNS-antiserum, followed by Western blotting with primary antibodies detecting μNS, μ1C, σ1, σ3 and λ1 (arrows). M = molecular weight standard.

Table 1. Identified *piscine orthoreovirus* (PRV) peptides following immunoprecipitation with anti-μNS and mass spectrometry (MS).

* Band Excised from SDS-PAGE (kDa)	Identified PRV Proteins	Unique Peptides	Theoretical PRV Protein Size (kDa)
140 (5 wpc)	μNS	1	83.5
130 (5 wpc)	λ3	2	144.5
	λ2	7	143.7
	λ1	14	141.5
	μNS	9	83.5
	σ1	1	34.6
80 (5 wpc)	λ1	11	141.5
	μNS	24	83.5
70 (4 wpc)	μNS	16	83.5
35 (5 wpc)	μNS	4	83.5
	δ [†]	3	37.7
	σNS	1	39.1
	σ1	2	34.6

* Approximate size of proteins excised from bands following IP with anti-μNS antisera at four and five weeks post challenge (wpc). [†] Proteolytic fragment of μ1 proposed in the present work.

3.9. μNS Exists in Two Forms

WB of infected blood cells consistently produced two μNS bands of approximately 83.5 and 70 kDa (Figures 5 and 6). Due to the presence of two translation initiation sites in MRV segment M3, the LC–MS data were analyzed to identify putative shorter variants of the PRV μNS. The peptide distribution along the full-length μNS sequence and their spectrum matches are shown in Figure S7a. The μNS peptides and total spectrum matches obtained from the two bands are shown in Figure S7b. Several N-terminal μNS peptides were identified from the 83.5 kDa band that were not observed in the 70 kDa band. Furthermore, the peptide spectrum matches from the 83.5 kDa and 70 kDa bands in the 200 amino acid N-terminus were 10 to 1, respectively. In contrast, for the remaining C-terminal μNS sequence, the 83.5 kDa and 70 kDa bands produced similar or identical peptide spectrum matches, with a ratio of 66 to 63 (Figure S7). These results point to the presence of a second translation initiation site in the 5'- region of the μNS ORF. Start sites at M_{85}, M_{94}, M_{115} or M_{169} would provide proteins with predicted sizes of 74.5, 73.6, 71.1 and 65.5 kDa, respectively. M_{115} is the most likely candidate due to its size and presence in all PRV strains.

3.10. μ1 Has Two Putative Proteolytic Cleavage Sites

WB targeting the μ1 protein showed that the protein is present in different forms during infection. The putative full-length μ1 (74.1 kDa) was detected at 4 wpc (Figures 6 and 7). In contrast, smaller versions, with estimated sizes of 70, 37 and 32 kDa, replaced the full-length variant at 5 wpc (Figure 6). The three size variants from 5 wpc were subjected to LC–MS analysis (Figure S8). The 70 kDa band most likely represents μ1C following pre-cleavage at $N_{42}P_{43}$ (MW 69.8 kDa). Of the fourteen peptide spectrum matches identified from the 70 kDa band, two were found to overlap $N_{42}P_{43}$ (Figure S8a). This is most likely due to carryover of slightly larger full-length μ1 (74.1 kDa) following gel excision. No peptides stretching N-terminal to $N_{42}P_{43}$ were identified from the 37 and 32 kDa bands. Additional semi-tryptic peptides, i.e., peptides generated by trypsin cleavage at one end but not the other, were identified from both the 37 and 32 kDa bands (Figure S8a). Among these is a peptide identified from the 32 kDa band harboring an N-terminal S_{388}. Cleavage of μ1C at $F_{387}S_{388}$ would yield N- and C-terminal fragments of 37.7 kDa and 32.1 kDa, respectively. The distribution of peptide sequences and peptide spectrum matches provides support for proteolytic cleavage at or close to $F_{387}S_{388}$ (Figure S8). The results suggest that the 37 kDa and 32 kDa bands represent the PRV homologues of MRV μ1 fragments δ and ϕ, respectively. Besides the μ1 peptide sequences, peptides originating from other PRV proteins with sizes close to the sizes of the three excised fragments, were also identified. Peptide sequences matching λ1 and μNS (one peptide spectrum match each) were identified from the 70 kDa band, sequences matching σ1, σ3 and σNS were identified from the 37 kDa band (2, 2 and 11 peptide spectrum matches, respectively) and σ2 sequences were identified from the 32 kDa band (four peptide spectrum matches).

4. Discussion

Screening of farmed Atlantic salmon has indicated that PRV is ubiquitous in seawater and causes a persistent infection [9,11,41,42]. The study of PRV pathogenesis has been hampered by the lack of susceptible cell lines, and is currently dependent upon in vivo experiments. The fish in this experiment were challenged by cohabitation, i.e., through a natural transmission route. To ensure coordinated onset of infection, a high ratio of shedder fish was used. We found that PRV infection of salmon blood cells is acute and transient, with a peak lasting for 1–2 weeks under these experimental conditions.

Erythrocytes are major target cells for PRV [19]. Piscine erythrocytes are nucleated and contain the transcriptional and translational machinery enabling virus replication both in vivo and ex vivo [16,19]. We detected various PRV proteins in blood cells from 4 wpc, and the amount of protein was reduced at 6 wpc. Innate antiviral gene expression also peaked at 4–6 wpc and all selected genes were significantly induced during the peak period, in line with PRV protein production. In contrast to the transient peak displayed by PRV proteins, the viral RNA levels in blood cells persisted. The viral RNA level though, varied for the targeted genomic segments; the level of M2 (μ1) and M3 (μNS) remained high throughout the trial, while S1 (σ3) transcripts decreased from 6 wpc. TEM analysis corresponded well with viral protein production, i.e., the lamellar structures observed at 4 wpc developed into inclusions containing reovirus-like particles at 5 wpc, while no virus particles could be observed at 7 wpc. The findings support PRV, causing an acute infection in blood cells where high PRV protein and particle production are sustained 1–2 weeks before the infection becomes persistent. Our study shows that, after the acute phase, the PRV RNA level as determined by RT-qPCR does not reflect the virus load in blood.

The salmon does not appear to be able to eliminate PRV. Challenge experiments have shown that PRV RNA can be detected at a steady level in heart and liver until 36 wpc (end of experiment) [41], and in blood for more than a year after challenge [9,19]. In an experiment where the infectious potential of persistently PRV infected Atlantic salmon was studied, sentinel fish were added at 59 wpc, but no transmission to the sentinel fish was observed [9]. This indicates that fish persistently infected with PRV do not continuously shed the virus. Viral persistence is common in fish and has been demonstrated for several RNA viruses [9,43–47]. The only PRV protein that could be detected after the peak of virus protein production was a fragment of μ1, suggesting a possible role for this protein in persistent infection. In farmed salmon, where the size of the population in a net pen may exceed a hundred thousand individuals, and in the whole farm be more than a million fish, viral persistence in the population, but not necessarily in the individual, is also a critical parameter.

PRV infection in erythrocytes has previously been shown to induce expression of type I interferon and interferon-regulated genes [16,18]. In this study, the level of viral RNA correlated with the innate antiviral response in individual fish, with the exception of M3 expression after the virus peak (6–8 wpc). The continuous production of M3 RNA indicates that the innate antiviral immune response primarily inhibits virus replication post transcriptionally, which is in line with the functions of PKR and ISG15 on translation and protein modification, respectively [48,49].

Orthoreoviruses generate viral factories in the cytoplasm of infected cells [21,27,50–52], and PRV forms cytoplasmic globular viral factories resembling the structures produced by MRV T3D [16,19,31]. Viral factories are structures where virus replication and assembly occur, and thus where the viral proteins co-localize. The secluded nature of the viral factories modulates the level of the innate antiviral immune response. The orthoreoviral protein μNS is orchestrating the construction of the factories and in this study and earlier studies we have found that λ1, λ2, λ3, μ1, σNS, σ1, σ2 and σ3 interact with μNS [31]. The σ3 protein co-precipitated with μNS but was not identified by MS, however WB can be more sensitive than LC–MS [52]. This suggests that μNS interacts directly or indirectly with all three λ-proteins, the μ1 protein, and possibly all four σ-proteins. The μNS protein was detected in different molecular sizes at specific time points. Further investigations led to the finding of four possible internal translation initiation sites in the μNS gene. The M_{115} residue was determined to be the best candidate as M_{94} is not conserved among all PRV isolates, and M_{85} and M_{169} are unlikely due to the sizes of the proteins generated. Post-translational cleavage to generate μNSC as shown for ARV μNS cannot be excluded, although the specific proteolytic cleavage site in the ARV protein is not conserved in PRV [2,33,53]. The different μNS size variants, i.e., full-length μNS and the 70 kDa variant with putative translation initiation at M_{115}, may differ in their interactions with other PRV proteins. At 4 wpc, when only the 70 kDa variant of μNS was detected following IP, all targeted structural proteins co-precipitated. However, at 5 wpc, when full-length μNS was dominant, only the σ3 protein and the assumed μ1 fragment δ co-precipitated. Studies previously performed on aquareoviruses and ARV indicate that recruitment of viral proteins into viral factories occurs in a predefined order through direct or indirect association with μNS [50,54].

Four different molecular sizes of the μ1 protein were observed in the infected blood cells. Previous multiple sequence alignments of the μ1 amino acid sequence showed absolute conservation of the G2-myristoylation site and the autolytic $N_{42}P_{43}$ cleavage site, both regarded as crucial for reovirus μ1-mediated membrane penetration [2]. The band observed at 4 wpc represents the full-length μ1 protein while the 70 kDa band at 5 wpc most likely represents μ1C.

Although peptides containing amino acid sequences overlapping the $N_{42}P_{43}$ site were observed from the 70 kDa band following LC–MS, peptides ending in P_{43} were present in equal amount. We conclude that the presence of the $N_{42}P_{43}$ overlapping peptides originate from carryover of the slightly larger full-length μ1 following gel excision. In addition, proteins can exhibit different abilities to separate in SDS-PAGE. This explains the presence of a minor fraction of peptides from the δ fragment (i.e., 37 kDa band) in the φ fragment (i.e., 32 kDa band) and vice versa. No peptide sequences overlapping $N_{42}P_{43}$ were identified from the 37 and 32 kDa bands. Rather, a higher number of peptides with an N-terminal P_{43} generated by non-tryptic cleavage were identified, providing additional support for cleavage at $N_{42}P_{43}$.

MRV μ1 contains a second cleavage site in its C-terminal region which, upon cleavage by exogenous proteases, generates the additional fragments δ and φ [55]. In the present study, we propose that the 37 kDa and 32 kDa bands represent the PRV homologues of the MRV δ and φ proteins. Hence, PRV φ contains a larger N-terminal portion of μ1 compared to MRV φ. Although there is only 28% identity at the amino acid level [2], the secondary structure of the PRV μ1 monomer predicted by

PSIPRED [39] (not shown) is very similar to that of MRV μl [56]. This includes the helix-rich region in the C-terminal end [57], which for MRV largely constitutes the φ fragment shown to be crucial for membrane penetration, apoptosis induction and intracellular localization [57,58]. An interesting observation is that the three PRV μ1 peptide sequences detected in the 35 kDa band following IP with anti-μNS were all N-terminal to the proposed φ region, suggesting that μNS-interacting sites on μ1 may be located in the proposed δ region, between P_{43} and F_{387}. From 7 wpc and onwards, the only PRV protein detected was a ~35 kDa protein which could represent the δ proteolytic fragment.

Production-related diseases are often multifactorial and the outcome of a PRV infection is influenced by viral strain, age of the fish, production and environmental factors. Recently, PRV was demonstrated to be the etiologic agent of EIBS, causing anemia and mass mortality in juvenile Coho salmon [7]. The level of anemia in EIBS corresponded well with the level of viral replication in blood and it is therefore tempting to suggest that EIBS is a consequence of acute PRV infection, i.e., the direct effect of virus PRV replication in erythrocytes. PRV is also the causative agent of HSMI [1,4], which appears 2–3 weeks after virus replication peaks in blood cells. The dominance of CD8 positive inflammatory cells found in the HSMI specific heart lesions indicates that immune mediated mechanisms are a major cause of the myocarditis.

In this study, we show that PRV infection has an acute phase in blood cells with high virus production before the infection subsides to a low persistent level. The continued transcription of viral RNA in the persistent phase suggests that the innate antiviral immune response may act to inhibit the virus infection post transcriptionally.

Supplementary Materials: The following are available online at www.mdpi.com/1999-4915/9/3/49/s1, Figure S1: Specificity of μNS and λ1 antisera; Figure S2: PRV RNA load in blood cells (second challenge experiment); Figure S3: Expression of immune genes in blood cells; Figure S4: Expression of EF1α during PRV infection; Figure S5: PRV σ1 positive blood cells detected by flow cytometry; Figure S6: Presence of PRV proteins in blood cells (second challenge experiment); Figure S7: LC–MS analyses of PRV μNS; Figure S8: LC–MS analyses of PRV μ1; Table S1: Primers and probes used for construction of plasmids and expression of viral RNA levels; Table S2: Primers used for analysis of antiviral gene expression.

Acknowledgments: The Research Council of Norway supported the research with grant #237315/E40 and #235788. We would also like to thank Elisabeth Furuseth Hansen and Ida Aksnes from the University of Life Sciences, and Elisabeth Dahl Nybø at the Norwegian Veterinary Institute (NVI) for technical and scientific assistance.

Author Contributions: H.M.H.: Study design (parts), experiments, analysis, interpretation of data, drafting, revising and approving the manuscript. Ø.W.: Flow cytometry experiment (parts), analysis and interpretation of data, revising and approving the manuscript. T.M.: Data analysis and interpretation, writing, revising and approving the manuscript. M.L.: Challenge experiments, revising and approving the manuscript. B.T.: LC–MS experiments and data analysis, revising and approving the manuscript. I.B.N.: Antibody production, revising and approving the manuscript. S.B.: Flow cytometry experiment (parts), interpretation of data, revising and approving the manuscript. M.K.D.: Analysis, interpretation of data, drafting (parts), revising and approving the manuscript. E.R.: Study design, analysis, interpretation of data, drafting, revising and approving the manuscript.

References

1. Palacios, G.; Lovoll, M.; Tengs, T.; Hornig, M.; Hutchison, S.; Hui, J.; Kongtorp, R.T.; Savji, N.; Bussetti, A.V.; Solovyov, A.; et al. Heart and skeletal muscle inflammation of farmed salmon is associated with infection with a novel reovirus. *PLoS ONE* **2010**, *5*, e11487. [CrossRef] [PubMed]
2. Markussen, T.; Dahle, M.K.; Tengs, T.; Lovoll, M.; Finstad, O.W.; Wiik-Nielsen, C.R.; Grove, S.; Lauksund, S.; Robertsen, B.; Rimstad, E. Sequence analysis of the genome of piscine orthoreovirus (PRV) associated with heart and skeletal muscle inflammation (HSMI) in Atlantic salmon (*Salmo salar*). *PLoS ONE* **2013**, *8*, e70075. [CrossRef]
3. Day, J.M. The diversity of the orthoreoviruses: Molecular taxonomy and phylogentic divides. *Infect. Genet. Evol.* **2009**, *9*, 390–400. [CrossRef] [PubMed]

4. Finstad, O.W.; Falk, K.; Lovoll, M.; Evensen, O.; Rimstad, E. Immunohistochemical detection of piscine reovirus (PRV) in hearts of Atlantic salmon coincide with the course of heart and skeletal muscle inflammation (HSMI). *Vet. Res* **2012**, *43*, 27. [CrossRef] [PubMed]

5. Olsen, A.B.; Hjortaas, M.; Tengs, T.; Hellberg, H.; Johansen, R. First description of a new disease in Rainbow Trout (*Oncorhynchus mykiss* (Walbaum)) similar to Heart and skeletal muscle inflammation (HSMI) and detection of a gene sequence related to Piscine Orthoreovirus (PRV). *PLoS ONE* **2015**, *10*, e0131638. [CrossRef] [PubMed]

6. Sibley, S.D.; Finley, M.A.; Baker, B.B.; Puzach, C.; Armien, A.G.; Giehtbrock, D.; Goldberg, T.L. Novel reovirus associated with epidemic mortality in wild Largemouth Bass (*Micropterus salmoides*). *J. Gen. Virol.* **2016**, *97*, 2482–2487. [PubMed]

7. Takano, T.; Nawata, A.; Sakai, T.; Matsuyama, T.; Ito, T.; Kurita, J.; Terashima, S.; Yasuike, M.; Nakamura, Y.; Fujiwara, A.; et al. Full-Genome sequencing and confirmation of the causative agent of Erythrocytic inclusion body syndrome in Coho Salmon identifies a new type of Piscine Orthoreovirus. *PLoS ONE* **2016**, *11*, e0165424. [CrossRef] [PubMed]

8. Kibenge, M.J.; Iwamoto, T.; Wang, Y.; Morton, A.; Godoy, M.G.; Kibenge, F.S. Whole-genome analysis of piscine reovirus (PRV) shows PRV represents a new genus in family Reoviridae and its genome segment S1 sequences group it into two separate sub-genotypes. *Virol. J.* **2013**, *10*, 230. [CrossRef] [PubMed]

9. Garver, K.A.; Johnson, S.C.; Polinski, M.P.; Bradshaw, J.C.; Marty, G.D.; Snyman, H.N.; Morrison, D.B.; Richard, J. Piscine Orthoreovirus from Western North America is transmissible to Atlantic Salmon and Sockeye Salmon but fails to cause Heart and skeletal muscle inflammation. *PLoS ONE* **2016**, *11*, e0146229. [CrossRef] [PubMed]

10. Ferguson, H.W.; Kongtorp, R.T.; Taksdal, T.; Graham, D.; Falk, K. An outbreak of disease resembling heart and skeletal muscle inflammation in Scottish farmed salmon, *Salmo salar* L., with observations on myocardial regeneration. *J. Fish Dis.* **2005**, *28*, 119–123. [CrossRef] [PubMed]

11. Lovoll, M.; Alarcon, M.; Bang Jensen, B.; Taksdal, T.; Kristoffersen, A.B.; Tengs, T. Quantification of Piscine reovirus (PRV) at different stages of Atlantic salmon *Salmo salar* production. *Dis. Aquat. Organ.* **2012**, *99*, 7–12. [CrossRef] [PubMed]

12. Bjorgen, H.; Wessel, O.; Fjelldal, P.G.; Hansen, T.; Sveier, H.; Saebo, H.R.; Enger, K.B.; Monsen, E.; Kvellestad, A.; Rimstad, E.; et al. *Piscine orthoreovirus* (PRV) in red and melanised foci in white muscle of Atlantic salmon (*Salmo salar*). *Vet. Res.* **2015**, *46*, 89. [CrossRef] [PubMed]

13. Kongtorp, R.T.; Halse, M.; Taksdal, T.; Falk, K. Longitudinal study of a natural outbreak of heart and skeletal muscle inflammation in Atlantic salmon, *Salmo salar* L. *J. Fish Dis.* **2006**, *29*, 233–244. [CrossRef] [PubMed]

14. Wiik-Nielsen, C.R.; Ski, P.M.; Aunsmo, A.; Lovoll, M. Prevalence of viral RNA from piscine reovirus and piscine myocarditis virus in Atlantic salmon, *Salmo salar* L., broodfish and progeny. *J. Fish Dis.* **2012**, *35*, 169–171. [CrossRef] [PubMed]

15. Johansen, L.H.; Dahle, M.K.; Wessel, O.; Timmerhaus, G.; Lovoll, M.; Rosaeg, M.; Jorgensen, S.M.; Rimstad, E.; Krasnov, A. Differences in gene expression in Atlantic salmon parr and smolt after challenge with *Piscine orthoreovirus* (PRV). *Mol. Immunol.* **2016**, *73*, 138–150. [CrossRef] [PubMed]

16. Wessel, O.; Olsen, C.M.; Rimstad, E.; Dahle, M.K. *Piscine orthoreovirus* (PRV) replicates in Atlantic salmon (*Salmo salar* L.) erythrocytes ex vivo. *Vet. Res.* **2015**, *46*, 26. [CrossRef] [PubMed]

17. Morera, D.; Roher, N.; Ribas, L.; Balasch, J.C.; Donate, C.; Callol, A.; Boltana, S.; Roberts, S.; Goetz, G.; Goetz, F.W.; et al. RNA-Seq reveals an integrated immune response in nucleated erythrocytes. *PLoS ONE* **2011**, *6*, e26998. [CrossRef] [PubMed]

18. Dahle, M.K.; Wessel, O.; Timmerhaus, G.; Nyman, I.B.; Jorgensen, S.M.; Rimstad, E.; Krasnov, A. Transcriptome analyses of Atlantic salmon (*Salmo salar* L.) erythrocytes infected with piscine orthoreovirus (PRV). *Fish Shellfish Immunol.* **2015**, *45*, 780–790. [CrossRef] [PubMed]

19. Finstad, O.W.; Dahle, M.K.; Lindholm, T.H.; Nyman, I.B.; Lovoll, M.; Wallace, C.; Olsen, C.M.; Storset, A.K.; Rimstad, E. *Piscine orthoreovirus* (PRV) infects Atlantic salmon erythrocytes. *Vet. Res.* **2014**, *45*, 35. [CrossRef] [PubMed]

20. Kongtorp, R.T.; Taksdal, T.; Lyngoy, A. Pathology of heart and skeletal muscle inflammation (HSMI) in farmed Atlantic salmon *Salmo salar. Dis. Aquat. Organ.* **2004**, *59*, 217–224. [CrossRef] [PubMed]

21. Netherton, C.; Moffat, K.; Brooks, E.; Wileman, T. A guide to viral inclusions, membrane rearrangements, factories, and viroplasm produced during virus replication. *Adv. Virus Res.* **2007**, *70*, 101–182. [PubMed]

22. Knipe, D.M.; Howley, P.M. *Fields Virology*, 5th ed.; Wolters Kluwer/Lippincott Williams & Wilkins Health: Philadelphia, PA, USA, 2007; pp. 1854–1858.

23. Lee, P.W.; Hayes, E.C.; Joklik, W.K. Protein sigma 1 is the reovirus cell attachment protein. *Virology* **1981**, *108*, 156–163. [CrossRef]

24. Nibert, M.L.; Fields, B.N. A carboxy-terminal fragment of protein mu 1/mu 1C is present in infectious subvirion particles of mammalian reoviruses and is proposed to have a role in penetration. *J. Virol.* **1992**, *66*, 6408–6418. [PubMed]

25. Thete, D.; Snyder, A.J.; Mainou, B.A.; Danthi, P. Reovirus mu1 protein affects infectivity by altering virus-receptor interactions. *J. Virol.* **2016**, *90*, 10951–10962. [CrossRef] [PubMed]

26. Becker, M.M.; Peters, T.R.; Dermody, T.S. Reovirus σNS and μNS proteins form cytoplasmic inclusion structures in the absence of viral infection. *J. Virol.* **2003**, *77*, 5948–5963. [CrossRef] [PubMed]

27. Schiff, L.A.; Nibert, M.L.; Tyler, K.L. Orthoreoviruses and their replication. In *Fields virology*, 5th ed.; Knipe, D.M., Howley, P.M., Fields, B.N., Eds.; Wolters Kluwer/Lippincott Williams & Wilkins: Philadelphia, PA, USA, 2007; Volume 2, pp. 1853–1915.

28. Jayasuriya, A.K.; Nibert, M.L.; Fields, B.N. Complete nucleotide sequence of the M2 gene segment of reovirus type 3 dearing and analysis of its protein product mu 1. *Virology* **1988**, *163*, 591–602. [CrossRef]

29. Duncan, R. The low pH-dependent entry of avian reovirus is accompanied by two specific cleavages of the major outer capsid protein mu 2C. *Virology* **1996**, *219*, 179–189. [CrossRef] [PubMed]

30. Wiener, J.R.; Joklik, W.K. Evolution of reovirus genes: A comparison of serotype 1, 2, and 3 M2 genome segments, which encode the major structural capsid protein mu 1C. *Virology* **1988**, *163*, 603–613. [CrossRef]

31. Haatveit, H.M.; Nyman, I.B.; Markussen, T.; Wessel, O.; Dahle, M.K.; Rimstad, E. The non-structural protein μNS of piscine orthoreovirus (PRV) forms viral factory-like structures. *Vet. Res.* **2016**, *47*, 5. [CrossRef] [PubMed]

32. Parker, J.S.; Broering, T.J.; Kim, J.; Higgins, D.E.; Nibert, M.L. Reovirus core protein mu2 determines the filamentous morphology of viral inclusion bodies by interacting with and stabilizing microtubules. *J. Virol.* **2002**, *76*, 4483–4496. [CrossRef] [PubMed]

33. Busch, L.K.; Rodriguez-Grille, J.; Casal, J.I.; Martinez-Costas, J.; Benavente, J. Avian and mammalian reoviruses use different molecular mechanisms to synthesize their microNS isoforms. *J. Gen. Virol.* **2011**, *92*, 2566–2574. [CrossRef] [PubMed]

34. Touris-Otero, F.; Martinez-Costas, J.; Vakharia, V.N.; Benavente, J. Avian reovirus nonstructural protein microNS forms viroplasm-like inclusions and recruits protein sigmaNS to these structures. *Virology* **2004**, *319*, 94–106. [CrossRef] [PubMed]

35. Wiener, J.R.; Bartlett, J.A.; Joklik, W.K. The sequences of reovirus serotype 3 genome segments M1 and M3 encoding the minor protein mu 2 and the major nonstructural protein mu NS, respectively. *Virology* **1989**, *169*, 293–304. [CrossRef]

36. McCutcheon, A.M.; Broering, T.J.; Nibert, M.L. Mammalian reovirus M3 gene sequences and conservation of coiled-coil motifs near the carboxyl terminus of the microNS protein. *Virology* **1999**, *264*, 16–24. [CrossRef] [PubMed]

37. Koehler, C.J.; Bollineni, R.C.; Thiede, B. Application of the half decimal place rule to increase the peptide identification rate. *Rapid Commun. Mass Spectrom.* **2016**, *31*, 227–233. [CrossRef] [PubMed]

38. ExPASy Bioinformatics Resource Portal. Available online: http://web.expasy.org/compute_pi/ (accessed on 1 April 2016).

39. UCL Department Of Computer Science. Available online: http://bioinf.cs.ucl.ac.uk/psipred/ (accessed on 1 April 2016).

40. Su, J.; Zhang, R.; Dong, J.; Yang, C. Evaluation of internal control genes for qRT-PCR normalization in tissues and cell culture for antiviral studies of grass carp (*Ctenopharyngodon idella*). *Fish Shellfish Immunol.* **2011**, *30*, 830–835. [CrossRef] [PubMed]

41. Lovoll, M.; Wiik-Nielsen, J.; Grove, S.; Wiik-Nielsen, C.R.; Kristoffersen, A.B.; Faller, R.; Poppe, T.; Jung, J.; Pedamallu, C.S.; Nederbragt, A.J.; et al. A novel totivirus and piscine reovirus (PRV) in Atlantic salmon (*Salmo salar*) with cardiomyopathy syndrome (CMS). *Virol. J.* **2010**, *7*, 309. [CrossRef] [PubMed]

42. Marty, G.D.; Morrison, D.B.; Bidulka, J.; Joseph, T.; Siah, A. Piscine reovirus in wild and farmed salmonids in British Columbia, Canada: 1974–2013. *J. Fish Dis.* **2015**, *38*, 713–728. [CrossRef] [PubMed]

43. Julin, K.; Johansen, L.H.; Sommer, A.I.; Jorgensen, J.B. Persistent infections with infectious pancreatic necrosis virus (IPNV) of different virulence in Atlantic salmon, *Salmo salar* L. *J. Fish Dis.* **2015**, *38*, 1005–1019. [CrossRef] [PubMed]

44. Gjessing, M.C.; Kvellestad, A.; Ottesen, K.; Falk, K. Nodavirus provokes subclinical encephalitis and retinochoroiditis in adult farmed Atlantic cod, *Gadus morhua* L. *J. Fish Dis.* **2009**, *32*, 421–431. [CrossRef] [PubMed]

45. Amend, D.F. Detection and transmission of infectious hematopoietic necrosis virus in rainbow trout. *J. Wildl. Dis.* **1975**, *11*, 471–478. [CrossRef] [PubMed]

46. Neukirch, M. Demonstration of persistent viral haemorrhagic septicaemia (VHS) virus in rainbow trout after experimental waterborne infection. *Zentralbl Veterinarmed B* **1986**, *33*, 471–476. [CrossRef] [PubMed]

47. Hershberger, P.K.; Gregg, J.L.; Grady, C.A.; Taylor, L.; Winton, J.R. Chronic and persistent viral hemorrhagic septicemia virus infections in Pacific herring. *Dis. Aquat. Organ.* **2010**, *93*, 43–49. [CrossRef] [PubMed]

48. Dalet, A.; Gatti, E.; Pierre, P. Integration of PKR-dependent translation inhibition with innate immunity is required for a coordinated anti-viral response. *FEBS Lett.* **2015**, *589*, 1539–1545. [CrossRef] [PubMed]

49. Durfee, L.A.; Lyon, N.; Seo, K.; Huibregtse, J.M. The ISG15 conjugation system broadly targets newly synthesized proteins: Implications for the antiviral function of ISG15. *Mol. Cell* **2010**, *38*, 722–732. [CrossRef] [PubMed]

50. Touris-Otero, F.; Cortez-San Martin, M.; Martinez-Costas, J.; Benavente, J. Avian reovirus morphogenesis occurs within viral factories and begins with the selective recruitment of sigmaNS and lambdaA to microNS inclusions. *J. Mol. Biol.* **2004**, *341*, 361–374. [CrossRef] [PubMed]

51. Carroll, K.; Hastings, C.; Miller, C.L. Amino acids 78 and 79 of Mammalian Orthoreovirus protein microNS are necessary for stress granule localization, core protein lambda 2 interaction, and de novo virus replication. *Virology* **2014**, *448*, 133–145. [CrossRef] [PubMed]

52. Aebersold, R.; Burlingame, A.L.; Bradshaw, R.A. Western blots versus selected reaction monitoring assays: time to turn the tables? *Mol. Cell. Proteomics* **2013**, *12*, 2381–2382. [CrossRef] [PubMed]

53. Rodriguez-Grille, J.; Busch, L.K.; Martinez-Costas, J.; Benavente, J. Avian reovirus-triggered apoptosis enhances both virus spread and the processing of the viral nonstructural muNS protein. *Virology* **2014**, *462–463*, 49–59. [CrossRef] [PubMed]

54. Yan, L.; Zhang, J.; Guo, H.; Yan, S.; Chen, Q.; Zhang, F.; Fang, Q. Aquareovirus NS80 initiates efficient viral replication by retaining core proteins within replication-associated viral inclusion bodies. *PLoS ONE* **2015**, *10*, e0126127. [CrossRef] [PubMed]

55. Nibert, M.L.; Odegard, A.L.; Agosto, M.A.; Chandran, K.; Schiff, L.A. Putative autocleavage of reovirus mu1 protein in concert with outer-capsid disassembly and activation for membrane permeabilization. *J. Mol. Biol.* **2005**, *345*, 461–474. [CrossRef] [PubMed]

56. Liemann, S.; Chandran, K.; Baker, T.S.; Nibert, M.L.; Harrison, S.C. Structure of the reovirus membrane-penetration protein, Mu1, in a complex with is protector protein, Sigma3. *Cell* **2002**, *108*, 283–295. [CrossRef]

57. Coffey, C.M.; Sheh, A.; Kim, I.S.; Chandran, K.; Nibert, M.L.; Parker, J.S. Reovirus outer capsid protein micro1 induces apoptosis and associates with lipid droplets, endoplasmic reticulum, and mitochondria. *J. Virol.* **2006**, *80*, 8422–8438. [CrossRef] [PubMed]

58. Danthi, P.; Coffey, C.M.; Parker, J.S.; Abel, T.W.; Dermody, T.S. Independent regulation of reovirus membrane penetration and apoptosis by the mu1 phi domain. *PLoS Pathog.* **2008**, *4*, e1000248. [CrossRef] [PubMed]

viruses

MDPI

Article

Nutrients and Other Environmental Factors Influence Virus Abundances across Oxic and Hypoxic Marine Environments

Jan F. Finke [1,2], Brian P. V. Hunt [1,2,3], Christian Winter [1,†], Eddy C. Carmack [4] and Curtis A. Suttle [1,2,5,6,*]

1   Department of Earth, Ocean and Atmospheric Sciences, University of British Columbia,
    Vancouver, BC V6T 1Z4, Canada; jfinke@eos.ubc.ca (J.F.F.); b.hunt@oceans.ubc.ca (B.P.V.H.);
    christian.winter@univie.ac.at (C.W.)
2   Institute for the Oceans and Fisheries, University of British Columbia, Vancouver, BC V6T 1Z4, Canada
3   Hakai Institute, P.O. Box 309, Heriot Bay, BC, Canada
4   Fisheries and Oceans Canada, Institute of Ocean Sciences, Sidney, BC V8L 4B2, Canada;
    Eddy.Carmack@dfo-mpo.gc.ca
5   Department of Botany, University of British Columbia, Vancouver, BC V6T 1Z4, Canada
6   Department of Microbiology and Immunology, University of British Columbia,
    Vancouver, BC V6T 1Z3, Canada
*   Correspondence: suttle@science.ubc.ca
†   Current Address: Department of Limnology and Bio-Oceanography, University of Vienna, Althanstrasse 14,
    1090 Vienna, Austria.

Academic Editors: Mathias Middelboe and Corina P. D. Brussaard
Received: 15 March 2017; Accepted: 13 June 2017; Published: 17 June 2017

Abstract: Virus particles are highly abundant in seawater and, on average, outnumber microbial cells approximately 10-fold at the surface and 16-fold in deeper waters; yet, this relationship varies across environments. Here, we examine the influence of a suite of environmental variables, including nutrient concentrations, salinity and temperature, on the relationship between the abundances of viruses and prokaryotes over a broad range of spatial and temporal scales, including along a track from the Northwest Atlantic to the Northeast Pacific via the Arctic Ocean, and in the coastal waters of British Columbia, Canada. Models of varying complexity were tested and compared for best fit with the Akaike Information Criterion, and revealed that nitrogen and phosphorus concentrations, as well as prokaryote abundances, either individually or combined, had significant effects on viral abundances in all but hypoxic environments, which were only explained by a combination of physical and chemical factors. Nonetheless, multivariate models of environmental variables showed high explanatory power, matching or surpassing that of prokaryote abundance alone. Incorporating both environmental variables and prokaryote abundances into multivariate models significantly improved the explanatory power of the models, except in hypoxic environments. These findings demonstrate that environmental factors could be as important as, or even more important than, prokaryote abundance in describing viral abundance across wide-ranging marine environments.

Keywords: viral abundance; environmental variables; multivariate model; Akaike Information Criterion

---

1. Introduction

Viruses play an important role in aquatic ecosystems, which includes influencing host diversity and the flux of nutrients and carbon through the viral shunt [1]. They are highly abundant, typically ranging in concentration across different environments from 10^6 mL^{-1} to as high as 10^8 mL^{-1} [2–4], with generally lower abundances in the deep sea and higher abundances at productive coastal sites.

Because contact rates between viruses and their potential hosts are proportional to viral abundance, higher densities of viruses generally lead to a greater impact on microbial host populations [5,6]. Given that the most abundant host cells for viruses in the oceans are prokaryotes, and that these are largely bacteria, prokaryotes will henceforth be referred to as bacteria.

Over the years, it has been established that viral abundance is about an order of magnitude higher than bacterial abundance [7], but the virus to bacteria ratio (VBR) varies greatly among host-virus systems and environments [8–11]. In a meta-analysis of 25 studies, Wigington et al. [10] found that the VBR ranged from 10.5 to 16. They also demonstrated the limitation of models using a fixed VBR ratio of 10:1, and applied non-linear power functions to relate viral and bacterial abundances. Conversely, Knowles et al. [9] showed a linear correlation between viral and bacterial abundances across a range of habitats, and that there was a relative decrease in the relationship with increasing bacterial abundance. From Wigington et al. [10] and Knowles et al. [9], it is apparent that the relationship between viral and bacterial abundances varies substantially among studies. Additionally, there were significant differences in correlations between bacterial and viral abundances in samples from lakes, the upper Pacific, deep Pacific and Arctic oceans [11]. Observations that the VBR varies under different conditions and among locations implies that it could be affected by environmental variables, with burst size, viral decay rates and photosynthetic host density potentially affecting the VBR [10–12]. Hence, while viral and bacterial abundances for specific studies or locations are typically highly correlated, deriving relationships that extend across biomes requires models that include environmental variables that affect the virus–host relationship.

Temperature and salinity are environmental variables that can directly affect virus–host interactions. For example, in the microalgae *Phaeocystis globosa* and *Heterosigma akashiwo*, lysis of infected cells occurred over a narrow temperature range and the different viruses were inactivated above temperatures ranging from 20 to 35 °C [13,14]. A similar pattern of inactivation at 40 °C was shown for a phage of the marine prokaryote *Pseudoalteromonas marina* [15]. Inactivation temperatures for marine viruses, however, are usually above 20 °C, which is higher than that which many virus–host systems are likely to encounter in temperate and Arctic waters, but can play a role in microenvironments in temperate waters. Furthermore, a rise in temperatures can favor the switch from a lysogenic to a lytic cycle in a marine phage–host system [16], which would affect the total community viral production. Salinity has also been shown to interfere with the initial step of viral infection; salt concentrations above 3 M NaCl lowered infectivity and adsorption in a marine bacteria–virus system in culture [17]. Additionally, marine phages can require salt for particle stability [18]. However, another study showed an increase in viral abundance and drastic change in the viral community composition at hypersaline conditions above 240 practical salinity units (PSU) [19].

Light can also influence virus-host interactions in both positive and negative ways. Photosynthetically active radiation (PAR) is required for phytoplankton growth, and is thus crucial for replication of phytoplankton viruses. Even adsorption of viral particles to their host can be light dependent [20], as can be the duration of the viral replication cycle and the burst size [21,22]. Yet, some viruses infecting phytoplankton, including those infecting *H. akashiwo*, appear to be less sensitive to changes in the light regime [23,24]. Nonetheless, the final stage of virus replication is very energy demanding, and can be especially vulnerable to light limitation in photosynthetic hosts [25]. Light can also have highly negative effects on viral replication. For example, UV radiation is a major factor causing viral decay, and decay rates for viruses of bacteria, cyanobacteria and eukaryotic phytoplankton increase in proportion to irradiance [5,26–28]. In the ocean, light effects are restricted to the upper photic zone, with PAR influencing interactions of viruses of photosynthetic hosts, and UV radiation causing decay of all viruses.

Nutrients also have profound effects on virus–host interactions. Since viral particles mainly consist of a genome and a capsid, they have a different stoichiometric composition than cellular organisms. A recent study [29] calculated that the C:N:P stoichiometry of viruses is about 17:6:1, which is very different from that of their cellular hosts, which is typically 69:16:1 for heterotrophs and

106:16:1 for phototrophs [29–31]. Moreover, up to 87% of cellular phosphorus can be assimilated into viral particles during replication, highlighting the relatively high demand of viruses for nitrogen and phosphorus, and the importance of these nutrients for viral replication [29]. For example, phosphorus depletion can result in reduced viral production for a variety of prymnesiophytes and their viruses [32,33], and production of viruses infecting *Emiliania huxleyi* were affected by phosphate and nitrate availability [34]. In turn, phosphate addition can increase viral production [35]. The limited available data indicate that nitrogen limitation either has no impact, or reduces viral production [32,36]. Moreover, there is mounting evidence that hosts and viruses adapt to environmental conditions [37]. In summary, environmental factors affect viral replication, and thus would be expected to affect the relationship between virus and bacterial abundances.

Despite the highlighted importance of environmental factors to virus–host interactions, their relationship to the relative abundances of viruses and bacteria in the environment has not been rigorously explored. This study addresses these influences by exploring which environmental variables influence the relative abundances of viruses and bacteria across a wide range of samples derived from diverse environments. This approach allows better predictions of how environmental differences affect the relative abundances of viruses and bacteria.

2. Materials and Methods

Data from 515 samples were compiled from several years of data collected in Saanich Inlet (SI; 48°35′ N, 123°30′ W) and Rivers Inlet (RI; 51°26′ N, 127°38′ W) [38], BC, Canada, as well as along a cruise track from the Labrador Sea to the coast of British Columbia through the Arctic Ocean as part of the Canada's Three Oceans project (C3O) [39] (Figure 1). Water samples from depth profiles were collected with Go-Flo bottles and subsampled for various analyses, as detailed below. Samples were taken from surface waters to a maximum depth of 1000 m.

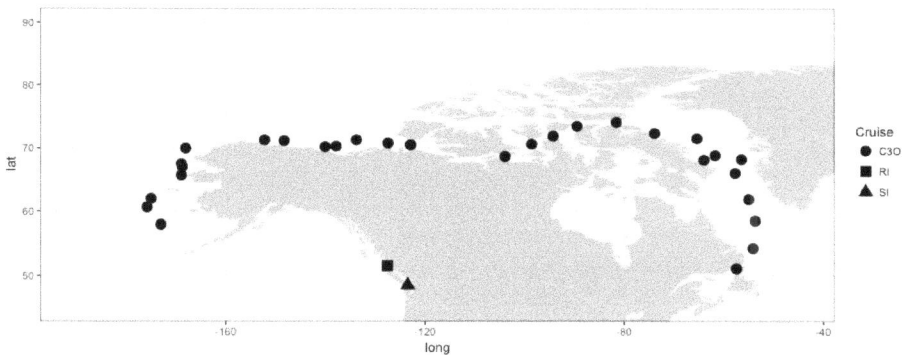

Figure 1. Map of sampling locations by project. Each location represents multiple depths and/or time points. C3O: Canada's Three Oceans project; lat: Latitude; long: Longitude; RI: Rivers Inlet; SI: Saanich Inlet.

Abundances of double-stranded DNA (dsDNA) viruses and bacteria were determined in duplicate water samples using a Beckton Dickinson FACSCalibur flow cytometer (Franklin Lakes, NJ, USA) with a 15 mW 488 nm air-cooled argon ion laser, as described in [40]. Briefly, samples were fixed for 15 min at 4 °C in the dark with electron microscopy-grade glutaraldehyde (25%; Sigma-Aldrich, Saint Louis, MO, USA), final concentration 0.5%, followed by snap-freezing in liquid nitrogen and storage at −80 °C. Right before analysis, the samples are thawed and diluted in 0.2 μm filtered, autoclaved 10:1 TE buffer (10 mM Tris HCl; 1 mM ethylenediaminetetraacetic acid (EDTA) pH 8.0) and stained with SYBR Green I (Invitrogen, Carlsbad, CA, USA) at a final concentration of 0.5×10^{-4} of the commercial stock, for 10 min at 80 °C in a water bath. Samples were diluted in TE buffer (pH 8.0), if necessary, to reach 100 to 1000 events s^{-1}. Viruses were discriminated by plotting green fluorescence against side scatter, and the results analyzed with CYTOWIN version 4.31 [41].

Nutrient samples were filtered through 0.22 μm pore-size polyvinylidene difluoride (PVDF) syringe filters and stored at −20 °C till analysis. Total nitrate (NO_3) (reduced to nitrite) and nitrite (referred to as the predominant nitrate hereafter), phosphate (PO_4) and silicate (SiO_4) were analyzed with a Bran & Luebbe AutoAnalyzer 3 (Norderstedt, Germany) using air-segmented continuous-flow analysis. Colorimetry was used to measure the concentrations of reduced nitrate [42] and silicate at 550 nm, and reduced orthophosphate [43] at 880 nm.

For physical data, in situ profiles of temperature, salinity and depth were measured with a SBE 25 (SI and RI) or SBE 911 (C3O) CTD (Seabird Electronics, Inc., Bellevue, WA, USA). Chlorophyll concentration was estimated by a fast-repetition-rate fluorometer (FRRF), for SI and RI a WetStar fluorometer (Seabird Electronics, Inc., Bellevue, WA, USA) for C3O a Seapoint Chlorophyll Fluorometer (Seapoint Sensors, Exeter, NH, USA), mounted to the CTD. Fluorescence data were converted to chlorophyll concentrations based on standard curves. These curves were derived from measurements of in situ fluorescence, as well as extracted chlorophyll concentrations made on samples from a range of environments. Oxygen was measured with a SBE 43 oxygen sensor and PAR was measured with a QSP-200PD (SI and RI) or QSP-2300 (C3O) profiling sensor (Biospherical Instruments, San Diego, CA, USA).

Of the 515 samples, 47 samples from Saanich Inlet were missing bacterial counts, and 211 samples from Rivers Inlet did not have PAR data; these were left out of the analysis when applicable. Other irregularly missing data points, with <10% missing per variable, were filled with weighted data by multiple imputation, a statistical technique to analyze data sets with missing values. The data were divided into the following three subsets: "Arctic", including sub-Arctic samples from the Atlantic and Pacific; "inlet"; and "hypoxic". Data from Saanich Inlet and Rivers Inlet comprised the inlet subset; data from C3O made up the Arctic subset; and all samples with an oxygen concentration below 1.5 mL·L^{-1} [44] were pooled into the hypoxic subset. Statistical analysis was done in the programming language, R [45]. A linear discriminant analysis (LDA) of the samples based on scaled environmental variables was performed with the MASS package (version 7.3-40) to confirm the prior classification of samples into environments. Input variables for the LDA were temperature, salinity, chlorophyll, nitrate, phosphate, silicate and oxygen. Samples for one sampling day and one site were removed from the inlet subset due to extremely high viral counts, exceeding 1.5 times the interquartile range, and were thus considered to be outliers. Temperature, salinity and chlorophyll were log transformed to compensate for outliers and approximate normal distribution. Viral and bacterial abundances were log_{10} transformed. Transformations were kept consistent across sub-sets of data so that the models were comparable. The data were explored for normal distributions in histogram plots and Pearson correlation coefficients were used to explore variables for patterns of collinearity (Figures S1–S3).

Single variable correlations were measured using linear models with log_{10} transformed viral and bacterial abundances, while nitrate and phosphate data were not transformed. The explanatory power of the models was expressed as the coefficient of determination (R^2) and significances in p-values; the slope of the regression is also given. Multivariate regressions were determined with generalized linear models (GLM), with a Gaussian distribution and an identity link function being run for log_{10} transformed viral abundance against environmental variables and/or log_{10} transformed bacterial abundance using the MASS package [46]. Models were run at a range of complexities, ranging from one input variable to all possible variables. For each complexity, the optimal combination of variables was selected based on the Akaike Information Criterion (AIC) with the Stats package [45]. Optimal models were then selected by comparing the AICs and considering improvements in explanatory power at different complexities; a relative drop in the AIC of two was considered relevant. Model fit was tested with a combined McFadden pseudo R^2, and significance was tested on z-values per coefficient. Pseudo R^2s were determined with the BaylorEdPsych (version 0.5) package [47]. The use of GLMs and model selection based on the AIC was done to account for deviations from a normal distribution in the variable and to reduce model complexity to significant predictors. Multicollinearity of predictors in the models was assessed by the Variance Inflation Factor (VIF), collinear predictors were then removed from the models, retaining only one. Models were assessed for their homogeneity of variance and the normal distribution of residuals, additionally the normal distribution of residuals was tested with the Shapiro–Wilk test.

3. Results

The data used in this study are categorized into "inlet" samples from Saanich and Rivers Inlets, "hypoxic" samples, mainly from deep inlet water, and "Arctic" samples from the Canadian Arctic and sub-Arctic; each environmental category has distinguishing environmental conditions.

Viral abundance data that went into models ranged from 4.83×10^5 to 1.40×10^8 viruses mL^{-1}, and bacterial abundances ranged from 7.31×10^4 to 7.40×10^7 bacteria mL^{-1} (Table 1). A set of outlier samples from June 2009 in Rivers Inlet had extraordinarily high viral abundances with 1.40×10^8 viruses mL^{-1} at 10 m, which remained above 4×10^7 viruses mL^{-1} until 320 m depth. Bacterial abundances were proportionally high and varied between 7.4×10^7 and 2.04×10^7 bacteria mL^{-1} over the same depths, but the environmental variables did not show a correlated pattern.

Table 1. Ranges, mean values and units of data included in the statistical analysis. PAR: Photosynthetically active radiation; PSU: Practical salinity units.

Variable	Min.	Max.	Mean	Unit
Temperature	−1.710	15	7	°C
Salinity	3.060	35	31	PSU
Chlorophyll	0.030	44	2	mg·m^{-3}
Oxygen	0.005	10	4	mL·L^{-1}
PAR	0.000	669	25	µmol quanta m^{-2}·s^{-1}
NO$_3$	0.010	54	15	µM
PO$_4$	0.006	7	2	µM
SiO$_4$	0.070	141	43	µM
Bacteria	7.31×10^4	7.40×10^7	1.66×10^6	Cells mL^{-1}
Viruses	4.83×10^5	1.40×10^8	8.35×10^6	Viruses mL^{-1}

The range in environmental data was also large. Temperature ranged from −2 to 15 °C and salinity from 3 to 35 PSU, while chlorophyll and oxygen ranged from 0.03 to 44 mg·m^{-3} and from 0.005 to 10 mL·L^{-1}, respectively. PAR data, which was only available for Saanich Inlet and C3O had a maximum of 669 µmol quanta m^{-2}·s^{-1} at the surface and was undetectable in hypoxic waters in Saanich Inlet. Nutrient values ranged from 0.01 to 54 µM for nitrate, 0.006 to 7 µM for phosphate and 0.07 to 141 µM for silicate. After classifying the data into the three environments and appropriate transformations, the data generally demonstrated normal distribution. However, even after log transformation, temperature and salinity in some environments were somewhat skewed (Figures S1–S3). Correlating all environmental variables, especially nutrient data in the inlet environment, showed some degree of collinearity based on the Pearson correlation coefficient (Figures S1–S3). Variables displaying collinearity in the multivariate models based on the VIF were subsequently reduced to one variable.

3.1. Samples Can Be Classified into Environments

Linear discriminant analysis (LDA) of all samples based on scaled environmental data, consisting of temperature, salinity, oxygen, nitrate, phosphate, silicate and chlorophyll, supported the classification of the data into three groups (Figure 2), reflecting Arctic, inlet and hypoxic environments. The first dimension LD1 describes 92.6% of the variation and the second dimension LD2 7.4%, with temperature and phosphate concentrations being the strongest components. The environments form well-defined clusters, with the Arctic and inlet samples partially overlapping and the hypoxic samples a clearly separated.

Besides their variability in temperature and salinity, the three environments varied markedly in the concentrations of nitrate and phosphate (Figure 3). Nitrate to phosphate ratios in the inlet and coastal environments co-varied with a ratio of about 12:1, higher than the average elemental N:P stoichiometry of 5:1 for viral particles, but lower than the ratio of 16:1 associated with phytoplankton in balanced growth or heterotrophic bacteria [29,31]. Nutrient concentrations also co-varied with depth, with surface samples generally being low in nutrients. Furthermore, coastal samples generally showed lower nitrate concentrations than inlet samples. The majority of samples had relatively low phosphate concentrations compared to nitrate concentrations. This trend was reversed in the hypoxic samples with nitrate and phosphate concentrations being negatively correlated.

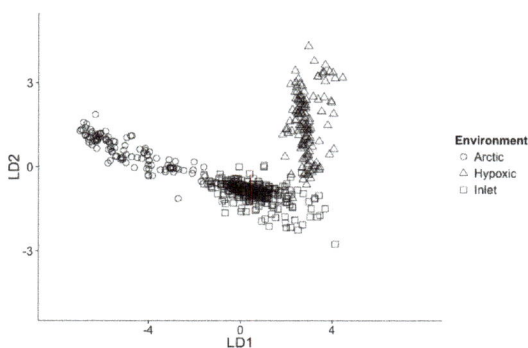

Figure 2. Linear discriminant analysis of samples used in models, based on temperature, salinity, nitrate, phosphate, silicate, chlorophyll and oxygen. Arctic samples (open circles); Inlet samples (open squares); Hypoxic samples (open triangles).

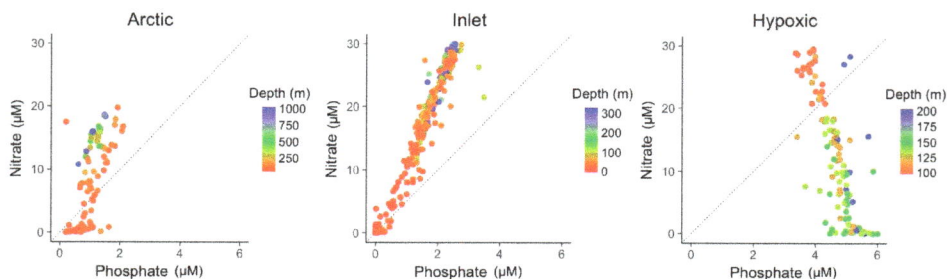

Figure 3. Nitrate to phosphate ratio for the samples from the three different environments. Colors indicate the sampling depth. The dashed line indicates the elemental 5:1 stoichiometric N:P ratio of viral particles.

3.2. Explanatory Power of Single Variable Linear Models

Linear models (LM) showing the distribution of direct relationships of $\log_{10}$ transformed viral abundances vs. $\log_{10}$ transformed bacterial abundances for the Arctic, inlet and hypoxic data sets are shown in Figure 4. For the inlet and Arctic data sets there were significant positive relationships between viral and bacterial abundances, explaining 48% of the variation in viral abundance in the inlet and 66% in the Arctic (Table 2). In the hypoxic samples, there was no discernable relationship between viral and bacterial abundances.

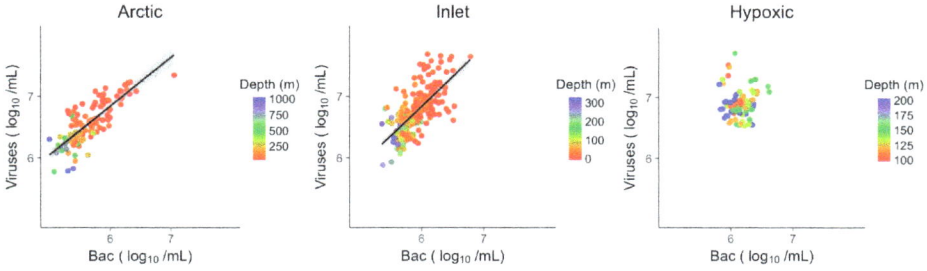

Figure 4. Linear models of log transformed viral abundances to log transformed bacterial abundances. Grey shading indicates the 95% confidence interval. Bac: Bacteria.

Nitrate and phosphate concentrations showed significant relationships with viral abundances in Arctic and inlet environments (Figures 5 and 6). However, these relationships varied in strength and only explained ~10 to 40% of the variation in viral abundances (Table 2). For nitrate, the R^2 values were 0.37 for Arctic samples and 0.33 for inlet samples, while for phosphate the values were 0.12 and 0.28, respectively. Relationships between viral abundances and nitrate or phosphate for the hypoxic samples were not significant. Generally, viral abundance and bacterial abundance were inversely correlated to depth, while nitrate and phosphate showed an opposite trend. However, this is not the case for the hypoxic samples. Based on the Shapiro–Wilk test, the residuals of the bivariate linear models were not normally distributed; however, the models displayed homogeneity of variance and the normal distribution of residuals, appropriate for large data sets (Figures S4–S9).

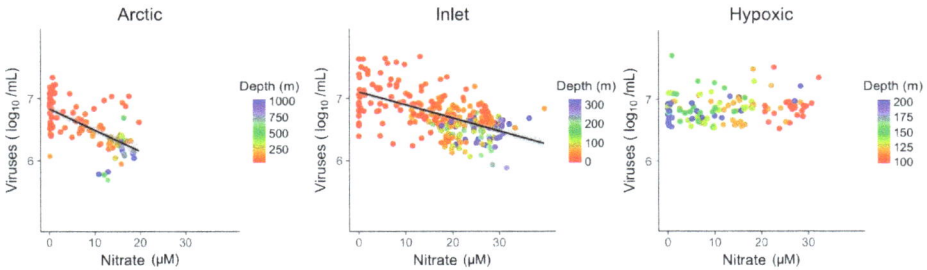

Figure 5. Linear models of log transformed viral abundance vs. nitrate (μM) concentration. Grey shading indicates the 95% confidence interval.

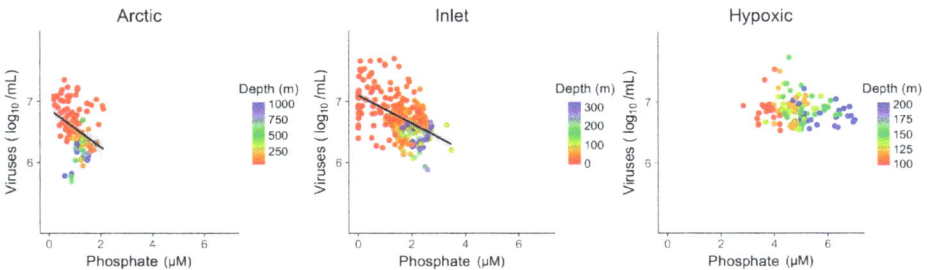

Figure 6. Linear models of log transformed viral abundance to phosphate (μM) concentration. Grey shading indicates the 95% confidence interval.

Table 2. Results for the significant linear models of viral abundance and bacterial abundance, nitrate and phosphate in the Arctic and inlet environments. Samples from the hypoxic environment did not show significant relationships and are not listed.

Variable	Parameter	Arctic	Inlet
Bacteria ($\log_{10}$)	R^2	0.66	0.48
	Slope	0.80	0.97
	p-value	2.5×10^{-27}	1.9×10^{-37}
NO_3	R^2	0.37	0.33
	Slope	−0.03	−0.02
	p-value	1.4×10^{-12}	1.2×10^{-24}
PO_4	R^2	0.12	0.28
	Slope	−0.31	−0.23
	p-value	1.0×10^{-04}	3.6×10^{-20}

3.3. Multivariate Models Show Increased Explanatory Power

Multivariate models of viral abundance were based on GLM of transformed data. For each environment, the best model was selected based on the AIC, and collinear predictors were reduced to one representative predictor. Combining only environmental variables and excluding bacterial abundance produced meaningful models in all three environments, matching or surpassing the explanatory power of bacterial abundance alone (Figure 7). The coefficient of determination for the three multivariate models was assessed by McFadden pseudo R^2. Pseudo R^2 of the GLMs and viral abundance in Artic, inlet and hypoxic environment were 0.56, 0.47 and 0.31, respectively. Significant predictors across all three environments were temperature and one of the nutrients (Table 3). Chlorophyll was a significant variable for the Arctic and hypoxic environments. Notably, for the inlet and hypoxic samples the models using combined environmental variables had an explanatory power that matched or exceeded the models based on bacteria only.

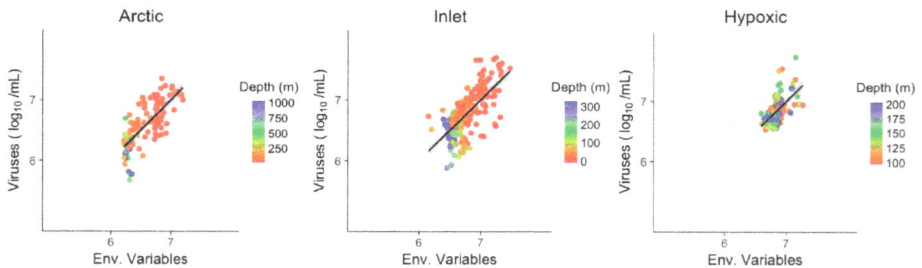

Figure 7. Generalized linear models of viral abundance and modeled abundance based on environmental variables for the Arctic, inlet and hypoxic environments, grey shading indicates the 95% confidence interval. Env.: Environmental variables.

The combined models of bacterial abundance and environmental variables substantially improved the relationship relative to bacterial abundances alone, for the Arctic and inlet environments (Figure 8). For the Arctic and inlet samples, pseudo R^2 values were high, at 0.73 and 0.59, respectively. Again, best models were identified by the AIC for each environment and only one representative of collinear predictors was retained. Besides bacterial abundance, the only significant predictor in the models for both environments was nitrate (Table 4). Chlorophyll was a significant explanatory variable for the Arctic samples, while temperature was only significant for the inlet samples. For the hypoxic samples, including bacterial abundance did not significantly improve the explanatory power of the combined environmental variables over viral abundance, and was left out.

Table 3. Results and significant predictors of generalized linear models based on environmental variables (Env.) per environment. Akaike Information Criterion (AIC), pseudo R^2, sample size (n) and degrees of freedom (df) shown with the effect sizes for predictors, fonts indicate the significance level.

Env.	Arctic	Inlet	Hypoxic
McFadden (R^2)	0.56	0.47	0.31
Slope	1.00	1.00	1.00
n/df	109/104	261/258	126/122
Intercept	5.545	3.75	13.721
Temperature	0.141	1.068	−2.763
Salinity	-	0.199	-
Chlorophyll	0.191	-	0.384
Oxygen	0.112	-	-
NO_3	−0.018	−0.013	-
PO_4	-	-	−0.078
SiO_4	0.009	-	0.004
PAR	-	-	-
Signif. level	<0.01	<0.05	<0.1

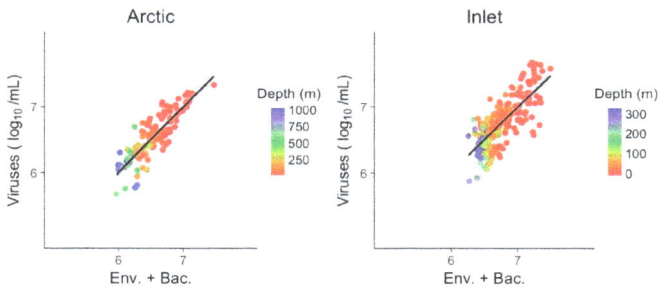

Figure 8. Generalized linear models of viral abundance and modeled abundance based on log-transformed bacterial abundances combined with environmental variables for the Arctic and inlet environments, grey shading indicates the 95% confidence interval. The model for the hypoxic environment did not improve by adding bacterial abundance relative to using environmental variables only in the model, and is not shown. Env.: Environmental variables; Bac.: Bacterial abundance.

Table 4. Results and significant predictors of combined generalized linear models based on environmental variables and bacterial abundance (Env. + Bac.) for the Arctic and inlet environment. AIC, pseudo R^2, sample size (n) and degrees of freedom (df) shown with the effect sizes for predictors, fonts indicate the significance level.

Env. + Bac.	Arctic		Inlet	
McFadden (R^2)	0.73		0.59	
Slope	1.00		1.00	
n/df	109/105		252/249	
Intercept	5.008		1.020	
Temperature	-		0.774	
Salinity	−0.523		-	
Chlorophyll	0.098		-	
Oxygen	-		-	
NO_3	−0.010		−0.003	
PO_4	-		-	
SiO_4	-		-	
PAR	-		-	
Bacteria (log_{10})	0.607		0.665	
Signif. level	<0.01		<0.05	<0.1

Using environmental variables, the improvement over models solely based on bacterial abundances was stronger for the inlet and hypoxic samples than for Arctic samples. GLMs for samples where PAR data were available showed that PAR was not a significant predictor and did not improve the explanatory power of the models. Additionally, based on the Shapiro–Wilk test, the residuals for the GLMs were not normally distributed. However, the residuals were centered around zero and the deviation from the normal distribution appeared random; all GLMs demonstrated homogeneity of variance to a level that can be expected for models of this size (Figures S10–S13). While other model approaches on these data sets produced higher explanatory power, this came at the expense of more pronounced heterogeneity of variance.

4. Discussion

As has been found in many previous studies, viruses are typically about ten times more abundant than bacteria in marine surface waters, although there is wide variation around this mean across environments [10,12] that is difficult to explain [11]. In this study, we used a series of models of varying complexity to investigate relationships between viral abundances and several environmental variables in an effort to explain the factors responsible for variation in viral abundances. We found that viral abundances across locations and time were related to a suite of environmental factors, but particularly nitrogen and phosphorus concentrations, as well as bacterial abundances. The exception was hypoxic environments, in which viral abundances were only explained by a combination of physical and chemical factors. These findings are discussed in detail below.

A database was compiled from samples collected from different depths, across a wide geographic range at different times of year. The values of environmental variables, including bacterial and viral abundances, were in the typical ranges for these habitats. One set of outlying data from Rivers Inlet was excluded from the models because of excessively high viral and bacterial abundances that could not be related to any of the environmental variables or explained in a model. Presumably, these data were due to high rates of bacterial growth and a lysis event during sampling, and show the difficulty in accounting for such extremes in models.

Samples were classified into Arctic, inlet and hypoxic environments. The LDA of the environmental variables for the three environments supported the approach to classifying samples based on the prevailing conditions, rather than by geographic location, cruise or project. The Arctic and inlet samples represent a continuum of environmental conditions. In contrast, the hypoxic samples were collected from depths below 100 m, had dissolved oxygen concentrations below 1.5 mL·L^{-1} and an altered nitrate to phosphate stoichiometry; thus, they represent a much different environment [44,48].

Given the stoichiometry of viral particles, nitrogen and phosphorus are key resources for viral replication and their availability would be expected to affect viral production. Nitrate to phosphate ratios averaged about 12:1 for the Arctic and inlet data, although in some cases reached much higher values for the inlet samples. This ratio was higher than the estimated elemental ratio of 5:1 for viral particles [29], but lower than the nitrate to phosphate ratio of ~15:1 previously found in marine samples [49]. The ratio of nitrate to phosphate was inverted to 1:12 in the hypoxic samples, as nitrate is used as an alternative electron acceptor by bacteria under anoxic conditions [48,50]. Arctic surface and hypoxic deep samples display the potential for nitrate limitation during viral replication in some virus–host systems with concentrations approaching zero. Nitrate and phosphate ratios in seawater show a similarity to the elemental nitrogen and phosphorus ratios in cells [31,51]. Consequently, shifts in the nitrate to phosphate ratio in seawater could link to the nitrogen and phosphorus supply to cells. When growing at relatively low phosphate concentrations, the high phosphorus accumulation of up to 87% of the cellular content in viral particles [29] could lead to a limitation in phosphorus supply during viral replication in autotrophic hosts.

The strength of relationships between viral abundance and single variables differed among the subsets of data. The explanatory power of bacterial abundance was higher for the Arctic data ($R^2 = 0.66$) than for the inlets data ($R^2 = 0.48$), although both were comparable to relationships reported for other surface and sub-surface studies [9,10]. Relationships of viral abundances to nitrate or phosphate were weaker than for bacterial abundance in the Arctic and inlet samples; however, the significant explanatory power of nitrate (R^2 = 0.37 and 0.33) in the Arctic and inlet environments comes close to that of bacterial abundance, highlighting the importance of nitrate. In the Arctic and inlet models, viral abundance and depth covaried; however, within the scope of this study, we treated depth as a co-variate for the environmental variables, e.g., salinity, temperature or light, rather than as an independent variable. That viral abundance was not significantly related to any of the three single variables in the hypoxic data implies that viral production is dependent on different processes in this environment.

Combining environmental variables into multivariate models showed high explanatory power of viral abundance in all environments. Based on the pseudo R^2 values, the models for the Arctic and the inlet data explained about 50% of the variation in viral abundance; for the inlet data, environmental variables surpassed the explanatory power of bacterial abundance alone. For the hypoxic data, the explanatory power of environmental variables was 31%, a substantial improvement compared to the absence of significant correlations with bacterial abundance, nitrate, or phosphate alone. After removing collinear nutrient variables, significant components of the models across data sets were temperature, chlorophyll and representative nutrients, nitrate, phosphate and silicate.

Phosphate was a significant component of the model for the hypoxic environment, but not for the Arctic or inlet samples, which generally had higher nitrate to phosphate ratios than the hypoxic samples. Phosphate is important to viral replication and infection, highlighted by reduced viral mortality of phytoplankton under phosphate limitation [33]. However, the collinearity of nitrate and phosphate data in the Arctic and inlet samples makes it difficult to identify which nutrient is eventually affecting viral replication. That phosphate was a statistically more significant variable than nitrate in the hypoxic model is presumably a result of the full depletion of nitrate by denitrification in samples that are truly anoxic [48,50].

The observation that chlorophyll was a significant variable in the Arctic but not in the inlet samples can be explained by phytoplankton blooms in the Arctic, which are associated with increases in viral abundance. For example, a seasonal study in the Beaufort Sea shelf showed a significant correlation between chlorophyll and viral abundance [52], as did another study in fresh waters [53]. The significance of chlorophyll in the deep hypoxic environment, however, must be related to phytoplankton cells sinking out of the photic zone, or is a statistical artefact. Based on the data presented, using chlorophyll as a proxy indicates that phytoplankton were not important in the inlet environments, where the majority of viruses are produced by and infect heterotrophic bacteria. Overall, it is remarkable that multivariate models built from environmental variables alone explain viral abundance as well as, or even exceed, the explanatory power of bacterial abundance.

Combining data for environmental variables and bacterial abundance further improved the explanatory power of the models for the Arctic and inlet data, with 73 and 59% of the variation in viral abundance explained by the multivariate models. In contrast, for the hypoxic data, including bacterial abundance did not increase the explanatory power from the multivariate model using environmental variables only. This suggests a strong effect on viral production by nutrient stoichiometry and other environmental conditions. Across these multivariate models, the consistent component besides bacterial abundance was nitrate. While temperature or salinity were significant variables in the models for the Arctic and inlet environments, again, chlorophyll was only a significant variable in the Arctic environment and can be explained by phytoplankton blooms [52,53]. The influence of environmental variables on the relationship between viral and bacterial abundances, and the differences among environments, is consistent with observations from marine and freshwater environments [10,12]. The data presented here show that much of this variation is likely explained by differences in nutrient availability.

In conclusion, the environmental variables examined here are associated with changes in viral abundance and the relationship between viruses and bacteria in diverse marine samples. We provide a first attempt at generalized statistical models that capture these relationships, and a first step towards a better

Viruses **2017**, *9*, 152

ecological understanding of the processes controlling virus abundance in the ocean. For the purpose of explanatory models, samples can be classified by their environment, rather than arbitrarily by project, cruise or station. While bacterial abundance is a well-established predictor for viral abundance, it fails in certain marine environments, and can be substantially improved by more complex models incorporating environmental variables. Individual environmental variables do not have great explanatory power for predicting viral abundances; yet, when combined in multivariate models they can produce explanatory power equal to or surpassing that of bacterial abundance. This study shows that the environmental variables explaining viral abundance vary among environments, but nutrient concentrations, as well as salinity and temperature, appear to be key factors. The relationships described here only apply to viruses that can be detected by flow cytometry. RNA viruses with small genomes can be difficult to detect and distinguish by flow cytometry and may have different relationships to environmental variables.

The three types of environments studied in this project are predicted to be strongly affected by climate change, with increased stratification in inlets, the North Atlantic, Arctic and Northeast Pacific, and associated changes in vertical nutrient fluxes and expanding oxygen minimum zones [54–57]. Understanding the interplay between viruses, hosts and environmental variables in these types of environments improves the potential of predicting how virus-host systems will respond to environmental changes.

Supplementary Materials: The following are available online at www.mdpi.com/1999-4915/9/6/152/s1, Table S1: Viral, bacterial and environmental data used in building the models. Figure S1: Data distribution and direct correlations (Pearson's) of data in the Arctic environment, Figure S2: Data distribution and direct correlations (Pearson's) of data in the inlet environment, Figure S3: Data distribution and direct correlations (Pearson's) of data in the hypoxic environment, Figure S4: Residual density for linear models of $\log_{10}$ viral abundance and $\log_{10}$ bacterial abundance in the three environments. Shapiro–Wilk test: Arctic, w = 0.99, p-value = 0.54; inlet, w = 0.96, p-value = 0.0002; hypoxic, 0.93, p-value = 0.0001, Figure S5: Residual distribution and qq-plots for linear models of $\log_{10}$ viral abundance and $\log_{10}$ bacterial abundance for the Arctic (a), inlet (b) and hypoxic (c) environments, Figure S6: Residual density for linear models of $\log_{10}$ viral abundance and nitrate in the three environments. Shapiro–Wilk test: Arctic, w = 0.98, p-value = 0.087; inlet, w = 0.99, p-value = 0.176; hypoxic, 0.93, p-value = 3.61×10^{-6}, Figure S7: Residual distribution and qq-plots for linear models of $\log_{10}$ viral abundance and nitrate for the Arctic (a), inlet (b) and hypoxic (c) environments, Figure S8: Residual density for linear models of $\log_{10}$ viral abundance and phosphate in the three environments. Shapiro–Wilk test: Arctic, w = 0.97, p-value = 0.031; inlet, w = 0.99, p-value = 0.034; hypoxic, 0.93, p-value = 7.69×10^{-6}, Figure S9: Residual distribution and qq-plots for linear models of $\log_{10}$ viral abundance and phosphate for the Arctic (a), inlet (b) and hypoxic (c) environments, Figure S10: Residual density for generalized linear models of $\log_{10}$ viral abundance and combined environmental variables in the three environments. Shapiro–Wilk test: Arctic, w = 0.99, p-value = 0.406; inlet, w = 0.95, p-value = 1.68×10^{-7}; hypoxic, 0.95, p-value = 6.24×10^{-5}, Figure S11: Residual distribution and qq-plots for linear models of $\log_{10}$ viral abundance and combined environmental variables for the Arctic (a), inlet (b) and hypoxic (c) environments, Figure S12: Residual density for generalized linear models of $\log_{10}$ viral abundance and combined $\log_{10}$ bacterial abundance and environmental variables in the Arctic and inlet environments. Shapiro–Wilk test: Arctic, w = 0.99, p-value = 0.685; inlet, w = 0.98, p-value = 0.002, Figure S13: Residual distribution and qq-plots for linear models of $\log_{10}$ viral abundance and combined $\log_{10}$ bacterial abundance and environmental variables for the Arctic (a) and inlet (b) environments.

Acknowledgments: Special thanks are due to Stilianos Louca (University of British Columbia, Canada) who was invaluable in advising on the analysis of the data. For Saanich Inlet sampling and logistical support we thank past and present members of the Hallam lab (University of British Columbia, Canada) for logistical support, especially Steven Hallam, Alyse Hawley and Monica Torres Beltran, as well as Chris Payne, Lora Pakhomova and Richard Pawlowicz from the University of British Columbia Department of Earth, Ocean & Atmospheric Sciences. We thank Julian Ho and Adrian Jones from the University of British Columbia Department of Statistics for advice. For C3O, we thank Jane Eert, Mike Dempsey, Sarah Zimmermann, Caroline Chenard and Amy M. Chan for sample collection and logistics, as well as the entire IPY C3O team. We thank the Hakai Institute team for Rivers Inlet sampling, and the captains and crews of the *HMS John Strickland*, *CCGS Wilfried Laurier* and the *CCGS Louis S. Saint-Laurent for their service in Saanich Inlet and C3O*. Saanich Inlet ship time support was provided by NSERC between 2007–2014 through grants awarded to Steven J. Hallam (University of British Columbia, Canada) and Philippe Tortell (University of British Columbia, Canada). Canada's Three Oceans project was funded by the Canadian Federal Program Office of the International Polar Year through an award to EC Carmack. This project was supported by awards to CAS from the Tula Foundation and NSERC-DG grant program.

Author Contributions: Jan F. Finke and Curtis A. Suttle conceived and designed the project; Jan F. Finke and Christian Winter measured viral and bacterial abundances; Brian P. V. Hunt and Eddy C. Carmack contributed samples and data, and advised on the analysis; Jan F. Finke performed the analysis and wrote the paper with input from the other authors.

Conflicts of Interest: The authors declare no conflict of interest. The founding sponsors had no role in the design of the study; in the collection, analyses, or interpretation of data; in the writing of the manuscript, and in the decision to publish the results.

References

1. Wilhelm, S.W.; Suttle, C.A. Viruses and nutrient cycles in the sea. *Bioscience* **1999**, *49*, 781–788. [CrossRef]
2. Ortmann, A.C.; Suttle, C.A. High abundances of viruses in a deep-sea hydrothermal vent system indicates viral mediated microbial mortality. *Deep Sea Res Part I* **2005**, *52*, 1515–1527. [CrossRef]
3. Paul, J.H.; Rose, J.B.; Jiang, S.C.; Kellogg, C.A.; Dickson, L. Distribution of viral abundance in the reef environment of Key Largo, Florida. *Appl. Environ. Microbiol.* **1993**, *59*, 718–724. [PubMed]
4. Proctor, L.M.; Fuhrman, J.A. Viral mortality of marine bacteria and cyanobacteria. *Nature* **1990**, *343*, 60–62. [CrossRef]
5. Murray, A.G.; Jackson, G.A. Viral dynamics: A model of the effects of size, shape, motion and abundance of single-celled planktonic organisms and other particles. *Mar. Ecol. Prog. Ser.* **1992**, *89*, 103–116. [CrossRef]
6. Mann, N.H. Phages of the marine cyanobacterial picophytoplankton. *FEMS Microbiol. Rev.* **2003**, *27*, 17–34. [CrossRef]
7. Wommack, K.E.; Colwell, R.R. Virioplankton: Viruses in aquatic ecosystems. *Microbiol. Mol. Biol. Rev.* **2000**, *64*, 69–114. [CrossRef] [PubMed]
8. Fuhrman, J.A.; Suttle, C.A. Viruses in marine planktonic systems. *Oceanography* **1993**, *6*, 51–63. [CrossRef]
9. Knowles, B.; Silveira, C.B.; Bailey, B.A.; Barott, K.; Coutinho, F.H.; Dinsdale, E.A.; Felts, B.; Furby, K.A.; George, E.E.; Green, K.T.; et al. Lytic to temperate switching of viral communities. *Nature* **2016**, *531*, 466–470. [CrossRef] [PubMed]
10. Wigington, C.H.; Sonderegger, D.; Brussaard, C.P.D.; Buchan, A.; Finke, J.F.; Fuhrman, J.A.; Lennon, J.T.; Middleboe, M.; Suttle, C.A.; Stock, C.; et al. Marine virus and microbial cell abundances. *Nat. Microbiol.* **2016**, *1*, 4–11. [CrossRef] [PubMed]
11. Parikka, K.J.; Le Romancer, M.; Wauters, N.; Jacquet, S. Deciphering the virus-to-prokaryote ratio (VPR): Insights into virus-host relationships in a variety of ecosystems. *Biol. Rev. Camb. Philos. Soc.* **2017**, *92*, 1081–1100. [CrossRef] [PubMed]
12. Clasen, J.L.; Brigden, S.M.; Payet, J.P.; Suttle, C.A. Evidence that viral abundance across oceans and lakes is driven by different biological factors. *Freshw. Biol.* **2008**, *53*, 1090–1100. [CrossRef]
13. Baudoux, A.C.; Brussaard, C.P.D. Characterization of different viruses infecting the marine harmful algal bloom species *Phaeocystis globosa*. *Virology* **2005**, *341*, 80–90. [CrossRef] [PubMed]
14. Nagasaki K, YM. Effect of temperature on the algicidal activity and the stability of HaV (*Heterosigma akashiwo* virus). *Aquat. Microb. Ecol.* **1998**, *15*, 211–216. [CrossRef]
15. Hardies, S.C.; Hwang, Y.J.; Hwang, C.Y.; Jang, G.I.; Cho, B.C. Morphology, physiological characteristics, and complete sequence of marine bacteriophage RIO-1 infecting *Pseudoalteromonas marina*. *J. Virol.* **2013**, *87*, 9189–9198. [CrossRef] [PubMed]
16. Williamson, S.J.; Paul, J.H. Environmental factors that influence the transition from lysogenic to lytic existence in the HSIC/*Listonella pelagia* marine phage-host system. *Microb. Ecol.* **2006**, *52*, 217–225. [CrossRef] [PubMed]
17. Kukkaro, P.; Bamford, D.H. Virus-host interactions in environments with a wide range of ionic strengths. *Environ. Microbiol. Rep.* **2009**, *1*, 71–77. [CrossRef] [PubMed]
18. Keynan, A.; Nealson, K.; Sideropoulos, H.; Hastings, J.W. Marine transducing bacteriophage attacking a luminous bacterium. *J. Virol.* **1974**, *14*, 333–340. [PubMed]
19. Bettarel, Y.; Bouvier, T.; Bouvier, C.; Carr, C.; Desnues, A.; Domaizon, I.; Jacquet, S.; Robin, A.; Sime-Ngando, T. Ecological traits of planktonic viruses and prokaryotes along a full-salinity gradient. *FEMS Micriobiol. Ecol.* **2011**, *76*, 360–372. [CrossRef] [PubMed]
20. Jia, Y.; Shan, J.; Millard, A.; Clokie, M.R.J.; Mann, N.H. Light-dependent adsorption of photosynthetic cyanophages to *Synechococcus* sp. WH7803. *FEMS Microbiol. Lett.* **2010**, *310*, 120–126. [CrossRef] [PubMed]
21. Baudoux, A.C.; Brussaard, C.P.D. Influence of irradiance on virus-algal host interactions. *J. Phycol.* **2008**, *44*. [CrossRef] [PubMed]

22. Brown, C.M.; Campbell, D.A.; Lawrence, J.E. Resource dynamics during infection of *Micromonas pusilla* by virus MpV-Sp1. *Environ. Microbiol.* **2007**, *9*, 2720–2727. [CrossRef] [PubMed]

23. Juneau, P.; Lawrence, J.E.; Suttle, C.A.; Harrison, P.J. Effects of viral infection on photosynthetic processes in the bloom-forming alga *Heterosigma akashiwo*. *Aquat. Microb. Ecol.* **2003**, *31*, 9–17. [CrossRef]

24. Lawrence, J.E.; Suttle, C.A. Effect of viral infection of sinking rates of *Heterosigma akashiwo* and its implications for bloom termination. *Aquat. Microb. Ecol.* **2004**, *37*, 1–7. [CrossRef]

25. Mojica, K.D.A.; Brussaard, C.P.D. Factors affecting virus dynamics and microbial host-virus interactions in marine environments. *FEMS Microbiol. Ecol.* **2014**, *89*, 495–515. [CrossRef] [PubMed]

26. Cottrell, M.T.; Suttle, C.A. Dynamics of a lytic virus infecting the photosynthetic marine picoflagellate *Micromonas pusilla*. *Limnol. Oceanogr.* **1995**, *40*, 730–739. [CrossRef]

27. Noble, R.T.; Fuhrman, J.A. Virus decay and its causes in coastal waters. *Appl. Environ. Microbiol.* **1997**, *63*, 77–83. [PubMed]

28. Garza, D.R.; Suttle, C.A. The effect of cyanophages on the mortality of *Synechoccocus* spp. and selection for UV resistant viral communities. *Microb. Ecol.* **1998**, *36*, 281–292. [CrossRef] [PubMed]

29. Jover, L.F.; Effler, T.C.; Buchan, A.; Wilhelm, S.W.; Weitz, J.S. The elemental composition of virus particles: Implications for marine biogeochemical cycles. *Nat. Rev. Microbiol.* **2014**, *12*, 519–528. [CrossRef] [PubMed]

30. Suttle, C.A. Marine viruses-major players in the global ecosystem. *Nat. Rev. Microbiol.* **2007**, *5*, 801–812. [CrossRef] [PubMed]

31. Redfield, A.C.; Ketchum, B.H.; Richards, F.A. The composition of seawater: Comparative and descriptive oceanography. In *The Sea: Ideas and Observations on Progress in the Study of the Seas*; Hill, N.M., Ed.; Interscience: New York, NY, USA, 1963; pp. 26–77.

32. Bratbak, G.; Jacobsen, A.; Heldal, M.; Nagasaki, K.; Thingstad, F. Virus production in *Phaeocystis pouchetii* and its relation to host cell growth and nutrition. *Aquat. Microb. Ecol.* **1998**, *16*, 1–9. [CrossRef]

33. Maat, D.S.; Crawfurd, K.J.; Timmermans, K.R.; Brussaard, C.P.D. Elevated CO_2 and phosphate limitation favor *Micromonas pusilla* through stimulated growth and reduced viral impact. *Appl. Environ. Microbiol.* **2014**, *80*, 3119–3127. [CrossRef] [PubMed]

34. Jacquet, S.; Heldal, M.; Iglesias-Rodriguez, D.; Larsen, A.; Wilson, W.H.; Bratbak, G. Flow cytometric analysis of an *Emiliania huxleyi* bloom terminated by viral infection. *Aquat. Microb. Ecol.* **2002**, *27*, 111–124. [CrossRef]

35. Motegi, C.; Kaiser, K.; Benner, R.; Weinbauer, M.G. Effect of P-limitation on prokaryotic and viral production in surface waters of the Northwestern Mediterranean Sea. *J. Plankton Res.* **2015**, *37*, 16–20. [CrossRef]

36. Bratbak, G.; Egge, J.K.; Heldal, M. Viral mortality of the marine alga *Emiliania huxleyi* (Haptophyceae) and termination of algal blooms. *Mar. Ecol. Prog. Ser.* **1993**, *93*, 39–48. [CrossRef]

37. Chow, C.T.; Kim, D.Y.; Sachdeva, R.; Caron, D.A. Top-down controls on bacterial community structure: Microbial network analysis of bacteria, T4-like viruses and protists. *ISME J.* **2013**, *8*, 816–829. [CrossRef] [PubMed]

38. Tommasi, D.; Hunt, B.P.V.; Pakhomov, E.A.; Mackas, D.L. Mesozooplankton community seasonal succession and its drivers: Insights from a British Columbia, Canada, fjord. *J. Mar. Syst.* **2013**, *116*, 10–32. [CrossRef]

39. Carmack, E.C.; Mclaughlin, F.A.; Vagle, S.; Melling, H.; Williams, W.J. Structures and property distributions in the three oceans surrounding Canada in 2007: A basis for a long-term ocean climate monitoring strategy. *Atmosphere-Ocean* **2010**, *48*, 211–224. [CrossRef]

40. Brussaard, C.P.D. Optimization of procedures for counting viruses by flow cytometry. *Appl. Environ. Microbiol.* **2004**, *70*, 1506–1513. [CrossRef] [PubMed]

41. Vaulot, D. CYTOPC: Processing software for flow cytometric data. *Signal Noise* **1989**, *2*, 292.

42. Armstrong, F.A.J.; Stearns, C.R.; Strickland, J.D.H. The measurement of upwelling and subsequent biological processes by means of the Technicon AutoAnalyzer^{TM} and associated equipment. *Deep Sea Res. Oceanogr. Abstr.* **1967**, *14*, 381–389. [CrossRef]

43. Murphey, J.; Riley, J.P. A modified single solution method for the determination of phosphate in natural waters. *Anal. Chim. Acta* **1962**, *27*, 31–36. [CrossRef]

44. Moffitt, S.E.; Moffitt, R.A.; Sauthoff, W.; Davis, C.V.; Hewett, K.; Hill, T.M. Paleoceanographic insights on recent oxygen minimum zone expansion: Lessons for modern oceanography. *PLoS ONE* **2015**, *10*, e0115246. [CrossRef] [PubMed]

45. R Core Team. *R: A Language and Environment for Statistical Computing*; R Foundation for Statistical Computing: Vienna, Austria, 2015.

46. Venables, W.N.; Ripley, B.D. *Modern Applied Statistics with S-Plus*, 4th ed.; Springer: New York, NY, USA, 2002.

47. Beaujean, A.A. BaylorEdPsych. 2012.

48. Zaikova, E.; Walsh, D.; Stilwell, C.P.; Mohn, W.W.; Tortell, P.D.; Hallam, S.J. Microbial community dynamics in a seasonally anoxic fjord: Saanich Inlet, British Columbia. *Environ. Microbiol.* **2010**, *12*, 172–191. [CrossRef] [PubMed]

49. Tyrrell, T. The relative influences of nitrogen and phosphorus on oceanic primary production. *Nature* **1999**, *400*, 525–531. [CrossRef]

50. Somes, C.J.; Schmittner, A.; Galbraith, E.D.; Lehmann, M.F.; Altabet, M.A.; Montoya, J.P.; Letelier, R.M.; Mix, A.C.; Bourbonnais, A.; Eby, M. Simulating the global distribution of nitrogen isotopes in the ocean. *Glob. Biogeochem. Cycles* **2010**, *24*, 1–16. [CrossRef]

51. Moore, C.M.; Mills, M.M.; Arragio, K.R.; Berman-Frank, I.; Bopp, L.; Boyd, P.W.; Galbraith, E.D.; Geider, R.J.; Jaccard, S.L.; Jickells, T.D.; et al. Processes and patterns of oceanic nutrient limitation. *Nat. Geosci.* **2013**, *6*, 701–710. [CrossRef]

52. Payet, J.P.; Suttle, C.A. Physical and biological correlates of virus dynamics in the southern Beaufort Sea and Amundsen Gulf. *J. Mar. Syst.* **2008**, *74*, 933–945. [CrossRef]

53. Maranger, R.; Bird, D. Viral abundance in aquatic systems: A comparison between marine and fresh waters. *Mar. Ecol. Prog. Ser.* **1995**, *121*, 217–226. [CrossRef]

54. Capotondi, A.; Alexander, M.A.; Bond, N.A.; Curchitser, E.N.; Scott, J.D. Enhanced upper ocean stratification with climate change in the CMIP3 models. *J. Geophys. Res. Ocean* **2012**, *117*, 1–23. [CrossRef]

55. Hordoir, R.; Meier, H.E.M. Effect of climate change on the thermal stratification of the baltic sea: A sensitivity experiment. *Clim. Dyn.* **2012**, *38*, 1703–1713. [CrossRef]

56. Keeling, R.E.; Körtzinger, A.; Gruber, N. Ocean deoxygenation in a warming world. *Ann. Rev. Mar. Sci.* **2010**, *2*, 199–229. [CrossRef] [PubMed]

57. Carmack, E.; Mclaughlin, F.; Whiteman, G.; Homer-dixon, T. Detecting and coping with disruptive shocks in Arctic marine systems: A resilience approach to place and people. *AMBIO* **2012**, *41*, 56–65. [CrossRef] [PubMed]

viruses

MDPI

Article

The Response of Heterotrophic Prokaryote and Viral Communities to Labile Organic Carbon Inputs Is Controlled by the Predator Food Chain Structure

Ruth-Anne Sandaa [1,*], Bernadette Pree [1], Aud Larsen [2], Selina Våge [1], Birte Töpper [1], Joachim P. Töpper [3], Runar Thyrhaug [1,†] and Tron Frede Thingstad [1]

[1] Department of Biology, University of Bergen, N-5020 Bergen, Norway; Bernadette.Pree@uib.no (B.P.); selina.vage@bio.uib.no (S.V.); Birte.topper@bio.uib.no (B.T.); frede.thingstad@bio.uib.no (T.F.T.)
[2] Uni Research Environment, Nygårdsgaten 112, 5008 Bergen, Norway; aud.larsen@bio.uib.no
[3] Norwegian Institute for Nature Research, Thormøhlensgate 55, N-5008 Bergen, Norway; joachim.topper@nina.no
[*] Correspondence: Ruth.Sandaa@uib.no; Tel.: +47-5558-4646
[†] Deceased.

Academic Editors: Mathias Middelboe and Corina P. D. Brussaard
Received: 27 June 2017; Accepted: 17 August 2017; Published: 23 August 2017

Abstract: Factors controlling the community composition of marine heterotrophic prokaryotes include organic-C, mineral nutrients, predation, and viral lysis. Two mesocosm experiments, performed at an Arctic location and bottom-up manipulated with organic-C, had very different results in community composition for both prokaryotes and viruses. Previously, we showed how a simple mathematical model could reproduce food web level dynamics observed in these mesocosms, demonstrating strong top-down control through the predator chain from copepods via ciliates and heterotrophic nanoflagellates. Here, we use a steady-state analysis to connect ciliate biomass to bacterial carbon demand. This gives a coupling of top-down and bottom-up factors whereby low initial densities of ciliates are associated with mineral nutrient-limited heterotrophic prokaryotes that do not respond to external supply of labile organic-C. In contrast, high initial densities of ciliates give carbon-limited growth and high responsiveness to organic-C. The differences observed in ciliate abundance, and in prokaryote abundance and community composition in the two experiments were in accordance with these predictions. Responsiveness in the viral community followed a pattern similar to that of prokaryotes. Our study provides a unique link between the structure of the predator chain in the microbial food web and viral abundance and diversity.

Keywords: marine viral diversity; viral–host interaction; high latitude microbes; minimum food web model; copepods; ciliates; nutrient limitation; trophic cascade

1. Introduction

Microorganisms are the main controllers of biomass and energy fluxes in the ocean. Together with their viruses, they form a tightly linked web of trophic interactions at the base of marine food webs, typically connected to the higher part of the food chain with multicellular organisms through copepod predation on microbes. Within the microbial part of this ecosystem, the composition and activity of the community of pelagic heterotrophic prokaryotes are presumably shaped both by bottom-up factors such as the availability of mineral and organic-C nutrients, and by the top-down mechanisms of predation [1] and viral lysis [2–4]. Nutrient availability and predator control have to a large extent been studied using a "black box" approach, treating the heterotrophic prokaryotes as one plankton functional type (PFT), disregarding internal community composition and differences in activity between community members [5,6], and leaving us with a limited understanding of how population dynamics at the two levels of resolution are connected.

Top-down control by viruses is believed to be more specific than predatory control, regulating the size of specific host groups, and thus acts more on community composition than on community size [7]. With host-specific viruses, host–virus interactions must, however, work both ways so that a change in host community composition will be reflected in a subsequent change in viral composition.

In the present study, we approach these interactions by revisiting the study by Larsen et al. [8], who demonstrated the explanatory power of a black-box modeling approach in their analysis of contrasting food web level responses observed in two similarly bottom-up perturbed mesocosm experiments (Polar Aquatic Microbial Ecology) (PAME)-I and PAME-II). Larsen et al. [8] were able to explain these contrasting responses using a "minimum" food web model, consisting of six PFTs (Figure S1). An essential conclusion of their study was that food web level responses to the bottom-up manipulations applied (glucose and mineral nutrients) were strongly modulated by the different states of the trophic cascade, from copepods via ciliates and heterotrophic nanoflagellates (HNF) to heterotrophic prokaryotes, essentially identifying the seasonal vertical migration of Arctic copepods as an ultimate cause in the cause–effect chain of such a trophic cascade. Larsen et al. [8] did not include any resolution of the prokaryote community, or any representation of viruses, in their analysis.

We here address the observation that the heterotrophic prokaryote (Figure 1A,B) and virus communities responded to labile C-addition in PAME-I, but lacked such a response in PAME-II. We hypothesize that these differences within the prokaryote and virus communities arose from the different states of the trophic cascade, as inferred by Larsen et al. [8]. Using a simplified steady state analysis of the "minimum" model, we explain how low versus high initial ciliate abundances drive model heterotrophic prokaryote growth towards mineral nutrient (MN) versus organic carbon (OC) limitation, respectively (Figure 1C). A C-limited community is presumably more responsive to the addition of glucose than an MN-limited community, replete in organic-C substrates, and the model predictions are thus in qualitative accordance with the observed differences in prokaryote community composition response. This interpretation extends the cascading effect of seasonally migrating Arctic copepods from the microbial food web level, as suggested by Larsen et al. [8], to the level of internal community structures of the prokaryote, and even the viral communities.

2. Materials and Methods

2.1. Model Framework

We used the minimum microbial food web model (Figure S1) to explain observed responses and derive a theory for different states of limitation (Figure 1). The dynamic version of the minimum model accurately predicts mesocosm responses in different environments [5,8]. Here we use a simplified steady-state analysis of the model. The trophic interactions between ciliates, heterotrophic nanoflagellates, prokaryotes, and autotrophic flagellates (Figures 1C,D and S1, left) reveal a link between ciliate abundance and bacterial carbon demand (Equations (4)–(6)), as described in the following. Under steady state, mass balance must be fulfilled such that "growth = loss" for each state variable. Assuming that food consumption is proportional to food concentration, heterotrophic flagellates balance their growth by their loss as follows (explanation of symbols is found in Table 1):

$$Y_H \alpha_H BH = \alpha_C HC. \text{ Solving this for B gives} \qquad B = \frac{\alpha_C}{Y_H \alpha_H} C. \qquad (1)$$

For autotrophic flagellate, growth = loss is given by:

$$\alpha_A PA = \alpha_C AC. \text{ Solving this for P gives} \qquad P = \frac{\alpha_C}{\alpha_A} C. \qquad (2)$$

For phosphate-limited growth of heterotrophic prokaryotes, growth is expressed as $\mu = \alpha_B P$, which, by insertion from Equation (2), gives:

$$\mu = \frac{\alpha_B \alpha_C}{\alpha_A} C. \tag{3}$$

Steady-state prokaryote carbon demand under P-limitation (BCD_P) is given by the ratio of production (μB) over yield (Y_{BC}). Assuming yield (Y_{BC}) to be a constant and independent of the growth rate μ, BCD_P can from inserting Equations (1) and (3) be expressed in terms of ciliate biomass as:

$$BCD_P = \frac{\mu B}{Y_{BC}} = \frac{\alpha_B \cdot \alpha_C^2}{Y_{BC} Y_H \alpha_H \alpha_A} C^2. \tag{4}$$

This gives a quadratic relationship between prokaryote carbon demand and ciliate abundance (Figure 1, MNL state). Prokaryote carbon demand can, however, not exceed the rate ψ, at which labile DOC is supplied by autochthonous and/or allochthonous sources. For high ciliate abundances giving $BCD_P > \psi$, the system thus shifts to C-limited prokaryote growth with prokaryote carbon demand (OCL state in Figure 1):

$$BCD_C = \Psi. \tag{5}$$

Equations (4) and (5) can be summarized as:

$$BCD = \min(\frac{\alpha_B \cdot \alpha_C^2}{Y_{BC} Y_H \alpha_H \alpha_A} C^2, \Psi), \tag{6}$$

where the left entry in the bracket describes BDC under mineral nutrient limitation, and the right entry describes BDC under carbon limitation (Figure 1). Using a conversion factor of P-content per ciliate (σ) and optimizing the model for the PAME experiments by assuming a temperature sensitivity of the rates with a $Q_{10} \approx 1.3$ [8], this adjusts Equation (6) (shown in Figure 1B) to:

$$BCD = \min(\frac{\alpha_B \cdot \alpha_C^2 \sigma^2}{Y_{BC} Y_H \alpha_H \alpha_A Q_{10}} C^2, \Psi), \tag{7}$$

where ciliate abundance C is in cells mL^{-1} and the parameters are listed in Table 1.

To obtain the estimate of σ, we assumed a specific carbon content of 0.13 pg-C μm^{-3}, Redfield stoichiometry (molar C:P = 106:1), and an equivalent spherical diameter of 20 μm [9].

Table 1. Symbols and parameter values used in drawing Figure 1D.

Symbol	Meaning	Numerical Value	Unit
Biomasses			
B	Heterotrophic prokaryotes		nmol-P L^{-1}
H	Heterotrophic flagellates		nmol-P L^{-1}
C	Ciliates		nmol-P L^{-1}
A	Autotrophic flagellates		nmol-P L^{-1}
P	Free phosphate		nmol-P L^{-1}
Affinities/clearance rates		**Value at 17 °C**	
α_B	Heterotrophic prokaryote affinity for phosphate	0.08	L nmol-P^{-1}h^{-1}
α_A	Autotrophic flagellate affinity for phosphate	0.04	L nmol-P^{-1}h^{-1}
α_H	Heterotrophic flagellate clearance rate for bacteria	0.0015	L nmol-P^{-1}h^{-1}
α_C	Ciliate clearance rate for flagellates	0.0005	L nmol-P^{-1}h^{-1}
Yields			
Y_H	Heterotrophic flagellate yield on heterotrophic prokaryotes	0.3	nmol-P nmol-P^{-1}
Y_{BC}	Heterotrophic prokaryote yield on DOC		nmol-P nmol-C^{-1}
Conversion factor			
α	P per ciliate	0.00043	nmol-P cell^{-1}
Temperature sensitivity of α-parameters			
Q_{10}		1.3	dimensionless

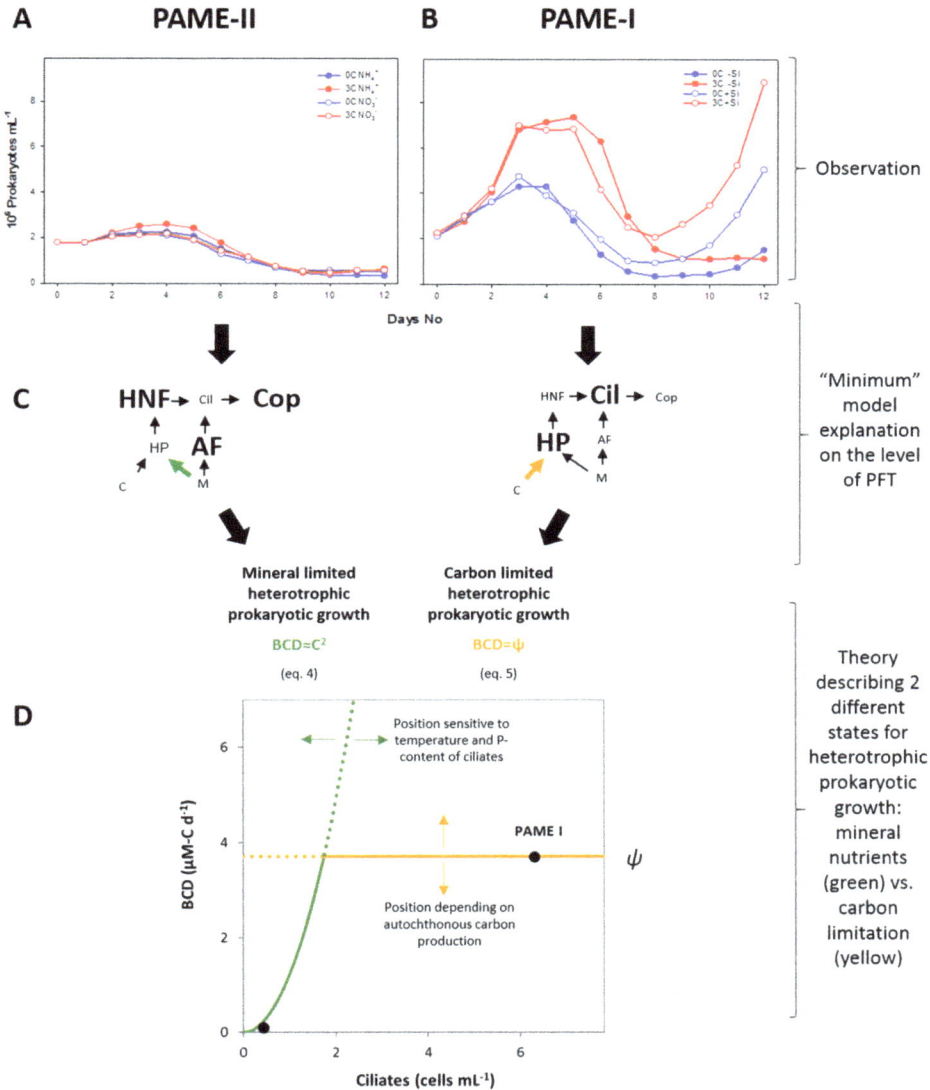

Figure 1. Different heterotrophic prokaryote (HP) responses to glucose additions ((**A,B**) data from [8],) are predicted qualitatively by the model (**C**) that links the type of growth rate limitation of bacteria to ciliate abundance (**D**) [8]. (**A**) Observed response in PAME-II was negligible, while (**B**) observed response in PAME I was a strong effect of glucose addition. (**C**) The pentagon food web structure (see also Figure S1, left) has two possible limiting states for heterotrophic prokaryotic growth: (**D**) either free mineral nutrients limitation (MNL), where bacterial carbon demand increases as the square of ciliate biomass increases (green curve); or organic carbon limitation (OCL), where bacterial carbon demand is equal to the supply (ψ) of degradable organic-C (yellow line). Since bacterial carbon demand (BCD) cannot sustainably exceed the supply, the shift between the two states occurs at the ciliate density where the green and yellow curves cross. HNF: heterotrophic nanoflagellates; AF: autothrophic flagellates; Cop: copepodes.

2.2. Experimental Design

2.2.1. Mesocosm Setup

In PAME-I eight units (each 700 L) formed two four-point gradients of additional DOC in the form of glucose (0, 0.5, 1 and 3 times Redfield ratio in terms of carbon relative to the nitrogen and phosphorus additions) (Figure 2). All eight units also got a daily dose of NH_4^+ and PO_4^{3-} in Redfield ratio. Two gradients were set up, one with silicate addition [8]. Samples for flow cytometric counts of viruses were collected from all treatments every day. Samples for DGGE and PFGE were collected from four of the tanks: 0 and 3C without Si ($-$Si) treatment; 0 and 3C with Si treatment (+Si).

In PAME-II all nine units (each 900 L) formed two four-point gradients of additional DOC in the form of glucose (0, 0.5, 1, 2 and 3 times Redfield ratio in terms of carbon relative to nitrogen and phosphorus additions) (Figure 2). The two gradients in glucose were kept silicate-replete. NH_4^+ was used as the dissolved inorganic nitrogen (DIN) source in one gradient (units 1 to 5) and NO_3^- in the other (units 6–9). All units got a daily dose of PO_4^{3-} in Redfield ratio. Flow cytometric counts of viruses and bacteria were performed on samples from all treatments, whereas DGGE and PFGE was performed on samples from four of the tanks: 0; 3C with NH_4^+ treatment; and 0, 3C with NO_3^- treatment. For a more detailed description of the PAME-I and PAME-II setup, see [8,10].

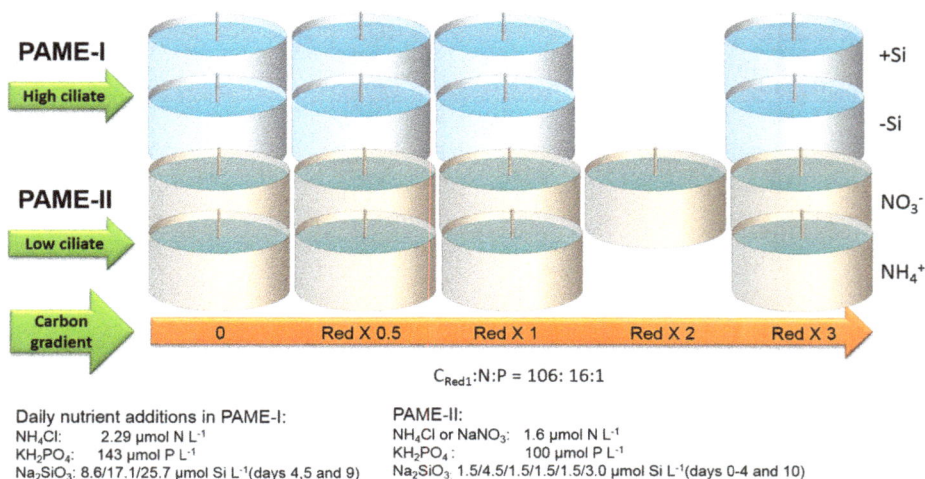

PAME-I +Si

High ciliate -Si

PAME-II NO₃⁻

Low ciliate NH₄⁺

Carbon gradient | 0 | Red X 0.5 | Red X 1 | Red X 2 | Red X 3

C_{Red1}:N:P = 106: 16:1

Daily nutrient additions in PAME-I:
NH_4Cl: 2.29 µmol N L⁻¹
KH_2PO_4:   143 µmol P L⁻¹
Na_2SiO_3: 8.6/17.1/25.7 µmol Si L⁻¹(days 4,5 and 9)

PAME-II:
NH_4Cl or $NaNO_3$: 1.6 µmol N L⁻¹
KH_2PO_4:   100 µmol P L⁻¹
Na_2SiO_3 1.5/4.5/1.5/1.5/1.5/3.0 µmol Si L⁻¹(days 0-4 and 10)

Figure 2. Experimental design of the two mesocosms, (Polar Aquatic Microbial Ecology) PAME-I (2007) and PAME-II (2008). Both experiments received the same dose of nitrogen (N) and phosphorus (P) in Redfield ratio (C:N:P = 106:16:1 molar). Glucose was added in two four-point carbon addition-gradients (0, 0.5, 1, 3 × Redfield in glucose) in PAME–I and one of the series in PAME-II. The other series in PAME-II was a five-point carbon addition-gradient (0, 0.5, 1, 2, 3 × Redfield in glucose C). Samples for this study were taken from the treatments with 0 and 3 × Redfield carbon additions. In PAME-I, nitrogen was added as NH_4Cl. In PAME-II, nitrogen was added as $NaNO_3$ in the NO_3^- gradient and as NH_4Cl in the NH_4^+ gradient. In PAME-I, silicate was added to the +Si units only. In PAME–II all tanks were kept silicate-replete. More information about the experimental setup can be found in [8].

2.2.2. Viral Counts

Total number of viruses was determined with a FACSCalibur flow cytometer (Becton–Dickinson, FACSCalibur, Biosciences, Franklin Lakes, NJ, USA) equipped with an air-cooled laser providing 15 mW at 488 nm and with standard filter set-up. Enumeration of virus-like particles was performed on samples fixed with glutaraldehyde (final concentration 0.5% v/v) prior to staining with 1 × SYBR

Green I (Molecular Probes, Eugene, OR, USA). A minimum of two different dilutions per sample were counted for 60 s each at a viral event rate between 100 and 1000 s^{-1}. Blanks (Tris EDTA (TE) buffer and SYBR Green) were run at regular intervals and subtracted from the total number in order to ensure instrument noise was not being counted as viral particles. The flow cytometer instrumentation and general methodology followed the recommendations of Marie et al. [11].

2.2.3. Viral Concentration and Pulse Field Gel Electrophoresis (PFGE)

Two liters of seawater per sample were pre-filtered through a 142 mm diameter 1.2 μm pore-size low-protein-binding Durapore membrane filter (Millipore Corp., Billerica, MA, USA) to remove larger particles/microorganisms from the sample. The virus-containing filtrates were further concentrated to a final volume of approx. 50 mL by tangential flow filtration using a QuixStand benchtop system equipped with a 100,000 pore size (NMWC) hollow fibre cartridge (GE Healthcare Bio-Sciences AB, Uppsala, Sweden) [12]. Recovery of the viruses using this approach has been measured to be 40–60% (Sandaa, personal observation). Viral particles were subsequently concentrated by ultracentrifugation (Beckman L8-M with SW-28 rotor) for 2 h at 28,000 rpm at 10 °C. The viral pellet was dissolved in 200 μL of SM buffer (0.1 M NaCl, 8 mM $MgSO_4 \times 7H_2O$, 50 mM Tris-HCl, 0.005% *w/v* glycerine).

PFGE was performed according to Sandaa et al. [13]. Four viral agarose plugs were prepared from the 200 μL viral concentrate for PFGE. The plugs were made immediately after viral concentration and stored in a TE buffer (20:50) at 4 °C and analyzed within a month after sampling. The samples were run on a 1% *w/v* SeaKem GTG agarose (FMC, Rockland, ME, USA) gel in 1 × TBE gel buffer using a Bio-Rad DR-II CHEF Cell (Bio-Rad, Richmond, CA, USA) electrophoresis unit. From each sample we used three of the plugs and ran them at three different pulse-ramp conditions in order to separate a large range of viral genome sizes [3,13]. Gels were visualized and digitized using the Fujifilm imaging system, LAS-3000.

2.2.4. DNA Isolation, PCR, and Denaturing Gradient Gel Electrophoresis (DGGE)

Depending on the biomass in the different units, 50–250 mL of water was filtered through a sterile 0.2 μm polycarbonate filter. The filters were flash-frozen in liquid nitrogen and stored for approx. two months before further analysis. DNA isolation, PCR, and DGGE were performed as described in Töpper, et al. [14]. The DGGE of samples from PAME-II was analyzed with a denaturing gradient of 20–60% and an internal standard consisting of 16S rDNA amplicons of four bacteria isolates: *Gelidibacter algens* (DSMZ: 12408), *Microbacterium maritypicum* (DSMZ: 12512), *Flexibacter aurantiacus* (DSMZ: 6792), and *Sulfitobacter mediterraneus* (DSMZ: 12244) [14].

2.2.5. Statistical Analysis

Correlation between ciliate abundance [8] and virus to prokaryote ratio (VPR) was investigated by calculating the Pearson correlation coefficient (ρ). Digitized images of DGGE (bacteria) and PFGE (virus) were analyzed using the programs Gel 2K (Svein Norland, Dept. of Biology, University of Bergen, Norway) and Bionumerics 4.5 (Applied Maths, Sint-Martens-Latem, Belgium), respectively, which determine the presence and absence of bands (fingerprints). Binary data of the DGGE analysis from PAME-I [15] and PAME-II (this study) together with the binary data of the PFGE analysis were compared using the program R 3.3.1 (R Development Core Team 2016). Non-metric multidimensional scaling (NMDS) was performed by using the function metaMDS in the R package vegan [16] applying the Jaccard similarity coefficient.

Correlation between bacteria and virus ordinations of each treatment was calculated using the Procrustes statistic (R package vegan) [16]. A permutation test with 1000 repetitions of the NMDS and Procrustes statistics was performed to obtain reliable *p*-values. Permutational Multivariate Analysis of Variance (PERMANOVA, 99999 permutations) of band patterns in the DGGE and PFGE analysis were used to test whether the bacterial and viral communities grouped according to nutrient treatment (function "Adonis" R package vegan).

3. Results

3.1. Theoretical Considerations

Combining the Lotka–Volterra equations for growth of prey communities of ciliates (i.e., AFs and HNFs, Figure 1) assuming steady-state and phosphate limitation, leads to a carbon demand for heterotrophic prokaryotes (BCD_P) that scales approximately to the second power of ciliate (C) abundance ($BCD_P \sim C^2$) (see Section 2.1 for assumptions and arguments). The relationship $BCD_P \sim C^2$ is illustrated by the green curve in Figure 1D. BCD_P increases rapidly with ciliate abundance, but since the consumption of labile organic material cannot sustainably exceed its autochthonous production (ψ, horizontal line in Figure 1D), heterotrophic prokaryotes will become C-limited when $BCD_P > \psi$. A closed system will therefore have a steady state with mineral nutrient-limited (MNL) bacterial growth at low ciliate abundances with a shift towards organic carbon limited (OCL) heterotrophic prokaryote growth at high ciliate abundances (Figure 1D). Supplying allochthonous labile organic-C to an OCL system is expected to evoke a marked response, compared to no response when adding it to a system in the MNL state. Although the slope of the quadratic function $BCD_P \sim C^2$, the position of ψ, and therefore the transition from MNL to OCL state are subject to several assumptions, they are consistent with the experimental results from Larsen et al. [8]. Interpreting the nutrient manipulated mesocosms as upscaled bioassays, the lack of response in prokaryote abundance to glucose addition indicates MNL bacterial growth in PAME-II (Figure 1A, data from [8]), in contrast to PAME-I (Figure 1B), where the strong positive response indicates OCL bacterial growth.

3.2. Experimental Results

3.2.1. Total Viral Abundance

In general, the total viral abundance was lower in all treatments in PAME-II than in PAME-I (Figure 3A,B). Viral abundance varied between 0.9 and 2.0×10^7 particles mL^{-1} in PAME-II in the start of the experiment, whereas the initial abundances in PAME-I were slightly higher (2.4×10^7 mL^{-1}). In PAME-II, we observed no major differences in viral abundance between the different tanks; and the abundance remained at the same level during the whole experimental period. In PAME-I, the viral abundance in tanks receiving glucose (3C−Si and 3C+Si) increased to peak values on day 6 (8.7 and 9.5×10^7 particles mL^{-1}, respectively). We observed a second increase in tank 3C+Si starting on day 9 and culminating on day 12 (1.7×10^8 particles mL^{-1}). In the two tanks without glucose addition, the viral abundance increased slightly throughout the entire experiment, reaching 6.5 and 3.9×10^7 particles mL^{-1}, respectively, on day 12.

The viral particles were grouped into four populations on the basis of side scatter signal vs. green fluorescent signal after staining with SYBR Green I in PAME-I: I, II, III, and IV (I = small viruses, II: medium viruses, III: large viruses, IV: huge viruses (Figures 4A and S2)). Both side- and fluorescent signals are indicators of size. Larger surface area will give a stronger scatter signal and bigger viral particles will have more DNA, which will be reflected in the more intense green fluorescence signal after staining with SYBR Green. The signals do not, however, give a basis for accurate size measurements as both granularity and surface properties affect signal strength. In PAME-II all viruses belonged to I + II (results not shown). In PAME-I, the dynamics of the various viral size groups all followed a pattern that was similar, but not identical, to that of total viruses, with low initial concentrations followed by a proliferation around day 3–5. This was most evident in the tanks receiving glucose, especially 3C+Si (Figure 4B). Group IV (huge viruses) peaked at day 6 in all tanks, with the highest concentrations in 3C+Si (1.7×10^6 particles mL^{-1}).

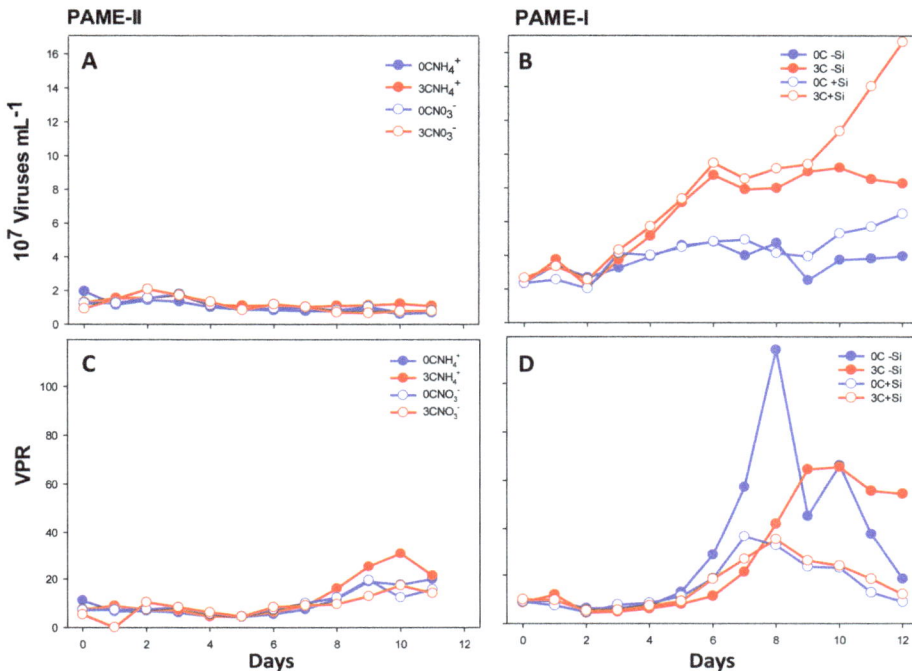

Figure 3. A and B, total abundance of viruses (all groups). C and D, virus to prokaryote ratio (VPR) determined by flow cytometry (FCM) of small viruses and prokaryotes. Blue circles show the data of tanks with no addition of carbon in the form of glucose, while red circles show three Redfield additions of carbon. Open and filled circles display perturbations with different nitrogen sources (open: NO_3; filled: NH_4) in PAME-II and with or without silicate (open: no Si addition; filled: Si addition) in PAME-I. See Figure 2 for more information about experimental design. Viral abundances during PAME-I have been previously published in Töpper et al. [15].

The abundance of group III (large viruses) increased from day 3 to day 6 in all tanks and continued to increase in tank 3C+Si, reaching 1.8×10^7 particles mL^{-1} on day 12. Group II (medium sized viruses) increased only slightly in numbers during the experimental period, except in tank 3C+Si, where we observed a pronounced increase starting at day 8, with 7.2×10^6 particles mL^{-1} reaching 4.2×10^7 particles mL^{-1} on day 12. The concentration of viruses belonging to group I (small viruses) was highest in the two glucose-amended tanks, with peaks on day 6. In 3C+Si, the concentration of this viral group increased again from day 9 to day 12, reaching a final concentration of 1.1×10^8 particles mL^{-1}.

Viral to prokaryote ratio (VPR) was calculated using the abundance of viruses in group I (small viruses mainly infecting prokaryotes [17]) and total prokaryotes, determined by FCM [8]. Initial VPR was approx. 10 in both experiments, and in general VPR varied more in PAME-I (Figure 3C,D) than in PAME-II. The highest ratio (114) was detected in 0C−Si of PAME-I on day 8. The VPR showed a large and significant positive correlation with ciliate numbers (Table 2) for all treatments except PAME-I tank 0C where a non-significant and slightly negative ($-\rho = 0.175$, $p = 0.678$) association was measured.

Table 2. Pearson correlation coefficient between ciliate numbers (data from [8]) and viral to prokaryote ratio (VPRs data form Figure 3C,D). $0 \times$ C: no addition of carbon in form of glucose, $3 \times$ C: three Redfield additions of carbon.

	PAME-I		PAME-II	
	$0 \times$ **C**	$3 \times$ **C**	$0 \times$ **C**	$3 \times$ **C**
Persons coefficient (ρ)	-0.175	0.835	0.630	0.783
p Value	$p = 0.678$	$p = 0.00982$	$p = 0.0281$	$p = 0.00258$

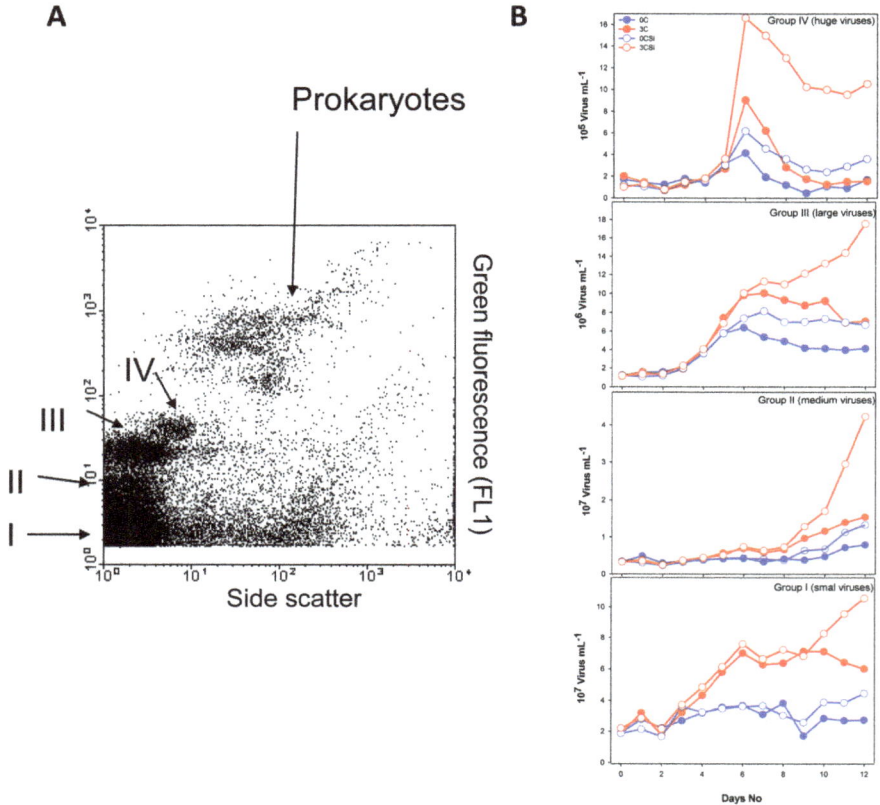

Figure 4. Biparametric flow cytometry virus plots. Four different viral populations were discriminated combining side scatter signal vs. green fluorescent signal after staining with SYBR Green I (reflecting the amount of DNA and hence genome sizes): I, Small viruses, II, medium viruses, III, large viruses, IV, huge viruses (**A**) (see also Figure S2). Abundance of small (I), medium (II), large (III), and huge (IV) viruses determined by FCM in PAME-I (**B**). Blue circles show no addition of carbon in the form of glucose, while red circles show three Redfield additions of carbon. Open and filled circles display perturbations with or without silicate (open: no Si addition; filled: Si addition). See Figure 2 for more information about experimental design.

3.2.2. Bacterial and Viral Community Structure

Thirty-one pulsed field gel electrophoresis (PFGE) bands, with genome sizes ranging from 14 to 502 kb (Figure 5), were detected in PAME-I. The number of bands was highest in the 3C+Si tank, with an increase from 11 on day 0 to 20 on day 12. Viruses with genomes larger than 350 kb were also more dominant (highest pixel value) in tank 3C+Si than in the others. During PAME-II, a total of 11 unique PFGE bands were detected ranging in size from 22 to 352 kb, with a dominance of bands between 42 and 54 kb. For PAME-II, the non-metric multidimensional scaling (NMDS) analysis based on the denaturing gradient gel electrophoresis (DGGE) and PFGE band patterns (Figure 6A,C) did not reveal any grouping. Furthermore, there was no significant support ($p > 0.05$) (Table 3) of grouping in either the DGGE nor PFGE patterns due to glucose addition, nor for co-occurring changes in bacterial host and viral communities ($p > 0.05$, Table 4. In PAME-I, both the bacterial and viral community structure changed over time and responded to different treatments (Figure 6B,D). Grouping of DGGE and PFGE patterns due to glucose and silicate addition was supported statistically ($p < 0.05$, Table 3). Positive correlations (median *p*-value < 0.05) were detected between the bacterial host and the viral communities (0C, 3S, 0CSi), except for the treatment with both carbon and silicate addition (3CSi, median *p*-value 0.08) (Table 4). Correlations were tested with 1000 NMDS permutations of DGGE and PFGE patterns and resulted in a % of *p*-value < 0.05 of 100% (0C), 97.1% (3C), 99.6%, and 27.5% (3CSi) (Table 4). The correlation between the bacterial and viral communities was paralleled by both communities showing significant responses to carbon and silicate treatments (Figure 6B,D, Table 3). However, viral communities responded less to carbon than to silicate, while the opposite was true for the bacterial community (Table 3), with the result that within the mesocosms treated with both C and Si bacterial and viral communities were no longer correlated (Table 4).

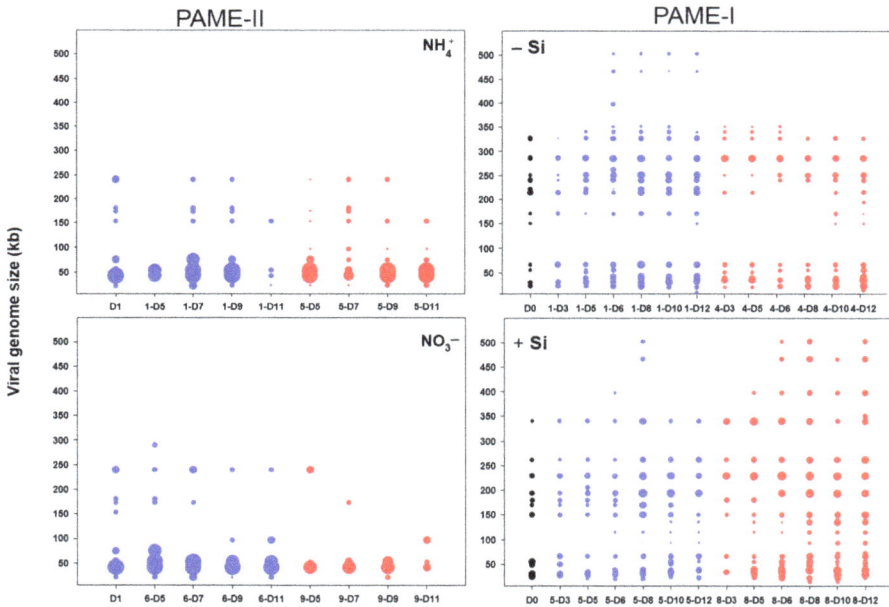

Figure 5. Schematic outline of the PFGE bands sorted by genome size during the mesocosm experiment in PAME-II and PAME-I. D: sampling day. The outline is based on three different electrophoresis runs for each viral sample. Blue circles show samples from treatments with no addition of carbon in the form of glucose, while red circles show samples from treatments with three Redfield additions of carbon. See Figure 2 for more information about experimental design.

Table 3. Permutational multivariate analysis of variance of band patterns in the denaturing gradient gel electrophoresis (DGGE) and pulsed field gel electrophoresis (PFGE) analysis for bacteria and virus, respectively, from PAME-I and PAME-II mesocosm experiments. Glucose treatments in PAME-I and PAME-II included no addition of carbon in form of glucose ($0 \times C$) and three Redfield additions of carbon ($3 \times C$). Silicate treatment in PAME-I comprised of silicate addition and ambient silicate concentrations. Nitrogen treatment in PAME-II comprised of NH_4^+ and NO_3^- addition. All treatments in PAME-II were kept silicate replete. Number of permutations in all analyses: 99999. Residual degrees of freedom: 20 in PAME-I and 12 in PAME-II.

Experiments	Treatment	Bacteria Community			Viral Community		
		F	R^2	*p*	F	R^2	*p*
PAME-I	Glucose	11.86	0.31	<0.001	7.33	0.11	0.002
	Silicate	5.17	0.14	0.003	38.16	0.56	<0.001
PAME-II	Glucose	2.23	0.13	0.07	1.32	0.06	0.3
	Nitrogen	0.63	0.04	0.67	3.53	0.17	0.048

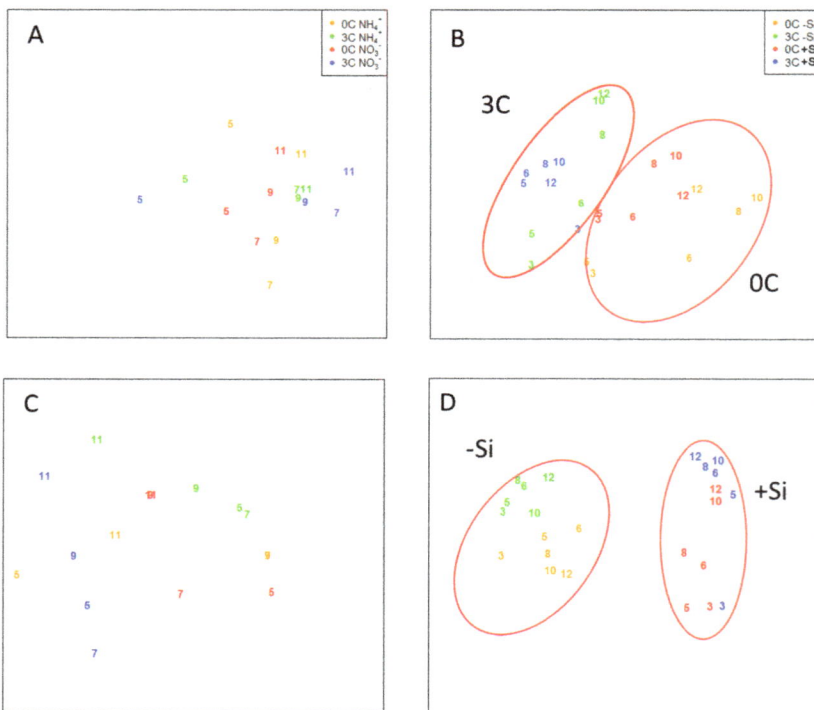

Figure 6. Nonmetric multidimensional scaling analysis (NMDS) based on the band pattern in the DGGE (bacteria (**A**,**B**)) and PFGE (virus (**C**,**D**)) analysis. Binary data for NMDS analysis from PAME-I is redrawn from [15]. See Figure 2 for more information about experimental design.

Table 4. Median *p*-values, the percentage of significant *p*-values (<0.05) and median correlation values issued from 1000 non-metric multidimensional scaling (NMDS) permutations of denaturing gradient gel electrophoresis (DGGE) and pulsed field gel electrophoresis (PFGE) band patterns are presented for the two different enrichment experiments (PAME-I and -II) and the different tanks. $0 \times C$: no addition of carbon, $3 \times C$: three Redfield additions of carbon in form of glucose. Si = Silicate.

	PAME-I				PAME-II			
Statistics	0C	3C	0CSi	3CSi	0CNH$_4$	3CNH$_4$	0CNO$_3$	3CNO$_3$
median *p*-value *	0.013	0.01	0.007	0.08	0.083	0.542	0.085	0.125
p-values < 0.05 [%] *	100	97.1	99.6	27.5	0	0	0	0
median Procrustes correlation *	0.852	0.878	0.906	0.723	0.963	0.491	0.858	0.945

* Issued from 1000 non-metric multidimensional scaling NMDS permutations.

4. Discussion

Key differences in responses on the level of PFTs in two similar mesocosm experiments (PAME-I and -II) were previously traced to a central role of ciliates, which themselves were top-down controlled by the copepod standing stock [8]. Here, we expand the analysis and demonstrate, first theoretically, how ciliate abundance can be connected to different states of growth limitation of heterotrophic prokaryotes, then experimentally how this is reflected not only in the abundance and activity of heterotrophic prokaryotes but also in the viral communities.

Our model predicts that high abundance of ciliates (PAME-I) creates a prokaryote community that is organic carbon-limited (OCL) (Figure 1D), which in turn leads to the expectation that glucose addition has a great effect on both abundance and structure of the virus host community. Low ciliate abundances (PAME-II) (Figure 1D), on the other hand, should, according to the model, promote a mineral nutrient-limited (MNL) prokaryote community, giving no expectations of a similar effect when carbon is added. Our experimental results supported our model predictions, with an increase in heterotrophic prokaryote abundance and a statistical supported effect on the bacterial community structure that received additional carbon in PAME-I (Figures 1B and 6B, Table 3). No similar effect or statistical support for changes in the prokaryote abundance or bacterial community structure was seen in PAME-II when carbon was added (Figures 1A and 6A, Table 3). In PAME-I, positive correlations between changes in the bacterial and viral community structure were observed, with one exception (Table 4). A corresponding link was not detected in PAME-II (Table 4). The significant positive correlation between changes in ciliate abundance and virus to prokaryote ratio (VPR) due to organic carbon load (Table 2) further supports our model predictions and demonstrates the strong link between the predator food chain in the marine microbial food web and the activity of viruses.

Glucose was the key mediator shaping the heterotrophic prokaryote community structure in PAME-I (Table 3, Figure 6B). We also observed, however, an enigmatic effect of silicate on prokaryotes, visible both as an increase in abundance in Si treatments towards the end of the experiment (Figure 1B), and small differences between similar carbon treatments with and without silicate in the community structure of the bacterial community (Figure 6B). Virus abundance increased concomitantly with the prokaryotes (3C+Si) and is indicative of a link between these two populations. Prokaryotes are not expected to be directly influenced by changes in Si concentrations. We believe that the explanation of the enigmatic Si effect on prokaryotes is indirect and linked to the total dominance of the autotrophic community of a single-celled diatom (*Thalassiosira* sp. 5–10 μm; [10]) small enough to be preyed upon by ciliates in the Si treatments. This led to an increasing ciliate population, which also feed on HNFs and release prokaryotes from predation pressure by HNF, allowing for the observed proliferation of prokaryotes at the end of the experiment [8].

As demonstrated in other studies [3,18,19], a statistically supported link was detected between changes in the bacterial host community and changes in the viral community structure (Figure 6B,D, Table 4) in PAME-I. Nevertheless, glucose, which was the main driver for changes in the bacterial community structure, did not seem to influence the viral community to the same extent as silicate (Figure 6D, Table 3). A plausible explanation is that PFGE captures viruses infecting both prokaryotes and eukaryotes [20]. The high abundance of huge viruses (FCM, Figure 4B) and double stranded DNA (dsDNA) viral genomes larger than 250 kb (PFGE, Figure 5) is probably linked to eukaryotic populations, as these generally host bigger viruses than prokaryotes [21–25]. This then explains the low correlation between the bacterial and viral community structure in the 3C+Si tanks (Table 4). *Thalassiosira* sp. diatoms could theoretically host the biggest viruses since they dominated the Si mesocosms [8], but there are presently no reports of huge dsDNA viruses infecting diatoms [26]. We thus deem it more likely that HNFs were hosts to these huge viruses, since the decline in the HNF population [8] coincided with the increase in the huge viral group. This would also be consistent with already characterized giant HNF viruses [24,27].

Although the viral community structure in general was most affected by silicate, viral groups I–III were most abundant when carbon was added (Figure 4B). These groups probably comprise the seven PFGE bands with smaller genome size than 100 kb, which were mainly detected at high glucose levels in the period when prokaryote abundance strongly declined (day 5–8) (Figure 5) and may represent viruses able to lyse heterotrophic prokaryotes.

As important regulators of host diversity and nutrient recycling, viruses are central components of the marine environment and the marine food web. Our results show that the activity of this important regulator in the microbial food web is strongly linked to another top-down mechanism, namely predation and bottom-up factors such as availability of limiting nutrients. To our knowledge, this is the first study that links external trophic interactions in the microbial food web to the internal structure and function of viral and prokaryote communities, using a combination of theoretical modeling and experimental data. Although challenging, understanding these relationships is of particular importance in the rapidly changing Arctic, where growth-limiting factors for prokaryotes have been found to vary [28] and allochthonous import of terrestrial dissolved organic carbon (DOC) from rivers [29], as well as seasonal migration of copepods [30], are central ecosystem characteristics. The present study provides a preliminary understanding of the intricate links between these processes and encourages further cross-scale studies.

Supplementary Materials: The following are available online at www.mdpi.com/1999-4915/9/9/238/s1, Figure S1: title; Minimum model of the photic zone microbial food-web Figure S2: title; Histogram representation of four different viral populations and prokaryotes detected by flow cytometry Supplementary raw data is alsoavailable in the Pangea repository doi:10.1594/PANGAEA.865353.

Acknowledgments: The authors thank all the participants in the PAME-I and II experiment. Financial support was provided by the Norwegian Research Council, the International Polar Year project 175939/S30 'PAME-Nor' (IPY activity ID No. 71), the European Research Council Advanced Grant ERC-AG-LS8 "Microbial Network Organization" (MINOS, project number 250254) and the Norwegian Research Council project MicroPolar (225956/E10). Further, we thank Jorun Egge, University of Bergen, Norway, for providing the ciliate numbers and Gunnar Gerdts from the Alfred Wegener Institute, Helgoland, Germany for providing analysis software.

Author Contributions: T.F.T. designed the experiment, which was conducted together with R.-A.S., B.T., A.L., and R.T., J.P.T. was responsible for the statistics. R.-A.S. wrote the manuscript with input from all co-authors.

Conflicts of Interest: The authors declare no conflict of interest.

References

1. Zöllner, E.; Hopp, H.-G.; Sommer, U.; Juergens, K. Effect of zooplankton-mediated trophic cascades on marine microbial food web components (bacteria, nanoflagellates, ciliates). *Limnol. Oceanogr.* **2009**, *54*, 262–275. [CrossRef]

2. Bouvier, T.; Del Giorgio, P.A. Key role of selective viral-induced mortality in determining marine bacterial community composition. *Environ. Microbiol.* **2007**, *9*, 287–297. [CrossRef] [PubMed]

3. Sandaa, R.-A.; Gomez-Consarnau, L.; Pinhassi, J.; Riemann, L.; Malits, A.; Weinbauer, M.G.; Gasol, J.M.; Thingstad, T.F. Viruses control of bacterial biodiversity—Linkages between viral and bacterial community

structure in a nutrient enriched mesocosm experiment. *Environ. Microbiol.* **2009**, *11*, 2585–2595. [CrossRef] [PubMed]

4. Weinbauer, M.G. Ecology of prokaryotic viruses. *FEMS Microbiol. Rev.* **2004**, *28*, 127–181. [CrossRef] [PubMed]

5. Thingstad, T.F.; Havskum, H.; Zweifel, U.L.; Berdalet, E.; Sala, M.M.; Peters, F.; Alcaraz, M.; Scharek, R.; Perez, M.; Jacquet, S.; et al. Ability of a "minimum" microbial food web model to reproduce response patterns observed in mesocosms manipulated with N and P, glucose, and Si. *J. Mar. Syst.* **2007**, *64*, 15–34. [CrossRef]

6. Beninca, E.; Huisman, J.; Heerkloss, R.; Jöhnk, K.D.; Branco, P.; Van Nes, E.H.; Scheffer, M.; Ellner, S.P. Chaos in a long-term experiment with a plankton community. *Nature* **2008**, *451*, 822–825. [CrossRef] [PubMed]

7. Våge, S.; Pree, B.; Thingstad, T.F. Linking internal and external bacterial community control gives mechanistic framework for pelagic virus-to-bacteria ratios. *Environ. Microbiol.* **2016**, *11*, 3932–3948. [CrossRef] [PubMed]

8. Larsen, A.; Egge, J.K.; Nejstgaard, J.C.; Di Capua, I.; Thyrhaug, R.; Bratbak, G.; Thingstad, T.F. Contrasting response to nutrient manipulation in Arctic mesocosms are reproduced by a minimum microbial food web model. *Limnol. Oceanogr.* **2015**, *60*, 360–374. [CrossRef] [PubMed]

9. Banchetti, R.; Nobili, R.; Esposito, F. An experimentally determined carbon: Volume ratio for marine "oligotrichous" ciliates from estuarine and coastal waters. *Limnol. Oceanogr.* **1989**, *34*, 1097–1103.

10. Thingstad, T.F.; Bellerby, R.G.J.; Bratbak, G.; Borsheim, K.Y.; Egge, J.K.; Heldal, M.; Larsen, A.; Neill, C.; Nejstgaard, J.; Norland, S.; et al. Counterintuitive carbon-to-nutrient coupling in an Arctic pelagic ecosystem. *Nature* **2008**, *455*, 387–390. [CrossRef] [PubMed]

11. Marie, D.; Brussaard, C.P.D.; Thyrhaug, R.; Bratbak, G.; Vaulot, D. Enumeration of marine viruses in culture and natural samples by flow cytometry. *Appl. Environ. Microbiol.* **1999**, *65*, 45–52. [PubMed]

12. Ray, J.; Dondrup, M.; Modha, S.; Steen, I.H.; Sandaa, R.-A.; Clokie, M. Finding a Needle in the Virus Metagenome Haystack-Micro-Metagenome Analysis Captures a Snapshot of the Diversity of a Bacteriophage Armoire. *PLoS ONE* **2012**, *7*, e34238. [CrossRef] [PubMed]

13. Sandaa, R.-A.; Short, S.M.; Schroeder, D.C. Fingerprinting aquatic virus communities. In *Manual of Aquatic Viral Ecology*; Wilhelm, S.W., Weinbauer, M.G., Suttle, C.A., Eds.; American Society of Limnology and Oceanography: Wasco, TX, USA, 2010; Volume 2, pp. 9–18.

14. Töpper, B.; Thingstad, T.F.; Sandaa, R.-A. Effects of differences in organic supply on bacterial diversity subject to viral lysis. *FEMS Microbiol. Ecol.* **2012**, *83*, 202–213. [CrossRef] [PubMed]

15. Töpper, B.; Larsen, A.; Thingstad, T.; Thyrhaug, R.; Sandaa, R.-A. Bacterial community composition in an Arctic phytoplankton mesocosm bloom: The impact of silicate and glucose. *Polar Biol.* **2010**, *33*, 1557–1565. [CrossRef]

16. Oksanen, J.; Blanchet, F.G.; Kindt, R.; Legendre, P.; Minchin, P.R.; O'Hara, R.B.; Simpson, L.G.; Solymos, P.; Stevens, M.H.H.; Wagner, H. Vegan: Community Ecology Package. R Package Version 2.3-0. 2015. Available online: https://cran.r-project.org/package=vegan (accessed on 19 March 2016).

17. Larsen, J.B.; Larsen, A.; Thyrhaug, R.; Bratbak, G.; Sandaa, R.-A. Marine viral populations detected during a nutrient induced phytoplankton bloom at elevated pCO_2 levels. *Biogeosciences* **2008**, *5*, 523–533. [CrossRef]

18. Ovreas, L.; Bourne, D.; Sandaa, R.A.; Casamayor, E.O.; Benlloch, S.; Goddard, V.; Smerdon, G.; Heldal, M.; Thingstad, T.F. Response of bacterial and viral communities to nutrient manipulations in seawater mesocosms. *Aquat. Microb. Ecol.* **2003**, *31*, 109–121. [CrossRef]

19. Riemann, L.; Steward, G.F.; Azam, F. Dynamics of bacterial community composition and activity during a mesocosm diatom bloom. *Appl. Environ. Microbiol.* **2000**, *66*, 578–587. [CrossRef] [PubMed]

20. Sandaa, R.-A. Burden or benefit? Virus-host interactions in the marine environment. *Res. Microbiol.* **2008**, *159*, 374–381. [CrossRef] [PubMed]

21. Campillo-Balderas, J.A.; Lazcano, A.; Becerra, A. Viral Genome Size Distribution Does not Correlate with the Antiquity of the Host Lineages. *Front. Ecol. Evol.* **2015**, *3*. [CrossRef]

22. Sandaa, R.A.; Heldal, M.; Castberg, T.; Thyrhaug, R.; Bratbak, G. Isolation and characterization of two viruses with large genome size infecting *Chrysochromulina ericina* (Prymnesiophyceae) and *Pyramimonas orientalis* (Prasinophyceae). *Virology* **2001**, *290*, 272–280. [CrossRef] [PubMed]

23. Johannessen, T.V.; Bratbak, G.; Larsen, A.; Ogata, H.; Egge, E.S.; Edvardsen, B.; Eikrem, W.; Sandaa, R.-A. Characterisation of three novel giant viruses reveals huge diversity among viruses infecting Prymnesiales (Haptophyta). *Virology* **2015**, *476*, 180–188. [CrossRef] [PubMed]

24. Fischer, M.G.; Allen, M.J.; Wilson, W.H.; Suttle, C.A. Giant virus with a remarkable complement of genes infects marine zooplankton. *Proc. Natl. Acad. Sci. USA* **2010**, *107*, 19508–19513. [CrossRef] [PubMed]

25. Massana, R.; del Campo, J.; Dinter, C.; Sommaruga, R. Crash of a population of the marine heterotrophic flagellate *Cafeteria roenbergensis* by viral infection. *Environ. Microbiol.* **2007**, *9*, 2660–2669. [CrossRef] [PubMed]

26. Nagasaki, K. Dinoflagellates, diatoms, and their viruses. *J. Microbiol.* **2008**, *46*, 235–243. [CrossRef] [PubMed]

27. Weinbauer, M.G.; Dolan, J.R.; Šimek, K. A population of giant tailed virus-like particles associated with heterotrophic flagellates in a lake-type reservoir. *Aquat. Microb. Ecol.* **2015**, *76*, 111–116. [CrossRef]

28. Vadstein, O. Large variation in growth-limiting factors for marine heterotrophic bacteria in the Arctic waters of Spitsbergen (78 degrees N). *Aquat. Microb. Ecol.* **2011**, *63*, 289–297. [CrossRef]

29. Matsuoka, A.; Babin, M.; Doxaran, D.; Hooker, S.B.; Mitchell, B.G.; Belanger, S.; Bricaud, A. A synthesis of light absorption properti,es of the Arctic Ocean: Application to semianalytical estimates of dissolved organic carbon concentrations from space. *Biogeosciences* **2014**, *11*, 3131–3147. [CrossRef]

30. Falk-Petersen, S.; Mayzaud, P.; Kattner, G.; Sargent, J.R. Lipids and life strategy of Arctic Calanus. *Mar. Biol. Res.* **2009**, *5*, 18–39. [CrossRef]

viruses

MDPI

Article

Seasonal Dynamics of Haptophytes and dsDNA Algal Viruses Suggest Complex Virus-Host Relationship

Torill Vik Johannessen [1], Aud Larsen [2], Gunnar Bratbak [3], António Pagarete [3], Bente Edvardsen [4], Elianne D. Egge [4] and Ruth-Anne Sandaa [3,*]

[1] Vaxxinova Norway AS, Kong Christian Frederiks plass 3, 5006 Bergen, Norway; torillvjohannessen@gmail.com
[2] Uni Research Environment, N-5008 Bergen, Norway; aud.larsen@bio.uib.no
[3] Department of Biology, University of Bergen, N-5020 Bergen, Norway; gunnar.bratbak@bio.uib.no (G.B.); antonio.pagarete@uib.no (A.P.)
[4] Department of Biosciences, University of Oslo, 0316 Oslo, Norway; bente.edvardsen@ibv.uio.no (B.E.); elianne.egge@gmail.com (E.D.E.)
*   Correspondence: ruth.sandaa@bio.uib.no

Academic Editors: Mathias Middelboe and Corina P.D. Brussaard
Received: 31 January 2017; Accepted: 13 April 2017; Published: 20 April 2017

Abstract: Viruses influence the ecology and diversity of phytoplankton in the ocean. Most studies of phytoplankton host–virus interactions have focused on bloom-forming species like *Emiliania huxleyi* or *Phaeocystis* spp. The role of viruses infecting phytoplankton that do not form conspicuous blooms have received less attention. Here we explore the dynamics of phytoplankton and algal viruses over several sequential seasons, with a focus on the ubiquitous and diverse phytoplankton division Haptophyta, and their double-stranded DNA viruses, potentially with the capacity to infect the haptophytes. Viral and phytoplankton abundance and diversity showed recurrent seasonal changes, mainly explained by hydrographic conditions. By 454 tag-sequencing we revealed 93 unique haptophyte operational taxonomic units (OTUs), with seasonal changes in abundance. Sixty-one unique viral OTUs, representing Megaviridae and *Phycodnaviridae*, showed only distant relationship with currently isolated algal viruses. Haptophyte and virus community composition and diversity varied substantially throughout the year, but in an uncoordinated manner. A minority of the viral OTUs were highly abundant at specific time-points, indicating a boom-bust relationship with their host. Most of the viral OTUs were very persistent, which may represent viruses that coexist with their hosts, or able to exploit several host species.

Keywords: Haptophyta; *Phycodnaviridae*; Megaviridae; viral–host interactions; metagenomics; marine viral ecology

1. Introduction

Marine phytoplankton account for approximately 50% of global primary production and have a strong impact on global nutrient cycling [1]. As key components within the phytoplankton community in both coastal and open oceans, and at all latitudes [2], haptophytes play important roles both as primary producers but also as mixotrophs, grazing on bacteria and protist [3]. Blooms of haptophytes can have significant ecological and economic impacts both through the amount of organic matter being produced and through production of toxins harmful to marine biota [4]. Most haptophyte species, however, do not usually form blooms, but rather appear at low concentrations at all times [5–7].

Phytoplankton diversity, abundance, and community composition change through the seasons, driven by variations in environmental conditions and biological processes. Viruses can, in theory, significantly condition those dynamics. Viral-based phytoplankton lysis can be at least as significant as grazing [8,9] and have the potential to drastically change host community structure [10]. Viral activity related to bloom forming haptophytes like *Emiliania huxleyi*, *Phaeocystis pouchetii*, and *Phaeocystis globosa* has been well studied [9,11–14]. During such blooms, viruses exhibit a strong regulatory role, and contribute to the termination of the bloom in what may be referred to as a "boom and bust" relationship [11,15,16]. Viruses may also prevent bloom formation by keeping host population at non-blooming levels [16–18]. Such interactions between host and virus have been described as a stable coexistence and explained by viral resistance, immunity and/or strain specificity [17,19–23].

The low diversity and high abundance, which characterize phytoplankton blooms, give species of specific viruses ample possibilities to find susceptible hosts. Most haptophyte species, such as species belonging to the Prymnesiales, however, are part of highly-diverse communities and occur at low concentrations [5–7], which decrease their chance of being infected by viruses with specific host requirements. Nevertheless, viruses infecting both *Prymnesium* and *Haptolina* species (order Prymnesiales) have been isolated, but have several characteristics that distinguish them from viruses infecting bloom-forming haptophytes like *E. huxleyi* [24,25]. Studies describing the seasonal diversity and abundances of these viruses and their potential host communities (haptophytes) in the environment are scarce.

All known haptophyte viruses have double-stranded DNA (dsDNA) genomes and belong to two related viral families, the *Phycodnaviridae* and Megaviridae, within a monophyletic group of nucleocytoplasmic large DNA viruses (NCLDV) [26]. Phycodnaviruses infect prasinophytes, chlorophytes, raphidophytes, phaeophytes, and haptophytes [27]. The Megaviridae family, not yet recognized as a taxon by the International Committee on Taxonomy of Viruses (ICTV), consists of NCLDVs infecting both non-photosynthetic protists such as *Acanthamoeba* and *Cafeteria roenbergensis* [28,29], as well as photosynthetic ones including prasinophytes, pelagophytes and prymnesiophytes (haptophytes) [14,30,31]. Both *Phycodnaviridae* and Megaviridae are abundant in aquatic environments but the majority are uncultured and not yet described [31–37]. The diversity within these two families is high, and available primers only match a fraction of its representatives [32, 38,39]. Moreover, only few polymerase chain reaction (PCR) primer-sets that target *Phycodnaviridae* and Megaviridae families are currently available [32,38]. The DNA polymerase B primers (polB) capture a wide diversity within the *Phycodnaviridae* family including the prasinoviruses and chloroviruses [36–39] whereas the major capsid protein (MCP) primers are better suited for capturing the diversity of the Megaviridae family including prymnesioviruses that infect various haptophytes ([32,39], this study). Coccolithoviruses (e.g., *Emiliania huxleyi virus* (EhV)), a diverged group in the *Phycodnaviridae* family, are not targeted by any of these primer-sets.

In previous studies, we have described the microbial community dynamics of the seasonal spring- and fall-blooms in a West Norwegian open fjord system (Raunefjorden) [40,41]. Virus infection seems to be one of the factors that drive the succession in the haptophyte community after the typical diatom spring bloom [40]. In the present study, we follow up on these investigations using methods with higher taxonomic resolution that enable a more specific focus on haptophytes and their potential viruses. By following dynamics and diversity in virus and haptophyte communities over a two-year period, we aimed at revealing the possible regulatory role of viruses, not only during blooms but also during periods with higher community diversity and lower productivity such as late fall and winter.

2. Results

2.1. Microbial Abundance and Abiotic Factors

The phytoplankton spring bloom, identified as elevated chlorophyll *a* (Chl *a*) fluorescence, started in late February before any stratification of the water masses, and lasted longer in 2011 than in 2010 (Figure 1). The water masses in Raunefjorden started to stratify in March–April, and the stratification was more pronounced and deeper in 2011 than in 2010 (Figure 1).

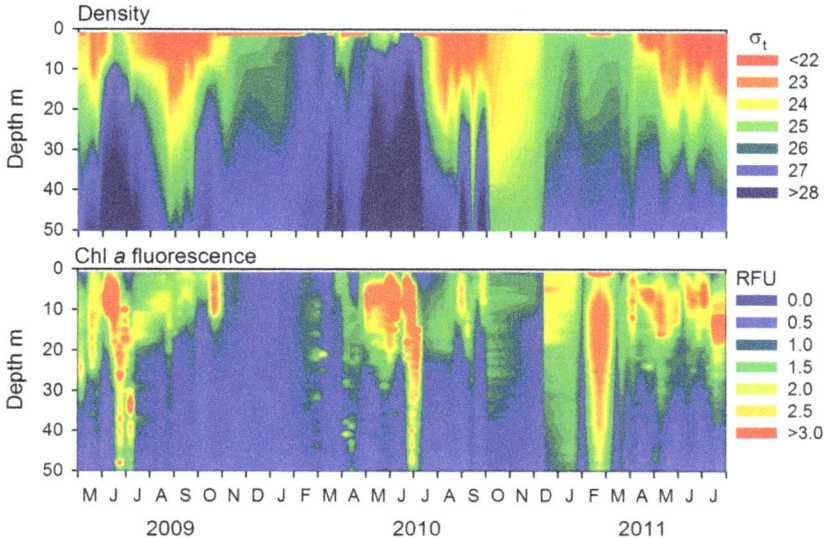

Figure 1. Isopleth diagrams showing seawater density (σt) and chlorophyll *a* (Chl *a*) fluorescence (RFU = relative fluorescence units) at the sampling station in Raunefjorden, respectively.

Several minor upwelling events and exchange of water masses were evident in spring (e.g., in June 2009, April 2010 and June 2010); concurrently May and June were characterized by several successive blooms with high Chl *a* levels (2–8 µg per L). The pycnocline deepened throughout summer and fall before the seasonal inflow and upwelling caused deep mixing in late fall, which corresponded to a temporary, slight increase in Chl *a* concentrations in fall each year (October–November). The water masses were well mixed through fall and winter.

The first increase in pico- and nano-eukaryote abundance, as measured by flow cytometry, was observed in late February (Figure 2A). The cell numbers increased throughout spring and summer with maximum abundance of both groups in August 2010 and May/June 2011. Total bacterial abundance was variable with a decreasing trend in fall and winter and an increasing trend in spring and summer-fall (Figure 2B), while the *Synechococcus* (cyanobacteria) abundance peaked once each year in late summer-fall (Figure 2B).

Viral abundance increased in the spring and summer. The highest values were found during summer and fall. The abundance of all three viral groups varied in synchrony (V2 vs. V1: $r = 0.603$, df = 26, $p < 0.0007$, V3 vs. V1: $r = 0.483$, df = 26, $p < 0.0091$) and the smaller viruses (V1) outnumbered the larger viruses (V2 and V3) by a factor of approximately 5–20 and 50–300, respectively (Figure 2C). Correlation analysis showed that the viral abundance was correlated ($p < 0.01$) with the abundance of bacteria, cryptophytes and *Synechococcus* (Table S1). The abundance of small-sized viruses (V1) also correlated with Chl *a* and abundance of nanoeukaryotes, while the abundance of intermediate-sized viruses (V2) correlated with abundance of picoeukaryotes and nanoeukaryotes.

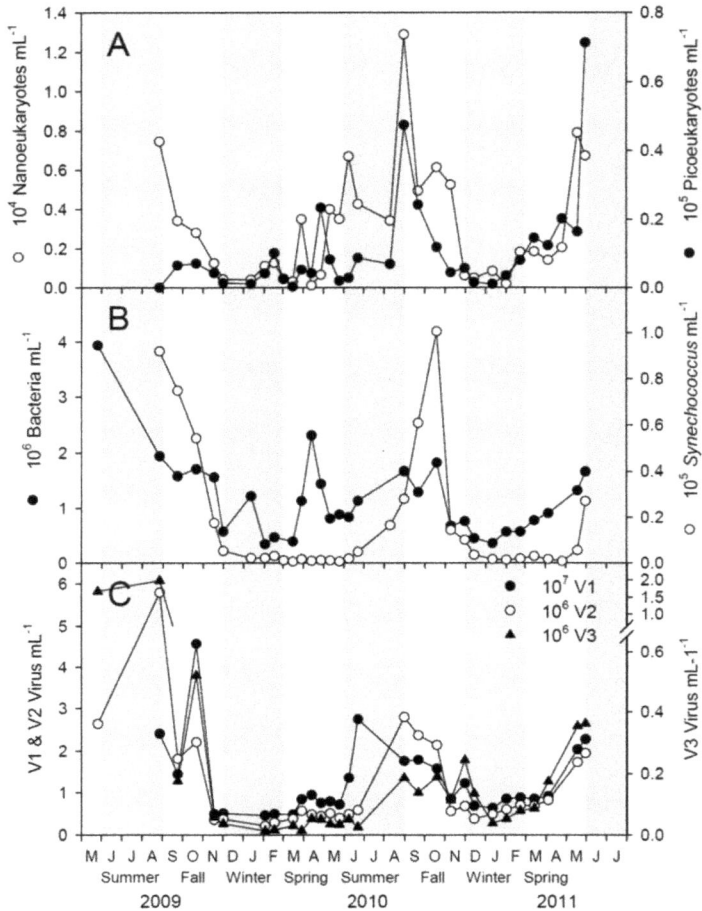

Figure 2. Abundance of microbial plankton at the sampling station in Raunefjorden measured by flow cytometry. (**A**) Phototrophic picoeukaryotes (filled circles) and nanoeukaryotes (open circles); (**B**) *Synechococcus* (open circles) and total bacteria (filled circles); and (**C**) V1 (low fluorescence viruses), V2 (intermediate fluorescence viruses) and V3 (high fluorescence viruses).

2.2. Haptophytes

Haptophyte reads (sequences), clustered based on 98% nucleotide sequence similarity, formed a total of 93 operational taxonomic units (OTUs) (Table S2). OTUs were classified against a curated Haptophyta reference sequence database [42] to the lowest reliable taxonomic level (Table S2). The classified OTUs were placed into one of seven haptophyte orders: Pavlovales, Phaeocystales, Zygodiscales, Syracosphaerales, Isochrysidales, Coccolithales and Prymnesiales (Figure 3). A number of the reads could not be classified to these formally-accepted taxa, and were assigned to defined clades without cultured representatives according to [42], here named Haptophyta unclassified (Clades HAP2, HAP3, HAP4, and HAP5) and Prymnesiophyceae unclassified (Clades B3, B4, D, E and F) (Table S2). Prymnesiales is, in Figure 3, divided into the families Chysochromulinaceae and Prymnesiaceae. OTUs assigned to the order Isochrysidales all belonged to the *E. huxleyi* cluster. A more detailed classification of the 93 haptophyte OTUs to is shown in Table S2.

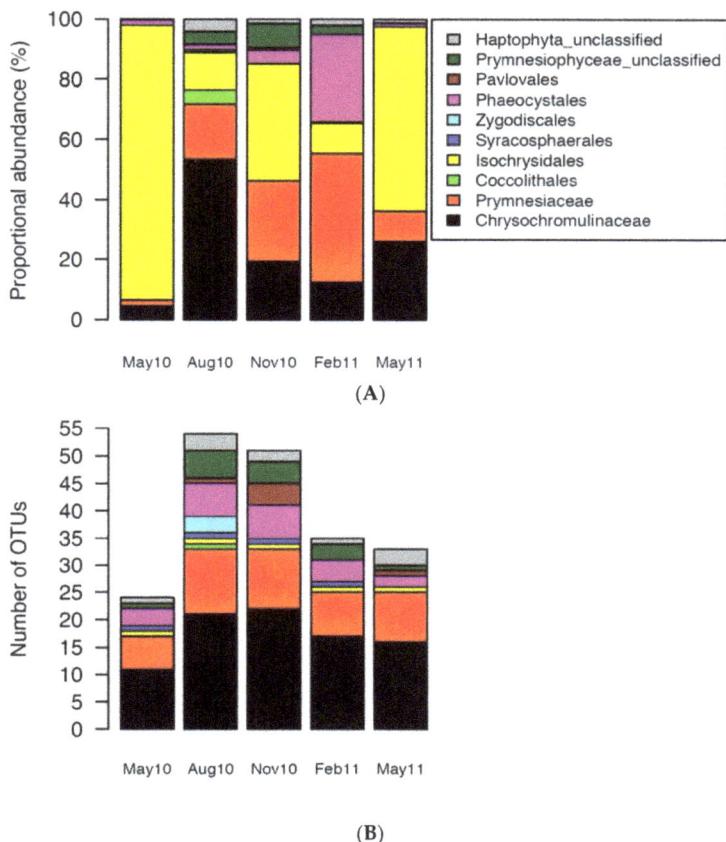

Figure 3. (A) Relative abundance of operational taxonomic units (OTUs) in the different haptophyte orders or clades. The OTUs represented seven accepted haptophyte orders: Pavlovales, Phaeocystales, Coccolithales, Isochrysidales, Syracosphaerales, Zygodiscales and Prymnesiales. The latter is represented by the families Chysochromulinaceae and Prymnesiaceae. The reads that did not belong to any of these orders were assigned to class Prymnesiophyceae (unclassified) or Haptophyta (unclassified); (B) Number of unique OTUs (richness) in each sample after rarifying to lowest read abundance (i.e., subsampling to obtain equal sample size).

Diversity and community composition varied between the samples, with highest diversity in August and lowest in May (Shannon diversity values of 2.97 and 0.50 respectively, Table S3). Based on the Bray-Curtis dissimilarity analysis we found that the August 2010 sample differed most from the rest (Figure 4).

OTUs assigned to Isochrysidales (only *E. huxleyi*) were present in all samples, with particularly high relative abundance in May. OTUs assigned to Prymnesiales, i.e., the families Prymnesiaceae and Chrysochromulinaceae, dominated the samples from August, November and February, while OTUs belonging to Phaeocystales occurred in high relative abundance only in February. Diversity was highest within Prymnesiales, with 35 and 21 unique OTUs assigned to Chrysochromulinaceae and Prymnesiaceae, respectively (Figure 3B). Ten different haptophyte OTUs were present in all samples, one was classified to *E. huxleyi*, one to Clade F, and eight to Prymnesiales (Table S2).

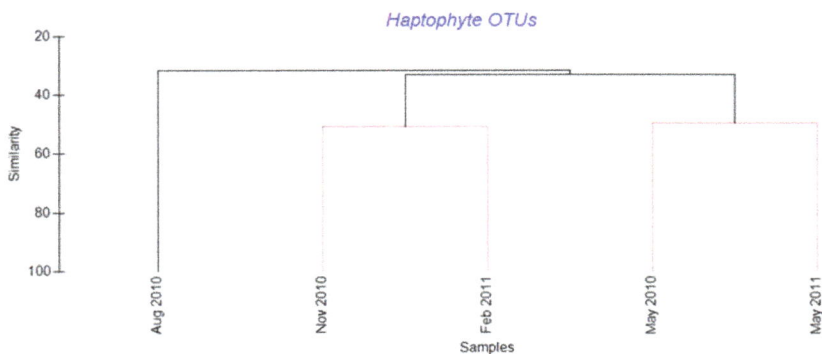

Figure 4. Cluster dendrogram illustrating Bray-Curtis dissimilarity in the haptophyte OTU compositions between the five samples from Raunefjorden. OTUs were defined as reads with ≥98% nucleotide similarity. Sequences were normalized to 100 in each sample and log-transformed prior to similarity calculations. Samples connected by red lines were not significantly differentiated (SIMPER permutation test). Black lines indicate significant differentiation ($p < 0.05$, SIMPER permutation test).

2.3. Megaviridae and Phycodnaviridae

All the quality-trimmed viral reads showed similarity to algal viruses in the Megaviridae and *Phycodnaviridae* families, with BLAST scores between 50 and 90% amino acid sequence identity. OTU clustering based on 95% amino acid identity gave a total of 161 OTUs containing 10 or more sequences (Table S4), with 61 being unique (Table S5). Forty-one and 20 of these OTUs showed highest similarity to the Megaviridae and *Phycodnaviridae* families, respectively (Table S5). Half of the OTUs (53%) were rare, each comprising less than 1% of the total reads (Table S5). The diversity was highest in May 2010 and lowest in February 2011 (Shannon diversity values of 2.66 and 1.45, respectively) (Table S4). Based on Bray-Curtis dissimilarity, the May 2010 sample differed significantly from the other 4 samples (Figure 5).

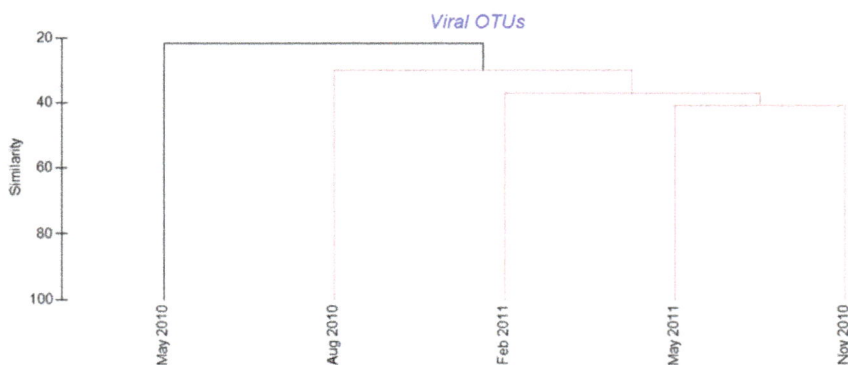

Figure 5. Cluster dendrogram illustrating Bray-Curtis dissimilarity between the virus OTU compositions in five samples from Raunefjorden. OTUs were defined as sequences with >95% amino acid similarity. The sequence data were normalized to 100 in each sample and log-transformed prior to similarity calculations. Samples connected by red lines could not be significantly differentiated (SIMPER permutation test). Black lines indicate significant differentiation ($p < 0.05$, SIMPER permutation test).

Fifty-eight percent of the viral OTUs were found in more than two of the samples (Table S5). Five viral OTUs (OTU009, OTU002, OTU001, OTU003, OTU008) were present in all the samples and dominated the samples from November, February2010 and May 2011. Four of these clustered within the *Phycodnaviridae* family. Others, such as OTU006, OTU373 and OTU010 that dominated the samples taken in May and August 2010, respectively, were almost absent or undetectable the rest of the year (Table S5, Figure 6).

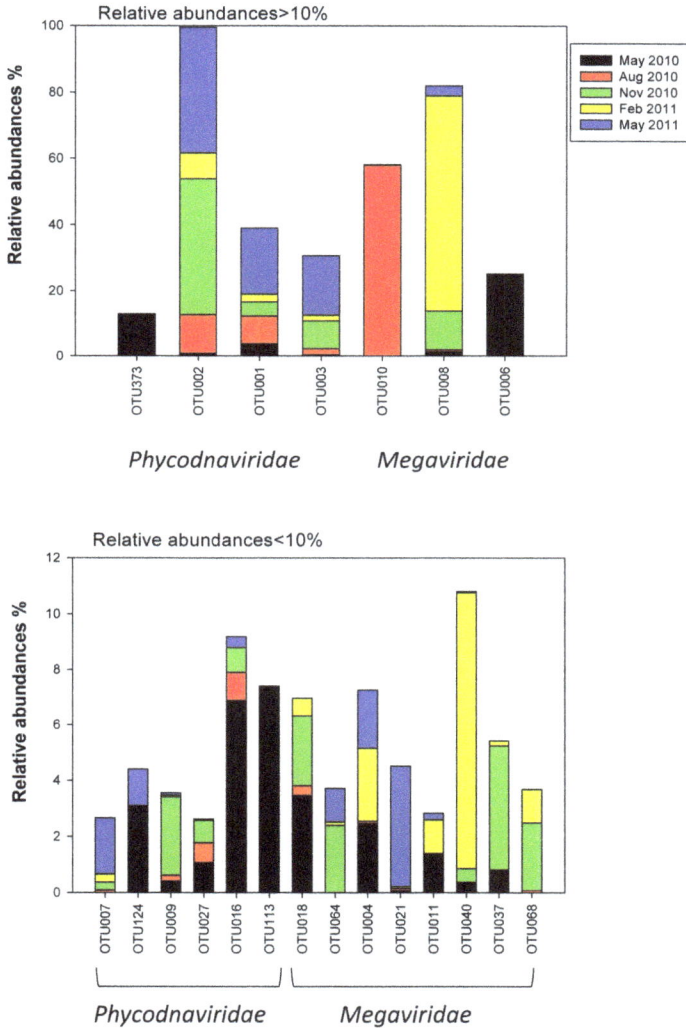

Figure 6. Relative abundances of 21 different OTUs containing more than 50 reads in the different samples.

The Megaviridae-like OTUs dominated the samples from August and February (Figure 6, Table S5). Three OTUs occurred in relative abundances over 10% (OTU010, OTU008 and OTU006). Two of them dominated the samples in August and February, the third dominated in May 2010. Further, eight OTUs occurred in low relative abundances (<10%) and were present in three or more samples. The most abundant algal Megaviridae-like OTU, OTU008, was distantly related to a virus infecting the haptophyte *Prymnesium kappa* (RF01) [25], with low bootstrap support (41%) (Figure 7). OTU008 dominated the sample from February (65%) and was present in all samples (relative abundances between 0.32–65%) (Figure 6, Table S5). The single most abundant OTU in May samples (OTU006) was also grouped within Megaviridae and clustered together with viruses infecting different haptophytes such as *Chrysochromulina*, *Phaeocystis* and *Prymnesium* (Figures 6 and 7). Three other OTUs (OTU037, OTU068 and OTU040) clustered within this clade as well. OTU037 and OTU040 were present at relatively low abundances (0.04–9.9%) in May and November 2010, and February 2011, while OTU068 were present at relative low abundances (0.09–2.4%) in samples from August and November 2010, and February 2011 (Figures 6 and 7, Table S5). Three new branches were made next to the Megaviridae family consisting of 4 OTUs from this study (Figure 7) where OTU010 and OTU18 clustered together with an environmental sequence from an earlier study at the same site [32].

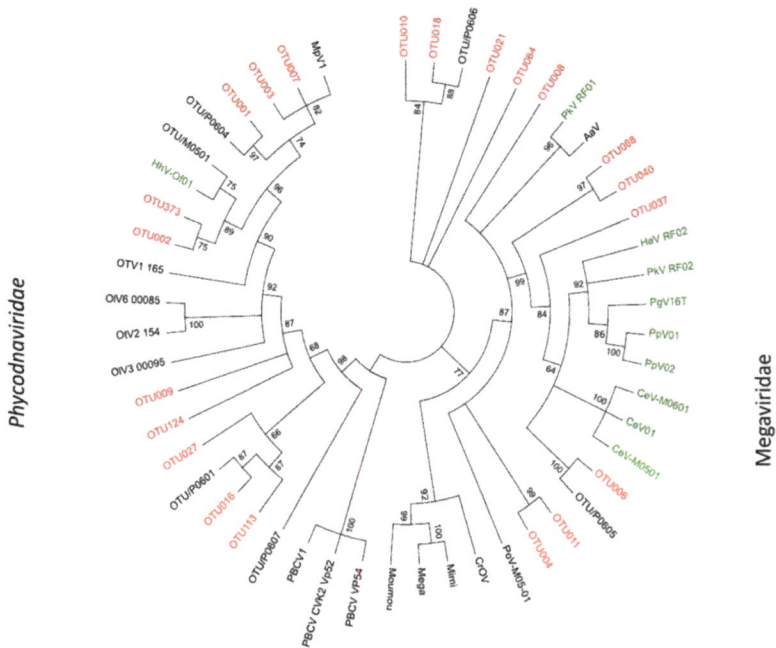

Figure 7. Midpoint rooted phylogenetic tree of the most abundant OTUs (>50 reads, marked in red) with similarities to the Megaviridae and *Phycodnaviridae* families, respectively. The tree was calculated based on the DNA-sequences encoding partial MCP-genes (FastTree v2.1.8 with default parameters). Bootstrap values form 100 replicates and aLRT- likelihood-values >0.5 are shown on nodes. Abbreviations: CroV; *Cafeteria roenbergensis virus*, Moumou; *Moumouvirus goulette*, Mimi; *Mimivirus*, Mega; *Megavirus chiliensis*, AaV; *Aureococcus anophagefferens virus*, PoV; *Pyramimonas orientalis virus*, PkV; *Prymnesium kappa virus*, HeV; *Haptolina ericina virus*, HhV; *Haptolina hirta virus*, CeV; *Chrysochromulina ericina virus*, PgV; *Phaeocystis globosa virus*, PpV; *Phaeocystis pouchetii virus*, PBCV; *Paramecium bursaria Chlorella virus*, MpV; *Micromonas pusilla virus*, OsV; *Ostreococcus* sp. *virus*, OlV; *Ostreococcus lucimarinus virus*. OTU/P0605, OTU/P0606, OTU/P0604, OTU/P0607, OTU/P0601, OTU/M0501 are all sequences from an earlier study in Raunefjorden [32]. Reference strains marked in green are haptophyte-infecting viruses maintained in culture.

OTU010 dominated in the sample from August (58%) and occurred at low relative abundances (0.04–0.05%) in the samples from November 2010 and May 2011 (Table S5). OTU18 occurred at low abundances in samples from May 2010, August, November and February. The two other collapsed branches comprised OTU064 and OTU021, both occurring at low abundances at three and four sample periods, respectively (Figure 6, Table S5). The two last OTUs in the Megaviridae family (OTU011, OTU004) showed highest similarity to a virus infecting the chlorophyte *Pyromimonas orientalis*. Both occurred at less than 10% in samples from May 2010, February 2011 and May 2011.

The *Phycodnaviridae* OTUs (Figure 6, Table S5) consisted of four OTUs (OTU373, OTU002, OTU001, OTU003) with relative abundances over 10% and six OTUs (OTU007, OTU124, OTU009, OTU027, OTU016, OTU113) with relative abundances below 10%. Three OTUs (OTU001, OTU002 and OTU003) dominated the samples from November and May 2011. They cluster within two subclades together with two cultured viruses, the prasinovirus *Micromonas pusilla virus*, and an unclassified virus shown to infect the haptophyte *Haptolina hirta* (HhV-Of01) [43] (Figure 7). Both clades also included environmental clones previously obtained at the same location [32]. The OTU016 and OTU113 also grouped together with an environmental clone from Raunefjorden [32]. OTU113 occurred only in May 2010 while OTU016 in addition were present on three other occasions (Figure 6, Table S5). OTU009, OTU124 and OTU027 did not match any viral sequences in GenBank. They were never abundant (relative abundances between 0.04–3.1%), but some were frequently observed (e.g., OTU009 and OTU027) (Figure 6, Table S5).

3. Discussion

3.1. Seasonal Patterns in Microbial Dynamics and Environmental Factors

The biological variables (Chl *a* concentration, phytoplankton, bacterial, and viral abundances) followed a seasonal and recurrent pattern that corroborated earlier descriptions of the microbial community in Raunefjorden [40,41,44–46]. The conditions in late winter and spring were nevertheless quite different in 2010 compared to 2011. The Chl *a* values were much higher in 2011 and nanoplankton, bacteria and virus abundances were stable or steadily increasing, although not fluctuating as in 2010. A deeper and more pronounced pycnocline in 2011 than in 2010 resulted in a more stable water column that sustained a longer-lasting bloom in 2011 than in 2010. In summer, fall and early winter, hydrographic conditions for the two years resembled each other, as did Chl *a* level and variation. High abundances of phototropic pico- and nanoeukaryotes and *Synechococcus* (Figure 2) matched peaks in Chl *a* concentration in August 2009, in June, September and October 2010, and in May 2011 (Figure 1). The concentration of Chl *a* in Raunefjorden is, however, largely determined by the abundance of larger phytoplankton forms like diatoms that were not counted in this study [40,44]. On several occasions (e.g., June 2009, June and September 2010, and March–May 2011) the decrease in Chl *a* concentrations may be related to a concurrent deep mixing and exchange of water masses (Figure 1).

3.2. Succession of Haptophytes and Co-Occurring DNA Viruses

454 pyrosequencing revealed a high diversity of haptophyte OTUs as well as of algal viruses. The diversity of haptophyte OTUs found in this study was larger than reported in earlier studies using microscopy (summarized in [47]). This demonstrates how high throughput sequencing of amplicon libraries is a powerful tool for detecting haptophyte species not yet morphologically and genetically characterized [48,49]. The level of haptophyte richness measured in Raunefjorden (93 different haptophyte OTUs) was at the same order of magnitude as the level previously found in Oslofjorden (156 haptophyte OTUs), a study for which the same sequencing technology and primers were used [47]. The loss of reads in the filtering process (Table S3) was high but can be explained by the fact that the multiplex identification *tag* was only present on the forward primer, leading to loss of nearly half of the reads.

Both haptophyte and viral communities varied substantially throughout the seasonal cycle. However, we could not distinguish a synchronization between their respective compositions. This may be explained by a complex virus-host relationship, or that our molecular approach was not sensitive enough to capture specific haptophyte viral-host pairs. Most empirical data on phytoplankton diversity resolve the host community diversity on a relatively coarse level, namely with host subgroups or species, typically defined by small subunit ribosomal RNA (SSU rRNA) gene sequences. Viral host-range, however, commonly dwells within strain or sub-species diversity levels [50]. Hence, 18S rRNA gene marker, as used in this and most other studies, might not be sensitive enough to capture the interaction dynamics between the viral sequences here observed and the true host to which they correspond.

Some viral OTUs resembled cultured haptophyte viruses but they were, in most cases, only distantly related. Others were more similar to viruses infecting other host groups. Due to the large diversity of haptophyte viruses and the paucity of isolated viruses infecting this important host group, there is, at present, no molecular approach available that allows us to target these viruses with specificity. Moreover, viral phylogeny does not necessarily reflect host phylogeny. Several algal viruses have been shown to cross host-species borders, and some infect hosts that are only distantly related [25]. Despite these challenges, our viral data did enable detection of successional patterns providing new insight into the interaction between viruses and their hosts. Some viral OTUs were highly abundant only at specific time-points, indicating a boom-bust relationship with their host, a pattern normally described for lytic viruses [11–13,15]. Surprisingly though, most of the viral OTUs were persistent indicating coexistence with their hosts, or alternatively an ability to exploit several host species.

Bloom communities normally comprise a few, and often recurrent, species [40,44]. Therefore, we were not surprised to find low haptophyte diversity, dominated by *E. huxleyi*, the common bloom-forming coccolithophores, and Chrysochromulinaceae (OTU001 and OTU004, respectively) in May both years. More to our surprise though, the diverse community of *Phycodneaviridae* and Megaviridae co-occurred with this recurrent, low-diversity haptophyte community, and the observed viral OTUs varied substantially between the two years. One possible explanation may be that genetically different viruses are exploiting the same hosts [25,51,52]. Even more surprising was the absence of the EhV in our flow cytometry (FCM) analysis. Blooms of *E. huxleyi* are frequently succeeded by large increases in this specific virus [53–55] which have a FCM characteristics that make them easy to distinguish and recognize [11,56] even without primers that capture their presence at the molecular level.

Our observation that the abundance of pico- and nanoeukaryotes and the diversity of haptophytes peaked in August is in accordance with the general narrative that relatively small forms typically dominate the diverse phytoplankton communities thriving in the temperate, stable, and nutrient-depleted summer water masses. The abundance of larger viruses (V2 and V3), i.e., viruses having a size typical of many dsDNA phytoplankton viruses [40,57], were also at their highest in the summer period. The *Phycodnaviridae* and Megaviridae communities were, however, dominated by a single OTU (OTU010), which was only distantly related to any known viral sequence. Thus, the diverse phytoplankton community in August seems to have sustained a high virus abundance with low diversity. One interpretation of this is that OTU010 represents a generalist virus type, able to infect several different species, a feature known for many viruses infecting members of Prymnesiales [25]. Other interpretations may be that phytoplankton viruses that are not targeted by our primers prevailed [25,32] or that the large viruses are related to other hosts groups.

Several of the haptophytes that we detected in our study are known to be susceptible to characterized algal viruses [24,25,58]. Viruses infecting *P. pouchetii, P. globosa, Haptolina ericina and Prymnesium kappa* have previously been isolated from Norwegian coastal waters and/or the North Sea [24,25,58,59], but none of the viral OTUs we found was similar to any of these characterized viruses within 95% aa similarity. Despite the low similarity, five of the OTUs clustered within the Megaviridae group together with several cultured representatives of viruses infecting the orders Prymnesiales and Phaeocystales. As very few algal viruses are cultured, our results may suggest that the diversity within this viral group is large. Alternatively, cultivated viruses may not represent the most abundant viral strains present in natural systems as current procedures for virus isolation [60] entail a strong selective pressure favoring lytic viruses with short replication cycles, a strategy perhaps not very common strategy in nature.

Some haptophyte and viral OTUs were remarkably persistent, considering that the samples were collected at different seasons, interspersed by mixing of the water-masses and changing environmental conditions. Haptophyte OTUs that were present in all samples may represent species that are able to tolerate a wide range of environmental conditions, either as actively dividing cells, or surviving periods of low activity. Most of these persistent OTUs were classified to Prymnesiales (eight OTUs), an order including several mixotrophic species known to survive even when light conditions are too low for photosynthesis [3]. A high degree of preservation and recurrence of virus-genotypes through the years have previously been observed for myovirus-like viruses [46], but this is the first observation for algal viruses. These year-long observations are contrasting and complementary to previous studies demonstrating clear boom and bust patterns for abundant algae and their viruses [15,53,61]. Viral particles are estimated to quickly degrade in seawater (inactivation rates of 5%–20% per h) [62–64], and should quickly disappear without co-occurring susceptible and active hosts. Persistent viral OTUs thus indicate either that they propagate and co-exist with a persisting host, or that they are able to infect various species. Virus-host coexistence [19,21,65] is regarded to be a paradox, since most cultured viruses quickly induce resistance in their hosts [21,66] and may be attributed to partial host resistance (strain specificity), low virus infectivity [21], or to chronic infections where only few cells in the host population produce the virus, while the rest grow normally [9,67]. Another possibility is that the persistent viruses have wide host ranges, which would allow them to proliferate on different host species [25]. The ability to infect several species may be especially beneficial in times when the phytoplankton community is very diverse or at low phytoplankton abundance and activity.

This inter-annual study of microbial communities in Raunefjorden is the first to apply high throughput sequencing to study seasonal variation in marine uncultured algal and viral communities. Five "snapshots" of the haptophyte and algal virus (*Phycodnaviridae* and Megaviridae) communities covering one year revealed a large diversity with many uncultured and unknown forms although we identified a stable "core" community of haptophyte and viral OTUs as well. Some abundant viral OTUs showed high relative abundance in several samples indicating virus-host coexistence or wider host-range than what we would expect from the existing isolates. The diversity varied a lot, and low haptophyte diversity in May was accompanied by high algal virus diversity whereas high haptophyte diversity in August co-occurred with low *Phycodnaviridae* and Megaviridae diversity. We suggest that several viruses may exploit the same hosts in the low-diversity spring communities, while a few viruses may be able to exploit several of the haptophytes in the high-diversity community in late summer. Notably, measured virus and host abundance illustrates the importance of viral caused mortality in the diverse late summer community.

4. Material and Methods

4.1. Sample Collection

Seawater samples were collected from 5 m depth in Raunefjorden (60°16.2′ N, 5°12.5′ E) Western Norway, between May 2009 and May 2011. The sampling interval was, with a few exceptions, 2–4 weeks. Temperature, salinity, density and Chl a fluorescence were determined using a CTD (Conductivity-Temperature-Depth) equipped with an in situ fluorometer (SD204 SAIV, SAIV A/S Environmental Sensors & Systems, Bergen, Norway). A 20 L aliquot of sampled water was filtered by peristaltic pumping through 3.00 μm and then through 0.45 μm pore-sized low-protein-binding filters (145 mm, Durapore, Millipore Corp., Billerica, MA, USA), within 3 h of collection. The filters were cut in two and immediately frozen in liquid N2 and thereafter stored at $-80\ °C$ until DNA-extraction for later use in PCR and 454 sequencing of the haptophyte community. Viruses in the 20 L 0.45 μm filtrate were concentrated 400 times (approximately 50 mL) using a tangential flow filtration system equipped with a 100,000 pore size (NMWC) hollow-fiber cartridge (QuixStand, GE Healthcare Bio-Sciences AB, Uppsala, Sweden). Aliquots of these viral concentrates were frozen at -80 °C for later use in PCR and 454 sequencing.

4.2. Microbial Abundance Measured by Flow Cytometry (FCM)

Phototrophic pico- and nano- plankton were counted in triplicate by FCM (Becton, Dickinson and Company, BD Biosciences, San Jose, CA, USA), using fresh, unpreserved samples, with the trigger set on red fluorescence and a flow rate giving 50–800 events per sec. Five different populations of phototrophs were defined in FCM-plots based on differences in side scatter, red and orange fluorescence: Synechococcus, picoeukaryotes, nanoeukaryotes, cryptophytes and E. huxleyi ([18,40,56,68], (Table S1)). Most haptophytes fall within the size class pico-and nanoeukaryotes [47,69,70].

Viruses and bacteria were counted in samples preserved with 1% glutaraldehyde (30 min at 4 °C) and snap frozen in liquid N_2. The samples were thawed, diluted, and stained with $1\times$SYBR Green I (Invitrogen, Carlsbad, CA, USA 10,000 $\times$ conc. in dimethyl sulfoxide (DMSO)) for 10 min at 80 °C [57], immediately before counting in triplicates. Bacteria and three different virus populations were defined based on side scatter properties and green fluorescence: low-, medium- and high-fluorescence viruses (V1, V2 and V3 respectively, [40]). Spearman rank order correlation analyses were calculated in Statistica 12 (StatSoft, Tulsa, OK, USA), to assess the relationship between abundance of different virus and algal groups, as well as their relationship with the measured abiotic factors. Missing values were pairwise deleted.

4.3. DNA Extraction, PCR and 454-Pyrosequencing

Five samples, collected on May 25, August 31, and November 30 (2010) and on February 22 and May 31 (2011), were used for targeted 454 pyrosequencing of haptophyte and *Phycodnaviridae*/Megaviridae communities. The samples were chosen based on pulsed field gel electrophoresis (PFGE) analysis to represent different seasonal community stages (Figure S1).

For haptophytes, DNA was extracted from $\frac{1}{2}$ of each 3 μm and 0.45 μm pore-sized filters (representing approximately 10 L of sea water) using DNeasy®Plant Mini kit (Qiagen, Hilden, Germany) according to the manufacturer's instructions. Initial re-suspension of cells was done by transferring the frozen filters into falcon tubes with 1 ml AP1 buffer (from the DNeasy®kit) and vortexing for 60 s. Extracted DNA from the two different size fractions was subjected to separate PCRs with tagged primers. The V4 region of the 18S rRNA gene (position 640–1060) was amplified using haptophyte-specific primers: 528Flong (5'GCGGTAATTCCAGCTCCAA3') and PRYM01+7 (5'-GATCAGTGAAAACATCCCTGG-3') [71]. Each PCR mixture contained 1 μL DNA template, 5 μL Phusion GC buffer (NEB Inc., Ipswich, MA, USA), 0.2 mM each deoxynucleoside triphosphate (dNTP), 400 nM each primer, 0.75 μL DMSO, 0.5 units Phusion®High-Fidelity DNA Polymerase (NEB Inc.), adjusted for a final volume of 25 μL. The cycling parameters were 98 °C for 30 s, and 30 cycles of 98 °C for 10 s, 55 °C for 30 s, and 72 °C for 30 s, with a final extension at 72 °C for 10 min [71].

For viruses, DNA was extracted from 0.5–1 mL of frozen viral concentrate (representing 200–400 mL of sea water). The concentrates were alternately heated to 90 °C and cooled on ice twice for 2 min. Ethylenediaminetetraacetic acid (EDTA, 20 mM) and proteinase K (100 μg/mL) were subsequently added before the samples were incubated for 10 min at 55 °C. Sodium dodecyl sulfate (SDS, final concentration 0.5%) was added and the samples incubated for 1 h at 55 °C. The extracted DNA was then purified with Zymo DNA Clean and Concentrator^{TM} kit (Zymo Research, Irvine, CA, USA) following the manufacturer's protocols.

A segment of the major capsid protein (MCP) gene was amplified by the primers: MCPforwd (5′-GGY GGY CAR CGY ATT GA-3′) and MCPrev (5′-TGI ARY TGY TCR AYI AGG TA-3′) developed by Larsen et al. [32]. The PCR reactions (25 μL) contained: 0.625 U of HotStarTaq DNA polymerase (Qiagen), 1×PCR buffer, 0.2 mM of each dNTP, 0.5 μM of each of the primers and 1 μL template. The following PCR-program was used: Initial activation at 95 °C (15 min), a touchdown PCR of 20 cycles of denaturation at 94 °C (30 s), annealing at 60 °C and decreasing 0.5 °C per round (30 s) and extension at 72 °C (30 s), followed by 35 cycles with fixed annealing temperature at 45 °C, and a final elongation of 72 °C for 7 min. The PCR products were cleaned (Zymo DNA Clean and Concentrator™ kit) and amplified in new PCR reactions with tagged primers specific to each sample, 25 cycles with annealing at 45 °C, and otherwise as above.

For each haptophyte or viral sample, the products from eight replicate PCR reactions were cleaned, quantified and pooled before sequencing. The DNA amplicons were sent for Roche 454 (GS-FLX Titanium) library sequencing.

Two-directional amplicon sequencing using L chemistry was performed by LGC Genomics GmbH, Berlin, Germany. The following numbers of reads were obtained: haptophytes, 22588 (25 May 2010), 23261 (31 August 2010), 31959 (30 November 2010), 32540 (22 February 2011), 30380 (31 May 2011) (Table S2), viruses, 7909 (25 May 2010), 8195 (31 August 2010), 8558 (30 November 2010), 6791 (22 February 2011), 10353 (31 May 2011) (Table S4).

4.4. Sequence Analysis

All reads were filtered using AmpliconNoise [72], with default settings, and further analyzed using Mothur (www.mothur.org) [73] with the commands here provided in italics within brackets. Reads were trimmed (*trim.seqs*) and checked for chimeras by uchime (*chimera.uchime*) with Silva reference sequences for the haptophytes [74].

Prior to clustering, haptophyte reads were aligned to a reference alignment [47] (*align.seqs*) to ensure that the reads aligned in the targeted region, and to enable distance calculation by *dist.seqs*. Based on these distances, reads were clustered de novo into OTUs with 98% similarity (*cluster*), with the average neighbor-algorithm. An OTU definition of 98% nucleotide similarity was applied here in accordance with studies showing this to be a good threshold for delineating different species of most protists, while at the same time accounting for intra-species variation [75]. OTUs of different lengths, but which were otherwise identical, were clustered at 100% similarity by Uclust [76] and the longest sequence of each OTU was picked as representative. OTUs shorter than 250 bp were removed. OTUs that were represented by only a single read (singletons) were excluded from the analysis. Taxonomic classification was performed by MegaBLASTin Geneiuos v. 8.1.6 against the PIP Haptophyte 18S rDNA reference sequence database as described in [42] and available from figshare [77]. Diversity analyses were performed in Mothur (*collect.shared, summary.single*). To compare OTU richness between samples, all samples were subsampled to the number of reads in the smallest sample (1320), by the function *rrarefy* in the *vegan* package v. 2.4-1 [78] in R v. 3.3.2 (R Core Team 2016).

As many of the OTUs were found in both size fractions (>3 and 3–0.45 μm), the number of reads in the two size fractions were pooled, and the relative abundance of each OTU in the five samples was determined.

The filtered viral MCP reads were translated into the corresponding amino acid sequence in BioEdit [79]. Alignment of the amino acid reads was done with MAFFT v7) [80], with a gap opening penalty of 2.5, offset value of 0.1, and BLOSUM62 as the amino acid scoring matrix. Insertion/deletion errors were manually corrected. Reads that then did not align in the mid-conserved region (approx. position 100 in the alignment) or contained stop-codons, indicative of sequencing errors, were removed. The remaining reads were trimmed to equal length, i.e., position 117 in the alignment. A protein distance matrix was calculated by PROTDIST v3.5c (©1993 Joseph Felsenstein), and used to cluster the sequences in Mothur [73]. As the large number of PCR-cycles is prone to create artifacts, a 95% amino acid sequence identity threshold was applied [35]. To further decrease the number of spurious reads, only OTUs containing ten or more reads were used in the analysis. Mothur was used for downstream calculations of diversity indices. A representative sequence for each OTU containing more than 50 reads was used for phylogenetic analysis (together representing 84% of the reads). These OTUs were tentatively assigned a phylogenetic affiliation (BLAST-search closest hits) to the Megaviridae or *Phycodnaviridae* family. The tree was constructed, comprising the representative sequences together with reference sequences. Alignment and phylogenetic reconstructions were performed using the function "build" of ETE3 v3.0.0b32 [81] implemented on the GenomeNet, Tree [82] . The tree was constructed using FastTree v2.1.8 with default parameters [83]. Statistical support for the internal branches was calculated by an aLRTtest (SH-like), and through 100 bootstraps. Cluster diagrams were drawn for the haptophyte and virus samples separately. The cluster diagrams were based on Bray Curtis similarities of relative abundance of each virus or haptophyte OTU in the samples. A SIMPROF permutation test was applied to test if the samples could be differentiated at $p < 0.05$ (Primer 6, Primer-E Ltd., Ivybridge, UK).

Supplementary Materials: The follow supplementary materials can be found online at www.mdpi.com/1999-4915/9/4/84/s1, Table S1: Spearman Rank Order Correlations of the quantitative biological data, including the chl a measurements and the population abundances obtained by flow cytometry. Table S2: Heatmap and OTU table showing the relative abundance and percentage identity to nucleotide blast hits of the haptophyte OTUs in samples from Raunefjorden. Table S3: Results of the 454 sequencing and analysis of the V4-region of 18S rDNA 18S rDNA in haptophytes from Raunefjorden. Figure S1: Schematic representation of the relative abundance of distinct viral populations. Table S4: Result of the 454 sequencing of the viral MCP gene in samples from Raunefjorden. Table S5: Heatmap showing relative abundance of the different viral MCP OTUs in samples from Raunefjorden. Suplementary method and material describing viral diversity explored by pulsed-field gel electrophoresis (PFGE) and method precautions.

Data Accessibility: The MCP and haptophyte sequences have been submitted to NCBI SRA-database under the bioproject id PRJNA262844.

Acknowledgments: This work was funded by the Research Council of Norway through the project 190307/S54 "HAPTODIV" and project number 225956/E10 "MicroPolar", and by the European Research Council through the Advanced Grant project No. 250254 "MINOS". We are grateful to Knut Tomas Holden Sørlie for help with sampling, and to Hilde Marie Stabell, Jessica Ray and Jorunn Egge for help with seawater filtering

Author Contributions: Main contributor to analysis and writing has been Torill Vik Johannesen. Aud Larsen, Gunnar Bratbak, Bente Edvardsen and Ruth-Anne Sandaa have all contributed to scientific discussion of results, analysis, and writing. Elianne D. Egge has contributed to scientific discussion and analyses, while António Pagarete has performed some of the analysis included in the paper.

Conflicts of Interest: The authors declare no conflict of interest.

References

1. Field, C.B.; Behrenfeld, M.J.; Randerson, J.T.; Falkowski, P. Primary production of the biosphere: Integrating terrestrial and oceanic components. *Science* **1998**, *281*, 237–240. [CrossRef] [PubMed]

2. Eikrem, E.; Medlin, L.K.; Henderiks, J.; Rokitta, S.; Rost, B.; Probert, I.; Throndsen, J.; Edvardsen, B. Haptophyta. In *Handbook of the Protists*; Archibald, J.M., Simpson, A.G.B., Slamovits, C.H., Margulis, L., Melkonian, M., Chapman, D.J., Corliss, J.O., Eds.; Springer International Publishing: Cham, Switzerland, 2016; pp. 1–61.

3. Unrein, F.; Gasol, J.M.; Not, F.; Forn, I.; Massana, R. Mixotrophic haptophytes are key bacterial grazers in oligotrophic coastal waters. *ISME J.* **2014**, *8*, 164–176. [CrossRef] [PubMed]

4. Hallegraeff, G.M. Ocean climate change, phytoplankton community responses, and harmful algal blooms: A formidable predictive challenge. *J. Phycol.* **2010**, *46*, 220–235. [CrossRef]

5. Leadbeater, B.S.C. Identification, by means of electron microscopy, of flagellate nanoplankton from the coast of Norway. *Sarsia* **1972**, *49*, 107–124. [CrossRef]

6. Thomsen, H.A.; Buck, K.R.; Chavez, F.P. Haptophytes as components of marine phytoplankton. *Syst. Assoc. Spec. Vol. Ser.* **1994**, *51*, 187–208.

7. Egge, E.S.; Johannessen, T.V.; Andersen, T.; Eikrem, W.; Bittner, L.; Larsen, A.; Sandaa, R.-A.; Edvardsen, B. Seasonal diversity and dynamics of haptophytes in the Skagerrak, Norway, explored by high-throughput sequencing. *Mol. Ecol.* **2015**, *24*, 3026–3042. [CrossRef] [PubMed]

8. Fuhrman, J.; Noble, R. Viruses and protists cause similar bacterial mortality in coastal seawater. *Limnol. Oceanogr.* **1995**, *40*, 1236–1242. [CrossRef]

9. Short, S.M. The ecology of viruses that infect eukaryotic algae. *Environ. Microbiol.* **2012**, *14*, 2253–2271. [CrossRef] [PubMed]

10. Bouvier, T.; Del Giorgio, P.A. Key role of selective viral-induced mortality in determining marine bacterial community composition. *Envir. Microbiol.* **2007**, *9*, 287–297. [CrossRef] [PubMed]

11. Castberg, T.; Larsen, A.; Sandaa, R.A.; Brussaard, C.P.D.; Egge, J.K.; Heldal, M.; Thyrhaug, R.; van Hannen, E.J.; Bratbak, G. Microbial population dynamics and diversity during a bloom of the marine coccolithophorid *Emiliania huxleyi* (Haptophyta). *Mar. Ecol. Prog. Ser.* **2001**, *221*, 39–46. [CrossRef]

12. Brussaard, C.P.D.; Bratbak, G.; Baudoux, A.C.; Ruardij, P. *Phaeocystis* and its interaction with viruses. *Biogeochemistry* **2007**, *83*, 201–215. [CrossRef]

13. Martinez, J.M.; Schroeder, D.C.; Larsen, A.; Bratbak, G.; Wilson, W.H. Molecular Dynamics of *Emiliania huxleyi* and Cooccurring Viruses during Two Separate Mesocosm Studies. *Appl. Environ. Microbiol.* **2007**, *73*, 554–562. [CrossRef] [PubMed]

14. Santini, S.; Jeudy, S.; Bartoli, J.; Poirot, O.; Lescot, M.; Abergel, C.; Barbe, V.; Wommack, K.E.; Noordeloos, A.A.M.; Brussaard, C.P.D.; et al. Genome of *Phaeocystis globosa virus* PgV-16T highlights the common ancestry of the largest known DNA viruses infecting eukaryotes. *Proc. Natl. Acad. Sci. USA* **2103**, *110*, 10800–10805. [CrossRef] [PubMed]

15. Brussaard, C.P.D.; Kuipers, B.; Veldhuis, M.J.W. A mesocosm study of *Phaeocystis globosa* population dynamics: I. Regulatory role of viruses in bloom control. *Harmful Algae* **2005**, *4*, 859–874. [CrossRef]

16. Tomaru, Y.; Hata, N.; Masuda, T.; Tsuji, M.; Igata, K.; Masuda, Y.; Yamatogi, T.; Sakaguchi, M.; Nagasaki, K. Ecological dynamics of the bivalve-killing dinoflagellate *Heterocapsa circularisquama* and its infectious viruses in different locations of western Japan. *Environ. Microbiol.* **2007**, *9*, 1376–1383. [CrossRef] [PubMed]

17. Suttle, C.A.; Chan, A.M. Dynamics and distribution of cyanophages and their effect on marine *Synechococcus* spp. *Appl. Environ. Microbiol.* **1994**, *60*, 3167–3174. [PubMed]

18. Larsen, A.; Castberg, T.; Sandaa, R.A.; Brussaard, C.P.D.; Egge, J.; Heldal, M.; Paulino, A.; Thyrhaug, R.; van Hannen, E.J.; Bratbak, G. Population dynamics and diversity of phytoplankton, bacteria and viruses in a seawater enclosure. *Mar. Ecol. Prog. Ser.* **2001**, *221*, 47–57. [CrossRef]

19. Cottrell, M.T.; Suttle, C.A. Dynamics of a lytic virus infecting the photosynthetic marine picoflagellate *Micromonas pusilla*. *Limnol. Oceanogr.* **1995**, *40*, 730–739. [CrossRef]

20. Tarutani, K.; Nagasaki, K.; Itakura, S.; Yamaguchi, M. Isolation of a virus infecting the novel shellfish-killing dinoflagellate *Heterocapsa circularisquama*. *Aquat. Microb. Ecol.* **2001**, *23*, 103–111. [CrossRef]

21. Thyrhaug, R.; Larsen, A.; Thingstad, T.F.; Bratbak, G. Stable coexistence in marine algal host-virus systems. *Mar. Ecol. Prog. Ser.* **2003**, *254*, 27–35. [CrossRef]

22. Brussaard, C.P.D. Viral control of phytoplankton populations—A review. *J. Eukaryot. Microbiol.* **2004**, *51*, 125–138. [CrossRef] [PubMed]

23. Demory, D.; Arsenieff, L.; Simon, N.; Six, C.; Rigaut-Jalabert, F.; Marie, D.; Ge, P.; Bigeard, E.; Jacquet, S.; Sciandra, A.; et al. Temperature is a key factor in Micromonas-virus interactions. *ISME J.* **2017**. [CrossRef] [PubMed]

24. Sandaa, R.A.; Heldal, M.; Castberg, T.; Thyrhaug, R.; Bratbak, G. Isolation and characterization of two viruses with large genome size infecting *Chrysochromulina ericina* (Prymnesiophyceae) and *Pyramimonas orientalis* (Prasinophyceae). *Virology* **2001**, *290*, 272–280. [CrossRef] [PubMed]

25. Johannessen, T.V.; Bratbak, G.; Larsen, A.; Ogata, H.; Egge, E.S.; Edvardsen, B.; Eikrem, W.; Sandaa, R.-A. Characterisation of three novel giant viruses reveals huge diversity among viruses infecting Prymnesiales (Haptophyta). *Virology* **2015**, *476*, 180–188. [CrossRef] [PubMed]

26. Iyer, L.; Balaji, S.; Koonin, E.; Aravind, L. Evolutionary genomics of nucleo-cytoplasmic large DNA viruses. *Virus Res.* **2006**, *117*, 156–184. [CrossRef] [PubMed]

27. Wilson, W.H.; Etten, J.L.; Allen, M.J. The *Phycodnaviridae*: The story of how tiny giants rule the world. In *Lesser Known Large dsDNA Viruses*; Papers in Plant Pathology; Springer: Heidelberg, Germany, 2009; pp. 1–42.

28. La Scola, B.; Audic, S.; Robert, C.; Jungang, L.; de Lamballerie, X.; Drancourt, M.; Birtles, R.; Claverie, J.-M.; Raoult, D. A giant virus in amoebae. *Science* **2003**, *299*, 2033. [CrossRef] [PubMed]

29. Fischer, M.G.; Allen, M.J.; Wilson, W.H.; Suttle, C.A. Giant virus with a remarkable complement of genes infects marine zooplankton. *Proc. Natl. Acad. Sci. USA* **2010**, *107*, 19508–19513. [CrossRef] [PubMed]

30. Moniruzzaman, M.; LeCleir, G.R.; Brown, C.M.; Gobler, C.J.; Bidle, K.D.; Wilson, W.H.; Wilhelm, S.W. Genome of brown tide virus (AaV), the little giant of the Megaviridae, elucidates NCLDV genome expansion and host-virus coevolution. *Virology* **2014**, *466–467*, 59–69. [CrossRef] [PubMed]

31. Moniruzzaman, M.; Gan, E.R.; LeCleir, G.R.; Kang, Y.; Gobler, C.J.; Wilhelm, S.W. Diversity and dynamics of algal Megaviridae members during a harmful brown tide caused by the pelagophyte, *Aureococcus anophagefferens*. *FEMS Microbiol. Ecol.* **2016**. [CrossRef] [PubMed]

32. Larsen, J.B.; Larsen, A.; Bratbak, G.; Sandaa, R.A. Phylogenetic analysis of members of the *Phycodnaviridae* virus family, using amplified fragments of the major capsid protein gene. *Appl. Environ. Microbiol.* **2008**, *74*, 3048–3057. [CrossRef] [PubMed]

33. Monier, A.; Larsen, J.B.; Sandaa, R.-A.; Bratbak, G.; Claverie, J.M.; Ogata, H. Marine mimivirus relatives are probably large algal viruses. *Virol. J.* **2008**, *5*, 12. [CrossRef] [PubMed]

34. Kristensen, D.M.; Mushegian, A.R.; Dolja, V.V.; Koonin, E.V. New dimensions of the virus world discovered through metagenomics. *Trends Microbiol.* **2010**, *18*, 11–19. [CrossRef] [PubMed]

35. Park, Y.; Lee, K.; Lee, Y.S.; Kim, S.W.; Choi, T.J. Detection of diverse marine algal viruses in the South Sea regions of Korea by PCR amplification of the DNA polymerase and major capsid protein genes. *Virus Res.* **2011**, *159*, 43–50. [CrossRef] [PubMed]

36. Short, S.M.; Rusanova, O.; Staniewski, M.A. Novel phycodnavirus genes amplified from Canadian freshwater environments. *Aquat. Microb. Ecol.* **2011**, *63*, 61–67. [CrossRef]

37. Rozon, R.M.; Short, S.M. Complex seasonality observed amongst diverse phytoplankton viruses in the Bay of Quinte, an embayment of Lake Ontario. *Freshw. Biol.* **2013**, *58*, 2648–2663. [CrossRef]

38. Short, S.M.; Suttle, C.A. Sequence analysis of marine virus communities reveals that groups of related algal viruses are widely distributed in nature. *Appl. Environ. Microbiol.* **2002**, *68*, 1290–1296. [CrossRef] [PubMed]

39. Wang, M.-N.; Ge, X.-Y.; Wu, Y.-Q.; Yang, X.-L.; Tan, B.; Zhang, Y.-J.; Shi, Z.-L. Genetic diversity and temporal dynamics of phytoplankton viruses in East Lake, China. *Virol. Sin.* **2015**, *30*, 290–300. [CrossRef] [PubMed]

40. Larsen, A.; Flaten, G.A.F.; Sandaa, R.A.; Castberg, T.; Thyrhaug, R.; Erga, S.R.; Jacquet, S.; Bratbak, G. Spring phytoplankton bloom dynamics in Norwegian coastal waters: Microbial community succession and diversity. *Limnol. Oceanogr.* **2004**, *49*, 180–190. [CrossRef]

41. Sandaa, R.A.; Larsen, A. Seasonal variations in viral-host populations in Norwegian coastal waters: Focusing on the cyanophage community infecting marine Synechococcus species. *Appl. Environ. Microbiol.* **2006**, *72*, 4610–4618. [CrossRef] [PubMed]

42. Edvardsen, B.; Egge, E.S.; Vaulot, D. Diversity and distribution of haptophytes revealed by environmental sequencing and metabarcoding—A review. *Perspect. Phycol.* **2016**, *3*, 77–91. [CrossRef]

43. Johannesen, T.V. Marine Virus-phytoplankton Interactions. Ph.D. Thesis, University of Bergen, Bergen, Norway, 2015.

44. Erga, S.R.; Heimdal, B.R. Ecological studies on the phytoplankton of Korsfjorden, western Norway. The dynamics of a spring bloom seen in relation to hydrographical conditions and light regime. *J. Plankton. Res.* **1984**, *6*, 67–90. [CrossRef]

45. Bratbak, G.; Heldal, M.; Norland, S.; Thingstad, T.F. Viruses as partners in spring bloom microbial trophodynamics. *Appl. Environ. Microbiol.* **1990**, *56*, 1400–1405. [PubMed]

46. Pagarete, A.; Chow, C.-E.T.; Johannessen, T.; Fuhrman, J.A.; Thingstad, T.F.; Sandaa, R.A. Strong Seasonality and Interannual Recurrence in Marine Myovirus Communities. *Appl. Environ. Microbiol.* **2013**, *79*, 6253–6259. [CrossRef] [PubMed]

47. Egge, J.S.; Eikrem, W.; Edvardsen, B. Deep branching novel lineages and high diversity of haptophytes in Skagerak (Norway) uncovered by 454-pyrosequencing. *J. Eukaryot. Microbiol.* **2015**, *62*, 121–140. [CrossRef] [PubMed]

48. Moon-van der Staay, S.Y.; van der Staay, G.W.M.; Guillou, L.; Vaulot, D.; Claustre, H.; Medlin, L.K. Abundance and diversity of prymnesiophytes in the picoplankton community from the equatorial Pacific Ocean inferred from 18S rDNA sequences. *Limnol. Oceanogr.* **2000**, *45*, 98–109. [CrossRef]

49. Liu, H.; Probert, I.; Uitz, J.; Claustre, H.; Aris-Brosou, S.; Frada, M.; Not, F.; de Vargas, C. Extreme diversity in noncalcifying haptophytes explains a major pigment paradox in open oceans. *Proc. Natl. Acad. Sci. USA* **2009**, *106*, 12803–12808. [CrossRef] [PubMed]

50. Thingstad, T.F.; Våge, S.; Storesund, J.E.; Sandaa, R.-A.; Giske, J. A theoretical analysis of how strain-specific viruses can control microbial species diversity. *Proc. Natl. Acad. Sci. USA* **2014**, *111*, 7813–7818. [CrossRef] [PubMed]

51. Baudoux, A.; Brussaard, C. Characterization of different viruses infecting the marine harmful algal bloom species *Phaeocystis globosa*. *Virology* **2005**, *341*, 80–90. [CrossRef] [PubMed]

52. Nagasaki, K. Dinoflagellates, diatoms, and their viruses. *J. Microbiol.* **2008**, *46*, 235–243. [CrossRef] [PubMed]

53. Bratbak, G.; Egge, J.K.; Heldal, M. Viral mortality of the marine alga *Emiliania-huxleyi* (Haptophyceae) and termination of algal blooms. *Mar. Ecol. Prog. Ser.* **1993**, *93*, 39–48. [CrossRef]

54. Bratbak, G.; Levasseur, M.; Michaud, S.; Cantin, G.; Fernandez, E.; Heimdal, B.R.; Heldal, M. Viral activity in relation to *Emiliana huxleyi* blooms: A mechanism of DSMP release? *Mar. Ecol. Prog. Ser.* **1995**, *128*, 133–142. [CrossRef]

55. Wilson, W.H.; Tarran, G.A.; Schroeder, D.; Cox, M.; Oke, J.; Malin, G. Isolation of viruses responsible for the demise of an *Emiliania huxleyi* bloom in the English Channel. *J. Mar. Biol. Assoc. UK* **2002**, *82*, 369–377. [CrossRef]

56. Jacquet, S.; Heldal, M.; Iglesias-Rodriguez, D.; Larsen, A.; Wilson, W.; Bratbak, G. Flow cytometric analysis of an *Emiliana huxleyi* bloom terminated by viral infection. *Aquat. Microb. Ecol.* **2002**, *27*, 111–124. [CrossRef]

57. Brussaard, C.P.D. Optimization of procedures for counting viruses by flow cytometry. *Appl. Environ. Microbiol.* **2004**, *70*, 1506–1513. [CrossRef] [PubMed]

58. Jacobsen, A.; Bratbak, G.; Heldal, M. Isolation and characterization of a virus infecting *Phaeocystis pouchetii* (Prymnesiophyceae). *J. Phycol.* **1996**, *32*, 923–927. [CrossRef]

59. Brussaard, C.P.D.; Short, S.M.; Frederickson, C.M.; Suttle, C.A. Isolation and phylogenetic analysis of novel viruses infecting the phytoplankton *Phaeocystis globosa* (Prymnesiophyceae). *Appl. Environ. Microbiol.* **2004**, *70*, 3700–3705. [CrossRef] [PubMed]

60. Nagasaki, K.; Bratbak, G. Isolation of viruses infecting photosynthetic and nonphotosynthetic protists. In *Manual of Aquatic Viral Ecology*; Wilhelm, S.W., Weinbauer, M.G., Suttle, C.A., Eds.; ASLO: Waco, TX, USA, 2010; pp. 92–101.

61. Sandaa, R.-A. Burden or benefit? Virus-host interactions in the marine environment. *Res. Microbiol.* **2008**, *159*, 374–381. [CrossRef] [PubMed]

62. Suttle, C.A.; Chen, F. Mechanisms and Rates of Decay of Marine Viruses in Seawater. *Appl. Environ. Microbiol.* **1992**, *58*, 3721–3729. [PubMed]

63. Noble, R.T.; Fuhrman, J.A. Rapid virus production and removal as measured with fluorescently labeled viruses as tracers. *Appl. Environ. Microbiol.* **2000**, *66*, 3790–3797. [CrossRef] [PubMed]

64. Mojica, K.D.A.; Brussaard, C.P.D. Factors affecting virus dynamics and microbial host-virus interactions in marine environments. *FEMS Microbiol. Ecol.* **2014**, *89*, 495–515. [CrossRef] [PubMed]

65. Needham, D.M.; Chow, C.E.T.; Cram, J.A.; Sachdeva, R.; Parada, A.; Fuhrman, J.A. Short-term observations of marine bacterial and viral communities: Patterns, connections and resilience. *ISME J.* **2013**, *7*, 1274–1285. [CrossRef] [PubMed]

66. Ray, J.L.; Haramaty, L.; Thyrhaug, R.; Fredricks, H.F.; Van Mooy, B.A.S.; Larsen, A.; Bidle, K.D.; Sandaa, R.-A. Virus infection of *Haptolina ericina* and *Phaeocystis pouchetii* implicates evolutionary conservation of programmed cell death induction in marine haptophyte–virus interactions. *J. Plankton Res.* **2014**, *36*, 943–955. [CrossRef] [PubMed]

67. Rozenn, T.; Grimsley, N.; Escande, M.L.; Subirana, L.; Derelle, E.; Moreau, H. Acquisition and maintenance of resistance to viruses in eukaryotic phytoplankton populations. *Environ. Microbiol.* **2011**, *13*, 1412–1420.

68. Marie, D.; Brussaard, C.P.D.; Thyrhaug, R.; Bratbak, G.; Vaulot, D. Enumeration of marine viruses in culture and natural samples by flow cytometry. *Appl. Environ. Microbiol.* **1999**, *65*, 45–52. [PubMed]

69. Edvardsen, B.; Eikrem, W.; Throndsen, J.; Sáez, A.G.; Probert, I.; Medlin, L.K. Ribosomal DNA phylogenies and a morphological revision provide the basis for a revised taxonomy of the Prymnesiales (Haptophyta). *Eur. J. Phycol.* **2011**, *46*, 202–228. [CrossRef]

70. Not, F.; Siano, R.; Kooistra, W.H.C.F.; Simon, N.; Vaulot, D.; Probert, I. Diversity and Ecology of Eukaryotic Marine Phytoplankton. In *Genomic Insights into the Biology of Algae*; Piganeau, G., Ed.; Academic Press: London, UK, 2012; Volume 64, pp. 1–53.

71. Egge, E.; Bittner, L.; Andersen, T.; Audic, S.; de Vargas, C.; Edvardsen, B. 454 Pyrosequencing to Describe Microbial Eukaryotic Community Composition, Diversity and Relative Abundance: A Test for Marine Haptophytes. *PLoS ONE* **2013**, *8*, e74371. [CrossRef] [PubMed]

72. Quince, C.; Lanzen, A.; Davenport, R.J.; Turnbaugh, P.J. Removing Noise From Pyrosequenced Amplicons. *BMC Bioinform.* **2011**, *12*, 38. [CrossRef] [PubMed]

73. Schloss, P.D.; Westcott, S.L.; Ryabin, T.; Hall, J.R.; Hartmann, M.; Hollister, E.B.; Lesniewski, R.A.; Oakley, B.B.; Parks, D.H.; Robinson, C.J.; et al. Introducing mothur: Open-Source, Platform-Independent, Community-Supported Software for Describing and Comparing Microbial Communities. *Appl. Environ. Microbiol.* **2009**, *75*, 7537–7541. [CrossRef] [PubMed]

74. Quast, C.; Pruesse, E.; Yilmaz, P.; Gerken, J.; Schweer, T.; Yarza, P.; Peplies, J.; Glöckner, F.O. The SILVA ribosomal RNA gene database project: Improved data processing and web-based tools. *Nucleic Acids Res.* **2013**, *41*, D590–D596. [CrossRef] [PubMed]

75. Caron, D.A.; Countway, P.D. Hypotheses on the role of the protistan rare biosphere in a changing world. *Aquat. Microb. Ecol.* **2009**, *57*, 227–238. [CrossRef]

76. Edgar, R.C. Search and clustering orders of magnitude faster than BLAST. *Bioinformatics* **2010**, *26*, 2460–2461. [CrossRef] [PubMed]

77. Figshare. Available online: https://dx.doi.org/10.6084/m9.figshare.2759983.v1 (accessed on 22 March 2017).

78. Oksanen, J.; Blanchet, E.G.; Friendly, M.; Kindt, R.; Legendre, P.; McGlinn, D.; Minchin, P.R.; O'Hara, R.B.; Simpson, G.L.; Solymos, P.; et al. Vegan: Community Ecology Package, R package version 2.4-1. 2016. Available online: https://CRAN.R-project.org/package=vegan (accessed on 2 March 2017).

79. Hall, T. BioEdit: A user friendly biologicla sequence alignment editor and analysis program for windows 95/98/NT. *Nucl. Acids. Symp. Ser.* **1999**, *41*, 95–98.

80. Katoh, K.; Standley, D.M. MAFFT Multiple Sequence Alignment Software Version 7: Improvements in Performance and Usability. *Mol. Biol. Evol.* **2013**, *30*, 772–780. [CrossRef] [PubMed]

81. Huerta-Cepas, J.; Serra, F.; Bork, P. ETE 3: Reconstruction, Analysis, and Visualization of Phylogenomic Data. *Mol. Biol. Evol.* **2016**, *33*, 1635–1638. [CrossRef] [PubMed]

82. GenomeNet, Tree. Available online: http://www.genome.jp/tools/ete/ (accessed on 18 November 2016).

83. Price, M.N.; Dehal, P.S.; Arkin, A.P. FastTree: Computing Large Minimum Evolution Trees with Profiles instead of a Distance Matrix. *Mol. Biol. Evol.* **2009**, *26*, 1641–1650. [CrossRef] [PubMed]

Article

Change in *Emiliania huxleyi* Virus Assemblage Diversity but Not in Host Genetic Composition during an Ocean Acidification Mesocosm Experiment

Andrea Highfield [1,†], Ian Joint [1,†], Jack A. Gilbert [2,3], Katharine J. Crawfurd [4] and Declan C. Schroeder [1,*]

[1] The Marine Biological Association, The Laboratory, Citadel Hill, Plymouth PL1 2PB, UK; ancba@mba.ac.uk (A.H.); ianjoi@mba.ac.uk (I.J.)
[2] The Microbiome Centre, Department of Surgery, University of Chicago, Chicago, IL 60637, USA; gilbertjack@uchicago.edu
[3] Division of Bioscience, Argonne National Laboratory, 9700 South Cass Avenue, Argonne, IL 60439, USA
[4] Department of Biological Oceanography, NIOZ–Royal Netherlands Institute for Sea Research, P.O. Box 59, 1790 AB Den Burg, Texel, The Netherlands; kate.crawfurd@gmail.com
* Correspondence: dsch@mba.ac.uk; Tel.: +44-1752-426-6484
† These authors contributed equally to this work.

Academic Editors: Corina P.D. Brussaard and Mathias Middelboe
Received: 12 January 2017; Accepted: 2 March 2017; Published: 8 March 2017

Abstract: Effects of elevated pCO_2 on *Emiliania huxleyi* genetic diversity and the viruses that infect *E. huxleyi* (EhVs) have been investigated in large volume enclosures in a Norwegian fjord. Triplicate enclosures were bubbled with air enriched with CO_2 to 760 ppmv whilst the other three enclosures were bubbled with air at ambient pCO_2; phytoplankton growth was initiated by the addition of nitrate and phosphate. *E. huxleyi* was the dominant coccolithophore in all enclosures, but no difference in genetic diversity, based on DGGE analysis using primers specific to the calcium binding protein gene (*gpa*) were detected in any of the treatments. Chlorophyll concentrations and primary production were lower in the three elevated pCO_2 treatments than in the ambient treatments. However, although coccolithophores numbers were reduced in two of the high-pCO_2 treatments; in the third, there was no suppression of coccolithophores numbers, which were very similar to the three ambient treatments. In contrast, there was considerable variation in genetic diversity in the EhVs, as determined by analysis of the major capsid protein (*mcp*) gene. EhV diversity was much lower in the high-pCO_2 treatment enclosure that did not show inhibition of *E. huxleyi* growth. Since virus infection is generally implicated as a major factor in terminating phytoplankton blooms, it is suggested that no study of the effect of ocean acidification in phytoplankton can be complete if it does not include an assessment of viruses.

Keywords: *Emiliania huxleyi*; CO_2; ocean acidification; climate change; *Coccolithovirus*; EhV

1. Introduction

The rise in anthropogenic CO_2 in the atmosphere and subsequent dissolution in the oceans has changed the carbonate: bicarbonate: dissolved CO_2 equilibrium, lowering seawater pH—a trend that is predicted to continue [1]. Change of pH is of particular significance for marine organisms that have calcium carbonate structures, such as corals and coccolithophores, because less alkaline conditions and pH-dependent shifts in equilibrium of the carbonate system will lead to higher dissolution rates of carbonate. Coccolithophores are ubiquitous and have global significance in regulating the carbon cycle in the oceans [2]. They form massive blooms, whose wide distribution and abundance is readily detected in satellite imagery. Given this wide distribution, it is important to determine if the lower pH of a future ocean will affect the success of coccolithophores and if there will be an impact on marine food webs and biogeochemical cycles.

The effect of changing pH on the important coccolithophore, *Emiliania huxleyi*, has been the focus of much research in recent years. However, results have been variable and consensus has been difficult to reach. In laboratory experiments, both negative and positive effects of increasing pCO_2 have been described (see, for example, [3–5]). Another important approach has been to use large volume enclosures—mesocosms—to investigate a range of conditions that might apply to the future ocean. Unlike laboratory-based experiments, which usually focus on a single organism in the experimental design, mesocosms include all components of the pelagic system from viruses to zooplankton. By maintaining the possibility of complex interactions between different components of the food web, it has been assumed that mesocosms should offer advantages over single-organism culture experiments. However, results have also been rather variable. Early experiments suggested negative effects of higher pCO_2 on production and calcification in *E. huxleyi* [6], but other studies have indicated that the effect of increased pCO_2 is minimal for other coccolithophore species [7]. Time series analysis of natural populations has been another approach and a recent analysis of coccolithophore abundance in the North Sea concluded that increasing pCO_2 on decadal scales has resulted in larger coccolithophore populations [8]. The contradictory results make it difficult to robustly predict how natural populations will respond to pH change in a future ocean.

We suggest that real understanding of the effect of pH change/higher pCO_2 requires more detailed information than has been obtained to date, particularly in relation to phytoplankton genetic variability and virus infection. In this study, the response of a population of *E. huxleyi* to increased pCO_2 at the early stages of a phytoplankton bloom in a mesocosm experiment has been investigated. In addition, changes in diversity of the viruses that infect *E. huxleyi* (EhVs) were followed during the experiment with diversity distinguished on the basis of a major capsid protein (*mcp*) gene as a molecular marker. Virus diversity is known to be high [9,10], and viruses are important components of the pelagic system that require attention in both laboratory and mesocosm experiments. All *E. huxleyi*-infecting viruses that have been characterised to date are dsDNA viruses, classified in the family *Phycodnaviridae*. They are significant mortality agents of *E. huxleyi*, implicated in the termination of large-scale blooms. Viruses have a proven role in structuring and maintaining host population diversity [11–14] and virus infection can have significance for the cycling of carbon and trace elements. The 'viral shunt' releases nutrients as well as dissolved and particulate organic matter from lysed organisms into the organic carbon pool [15,16]. This material, and the rate of supply by viral lysis, of substrates for heterotrophic microbial communities, has implications for species succession, biogeochemical cycles and feedback mechanisms.

Given that diversity of both *E. huxleyi* and EhV assemblages can be variable, and that different *E. huxleyi* and EhV assemblages may come to eventually dominate natural communities, it is important to know the impact of elevated pCO_2. *E. huxleyi* blooms are typically dominated by certain alleles/genotypes, and by asexual reproduction, with rarer alleles/genotypes tending to fluctuate [17]. As such, the impact of elevated CO_2 on the composition of *E. huxleyi* populations can easily be monitored by studying these entities. Virus infection may be an explanation for some of the variability reported from different mesocosm experiments that were designed to investigate potential effects of higher pCO_2. In this study, the aim was to understand how pCO_2 change might influence *E. huxleyi* and EhV population structure and the diversity of host and virus. We suggest that viral infection can result in variability between replicate enclosures.

2. Materials and Methods

2.1. Experimental Set-Up and Sampling

The mesocosm experiment was done in the Raunefjorden at the University of Bergen Espegrand field station, Norway (latitude: 60°16' N; longitude: 5°13' E) during May 2006. The experiment had two phases. The first phase, until 15 May, followed the development of a phytoplankton bloom and the second phase studied the decline of the bloom; only the first phase of the experiment is considered here. Six polyethylene enclosures of 2 m diameter and 3.5 m depth containing 11 m^3 water were moored ca. 200 m from the shore and filled simultaneously with fjord water, salinity 31.4, and temperature 10.4 °C. Over a 40 h period from 4–6 May, 3 enclosures were bubbled with air enriched with CO_2 to 760 ppmv whilst the other 3 enclosures were bubbled with air at ambient pCO_2. The pCO_2 in the air mixture was measured inline with a LI-COR 6262 CO_2/H_2O analyser (LI-COR, Inc., Lincoln, NE, USA). After equilibration, the pH of each of the high pCO_2 treatments was 7.8 and the ambient treatment mesocosms were all pH 8.15. High precision alkalinity and pCO_2 measurements were made throughout the experiment and pH was calculated [18]. All mesocosms were covered with UV-transparent polyethylene to maintain the appropriate CO_2 concentration in the headspace above the enclosures, whilst allowing transmission of the complete spectrum of light and the exclusion of rainwater. Phytoplankton blooms were initiated on 6 May by the addition of 15 $\mu mol \cdot L^{-1}$ $NaNO_3$ bringing the initial nitrate concentration to 16.1 $\mu mol \cdot N \cdot L^{-1}$ and 1 $\mu mol \cdot L^{-1}$ NaH_2PO_4 to give an initial phosphate concentration 1.19 $\mu mol \cdot P \cdot L^{-1}$. Silicate was not added because the aim was to test the effects of pH change on coccolithophores, but rather to stimulate diatom growth; the initial silicate concentration was 0.25 $\mu mol \cdot Si \cdot L^{-1}$.

2.2. Water Sampling

The majority of measurements were made on water samples taken at the same time each day, between 10 a.m. and 11 a.m. Water samples were collected in 5 L carboys and transported to the shore laboratory where they were processed in a temperature controlled room at ambient seawater temperature.

2.3. Nutrient and Phytoplankton Analysis

Nutrient concentrations were determined on duplicate water samples by colorimetric analysis using the methods of Brewer and Riley [19] for nitrate, Grasshoff [20] for nitrite, and Kirkwood [21] for phosphate. Chlorophyll concentration was determined fluorometrically each day during the experiment, using the method of Holm-Hansen et al. [22] on water samples filtered through GFF glass fibre filters to monitor phytoplankton development. Samples were also taken for HPLC analysis of phytoplankton pigments, with GFF filters being stored at −80 °C between the period of sampling and laboratory analysis. Coccolithophore numbers were enumerated by analytical flow cytometry.

The rate of carbon fixation was estimated from the incorporation of ^{14}C-bicarbonate following the method of Joint and Pomroy [23]. Surface water samples were collected from each mesocosm at dawn and transferred into five 60 mL clear polycarbonate bottles and a single black polycarbonate bottle; all bottles were cleaned following JGOFS protocols [24] to reduce trace metal contamination. Each bottle was inoculated with 37 kBq (1 μCi) $NaH^{14}CO_3$; bottles were incubated at the surface and depths of 0.5, 1, 2 and 3 m in the fjord adjacent to the mesocosm facility for 24 h. Samples were filtered through 0.2 μm pore-size polycarbonate filters, dried, and treated with fuming HCl to remove unfixed ^{14}C and the assimilated ^{14}C fraction was measured in a liquid scintillation counter. The efficiency of the LSC was determined with an external standard, channels ratio method. The quantity of ^{14}C added to the experimental bottles was determined by adding aliquots of the stock ^{14}C solution to a CO_2-absorbing scintillation cocktail, which was counted immediately in the LSC.

2.4. Extraction of DNA

Collected water was stored at 4 °C until it was filtered, which occurred within several hours. Five litres of water from each mesocosm were filtered through a Sterivex-GP Sterile Vented Filter Unit, 0.22 μm (Millipore, Merck KGaA, Darmstadt, Germany). Filters were snap frozen in liquid nitrogen and maintained at −80 °C until they were processed. In addition, 2 mL 1× PBS was applied to the filters to wash off biomass and this was pelleted by centrifugation. DNA was extracted using the Qiagen DNeasy blood and tissue kit (Qiagen, Valencia, CA, USA) according to the manufacturer's instructions.

2.5. Polymerase Chain Reaction (PCR) and Denaturing Gradient Gel Electrophoresis (DGGE) of E. huxleyi and EhV Populations

PCR/DGGE analyses of extracted DNA from the 6 mesocosms were carried out according to the protocol for *E. huxleyi* and *E. huxleyi* viruses (EhV), as detailed in Schroeder et al. [25] and Schroeder et al. [13], respectively, using primers specific to the calcium binding protein gene (*gpa*) for *E. huxleyi* and the major capsid protein (*mcp*) gene for EhV. PCR products for *gpa* and *mcp* were run on a 30%–50% denaturing gel according to Schroeder et al. in order to visualise the respective community structures [13]. DGGE profiles for EhV were analysed using Genetools (Syngene, Cambridge, UK) using rolling disk baseline correction and minimum peak detection; width 7, height 3, volume 2% and Savitsky–Golay filter 3 to discriminate and quantify different bands/peaks.

2.6. Statistical Analysis

Ambient and high CO_2 multi-dimensional analysis (MDA) ordinations were calculated using Primer (v6) [26] using Bray–Curtis resemblance matrices produced from the DGGE profiles where bands were detected according to their migration distance down the tracks using Genetools (Syngene, Cambridge, UK). Principal component analysis (PCA) were calculated in Primer using all data obtained in the experiment to investigate which components might define differences/similarities between samples.

3. Results

3.1. Bloom Evolution—pH, Nutrients and Primary Production

Following bubbling to achieve the target pHs in all mesocosms, the experimental phase was initiated on 6 May, by the addition of nitrate and phosphate. Initial pH of the non-modified treatment mesocosms was 8.14. Figure 1a shows the values of pH during the first nine days of the experiment that were calculated from high precision pCO_2 data (Figure 1b) [18]. For four days, pH and pCO_2 remained constant with little variation between replicate treatments. After 10 May, pH began to increase in all mesocosms, with declining pCO_2 values as the phytoplankton bloom developed. Figure 1c,d record the changes in nitrate and phosphate concentration, including the initial nutrient addition. Both nutrients declined in concentration after 10 May as phytoplankton biomass increased (Figure 1e). Chlorophyll a concentration increased rapidly in all enclosures (Figure 1f), reaching a maximum on 13 May. However, there were differences in the maximum concentrations attained; the three high pCO_2-treatment mesocosms had maximum chlorophyll concentrations of 6.23, 4.51 and 6.08 $\mu g \cdot L^{-1}$, but chlorophyll concentrations were higher (10.71 and 11.22 $\mu g \cdot L^{-1}$) in two of the ambient high pCO_2-treatment mesocosms (4 and 6). A slightly lower phytoplankton biomass developed in enclosure M5—one of the ambient pCO_2-treatment mesocosm—with a chlorophyll a concentration of 9.60 $\mu g \cdot L^{-1}$. The chlorophyll concentration in this mesocosm also declined more rapidly after 13 May than in the other treatments.

Primary production rates were very consistent in the three high pCO_2-treatment mesocosms (Figure 1f), reaching maximum values on 12 May, with little variation between enclosures. In all the ambient pCO_2 mesocosms, primary production was >900 mg C m$^{-2} \cdot$d^{-1} on 12 May and remained at this value for two days in M4 and M6. However, production in M5 was less than in the other two ambient pCO_2-treatments, which is consistent with the lower chlorophyll concentration in this enclosure.

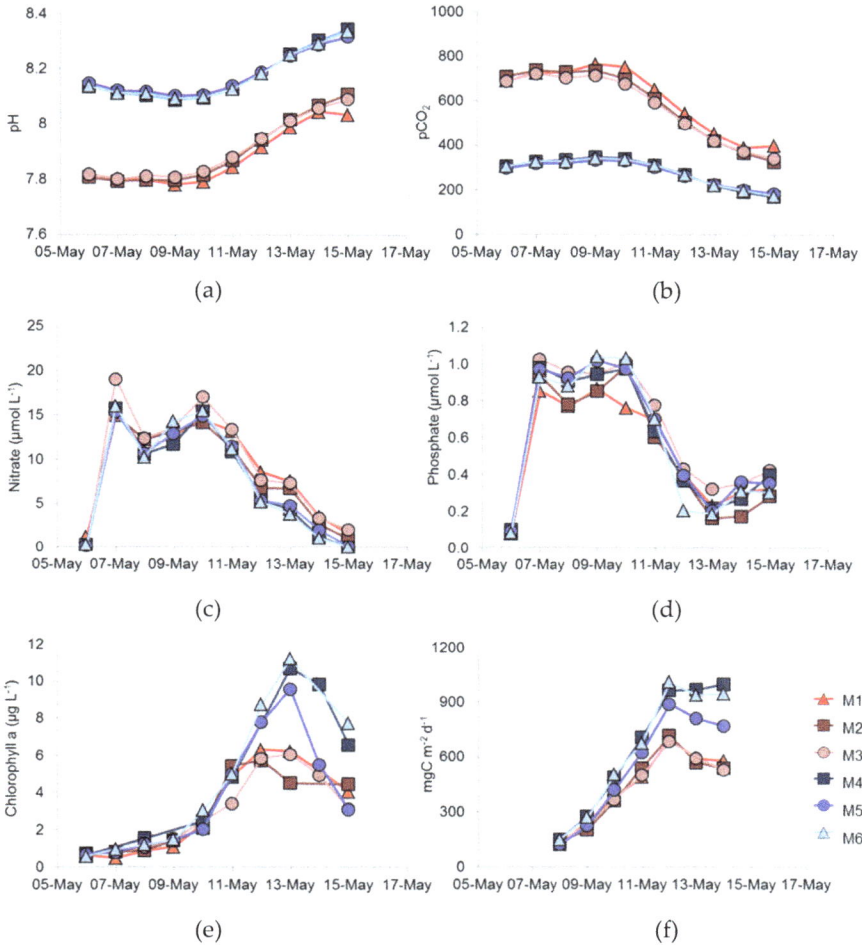

Figure 1. Temporal changes over the period of the experiment in (**a**) pH, which was calculated from measurements of (**b**) pCO_2 in μatmospheres; (**c**) nitrate concentration, $\mu mol \cdot N \cdot L^{-1}$; (**d**) phosphate concentration $\mu mol \cdot P \cdot L^{-1}$; (**e**) chlorophyll concentration $\mu g \cdot L^{-1}$; and (**f**) depth-integrated primary production as $mg \cdot C \cdot m^{-2} \cdot d^{-1}$. Enclosures M1 (▲), M2 (■), M3 ( ), M4 (■), M5 (●), M6 ( ).

3.2. E. huxleyi Genetic Composition during the Mesocosm Experiment

E. huxleyi was a significant component of the phytoplankton assemblage that developed in each enclosure. Diatom numbers were insignificant because silicate was not added to the initial nutrient addition, being three orders of magnitude less abundant in light microscope analysis than the total flagellate fraction, which includes coccolithophores. Hopkins et al. [18] reported the dominance of large picoeukaryotes in each mesocosm assemblage but with the flagellates contributing greatest to phytoplankton biomass.

All enclosures showed steady increases in coccolithophore numbers (as assessed by flow cytometry) immediately after nutrient addition. Numbers reached 600–1000 cells mL^{-1} on 9 May, which is typical of numbers seen during the early-stages of *E. huxleyi* blooms (Figure 2) [12]. In the ambient-pCO_2 treatments (M4, M5, M6) coccolithophore numbers increased until 12 May, but with a slight pause in growth, numbers increased further to a maximum of 2500–3000 cells mL^{-1} on 14 May. Cell numbers then declined to 1000–2000 cells mL^{-1} on 15 May. Coccolithophore biomass was different in the three replicate high pCO$_2$-treatment mesocosms. In two enclosures, numbers plateaued at about 1000 cells mL^{-1}, but, in the third mesocosm, numbers were higher and, indeed, very similar (3100 cells mL^{-1}) to the peak biomass in the ambient pCO$_2$-treatment mesocosms (Figure 2). Cell numbers declined in all six mesocosms after 14 May, even in those enclosures with lower cell numbers.

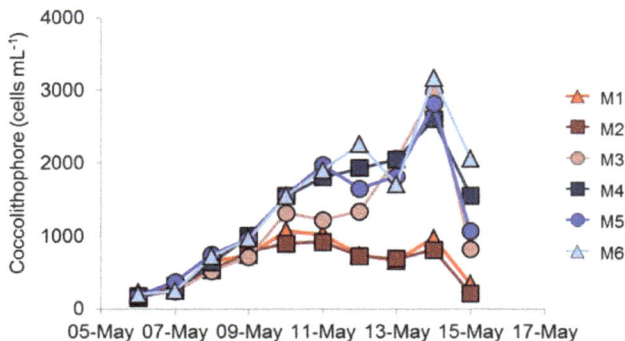

Figure 2. Total coccolithophore numbers assessed by flow cytometry. Enclosures M1 (▲), M2 (■), M3 (●), M4 (■), M5 (●), M6 (▲).

Traditional microscopy, neither light nor electron, is capable of distinguishing between *E. huxleyi* genotypes with Figure 3 showing that identical morphology (typical type A) was present in both pCO$_2$ treatments throughout the experiment.

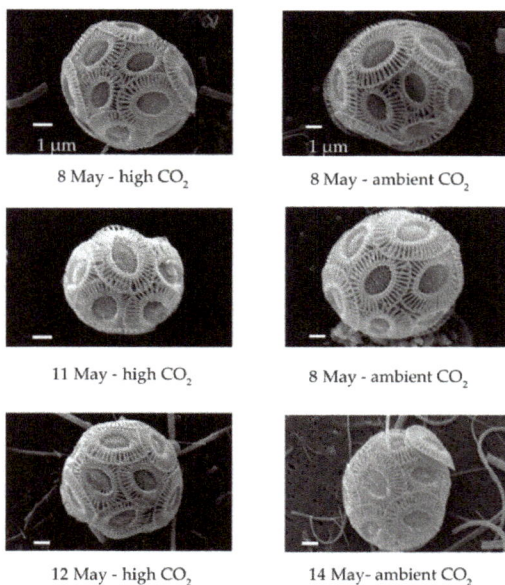

Figure 3. TEM images of identical *Emiliania huxleyi* morphologies (typical type A) present in both pCO$_2$ treatments throughout the experiment.

However, molecular analysis showed a large genetic diversity that was not revealed by microscopy. DGGE analysis of the *E. huxleyi* population using the *gpa* marker detected two to three dominant bands throughout the experiment as indicated by the arrows in Figure S1. This gene has been verified for *E. huxleyi* diversity analysis, with a limited number of genotypes known to exist [17] that can largely be separated by DGGE [25]. There was some small-scale variability in the *E. huxleyi* population between samples, as indicated by migration profiles. Overall, the *E. huxleyi* populations had similar genetic composition in all six mesocosms and no major differences were identified between treatments or replicates.

3.3. EhV Population Analysis

Flow cytometry revealed the presence of large DNA viruses in all enclosures (data not shown), indicating background levels less than 10^5 viruses per ml as described previously in other Bergen based mesocosm experiments [12,13]. Although there was little variation in *E. huxleyi* genotypes throughout the experiment, the virus (EhV) that infects this alga did show considerable variation. DGGE analysis (Figure 4) indicated that the EhV population was more diverse and had a more variable genetic structure than the host. Whilst DGGE has its limitations due to co-migration events meaning that a single band can be comprised of >1 OTU, it is still accepted as a useful tool for looking at changes in microbial communities. In particular, DGGE has proven to be a robust and reliable technique for the study of EhVs. Limitations often described in the literature, centre on PCR-DGGE designed to target a large taxonomic group where the scale of diversity is massive, e.g., 16S rRNA. Focussing on a smaller taxonomic unit improves resolution [27] as does careful design and optimisation of primers. A two-stage PCR has been well optimised for EhV with the use of a GC-clamp to improve resolution. Although not all EhVs can be discriminated from one another, such as EhV-201 and EhV-205 that only differ by 1 bp in the target *mcp* region, virus isolates, including EhVs 203, 201, 202, 163, 84 and 86, can clearly be separated on a DGGE gel, with EhV-84 and EhV-86 differing from each other by only 3 bp [13]. Furthermore, DGGE gels for EhV have been found to be highly reproducible with the samples being run on >1 separate gels, generating the same migration profile. This is also corroborated by previous work of Sorensen et al. [14] and Martinez-Martinez et al. [12], where replicate gels routinely produce the same profile.

In the early stages of the mesocosm bloom development before 9 May, when coccolithophore numbers were <800 cells mL^{-1}, there was high variability in the EhV population, both between replicates and on different days sampled, for example within mesocosm 4 on 7 May there were four bands, on 8 May seven bands and on 9 May eight bands, with four of these being unique; the percentage similarity was less than 40%. Whilst we cannot ascertain that each band represents a single out, we can still infer the changes observed and the overall temporal patterns of diversity. A Bray–Curtis similarity analysis (Figure 5) showed that, as coccolithophore numbers increased, exceeding 1.5×10^3 cells mL^{-1} in ambient enclosures, there was less variability in the EhV population in the ambient pCO$_2$-treatment mesocosms, which shared more similarity (>43%) between replicates. This was compared to as little as 1% in the early stages of the experiment when coccolithophore numbers were less than 1×10^3 cells mL^{-1} (Figure 5).

Within the 2 high-pCO$_2$ treatment enclosures that had the lowest coccolithophore cell number (M1 and M2), the DGGE profiles showed low similarity in the EhV population between dates and replicates (Figure 4). Stabilisation of the EhV population was not evident as the experiment proceeded in the high pCO$_2$ samples, mesocosms 1 and 2, and similarity between samples remained low (9%). In contrast, in mesocosm M3, all samples shared at least 55% similarity, indicating a stable EhV population across the time series. One of the dominant bands on the DGGE profile in M3 (marked with a triangle in Figure 4a) was an EhV genotype that also dominated in the ambient-pCO$_2$ treatments in enclosures M4, M5 and M6.

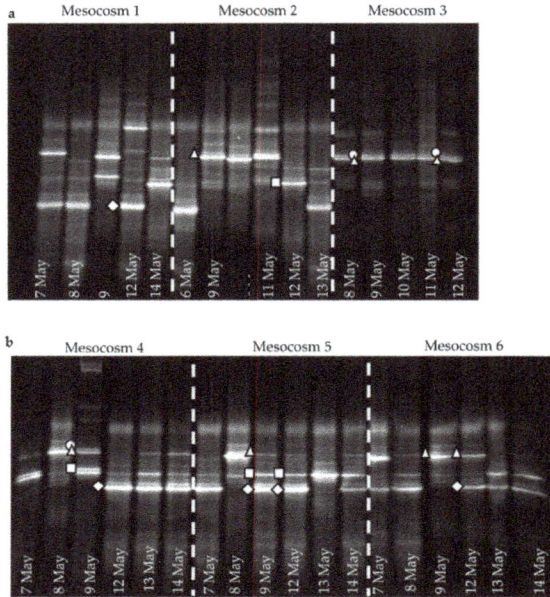

Figure 4. DGGE gels of EhV *mcp*-PCR products during the experiment from (**a**) high *p*CO₂-treatment mesocosms, 1, 2, and 3 and (**b**) ambient *p*CO₂-treatment mesocosms 4, 5, and 6. Bands that migrated at the same position when run on the same gel are indicated with the same symbol.

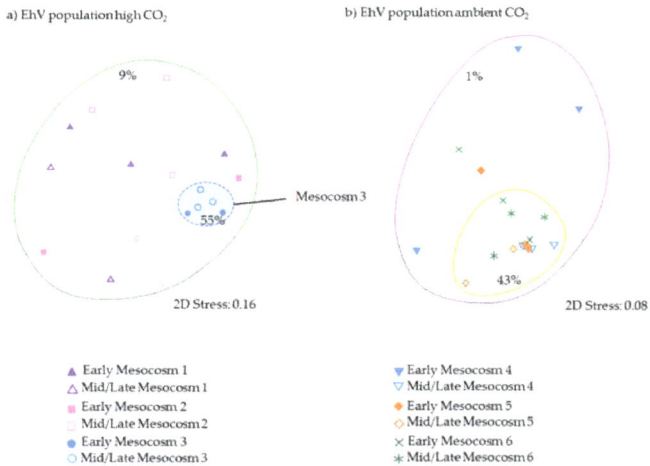

Figure 5. Bray–Curtis multidimensional plots based on the DGGE profiles (Figure 4) for EhV from (**a**) the high *p*CO₂-treatment mesocosms 1, 2, and 3 and (**b**) ambient *p*CO₂-treatment mesocosms 4, 5, and 6. "Early stage" corresponds to 7–9 May when coccolithophore numbers were <1000 cells mL⁻¹ in ambient enclosures and "mid/late stage" corresponds to 12–14 May when coccolithophore numbers exceeded 1500 cells mL⁻¹ in ambient enclosures. Contours indicate the percentage similarity, as indicated.

Samples from the 9 and 12 May from all six mesocosms were additionally run on the same gel. The shapes on the gels (Figure 4) indicate bands that migrated to the same position and hence can be inferred to be the same EhV sequence.

4. Discussion

A number of mesocosm experiments have been performed to investigate the potential effects of increased pCO_2 and reduced pH on complex pelagic assemblages—from bacteria to zooplankton. Mesocosm enclosures have the advantage of capturing more of the intrinsic complexity of a pelagic assemblage than is possible in a laboratory experiment because of the very large volumes of water (several thousand litres) that are involved. They also, by the nature of enclosure, eliminate problems of dispersal that make the study of variability in natural environments so complex. They are fundamentally attractive to experimentalists because they offer a means to manipulate large water volumes, with their associated planktonic assemblages, in order to test the effects of environmental problems such as eutrophication or ocean acidification.

In this study, we aimed to investigate how phytoplankton might respond in a future high-pCO_2 ocean by comparing natural phytoplankton assemblage development in enclosures at ambient pCO_2, (initial condition ca. 300 µatm.) and enriched pCO_2 conditions (initially ca. 700 µatm.)—Figure 1a,b; see also Hopkins et al. [18]. During the course of the experiment, utilisation of CO_2 by phytoplankton reduced pCO_2 and the pCO_2/pH values were continually changing. An obvious difference between the treatments was that less phytoplankton biomass, as indicted by chlorophyll *a* concentration, developed in the high- compared to the ambient-pCO_2 conditions (Figure 1e). Not only did less biomass develop, but primary production (Figure 1f) was also lower under high-, rather than under ambient-pCO_2, suggesting that increased pCO_2 might have a deleterious effect on the total phytoplankton biomass. In contrast, other experiments have suggested that dissolved inorganic carbon uptake would be enhanced under elevated pCO_2 conditions [28]. Different phytoplankton types are likely to respond differently to pCO_2 and Riebesell et al. suggested that diatoms showed enhanced uptake, whereas coccolithophores did not [28]. There was no suggestion in that study, though, that coccolithophore growth might be reduced under elevated pCO_2 conditions.

The dominant species of coccolithophore within the mesocosms was *E. huxleyi*, with coccolithophores being reported as contributing 6% and 12% to the total flagellate biomass for M1 and M6, respectively [18]. Manipulating the development of *E. huxleyi* blooms within mesocosm enclosures is well established at the mesocosm facility at Raunefjorden, and is well documented. Addition of nitrate and phosphate, with the omission of silicate, usually results in a bloom of *E. huxleyi* at this site, particularly during May/June [11,12,14] and an increase in *E. huxleyi* abundance happened in the present study. Previous mesocosm experiments of this nature have described a dominance of coccolithophores; however, differences in methodology most likely resulted in the dominance of large picoeukaryotes [18]. In the present study, a single nutrient enrichment was undertaken at the beginning of the study, whereas daily enrichments are often used e.g., Jacquet et al. [29]. Nevertheless, the maximum number of *E. huxleyi* cells that developed was significantly lower in the high- compared to ambient-pCO_2 conditions (Figure 2), suggesting that *E. huxleyi* might be particularly susceptible to variations in pCO_2.

Replication between enclosures was rather variable. The peak chlorophyll concentrations were very similar in the three ambient pCO_2 replicates (Figure 1e), but the bloom decayed more rapidly in M5 than in M4 and M6; there were similar differences in primary production (Figure 1f). However, the largest difference between replicates was in coccolithophore numbers (Figure 2), which showed significant differences between the three high-pCO_2 treatment mesocosms with such differences between replicates in other key phytoplankton groups not reported by Hopkins et al. [18]. Enclosure M3 had coccolithophore cell numbers at the peak of the bloom that were very similar to the three ambient pCO_2 mesocosms, unlike the lower numbers in M1 and M2. Although cell numbers were different in M3, no major differences in *E. huxleyi* genotype or phenotype were detected between treatments or over the course of the experiment, suggesting a stable community within all enclosures over the duration of the experiment. Stable *E. huxleyi* populations have been found in previous mesocosm experiments [12,14] and, indeed, in naturally occurring *E. huxleyi* blooms [9,17]. In this study, we have no evidence to support the hypothesis that higher pCO_2 conditions might benefit certain *E. huxleyi* genotypes; we could detect no restructuring of the *E. huxleyi* population in the different pCO_2 treatments.

Contradictory results are constant features of experiments to investigate the effect of pCO_2/pH change on coccolithophores. In laboratory culture experiments, Riebesell et al. [30], Zondervan et al. [31] and Richier et al. [32] all reported increased production by *E. huxleyi* under elevated pCO_2 conditions, but Sciandra et al. [33] and Langer et al. [4] found decreased production. In two different CO_2-manipulated mesocosm experiments in the Raunefjorden, Engel et al. [6] and Paulino et al. [34] found little difference in *E. huxleyi* cell concentrations over the course of their experiments. However, Engel et al. [6] calculated that the net specific growth rate of *E. huxleyi* was reduced at 710 μatm compared with 410 μatm. In a long-term batch culture experiment conducted over one year, Lohbeck et al. [5] found that *E. huxleyi* cultures maintained at ambient pCO_2 (400 μatm) went through 530 generations over the one year experimental period, but the same strain cultured at 1100 μatm achieved only 500 generations, and, at 2200 μatm, growth was even lower, with only 430 generations. *E. huxleyi* would appear to be more sensitive to pCO_2 change than other phytoplankton.

In the context of the present study, the reduced primary production in the high-pCO_2 treatment enclosures is consistent with the finding of Lohbeck et al. [5] that higher pCO_2/lower pH reduces primary production of an *E. huxleyi* dominated phytoplankton assemblage, although, in this study, other phytoplankton groups, e.g., picoeukaryotes, cryptophytes and cyanobacteria, will have also contributed to this. However, it is not consistent with the suggestion of Rivero-Calle et al. [8] that increasing pCO_2 is one of the factors most responsible for the decadal increase in abundance of coccolithophores in the North Atlantic. The relationship between pCO_2/pH change and success or, otherwise of cocccolithophores, remains confusing and requires clearer examination of mechanisms that might lead to phytoplankton changes in the future ocean.

Given that the coccolithophores numbers in enclosure M3 were very different from the other two high pCO₂ treatments, and yet the *E. huxleyi* genetic diversity was very similar in all three enclosures, could viral infection by EhV be a contributing factor to explain the observed variations within and between treatments? In the three ambient-pCO_2 enclosures, the EhV population followed a pattern that has been seen in other mesocosm experiments—high variability during the early stages of phytoplankton bloom development, with a smaller number of genotypes coming to dominate as the *E. huxleyi* numbers increase [12,14]. Although phytoplankton bloom development in the current study was short (<10 days), there was sufficient time for the virus populations to change because EhV populations are inherently dynamic [9] and known to change on very short time scales [14]. Daily changes in EhV composition can be expected since Sorensen et al. [14] showed that EhVs can appear/disappear from the water column in a matter of hours.

In the high-pCO_2 treatment enclosures M1 and M2, where *E. huxleyi* population densities did not exceed 1.1×10^3 mL^{-1}, the EhV population did not stabilise (Figures 4 and 5) and EhV diversity was typical of early or non-bloom conditions [10]; that is, it was a highly dynamic and diverse EhV population. EhVs are known to have different host ranges [35] as well as different characteristics of infection, such as burst size and latent period. Therefore, any changes in environmental conditions that directly affect these traits could ultimately select for different genotypes, hence restructuring the EhV population.

In contrast, the third high-pCO_2 treatment enclosure M3, where coccolithophore cell densities reached similar values to ambient enclosures, and the EhV population structure was very different. Two EhVs dominated over the course of the experiment (Figure 4) and the population was much less changeable compared to the other high-pCO_2 enclosures and, indeed, to the three ambient enclosures. The EhV assemblage in M3, right from the early stages of the experiment, reflected what would be expected in the later stages of a bloom. Even in the early stages of the bloom, the EhV assemblage was stable and clustered closely in the MDS plot (Figure 5) with samples taken later in the bloom.

Given that pH and pCO_2 were so similar throughout the experiment in the three high-pCO_2 enclosures, why did enclosure M3 have lower coccolithophore numbers and a lower and stable EhV diversity? Other studies have shown that, under non-bloom conditions, many different EhV genotypes are present and abundance fluctuates on short time scales [14]. In addition, during the initial phase of an *E. huxleyi* bloom, EhV populations remain diverse and are often highly dynamic. Sorensen et al. [14] showed that, as a bloom developed in a mesocosm experiment and viruses numbers proliferated, one viral genotype dominated, and suggested that this dominant virus caused the termination of the bloom. However, the dominant EhV is not always the same, even when host genotypes do not vary. Martínez-Martínez et al. [12] and Sorensen et al. [14] found that, although *E. huxleyi* populations were dominated by the same genotypes in different years at Raunefjorden (2000, 2003 and 2008), viruses changed and the dominant EhV in 2008 was different to the EhVs that dominated during the 2000 and 2003 mesocosm experiments. The reason is not known, but these authors suggest that a slight change in environmental conditions might have favoured dominance by a different virus genotype. Whilst the number of mesocosms sampled might be perceived as limited, the fact that Martinéz-Martinéz [35] described how the same DGGE profile was generated from four replicate mesocosms; in his studies, we can assume that the changes that we are observing are genuine.

Other mesocosm experiments have studied how altered pCO_2 influences natural virus communities. Larsen et al. [36], using flow cytometry analysis and pulsed field gel electrophoresis (PFGE), found slightly more (but statistically insignificant) EhVs under present-day pCO_2 mesocosms than in high-pCO_2 treatments; this was not a consequence of lower *E. huxleyi* cell densities in the high-pCO_2 treatments. The authors speculated that elevated-pCO_2 may affect host–virus interactions or influence viral replication, since a 26 kb genome virus was only detected in ambient conditions, and was absent from high-pCO_2 treatments, and a 105 kb genome virus was only detected in the highest pCO_2 treatment of 1050 ppm. Unfortunately, the taxonomic affiliation of the viruses was not verified, which limits comparison with the present study.

Some laboratory experiments have investigated the effect of higher pCO_2 on marine phytoplankton viruses. Carreira et al. [37] studied interaction between *E. huxleyi* and the virus EhV-99B1. *E. huxleyi* growth rate was not affected by the different pCO_2 treatments, but the burst size of EhV-99B1 was lower in present-day, compared with higher and lower (pre-industrial) pCO_2. In addition, release of EhVs was delayed in high-pCO_2 treatments. Other virus groups that have also been tested for sensitivity to elevated pCO_2. Chen et al. [38] found lower burst size of the *Phaeocystis globosa* virus (PgV) at high-pCO_2, and Traving et al. [39] found that the cyanophage S-PM2, which infects *Synechococcus* sp, had reduced burst size at lower pH. However, the extracellular phase, quantified as infectivity loss rates/decay, did not change. These experiments illustrate that pCO_2 can influence virus–host interactions, albeit to a relatively minor extent. However, extrapolation from these laboratory-scale experiments to natural virus communities would involve considerable uncertainty.

Another study considered a much longer time scale. Coolen [40] investigated *E. huxleyi* and EhV diversity in Black Sea sediments over a 7000 year period, showing that EhV diversity was highest during periods of change in hydrological and nutrient regimes. Shifts in EhV genotypic diversity typically coincided with Holocene environmental change with some viruses having limited persistence, yet others were found to persist for over a century. This study alluded to the impact that a change in CO_2/pH could have on future EhV populations.

Might the different EhV populations that dominated in each enclosure be an explanation for the differences observed? It is generally accepted that virus infection is a major reason why *E. huxleyi* cells stop growing and blooms are terminated [14]. Certainly, nutrients were still available, albeit at low concentrations when biomass peaked in each enclosure (Figure 1c,d). If viral mortality was the major limit on bloom development, then enclosures M1 & M2 must have been infected with more aggressive EhVs than in the other treatments. The effect of pCO_2 treatment, per se, on *E. huxleyi* cells cannot be responsible for the reduced growth in the two high-pCO_2 treatments because an identical pH/pCO_2 treatment in M3 did not reduce the size of the bloom. Therefore, pCO_2 change must have resulted in different viral diversity, if infection is indeed the main cause of lower cell numbers and smaller *E. huxleyi* bloom.

How might reduced EhV diversity in M3 have resulted in higher coccolithophore numbers developing than in the two other high pCO_2 treatments where growth was curtailed? One explanation would be that the EhVs that lead to rapid termination of *E. huxleyi* growth [14] were not present in sufficient numbers in M3 to suppress growth of the *E. huxleyi* population in this enclosure. Genetic diversity in natural populations of EhVs, especially in pre-bloom conditions [10], coupled with the rapid rate at which individual EhVs can come to dominate [9,14] means that the matrix of EhVs that could be selected for is large. It is not clear why only two EhVs were dominant in enclosure M3, but it is probably significant that viral infection in this enclosure did not suppress growth of *E. huxleyi* compared to M1 and M2.

In this study, it is difficult to distinguish whether pCO_2 change is affecting the viruses specifically, or their hosts independently, and/or the interactions between them. Both the external virus population (the virus particles present in the water used to fill the mesocosms) and internal virus population (present in infected *E. huxleyi* cells) are important for the ultimate progression of the *E. huxleyi* bloom and EhV population. By comparing data from Schroeder et al. [13] and Martínez Martínez [12] it can be seen that the diversity of EhVs amplified from water samples within a mesocosm bloom can be very different from that amplified from *E. huxleyi* cells, particularly at the early stage of the bloom—that is, external and internal EhV assemblages can have very different composition. Thus, studies that aim to explain the effect of elevated pCO_2 must incorporate into their experimental design ways to distinguish between a direct effect on external virus particles, or on *E. huxleyi* cells, or on EhVs that have already infected *E. huxleyi* cells. Alternatively, other unknown factors that are not related to pCO_2 cannot be dismissed and might be the cause of the different coccolithophore response in enclosure M3.

Our study demonstrates the need for further investigations on the effects of elevated pCO_2 on the *E. huxleyi* EhV system since there are ecological impacts of virus competition on biogeochemical cycles. Nissimov et al. [41] investigated competition between the two EhVs, finding that EhV-207 had a competitive advantage over EhV-86. It would thus be of value to determine how external factors, such as elevated CO_2, would affect relative competitive ability of EhVs; would EhV-207 still outcompete EhV-86?

In this study, we have shown that elevated pCO_2 can affect the structure of EhV assemblages. The data do not allow us to distinguish if this is a direct impact of pCO_2 on the viruses themselves, but it is clear that caution is required in interpreting results from manipulation experiments and that deep analysis is required to truly understand how complex assemblages respond. For example, analysis of only the chlorophyll concentration or primary production data would not have revealed that there were differences in the triplicate high-pCO_2 treatments: cell counts and microscopy would not have revealed the difference in *E. huxleyi* diversity in the replicate enclosures, and only analysis of virus genotypes could have revealed how different the viruses were in apparently identical replicate enclosures. We suggest that viruses cannot be ignored in any study of the potential effects of ocean acidification on phytoplankton productivity in the future high-CO_2 ocean.

Supplementary Materials: The following are available online at www.mdpi.com/1999-4915/9/3/41/s1, Figure S1. DGGE image of *E. huxleyi* amplified *gpa*-PCR products.

Acknowledgments: Twenty-seven people participated in the mesocosm experiment and we thank all of them for their contributions. In particular, we thank Dorothee Bakker for high precision measurements of the carbonate system, Isabel Mary and Andrew Whiteley for flow cytometer measurements of coccolithophore number and Cecilia Balestreri for laboratory support. Special thanks are due to the staff at the Espegrand field station for their assistance. The mesocosm experiment was supported by NERC, through Grant No. NE/C507902, as part of the Post-Genomics and Proteomics Programme.

Author Contributions: I.J. and D.C.S. conceived and designed the experiments; A.H. and K.J.C. performed the experiments; A.H., I.J. and D.C.S. analyzed the data; J.G. contributed reagents/materials/analysis tools; and A.H. and I.J. wrote the paper.

Conflicts of Interest: The authors declare no conflict of interest.

References

1. Orr, J.C.; Fabry, V.J.; Aumont, O.; Bopp, L.; Doney, S.C.; Feely, R.A.; Gnanadesikan, A.; Gruber, N.; Ishida, A.; Joos, F.; et al. Anthropogenic ocean acidification over the twenty-first century and its impact on calcifying organisms. *Nature* **2005**, *437*, 681–686. [CrossRef] [PubMed]

2. Balch, W.M. Re-evaluation of the physiological ecology of the coccolithophores. In *Coccolithophores: From Molecular Processes to Global Impact*; Thierstein, H.R., Young, J.R., Eds.; Springer: New York, NY, USA, 2004; pp. 165–190.

3. Iglesias-Rodriguez, M.D.; Halloran, P.R.; Rickaby, R.E.M.; Hall, I.R.; Colmenero-Hidalgo, E.; Gittins, J.R.; Green, D.R.H.; Tyrrell, T.; Gibbs, S.J.; von Dassow, P.; et al. Phytoplankton calcification in a high-CO_2 world. *Science* **2008**, *320*, 336–340. [CrossRef] [PubMed]

4. Langer, G.; Nehrke, G.; Probert, I.; Ly, J.; Ziveri, P. Strain-specific responses of *Emiliania huxleyi* to changing seawater carbonate chemistry. *Biogeosciences* **2009**, *6*, 2637–2646. [CrossRef]

5. Lohbeck, K.T.; Riebesell, U.; Reusch, T.B.H. Gene expression changes in the coccolithophore *Emiliania huxleyi* after 500 generations of selection to ocean acidification. *Proc. R. Soc. B* **2014**, *281*. [CrossRef] [PubMed]

6. Engel, A.; Zondervan, I.; Aerts, K.; Beaufort, L.; Benthien, A.; Chou, L.; Delille, B.; Gattuso, J.-P.; Harlay, J.; Heemann, C.; et al. Testing the direct effect of CO_2 concentration on a bloom of the coccolithophorid *Emiliania huxleyi* in mesocosm experiments. *Limnol. Oceanogr.* **2005**, *50*, 493–507. [CrossRef]

7. Meyer, J.; Riebesell, U. Reviews and Syntheses: Responses of coccolithophores to ocean acidification: A meta-analysis. *Biogeosciences* **2015**, *12*, 1671–1682. [CrossRef]

8. Rivero-Calle, S.; Gnanadesikan, A.; Del Castillo, C.E.; Balch, W.M.; Guikema, S.D. Multidecadal increase in North Atlantic coccolithophores and the potential role of rising CO_2. *Science* **2015**, *350*, 1533–1537. [CrossRef] [PubMed]

9. Highfield, A.; Evans, C.; Walne, A.; Miller, P.I.; Schroeder, D.C. How many *Coccolithovirus* genotypes does it take to terminate an *Emiliania huxleyi* bloom? *Virology* **2014**, *466*, 138–145. [CrossRef] [PubMed]

10. Rowe, J.M.; Fabre, M.F.; Gobena, D.; Wilson, W.H.; Wilhelm, S.W. Application of the major capsid protein as a marker of the phylogenetic diversity of *Emiliania huxleyi* viruses. *FEMS Microbiol. Ecol.* **2011**, *76*, 373–380. [CrossRef] [PubMed]

11. Bratbak, G.; Wilson, W.; Heldal, M. Viral control of *Emiliania huxleyi* blooms? *J. Mar. Syst.* **1996**, *9*, 75–81. [CrossRef]

12. Martínez Martínez, J.M.; Schroeder, D.C.; Larsen, A.; Bratbak, G.; Wilson, W.H. Molecular dynamics of *Emiliania huxleyi* and co-occurring viruses during two separate mesocosm studies. *Appl. Environ. Microbiol.* **2007**, *73*, 554–562. [CrossRef] [PubMed]

13. Schroeder, D.C.; Oke, J.; Hall, M.; Malin, G.; Wilson, W.H. Virus succession observed during an *Emiliania huxleyi* bloom. *Appl. Environ. Microbiol.* **2003**, *69*, 2484–2490. [CrossRef] [PubMed]

14. Sorensen, G.; Baker, A.C.; Hall, M.J.; Munn, C.B.; Schroeder, D.C. Novel virus dynamics in an *Emiliania huxleyi* bloom. *J. Plankton Res.* **2009**, *31*, 787–791. [CrossRef] [PubMed]

15. Gobler, C.J.; Hutchins, D.A.; Fisher, N.S.; Cosper, E.M.; Sanudo-Wilhelmy, S.A. Release and bioavailability of C, N, P, Se, and Fe following viral lysis of a marine chrysophyte. *Limnol. Oceanogr.* **1997**, *42*, 1492–1504. [CrossRef]

16. Wilhelm, S.W.; Suttle, C.A. Viruses and nutrient cycles in the sea—Viruses play critical roles in the structure and function of aquatic food webs. *Bioscience* **1999**, *49*, 781–788. [CrossRef]

17. Krueger-Hadfield, S.A.; Balestreri, C.; Schroeder, J.; Highfield, A.; Helaouet, P.; Allum, J.; Moate, R.; Lohbeck, K.T.; Miller, P.I.; Riebesell, U.; et al. Genotyping an *Emiliania huxleyi* (Prymnesiophyceae) bloom event in the North Sea reveals evidence of asexual reproduction. *Biosciences* **2014**, *11*, 5215–5234.

18. Hopkins, F.E.; Turner, S.M.; Nightingale, P.D.; Steinke, M.; Bakker, D.; Liss, P.S. Ocean acidification and marine trace gas emissions. *Proc. Natl. Acad. Sci. USA* **2010**, *107*, 760–765. [CrossRef] [PubMed]
19. Brewer, P.G.; Riley, J.P. The automatic determination of nitrate in seawater. *Deep Sea Res.* **1965**, *12*, 765–772.
20. Grasshoff, K. *Methods of Seawater Analysis*; Verlag Chemie: Weinheim, Germany, 1976.
21. Kirkwood, D.S. Simultaneous determination of selected nutrients in seawater. In *ICES*; National Marine Biological Library: Plymouth, UK, 1989.
22. Holm-Hansen, O.; Lorenzen, C.J.; Holmes, R.W.; Strickland, J.D.H. Fluorometric determination of chlorophyll. *ICES J. Mar. Sci.* **1965**, *30*, 3–15. [CrossRef]
23. Joint, I.; Pomroy, A. Phytoplankton biomass and production in the southern North Sea. *Mar. Ecol. Prog. Ser.* **1993**, *99*, 169–182. [CrossRef]
24. IOC; Paris, France. Protocols for the Joint Global Ocean Flux Study (JGOFS) core measurements. In *IOC Manuals and Guides No. 29*; JGOFS International Project Office: Bergen, Norway, 1994; p. 126.
25. Schroeder, D.C.; Biggi, G.F.; Hall, M.; Davy, J.; Matinez Martinez, J.; Richardson, A.J.; Malin, G.; Wilson, W.H. A genetic marker to separate *Emiliania huxleyi* (Prynesiophyceae) morphotypes. *J. Phycol.* **2005**, *41*, 874–879. [CrossRef]
26. Clarke, K.R.; Gorley, R.N. *PRIMER V6: User Manual/Tutorial*; PRIMER-E Ltd.: Plymouth, UK, 2006.
27. Marzorati, M.; Wittebolle, L.; Boon, N.; Daffonchio, D.; Verstraete, W. How to get more out of molecular fingerprints: Practical tools for microbial ecology. *Environ. Microbiol.* **2008**, *10*, 1571–1581. [CrossRef] [PubMed]
28. Riebesell, U.; Schulz, K.G.; Bellerby, R.G.J.; Botros, M.; Fritsche, P.; Meyerhöfer, M.; Neill, C.; Nondal, G.; Oschlies, A.; Wohlers, J.; et al. Enhanced biological carbon consumption in a high CO_2 ocean. *Nature* **2007**, *450*, 545–549. [CrossRef] [PubMed]
29. Jacquet, S.; Heldal, M.; Iglesias-Rodriguez, D.; Larsen, L.; Wilson, W.H.; Bratbak, G. Flow cytometric analysis of an *Emiliana huxleyi* bloom terminated by viral infection. *Aquat. Microb. Ecol.* **2002**, *27*, 111–124. [CrossRef]
30. Riebesell, U.; Zondervan, I.; Rost, B.; Tortell, P.D.; Zeebe, R.E.; Morel, F.M. Reduced calcification of marine plankton in response to increased atmospheric CO_2. *Nature* **2000**, *407*, 364–367. [CrossRef] [PubMed]
31. Zondervan, I.; Rost, B.; Riebesell, U. Effect of CO_2 concentration on the PIC/POC ratio in the coccolithophore *Emiliania huxleyi* grown under light-limiting conditions and different day lengths. *J. Exp. Mar. Biol. Ecol.* **2002**, *272*, 55–70. [CrossRef]
32. Richier, S.; Fiorini, S.; Kerros, M.E.; von Dassow, P.; Gattuso, J.P. Response of the calcifying coccolithophore *Emiliania huxleyi* to low pH/high pCO$_2$: From physiology to molecular level. *Mar. Biol.* **2011**, *158*, 551–560. [CrossRef] [PubMed]
33. Sciandra, A.; Harlay, J.; Lefèvre, D.; Lemée, R.; Rimmelin, P.; Denis, M.; Gattuso, J.P. Response of coccolithophorid *Emiliania huxleyi* to elevated partial pressure of CO_2 under nitrogen limitation. *Mar. Ecol. Prog. Ser.* **2003**, *261*, 111–122. [CrossRef]
34. Paulino, A.I.; Egge, J.K.; Larsen, A. Effects of increased atmospheric CO_2 on small and intermediate sized osmotrophs during a nutrient induced phytoplankton bloom. *Biogeosciences* **2008**, *5*, 739–748. [CrossRef]
35. Martínez, J.M. Molecular ecology of marine algal viruses. Ph.D. Thesis, University of Plymouth, Plymouth, UK, 2006.
36. Larsen, J.B.; Larsen, A.; Thyrhaug, G.; Bratbak, G.; Sandaa, R.A. Response of marine viral populations to a nutrient induced phytoplankton bloom at different *p*CO$_2$ levels. *Biogeosciences* **2008**, *5*, 523–533. [CrossRef]
37. Carreira, C.; Heldal, M.; Bratbak, G. Effect of increased *p*CO$_2$ on phytoplankton-virus interactions. *Biogeochemistry* **2013**, *114*, 391–397. [CrossRef]
38. Chen, S.; Gao, K. Viral attack exacerbates the susceptibility of a bloom-forming alga to ocean acidification. *Glob. Chang. Biol.* **2015**, *21*, 629–636. [CrossRef] [PubMed]
39. Traving, S.J.; Clokie, M.R.J.; Middelboe, M. Increased acidification has a profound effect on the interactions between the cyanobacterium *Synechococcus* sp. WH7803 and its viruses. *FEMS Microb. Ecol.* **2013**, *87*, 133–141. [CrossRef] [PubMed]

40. Coolen, M.J.L. 7000 years of *Emiliania huxleyi* viruses in the Black Sea. *Science* **2011**, *333*, 451–452. [CrossRef] [PubMed]
41. Nissimov, J.I.; Napier, J.A.; Allen, M.J.; Kimmance, S.A. Intragenus competition between *coccolithoviruses*: And insight on how a select few can come to dominate many. *Environ. Microbiol.* **2015**, *18*, 133–145. [CrossRef] [PubMed]

![viruses logo] *viruses*

MDPI

Article

Characterization and Temperature Dependence of Arctic *Micromonas polaris* Viruses

Douwe S. Maat [1,†], Tristan Biggs [1,†], Claire Evans [1,2], Judith D. L. van Bleijswijk [1], Nicole N. van der Wel [3], Bas E. Dutilh [4,5] and Corina P. D. Brussaard [1,*]

[1] Department of Marine Microbiology and Biogeochemistry, NIOZ Royal Netherlands Institute for Sea Research, and University of Utrecht, P.O. Box 59, 1790 AB Den Burg, Texel, The Netherlands; douwe.maat@nioz.nl (D.S.M.); tristan.biggs@nioz.nl (T.B.); clevans@noc.ac.uk (C.E.); judith.van.bleijswijk@nioz.nl (J.D.L.v.B.)
[2] Ocean Biogeochemistry & Ecosystems Research Group, National Oceanography Centre, Southampton, European Way, Southampton SO14 3ZH, UK
[3] Electron Microscopy Center Amsterdam, Department of Cell Biology and Histology, Academic Medical Center, University of Amsterdam, Meibergdreef 15, 1105 AZ Amsterdam, The Netherlands; n.n.vanderwel@amc.uva.nl
[4] Theoretical Biology and Bioinformatics, Utrecht University, 3584 CH Utrecht, The Netherlands; bedutilh@gmail.com
[5] Centre for Molecular and Biomolecular Informatics, Radboud University Medical Centre, 6525 GA Nijmegen, The Netherlands
*   Correspondence: corina.brussaard@nioz.nl; Tel.: +31-(0)222-369513
†   Joint first authors.

Academic Editor: Eric O. Freed
Received: 4 March 2017; Accepted: 25 May 2017; Published: 2 June 2017

Abstract: Global climate change-induced warming of the Artic seas is predicted to shift the phytoplankton community towards dominance of smaller-sized species due to global warming. Yet, little is known about their viral mortality agents despite the ecological importance of viruses regulating phytoplankton host dynamics and diversity. Here we report the isolation and basic characterization of four prasinoviruses infectious to the common Arctic picophytoplankter *Micromonas*. We furthermore assessed how temperature influenced viral infectivity and production. Phylogenetic analysis indicated that the putative double-stranded DNA (dsDNA) *Micromonas polaris* viruses (MpoVs) are prasinoviruses (Phycodnaviridae) of approximately 120 nm in particle size. One MpoV showed intrinsic differences to the other three viruses, i.e., larger genome size (205 ± 2 vs. 191 ± 3 Kb), broader host range, and longer latent period (39 vs. 18 h). Temperature increase shortened the latent periods (up to 50%), increased the burst size (up to 40%), and affected viral infectivity. However, the variability in response to temperature was high for the different viruses and host strains assessed, likely affecting the Arctic picoeukaryote community structure both in the short term (seasonal cycles) and long term (global warming).

Keywords: Arctic algal viruses; climate change; infectivity; *Micromonas* virus; prasinovirus; temperature; virus-host interactions

---

1. Introduction

Marine phycovirology, i.e., the study of viruses infecting marine eukaryotic algae, started with the lytic viruses infectious to the picophytoplankter *Micromonas pusilla* [1–5]. The genus *Micromonas* (class Mamiellophyceae) is ubiquitous, occurring from tropical to polar regions, and is readily infected by viruses [3,6–9]. The majority of *Micromonas* virus isolates belong to the double-stranded DNA (dsDNA) prasinoviruses [3–5,9], although a dsRNA *Micromonas* virus has also been reported [10,11].

The prasinoviruses are considered the most abundant group of marine phycodnaviruses [12] and virus abundances show synchrony with their hosts' temporal dynamics consistent with infection [13,14].

Micromonas is a globally important prasinophyte, which typically dominates the picophytoplankton fraction in marine Arctic waters [15–22]. Previous studies have shown that Arctic *Micromonas* forms a separate ecotype from lower latitude strains [16,21] adapted to grow at temperatures between 0 and 12 °C (with an optimum around 6–8 °C [16]). Considering Arctic sea surface temperature over the year to be in the range of −1 to a maximum 7 °C [23–25] and steadily increasing as a result of global warming (0.03–0.05 °C per year over the 21st century [24]), the *Micromonas* polar ecotype species (tentatively named *M. polaris*; [26]) can be expected to belong to the picophytoplankton predicted to benefit from a warming Arctic region [24,27–29]. Despite this predicted increase in abundance and relative share of picophytoplankton in the changing Arctic Ocean, it is still unclear how the viruses infecting the picophytoplankton are affected by changes in temperature. Little is known about Arctic phycoviruses in general, and to our knowledge, no viruses infectious to Arctic *Micromonas* species have yet been brought into culture [30–32].

Changes in an environmental variable, such as temperature, may directly affect virus infectivity and/or more indirectly impact virus proliferation due to alterations in the metabolic activity of the host [33]. Thus far the thermal stability of psychrophilic marine virus-host interactions has only been assessed for several phage-bacterium systems [34,35], despite the potential for special physiological adaptations by cold-adapted hosts and viruses [36,37]. It is likely that different viruses infecting the same host strain have distinct responses to shifting environmental factors and therefore environmental change may drive virus selection and host population dynamics. Nagasaki and Yamaguchi [38] found that the temperature ranges for successful infection were different for two virus strains infecting the raphidophyte *Heterosigma akashiwo* and that the host strain sensitivity to infection varied according to the temperature. Furthermore, temperature regulates growth by controlling cellular metabolic activity [39], which has been proportionally related to latent period length and burst sizes for *Vibrio natriegens* phages [40]. Recently, Demeroy and colleagues [41] demonstrated that temperature-regulated growth rates of *Micromonas* strains that originated from the English Channel were responsible for shortened latent periods and increased viral burst sizes upon infection. Ongoing change in the Arctic necessitates a better understanding of how Arctic phycoviruses are affected by temperature.

Here we report on the isolation of four *Micromonas* viruses from the Arctic. In addition to determining their viral characteristics (capsid morphology and size, genome type and size, latent period, phylogeny, host range, burst size, virion inactivation upon chloroform and freezing treatment), we investigated the impact of temperature change on virus infectivity and production. We hypothesize that (i) viral infectivity will increase with temperature, and (ii) increasing temperatures will stimulate virus production (shorter latent periods and higher burst sizes). For testing the latter hypothesis, we performed one-step virus growth experiments at a range of temperatures representative of the extremes over the polar growth season (0.5–7 °C) [23].

2. Materials and Methods

2.1. Isolation and Culturing

The *Micromonas* host TX-01 was isolated from Kongsfjorden, Spitsbergen, Norway (78°55.073′ N, 12°24.646′ E) on the 19 April 2014, by making an end-point, 10-fold dilution series of fjord water in F/4 medium (based on Whatman glass microfiber GF/F filtered, autoclaved fjord water; [42]). The other *Micromonas* species and strains used were obtained from the Bigelow National Center for Marine Algae and Microbiota (culture collection of marine phytoplankton (CCMP) coded strains; West Boothbay Harbor, ME, USA), the Culture Collection Marine Research Center of Göteborg University (LAC38; Göteborg, Sweden), and the Roscoff Culture Collection (RCC coded strains; Roscoff, France).

Micromonas TX-01 was classified based on its position in a Maximum-Likelihood dendrogram (Supplement Figure S1) of 18S rRNA sequences (1574 valid columns) of *Micromonas* strains with clade designations A–E after Slapeta et al. [43] and Ea after Lovejoy et al. [16]. Analysis was done using Randomized Axelerated Maximum Likelihood (RAxML) [44] implemented in the ARB software package [45]. *Micromonas* TX-01 (1051 Bp) was added to the tree using ARB Parsimony. Neighbor-Joining analysis gave a similar tree topology, whereby the tree was rooted using *Mantoniella squamata*. Primers 328F and 329R were used to amplify a part of the small subunit (SSU) ribosomal RNA gene of the *Micromonas* host according to Romari and Vaulot [46]. The same primers plus internal primer 528F were used for sequencing. Isolate TX-01 clustered in clade Ea which is composed of only Arctic clones [16]. Recently it has become clear that the genus *Micromonas* is not made up by solely *M. pusilla*, but instead consists of distinct genetic lineages and new species are described [6,26,47]. The strains in the Ea cluster are recently described as a new species of *Micromonas*, i.e., *M. polaris* [26], and with pending approval we consider TX-01 to be a putative *M. polaris* strain.

Micromonas species and strains (Table 1) were cultured in Mix-TX medium, a 1:1 mixture of f/2 medium [42] and artificial seawater [48] enriched with Tris-HCl and Na_2SeO_3 [3], under a light:dark cycle of 16:8 h. Light was supplied by 18W/965 OSRAM daylight spectrum fluorescent tubes (München, Germany) at intensities of 70–90 μmol quanta m^{-2} s^{-1}. Cultivation temperatures for the different *Micromonas* species and strains used for testing the host range of the virus isolates are listed in Table 1.

Table 1. Lytic activity of the four *Micromonas polaris* viruses (MpoV) against different *Micromonas* species and strains. Columns show from left to right: host strain code, origin of isolation, *Micromonas* species, culturing temperature, and the MpoV strain names. Grey cells mean that the virus from the column is able to infect and lyse the host from the row.

Host Code	Origin	*Micromonas*	Culture Temp.	Lytic Activity against *Micromonas*			
				MpoV-44T	MpoV-45T	MpoV-46T	MpoV-47T
TX-01	KF (2014)	*M. polaris*	3 °C	▓	▓	▓	▓
LAC 38	OFN (1998)	*M. commoda* [1]	3 °C	▓			
LAC 38	OFN (1998)	*M. commoda* [1]	15 °C	▓			
CCMP 1545	EC (1950)	*M. pusilla*	15 °C				
CCMP 2099	BB (1998)	*Micromonas* sp.	3 °C			▓	
RCC 461	EC (2001)	*M. pusilla*	15 °C				
RCC 834 *	EC (1950)	*M. pusilla*	20 °C				
RCC 2242	BzS (2009)	*M. polaris*	3 °C	▓		▓	▓
RCC 2246	BS (2009)	*M. polaris*	3 °C	▓		▓	▓
RCC 2257	BS (2009)	*M. polaris*	3 °C	▓		▓	▓
RCC 2258	BS (2009)	*M. polaris*	3 °C	▓		▓	▓
RCC 2306	BS (2009)	*M. polaris*	3 °C	▓		▓	▓
RCC 4298	GS (2014)	*M. polaris*	3 °C	▓		▓	
RCC 4778	GS (2014)	*M. polaris*	3 °C	▓		▓	
RCC 4779	GS (2014)	*M. polaris*	3 °C	▓		▓	

BB stands for Baffin Bay, BzS for Barents Sea, BS for Beaufort Sea, EC for English Channel, GS for Greenland Sea, KF for Kongsfjorden Spitsbergen, and OFN for Oslofjord Norway. [1] formerly known as *M. pusilla*; [47], * original CCMP1545.

The standard temperature at which the *M. polaris* strains used for the virus infection experiments were cultured was 3 °C. For investigating the effect of temperature on the viral growth cycle and virus infectivity, the host strains had been acclimated to various other temperatures (0.5, 2.5, 3.5, and 7 °C for TX-01; and 7 °C for RCC2257 and RCC2258) for several months prior to experimentation. Although the host strain LAC38 is not, the TX-01 and the RCC strains are obligate low-temperature strains, as they did not grow at 15 °C.

Four virus strains were isolated from the waters around Spitsbergen and were named MpoV as they infect Arctic *M. polaris* [26]. MpoV-44T was isolated during winter in 2006 using *Micromonas commoda* strain LAC38 (formerly known as *M. pusilla* [47]), and MpoV-45T to 47T during spring and summer in 2014 and 2015, respectively, using *M. polaris* TX-01 (Table 2). The reason that a low-temperature acclimated LAC38 culture was used for the isolation of MpoV-44T was due to the lack of available Arctic *Micromonas* host strains at the time of isolation. The lytic virus isolate MpoV-44T was isolated by adding whole seawater (15% *v/v*) to an exponentially growing culture of *M. commoda* LAC38 (acclimated to grow at 3 °C), and MpoV-45T, 46T, and 47T by adding 25% *v/v* 0.2 μm filtered (polyethersulfone membrane filter; Sartopore Midicap, Sartorius A.G. Goettingen, Germany) seawater to exponentially growing *M. polaris* TX-01 (standard culturing at 3 °C, but isolation was performed at 4 °C). MpoV-46T was the only one isolated from Storfjorden; the others came from Kongsfjorden (Table 2). Clearing of the infected algal cultures as compared to the non-infected control cultures was indicative of lysis. The lytic agents were confirmed as biological as the obtained lysates could be successfully propagated when 0.2 μm filtered, but not when autoclaved. The lysates were made clonal by end-point dilution (10-fold dilutions) and were maintained by regularly infecting exponentially growing host cultures with 10% *v/v* earlier produced lysates.

Table 2. Overview of the origin and basic characterization of the Arctic double-stranded DNA (dsDNA) virus isolates (MpoV-44T, 45T, 46T, and 47T) infecting *Micromonas*. Isolation coordinates Spitsbergen: Kongsfjorden 78°56′28.55″ N, 12°0′2.50″ E, Storfjorden: 77°37′35.26″ N, 20°46′3.74″ E.

MpoV Strain	Geographical Origin Spitsbergen	Date of Isolation	Host Strain of Isolation	Isolation Temperature (°C)	Genome Size (Kbp)	Lipid Membrane	Latent Period (h) *	Burst Size (Viruses Cell^{-1}) *
44T	Kongsfjorden	December 2006	LAC38	3	205 ± 2	+	30–51	267 ± 67
45T	Kongsfjorden	April 2014	TX-01	4	191 ± 2	+	16–24	296 ± 26
46T	Storfjorden	August 2015	TX-01	4	192 ± 3	+	16–24	233 ± 7
47T	Kongsfjorden	June 2014	TX-01	4	190 ± 6	+	16–24	256 ± 13

* Tested on *M. polaris* TX-01 at 3 °C.

2.2. Virus Growth Characteristics

To obtain (comparative) information about the latent period and viral burst size of the four MpoVs isolated, viral growth experiments were performed in triplicate at 3 °C in 100 mL Erlenmeyer flasks with an exponentially growing algal host culture of TX-01 and freshly made 0.2 μm filtered (polyethersulfone membrane filter; Sartopore Midicap, Sartorius A.G. Goettingen, Germany) viral lysates. Culture conditions were as described above for host culturing. The virus to host ratio was 10–60:1 (on average 25 ± 12), at all times sufficient to allow one-step viral growth curves. The host strain TX-01 was chosen as a model host system because it was indigenous to this Arctic region and isolated together with three of the four new MpoVs. Growth medium equal to the volume of the lysate was added to non-infected control cultures (in triplicate). Host cell and viral abundances were sampled every 6–24 h post infection (p.i.) at in situ temperatures. Flow cytometry was used to enumerate algae in unpreserved samples, which were kept chilled until analysis whereas samples for virus enumeration were fixed immediately after sampling.

Algal samples were counted using a Becton Dickinson (Becton Dickinson, Franklin lakes, NJ, USA) FACSCalibur benchtop flow cytometer (equipped with a 488 nm argon laser), with the trigger set on red chlorophyll autofluorescence [49]. Viral abundances were determined on fixed samples (final concentration 0.5% glutaraldehyde, EM-grade; Sigma-Aldrich, St. Louis, MO, USA) that were snap-frozen in liquid nitrogen and stored at −80 °C until analysis. Thawed virus samples were diluted in TE buffer (10 mM Tris-Base, 1 mM EDTA, pH 8.0), stained with the nucleic acid-specific green fluorescent dye SYBR Green-I (Invitrogen, Thermo Fisher, Waltham, MA, USA) and analyzed according to Brussaard [50]. *Micromonas* virus clusters were discriminated by a higher green fluorescence (similar to [50], and [9]).

2.3. Host Range

A range of *Micromonas* species and *M. polaris* strains were tested for susceptibility to infection by the four MpoVs (Table 1). Five hundred microliters of viral lysate was added to 4.5 mL of exponentially growing host, after which the lysis of the culture was monitored by visual inspection (clearing of the culture compared to non-infected control cultures). Cultures which had not lysed after 3 weeks were considered resistant to the lytic Arctic MpoVs. Lysed cultures were screened for virus production using flow cytometry.

2.4. Ultrastructure Analysis by Transmission Electron Microscopy (TEM)

For ultrastructural analysis by TEM, thin sectioned samples were prepared. Briefly, exponentially growing algal cells of *M. polaris* RCC2258 were infected with the respective virus, after which samples (3–6 tubes of 15 mL per sampling point) were taken at several time points within the latent period. Samples were prefixed with glutaraldehyde (EM-grade; 0.5% final concentration) for 30 min on ice. Algal cells were concentrated by low-speed centrifugation (3200× g, 10 min, 4 °C), after which the supernatant was decanted and the pellets were transferred to 1.5 mL Eppendorf tubes (three per tube) using a Pasteur pipet. These samples were further concentrated by centrifugation (3200× g, 10 min), followed by the transfer of two of the pellets into one Eppendorf tube and another round of centrifugation (3200× g, 10 min). Finally, these samples were fixed with glutaraldehyde (EM-grade; 2% final concentration) in 1 mL citrate-phosphate buffer (0.1 M $Na_2HPO_4 \cdot 12\ H_2O$, 9.7 mM citric acid, pH 7.2) containing 2.5 mM $CaCl_2$ on ice.

After fixation, the algae were washed in distilled water, osmicated for 60 min in 1% OsO_4 in water, and washed again in distilled water with centrifugation steps in between to spin down the algae. After the last spin down, the supernatant was removed and the algae were re-suspended in the remaining volume. An equal volume of 12% gelatin was added to the algal sample and centrifuged again to a non-compact pellet. The gelatin was solidified on ice and after 20 min a fixative (2% glutaraldehyde) was added to let the gelatin fixate overnight. The gelatin containing the algae was cut into small blocks of 1–2 mm² and dehydrated through a series of ethanols (70%, 80%, 90%, 96%). As a last dehydration step, propylene oxide was used before the samples were embedded in LX-112 resin. After polymerization at 60 °C, ultrathin sections of 90 nm were cut on a Reichert EM UC6 with a diamond knife, collected on Formvar coated grids and stained with uranyl acetate and lead citrate. Sections were examined with a FEI Tecnai-12 G2 Spirit Biotwin electron microscope (Fei, Eindhoven, The Netherlands), and images were taken with a Veleta camera using Radius software (EMSIS, Münster, Germany). The data analysis program within Radius was used to perform measurements of the virus particle size. The capsid diameter was measured for 100–400 virus particles per infection, discriminating intracellular and extracellular particles.

2.5. Sensitivity to Chloroform

Recently, a new group of *Micromonas* viruses has been reported to possess a lipid membrane [9]. To test for the presence of a viral lipid membrane in these Arctic strains, fresh viral lysates were exposed to chloroform. This organic solvent is an effective indicator of inner- and outer-viral lipid membranes [51,52]. Aliquots of 1 mL were incubated in 10% and 50% (v/v) chloroform for 10 min, after which the chloroform was separated by centrifugation (4000× g, 5 min) and the aqueous phase containing the viruses was recovered (in new 1.5 mL Eppendorf tubes). Tubes were left overnight at 3 °C to allow any remaining chloroform to evaporate. Treated lysates were added to exponentially growing cultures in 5 mL borosilicate tubes (10% v/v final concentration; total volume 5 mL) and incubated at standard light conditions and 3 °C. Non-infected negative controls received the same volume of media. Tubes were screened for lysis twice a week for three weeks.

2.6. Genome Size

MpoV lysates (~25 mL) were partially purified from cell debris and bacteria by centrifugation at 10,000× g for 30 min at 4 °C using a fixed angle rotor (type F 34-6-38) with conical adapters to fit the 30 mL Nalgene Oak Ridge centrifuge tubes in a Eppendorf 5810R centrifuge (Hamburg, Germany). Viral genome sizes were determined by Pulse Field Gel Electrophoresis (PFGE) according to Baudoux and Brussaard [53]. In short, the clarified supernatant was decanted and viral particles were concentrated by ultracentrifugation ($184,000 \times g$ for 2 h at 8 °C, using a fixed-angle rotor Beckman Coulter type 50.2Ti, in a Beckman Coulter Optima XPN-80 ultracentrifuge) (Pasadena, CA, USA). Pellets were resuspended in 150 µL SM buffer (0.1 M NaCl, 8 mM $MgSO_4 \cdot 7 H_2O$, 50 mM Tris-HCl, 0.0005% (*w/v*) glycerin), after which agarose plugs were prepared by mixing equal volumes of molten 1.5% (*w/v*) agarose (InCert; Lonza Group Ltd., Basel, Switzerland) with the virus concentrate in plastic molds. Plugs were incubated overnight at 30 °C in lysis buffer with proteinase K, followed by washing in TE buffer (10:1, pH 8.0) and storage in TE buffer (20:50, pH 8.0) at 4 °C until analysis. Plugged samples were loaded onto 1% SeaKem GTG agarose gels (InCert; Lonza Group Ltd., Basel, Switzerland) prepared in $1 \times$ TBE gel buffer (90 mM Tris-Borate and 1 mM EDTA, pH 8.0) and run in a PFGE Bio-Rad CHEF DR-II cell unit (Bio-Rad, Hercules, CA, USA), and corresponding CHEF DR-II chiller system, filled with 2 L $0.5 \times$ TBE buffer (45 mM Tris-Borate and 0.5 mM EDTA, pH 8.0), pre-cooled at 15 °C. Plugs were loaded with $0.5 \times$ TBE buffer (45 mM Tris-Borate and 0.5 mM EDTA, pH 8.0), at 6 V cm^{-1} with pulse ramps of 20 to 45 s at 14 °C for 22 h. Molecular size markers were included: DNA Lambda ladder plugs (Bio-Rad) and *Saccharomyces cerevisiae* DNA ladder plugs (Bio-Rad). Gels were visualized in a FluorS imager (Bio-Rad Instrument) after staining with SYBR Green I (1×10^4 of commercial solution, Invitrogen). Viral genome sizes were estimated in comparison to a molecular size marker ($n \geq 2$). Mean $\pm$ standard deviation were determined and the differences were tested by ANOVA (significance level $p = 0.05$) and Holm-Šidák multiple comparisons.

2.7. Virus Phylogeny

To determine the phylogenetic relationship between our new Arctic virus isolates and other *Micromonas* viruses, we amplified a part of the DNA polymerase B gene (*polB*) using the primer pair AVS1/AVS2 [54]. The viral lysate was diluted 1:5 in ultrapure water and sonicated (MSE Soniprep 150, London, UK) at an amplitude of 8 µm for 3×10 s with intervals of 30 s cooling on ice. Ten microliters of sonicated viral lysate was used as a template in a 50 µL PCR reaction containing 4.0 U of BiothermPlus DNA Polymerase (GeneCraft, Lüdinghausen, Germany), $1 \times$ buffer (including 1.5 mM $MgCl_2$), 0.25 mM of each dNTP, 0.8 µM of each primer, and 0.4 mg/mL BSA. Negative controls contained all reagents except the template. PCR cycling included an initial denaturation at 94 °C (4 min) followed by 37 cycles of denaturation at 94 °C (30 s), annealing at 45 °C (30 s), and extension at 72 °C (1 min), followed by a final extension at 72 °C (7 min). Sequencing was performed by BaseClear Ltd. (Leiden, The Netherlands). Based on 178 amino acid positions, a Maximum-Likelihood dendrogram was constructed with RAxML [44] implemented in ARB software [45]. The tree was rooted using *polB* sequences of *Bathycoccus* viruses (HM004432, FJ267515, FJ267518, KF501013, MEHZ011588827). These, and the *polB* sequences of *Ostreococcus* viruses (FJ267496, FJ267500, FJ267508, JN225873) were grouped to obtain a more compact tree. For comparison we added published *polB* sequences of the *Micromonas* virus isolates and contigs of an Arctic metagenome [22] that showed exact overlap with the *polB* fragment that we analyzed. Contig-95-10186 and contig-79-31207 were shorter (127 and 96 amino acid positions, respectively) and were later added to the dendrogram via ARB Parsimony.

2.8. Thermal Stability

We studied the effect of different temperatures on virus growth characteristics as well as on virus stability (loss of infectivity). To determine how the virus-host interaction might be affected over a range of different ecologically relevant temperatures, we used our model system of TX-01 with MpoV-45T, as both the host and virus were isolated in the same location and same period. Temperature sensitivity of the viral latent period and burst size were examined by one-step viral growth experiments at a range of growth temperatures, i.e., 0.5, 2.5, 3.5, and 7 °C. This range of temperatures represents natural water temperatures during the Arctic growth season [23–25]. Additionally, to test whether there are any species- and/or strain-specific responses to temperature, we furthermore tested both MpoV-44T and 45T on RCC 2257 and RCC2258 at 3 and 7 °C, whereby 3 °C represents the spring sea surface temperature around Spitsbergen (origin of TX-01) and southern Beaufort Sea (origin of RCC2257 and RCC2258; [55] and 7 °C represents the maximum Arctic summer temperatures (e.g., [23,56]). MpoV-44T and 45T were chosen as representative virus model systems because of their different viral growth characteristics with both being isolated from Kongsfjorden. Algal host and virus samples were taken regularly (every 6–8 h in the first 24 h and every 12–24 h for the rest of the experiment) and analyzed by flow cytometry as described above [49,50]. Other culture conditions were the same as for the virus growth experiments described above.

Viral infectivity was determined after exposure at −196, −80, −20, 0, 3, 7, and 15 °C and determined using the Most Probable Number (MPN) assay on the host strain RCC2258. This host strain was used instead of TX-01, because the latter did not grow well in the 5 mL tubes that we used for the assays (the large amount of dilutions and replicates did not allow the use of larger tubes or flasks). Aliquots of virus lysates (3 mL) were exposed to the different temperatures for 24 h when the exposure temperature was below zero and for 1 h when exposure temperature was above zero. Following exposure, viral lysates were added to the algal host ($n = 5$; 12×10-fold dilutions). In each MPN rack one additional row of tubes containing non-infected culture (also 5 mL per tube) served as a negative control. The MPN cultures were incubated at 3 °C under standard light conditions and were inspected at least once a week for 3 weeks for lysis. The titers were determined with the MPN Assay Analyzer [57] and data were normalized to the highest value, i.e., 9.8×10^8, 4.6×10^9, 2.1×10^1, and 4.6×10^9 mL^{-1} for MpoV-44T, 45T, 46T, and 47T, respectively. Statistics were carried out in SigmaPlot 13.0 (Systat Software Inc., Chicago, Il, USA). Differences between the viruses and temperature treatments were tested by ANOVA ($n = 3$, significance level $p = 0.05$) and Holm-Šidák multiple comparisons, either directly or after log transformation.

2.9. Diversity and Abundance in Metagenomes

To assess the diversity of MpoV in diverse marine environments, we searched the contigs generated by the Tara Oceans consortium [58], as well as KEGG Environmental sequences for MpoV homologs using blastn [59]. All hits had an E-value < 10^{-30}. Hit regions were excised from the contigs, aligned with the four MpoV sequences using Clustal Omega 1.2.0 [60], and converted into a phylogenetic tree using PhyML 3.0.1 [61] with the HKY85 model of substitution; four discrete gamma categories; shape parameter: 1.074; invariant proportion: 0.455; and transition/transversion ratio: 4.540. Finally, we assessed the ubiquity and abundance of all the MpoV-related sequences in the Tara Oceans samples [58]. Abundance was determined by mapping 2.5 billion metagenomic sequencing reads from 26 Tara Oceans metagenomes to the contig fragments using Burrows-Wheeler Aligner (BWA-MEM algorithm) with default parameters [62], with the number of mapped reads reflecting the relative abundance in the original samples. The IDs of the 26 screened metagenomes were: ERR594313.1, ERR598949.1, ERR598972.1, ERR598982.1, ERR599023.1, ERR599039.1, ERR599095.1, ERR594320.1, ERR598950.1, ERR598976.1, ERR598994.1, ERR599025.1, ERR599053.1, ERR599122.1, ERR594324.1, ERR598962.1, ERR598977.1, ERR599001.1, ERR599027.1, ERR599068.1, ERR594325.1, ERR598966.1, ERR598979.1, ERR599007.1, ERR599035.1, and ERR599078.1.

3. Results

3.1. Basic Virus Characteristics

The transmission electron microscope analysis of the four Arctic *Micromonas* viruses revealed virus-like particles in the cytoplasm of the host with a hexagonal shape (icosahedral symmetry) and a thick outer layer surrounding an electron-dense inner core (Figure 1). We detected no significant difference between the diameter of the different isolates nor for the intra- versus extracellular virus particles. The diameters of the virus particles were 119 ± 8, 121 ± 14, 120 ± 12, and 119 ± 9 nm respectively for MpoV-44T, 45T, 46T, and 47T. Furthermore, chloroform treatment revealed that all viruses lost their infectivity upon treatment with chloroform, indicative for the presence of a lipid membrane.

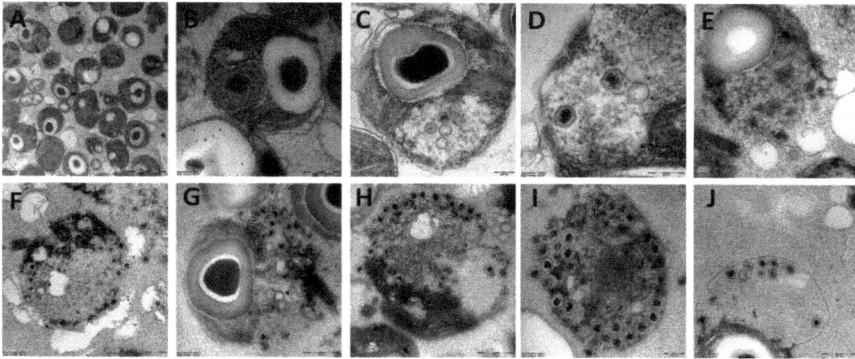

Figure 1. Transmission electron micrographs of thin sections of the uninfected *Micromonas* strain TX-01 (**A,B**), and infected with virus MpoV-44T (**C,D**), 45T (**E**), 46T (**F–H**), and 47T (**I,J**). Scale bar represents 200 nm (**A,C,D,I**) or 500 nm (**B,E–H,J**).

Cytograms of the viruses stained with the nucleic acid-specific dye SYBR Green I showed high green fluorescence signatures, similar to other known dsDNA *Micromonas* viruses (Supplement Figure S2; [9,50]). The dsDNA nature of the MpoV genomes was confirmed by the positive results from the PCR amplification of the partial DNA polymerase B gene (*polB*) using AVS1/AVS2 primers that were originally designed for dsDNA prasinoviruses [54]. The *polB* phylogeny based on inferred amino acid sequences indeed grouped the newly isolated MpoVs with other *Micromonas* viruses, but did not show a close match (i.e., >13 amino acid substitutions) to any of the other Arctic sequences (Figure 2). MpoV-45T and MpoV-47T were highly similar to each other (1 amino acid difference), but clearly distinct from MpoV-46T and 44T (>31 amino acids difference). MpoV-44T differed from MpoV-46T in 17 amino acid positions. The viral genome sizes of MpoV-45T, 46T, and 47T, (191 ± 3 Kb estimated by PFGE) were not significantly different from each other (one-way ANOVA $p > 0.818$, $n = 4$, 2, and 2, respectively) but were significantly smaller ($p < 0.008$) than the genome of MpoV-44T which displayed a genome size of 205 ± 2 Kb ($n = 3$; Table 2, Supplement Figure S4).

The virus isolates were specific for the genus *Micromonas* (Supplement Table S1), but were not species-specific. For example, MpoV-44T was isolated on *M. commoda* LAC38, but also infected TX-01, *M. pusilla*, and *M. polaris* strains (Roscoff Culture Collection; Table 1). Besides the host strain TX-01, RCC2257 and 2258 were also sensitive to infection by all four MpoVs. MpoV-44T displayed the broadest host range and was the only virus that could infect *Micromonas* strains LAC38, CCMP1545, RCC461, and 834 that grow at higher temperatures (8, 15, or 20 °C). No relationship could be established between virus infectivity or host susceptibility to infection based on the time of isolation, geographical origin, or host culture temperature.

One-step infection was observed for all the MpoV lytic virus growth cycles except for MpoV-44T when propagated on TX-01 (Figure 3A). TX-01 kept growing for the two days following virus addition to a higher extent than commonly observed [63,64]. However, this was not observed for other host strains, e.g., RCC2257, RCC2258, and LAC38 (Supplement Figure S3). MpoV45T, 46T, and 47T showed similar infection dynamics and latent periods, i.e., 16–24 h with a median of 18 h (Figure 3B; Table 2). The median latent period for MpoV-44T on host strain TX-01 was, at 39 h, twice as long. Viral burst sizes did not differ significantly for the four MpoVs on TX-01 (one-way ANOVA; $p = 0.317$) and the averages varied between 233 and 296 viruses produced per lysed host cell (Table 2).

Testing the sensitivity of the four Arctic MpoVs to temperature showed that all viruses lost most of their infectivity after 24 h exposure to temperatures < −20 °C (Figure 4). After a −20 °C treatment, only 1% of MpoV-44T remained infective while the other viruses retained over 25% of their infectivity when compared to the treatment at 3 °C. Moreover, the variability between the viruses in response to non-freezing temperatures was high. MpoV-45T and 46T displayed relatively narrow tolerance ranges, with maximal infectivity after the 0 °C and 3 °C treatments, respectively. MpoV-44T and 47T retained infectivity at the highest temperature tested (7 °C), with MpoV-47T being the least sensitive to temperature (consistent infectivity at 0–7 °C).

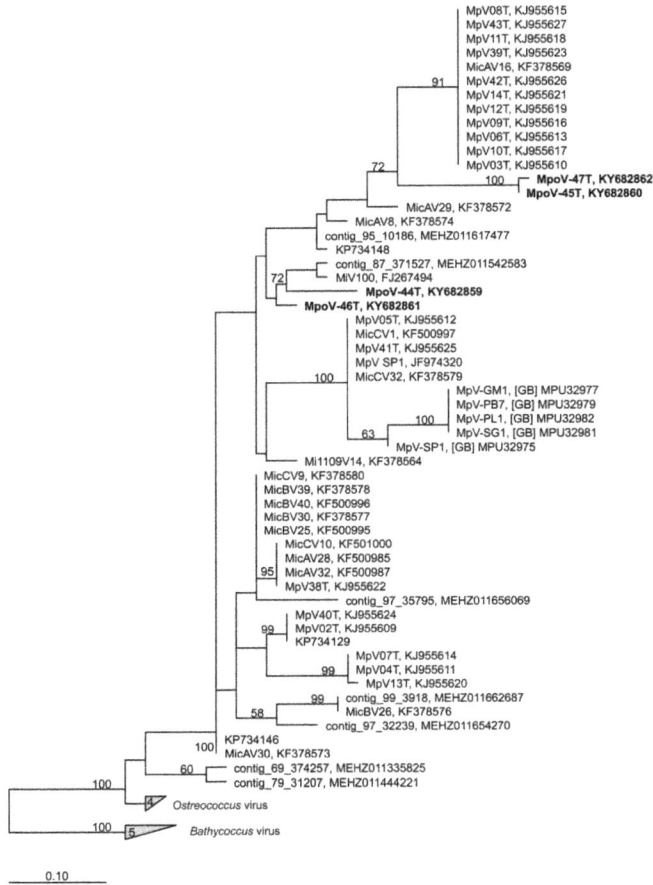

Figure 2. Position of the four *Micromonas polaris* viruses (MpoVs) (in bold) in a maximum likelihood dendrogram (100 bootstrap replicates), based on a multiple alignment of 178 amino acid positions of DNA polymerase B (*polB*). Only nodes with bootstrap values >50% are displayed. Virus strains and accession numbers are indicated. "Contigs" are *polB* sequences extracted from an Arctic marine metagenome [22]. The tree was rooted using *polB* sequences of *Bathycoccus* viruses.

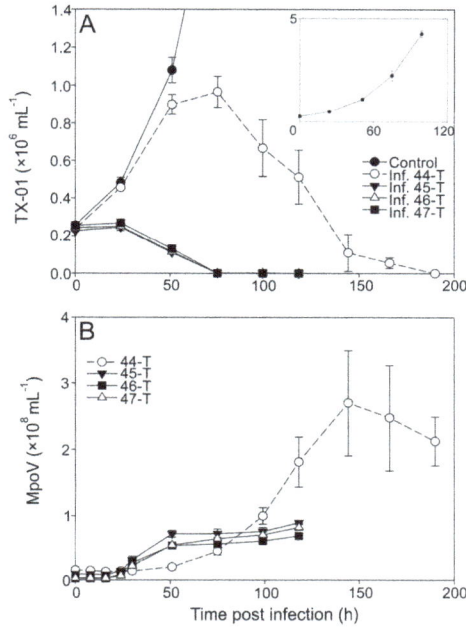

Figure 3. Abundances of *Micromonas* cells ($\times 10^6$ mL^{-1}) and viruses MpoV-44T, 45T, 46T, and 47T ($\times 10^8$ mL^{-1}) infecting host strain TX-01. Panel (**A**) shows the algal abundances (mean $\pm$ standard deviation (S.D.) over time, with the filled circles depicting the non-infected control cultures, open circles depicting the cultures infected with MpoV-44T, filled triangles depicting the ones infected with MpoV-45T, closed triangles depicting the ones infected with MpoV-46T, and the filled squares depicting the ones infected with MpoV-47T. The inlay panel shows the growth of the non-infected controls in detail. Panel (**B**) shows the viral abundances (mean $\pm$ S.D.) over time, with the symbols corresponding to panel (**A**), i.e., each virus is depicted by the same symbol as the culture it infected.

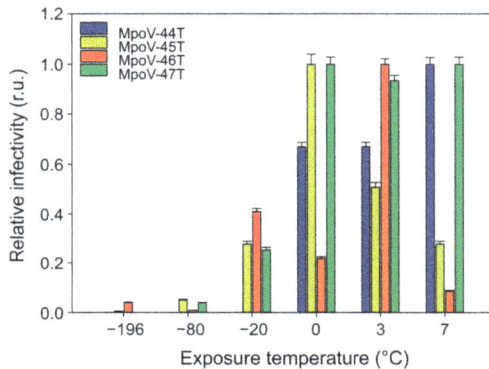

Figure 4. Effects of temperature exposure on the infectivity of MpoV-44T, 45T, 46T, and 47T (actual infection assay performed at 3 °C). The *x*-axis depicts the exposure temperature and the *y*-axis depicts the relative infectivity (normalized to highest infectivity) of the virus as determined by the most probable number (MPN) dilution assay. r.u. stands for relative units. Error bars show standard error ($n = 5$).

3.2. Temperature Dependent Virus Production

The lysis dynamics of the MpoV-45T infecting host TX-01 was similar for all four temperatures tested (0.5, 2.5, 3.5, 7 °C; Figure 5A) despite increasing exponential growth rates of the host (0.40 ± 0.05, 0.49 ± 0.06, 0.66 ± 0.01, 0.85 ± 0.02 d^{-1}, respectively). The latent period of MpoV-45T did not change with temperature (16–24 h; Figure 5B), but the viral burst sizes did show significant differences between 0.5, 2.5, and 3.5 °C (one-way ANOVA; $0.001 < p < 0.019$) and declined with lower temperatures by 15% and 28% for 2.5 and 0.5 °C, respectively, compared to 3.5 °C (Figure 5C).

Assessing the other virus-host combinations for temperature sensitivity (7 °C compared to 3 °C), the virus growth characteristics revealed host-specific effects. MpoV-45T showed a shorter latent period at higher temperature when RCC2257 was the host (from 12–18 h to 6–12 h), whereas no such effects were observed on hosts TX-01 and RCC2258; Figure 6A). Moreover, while TX-01 (infected with MpoV45T) did not show an increase in burst size from 3.5 to 7.0 °C (but did from 2.5 to 7.0 °C; see above), RCC2257 and RCC2258 showed increased burst sizes at 7 °C by respectively 150% and 140% (one-way ANOVAs; $p < 0.045$; Figure 6B, Supplement Table S2).

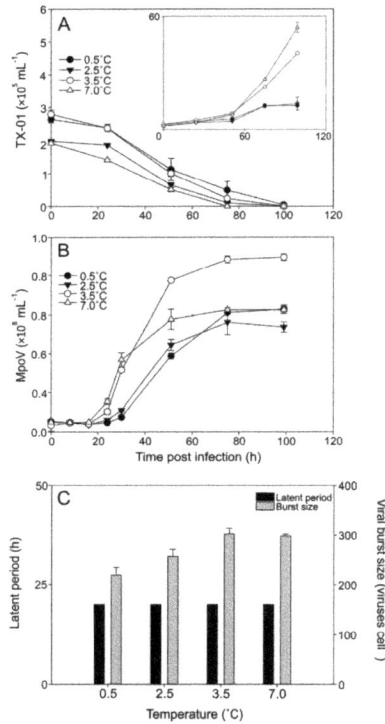

Figure 5. Abundances of *Micromonas* strain TX-01 ($\times 10^5$ mL^{-1}) and virus MpoV-45T ($\times 10^6$ mL^{-1}) tested at 0.5, 2.5, 3.5, and 7.0 °C. Panel (**A**) shows the algal abundances (mean ± S.D.) over time, with filled circles representing 0.5 °C, filled triangles representing 2.5 °C, open circles representing 3.5 °C, and open triangles representing 7.0 °C. The inlay panel shows the growth of the non-infected controls. Panel (**B**) shows the viral abundances (mean ± S.D.) over time, with the symbols corresponding to panel A, i.e., each virus is depicted by the same symbol as the culture it infected. Panel (**C**) depicts the median viral latent periods (black bars; determined with an 8 h sampling resolution) and viral burst sizes (grey bars; mean ± S.D.).

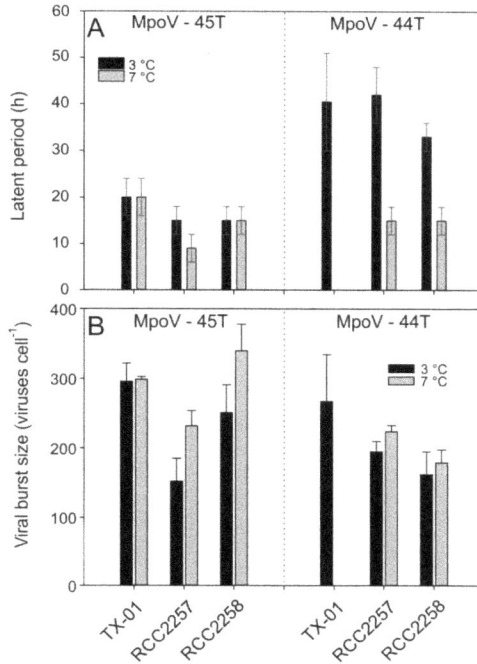

Figure 6. Median latent periods (**A**) and mean burst sizes (**B**) of MpoV-45T (left panels) and MpoV-44T (right panels) infecting host TX-01, RCC2257, and RCC2258 at 3 °C (black bars) and 7 °C (grey bars). Note that the TX-01 data are from the same as in Figure 5. The range bars in panel A depict the actual time interval on which the latent period is based. The error bars in panel B depict the standard deviation (S.D.). Statistical analysis of inter- and intra-strain differences are depicted in Supplement Table S2.

There were also virus-specific responses to temperature, as an increase from 3 °C to 7 °C showed a stronger effect on the latent periods of MpoV-44T rather than that of MpoV-45T, reducing it by roughly 50% from >30 h to 12–18 h on both hosts. Moreover, the viral burst sizes of MpoV-44T showed only a significant increase at 7 °C on host RCC2257 (one-way ANOVA; $p = 0.044$), which was also smaller than that for MpoV-45T (115% compared to 150%). Irrespective of temperature, on host RCC2258 the viral burst sizes of MpoV-45T were higher than those of MpoV-44T, but for host RCC2257 no such difference was observed (two-way ANOVA; two- $p < 0.001$ and $p < 0.916$, respectively).

3.3. Diversity and Abundance in Metagenomes

Finally, we assessed the diversity and abundance in the metagenomes by screening sequence databases for homologs of the amplified nucleotide region. The phylogenetic tree in Figure 7 shows that MpoVs are part of a family of viruses whose sequences were previously detected in marine samples from around the world from various studies including Tara Oceans [58] and the Ocean Sampling Day.

Figure 7. Unrooted maximum likelihood phylogeny of MpoV-related sequences from various studies, and their abundance in environmental metagenomes. Abundance is expressed as the total number of aligned reads out of 2.5 billion reads in 26 Tara Oceans datasets.

4. Discussion

To our knowledge this is the first report of the isolation and characterization of phycoviruses from polar marine waters. Similar to other reports of *Micromonas* virus isolates, and consistent with the *Phycodnaviridae*, the virus-like particles accumulated in the cytoplasm of the host cells [9]. The particles of the four MpoV were morphologically similar (no significant variance in virus particle size, i.e., on average 120 nm), and all contained a lipid membrane (sensitive to chloroform). Lipid-containing MpoVs were first reported by Martínez Martínez and colleagues [9]. These authors were able to clearly and convincingly divide nineteen newly isolated *Micromonas* viruses, across an area spanning the North Sea to the Mediterranean Sea, into two groups based on (i) their sensitivity to infection of LAC38 or CCMP1545, (ii) genome size (206 ± 6 Kb ($n = 12$) vs. 191 ± 4 Kb ($n = 8$)), and (iii) presence of a lipid membrane. Strikingly, all LAC38-infecting viruses with larger genomes contained a lipid membrane, whereas the smaller genome sized CCMP1545-infecting ones did not. Our data show a similar larger genome size for MpoV-44T which infects LAC38 (205 ± 2 Kb, in contrast to the 191 ± 3 Kb genomes of the other MpoVs), but in our case all MpoVs contained a lipid membrane.

Molecular phylogeny inferred from the amino acid sequences of the DNA polymerase gene B fragments established that the four Arctic MpoV isolates grouped distinctly with the other dsDNA *Micromonas* viruses belonging to the genus Prasinovirus. The genus Prasinovirus infects *Ostreococcus* and *Micromonas* species and is one of the six virus genera belonging to the *Phycodnaviridae* family; eukaryotic algal viruses with large dsDNA genomes (100–560 Kbp) [65,66]. A recent metagenomic survey (Tara Ocean Expedition) revealed that prasinoviruses are the most abundant group of phycodnaviruses in the oceans [12]. The newly isolated MpoVs did not group together; instead MpoV-44T and 46T were phylogenetically distinct, both from each other and from MpoV-45T and 47T. Furthermore, it is clear from the phylogenic analysis based on *polB* that the Arctic *Micromonas* viruses do not form a separate cluster. Screening the Tara Oceans' contigs and the KEGG Environmental database for homologs of the amplified nucleotide region of our MpoV isolates revealed a worldwide distribution and high diversity on thermal stability (i.e., from the Greenland Sea to the temperate regions to Antarctica and in waters from −1.6 to 17.3 °C, Table S3). These results confirm that the *Micromonas* viruses and their relatives are globally dispersed, show a high degree of genotypic diversity, and are ecologically relevant.

There was no general relationship between the phylogenies of the virus and host strains as revealed by Martínez Martínez et al. [9] for viruses infecting temperate *Micromonas* strains, but MpoV-44T could be distinguished from the other Arctic MpoV isolates based on its capability to virally infect *M. commoda* LAC38. Although LAC38 was being cultured at low temperature (3 °C) at the time of virus isolation, it is originally a temperate *Micromonoas* strain (Baltic Sea) [67]. We cannot exclude that the isolation of MpoV-44T on a different host is underlying its intrinsic differences to the other MpoVs isolated 8–9 years later using a local Arctic *Micromonas* host strain. These differences may also be due to MpoV-44T having been isolated during midwinter and years before the other MpoVs (isolated in summertime during two consecutive years). Successional patterns for marine virus communities with associations to temperature and host dynamics have been demonstrated (e.g., [68]). Even though Arctic *Micromonas* still grows well at low temperatures (0.4 d^{-1} at 0.5 °C, this study) and low light (0.2 d^{-1}, [16]), photosynthesis may not be possible during part of the Arctic winter. Several *Micromonas* species however exhibit phagotrophy (e.g., *M. polaris* CCMP2099; [69]) that could serve as an alternative energy source to maintain growth and/or virus production during the winter (see also [22]). Yet, our study shows that MpoV-44T is well adapted to relatively fast and high production of infective progeny at relatively higher temperatures, which makes it more likely that advection of relatively warm Atlantic water from the West Spitsbergen Current (WSC) [70] was responsible for being able to isolate MpoV-44T in winter. Water temperature in autumn of the year of isolation was in fact higher than the average of the preceding years [71]. Additionally, the ability of MpoV-44T to successfully infect *Micromonas* strains growing at higher temperatures up to 20 °C seems indicative that this specific virus has a high temperature tolerance. Still, a relatively high temperature optimum for a virus occurring in a cold environment could theoretically be an adaptation to be less virulent in order to avoid extinction of the host [14,72]. The relatively long latent periods and reduced infectivity of MpoV-44T at low temperatures would effectuate such low virulence for the slow growing hosts during the winter months. Then in the following more productive season, when the host growth rates increase, the latent period of MpoV-44T shortens and burst sizes increase to be able to keep in sync with host growth. Intriguingly, MpoV-47T also displayed thermostability, with an infectivity optimum at 7 °C. MpoV-47T was isolated only 2 months later than the temperature sensitive MpoV-45T (infectivity optimum at 0 °C), indicative of a high degree of diversity of virus thermostability in Arctic waters.

At temperatures above zero, the infectivity data do not confirm our first hypothesis that MpoV infectivity increases with temperature. The viral response to higher temperatures (0–7 °C) was highly variable and strain-specific. Two of the four virus isolates (MpoV-45 and 46T) even showed a loss of infectivity above 0 °C. All of the virus isolates, except for MpoV-44T, were able to maintain over 25% of their infectivity after being frozen at −20 °C. This suggests that these viruses are well able to withstand the freezing process during ice formation, a property which would maintain high titers during winter. Similar findings have been reported for Arctic marine bacteriophages [73] and dsDNA algal viruses during winter in a seasonally frozen pond [74]. Cottrell and Suttle [5] reported relatively high decay rates for MpV-SP1, however, this virus strain originated from subtropical waters and the decay rates were largely determined by sunlight (UV intensity). There is limited knowledge of dsDNA algal virus thermal stability and the mechanisms underlying the loss of infectivity have not been elucidated [9,41,53,75–77]. Only a few cold-active viruses (i.e., viruses that successfully infect hosts at 4 °C or below) have been brought into culture and all are bacteriophages [35,78,79]. Variability in MpoVs' temperature sensitivity demonstrates a specificity of infection efficiency related to temperature. Hypothesizing that our data are generally applicable, seasonal temperature shifts could regulate *Micromonas* host and virus succession. The infectivity loss at higher temperature (7 °C) for MpoV-45T and MpoV-46T shortens their window of optimal activity during the warmer summer months. Considering that the range of temperatures we tested are ecologically relevant for the Artic seas (water temperatures between −1 and 7 °C; [23]), our results indicate the need to determine the causal processes such as the means of virus entry and conformational changes in the virus particle (e.g., viral capsid proteins and lipid membrane properties).

When propagated on the putative *M. polaris* strain TX-01, MpoV-44T displayed a much longer median latent period than the other Arctic MpoVs (39 vs. 18 h, respectively). A comparably long latent period had, until recently, only been described for the dsRNA virus MpRV infecting *M. commoda* LAC38 (36 h; [10]). However, Baudoux and colleagues [14] reported a latent period of 27–31 h for a dsDNA virus infecting *Micromonas* isolates from the English Channel growing at 20 °C. The latent period of MpoV-44T displayed a strong temperature-dependence, i.e., with a temperature increase of 4 °C, the time of viral release decreased by >15 h (latent period 12–18 h; approximately 50% of the latent period at 3 °C). Although Baudoux et al. [14] did not find a correlation between differences in latent period (or burst size) with the host growth rates for the isolated MicVs, Demory et al. [41] reported for the virus-host model system Mic-B/MicV-B (virus infecting largely Clade B strains) an inverse relationship of the viral latent period with the host growth rate (whereby the growth rates were affected by the host culture temperature). We did not find such a relationship with growth rate for the latent periods of MpoV-44T growing on TX-01, but did find a similar significant linear relationship when infecting host RCC2258 (r^2 = 0.952, p = 0.018). Virus MpoV-45T did not show a dependency on host growth rate (TX-01, RCC2258, and RCC2257), but instead showed a shortened latent period on host RCC2257 at the highest temperature (7 °C compared to 3 °C). Furthermore, the latent period of MpoV-44T was strongly affected by temperature whereas increasing temperature only shortened the MpoV-45T latent period on RCC2257. On host TX-01, the latent period was unaffected over the whole range of 0.5 to 7 °C, however, we found a steeper increase of viruses with increasing temperature, i.e., virus production rates of 0.18, 0.37, 1.4, and 1.6 × 10^5 viruses h^{-1} between 16 and 30 h for 0.5, 2.5, 3.5, and 7.0 °C, respectively. While the data confirm our second hypothesis that the temperature increase stimulates MpoV production (through shortened latent periods, enhanced production rate and/or higher burst sizes), there is nonetheless a virus-specific response for the range of temperatures tested. Instead we found a high variability in response for the different virus isolates and host strains. When looking at a temperature increase from 3 to 7 °C, most virus-host combinations in our study showed enhanced viral burst sizes with a temperature-regulated increase in the host growth rates (0.53–0.59 d^{-1} at 3 °C and approximately 1.2-fold higher at 7 °C). However, MpoV-45T infecting TX-01 did not show this increase from 3 to 7 °C, but did exhibit an increasing burst size with temperatures from 0.5 up to 3.5 °C (growth rate TX-01 increased from 0.40 to 0.66 d^{-1}, respectively). Hence, the optimum temperature for virus production was not the same as the optimum host growth temperature. Wells and Deming [78] showed a similar situation in which phage 9A, infecting the psychrophile *Colwellia psychrerythraea* strain 34H, had a burst size optimum at −1 °C while the host's growth rate optimum was at 8 °C. These authors suggested that specific virally encoded enzymes have their own optimum temperatures.

Temperature strongly regulates Arctic *Micromonas* growth rates with increasing growth rates up to 7 °C (this study; [16,29]). These temperatures are at or above the present summer sea surface temperatures in the lower latitude regions of the Arctic [23,25]. At the time that TX-01 was isolated (half April), in situ picophotoeukaryotic gross growth rates were 0.58 d^{-1} at temperatures between 1 and 2 °C (Maat and Brussaard, unpublished data). By the end of May, the temperatures and consequently the growth rates had increased to 2–3 °C and 1.1 d^{-1}, respectively. This 20–50% growth rate increase is similar to TX-01 in the present study over the same temperature range. Our results indicate that over an Arctic growing season with increasing temperatures and host growth rates, viral activity can be expected to increase as a result of the decreasing latent periods and increasing burst sizes. Our data imply that temperature could affect host and virus diversity (strain dynamics), as for MpoV-45T with host TX-01 the latent period and viral burst size did not further change above 3 °C, but with other hosts (RCC2257 and 2258) and other viruses (MpoV-44T) the latent periods shortened and/or burst sizes increased to several extents. The different susceptibilities of viral infectivity to temperature, with a tolerance for the highest temperatures for MpoV-44T, strengthened this idea even more. Tarutani et al. [80] showed how within the observed abundances of *Heterosigma akashiwo* and the lytic virus HaV, a successional shift in clonal composition occurred due to differences in susceptibility/resistance of the host to the viruses. Based on our data a similar shift in strain and clonal composition of *Micromonas* and MpoV may occur, not only due to differences in susceptibility to the viruses but also because of differences in virus proliferation success at different temperatures [33].

In summary, the present study describes the first isolation and characterization of viruses infecting a cold-adapted polar phytoplankter. The relevance seems high, as it concerns the ubiquitous genus *Micromonas* which belongs to the picophytoplankton fraction and is expected to be favored under future Arctic conditions (due to warming and freshening induced vertical stratification; [24,27–29]). The Arctic region is warming to a greater extent than lower latitudinal marine waters [23,24] and current summer sea surface temperatures (August 2016) as high as 5 °C above the 1982–2010 mean [25] have been observed. *Micromonas* growth rates will enhance faster and earlier in the season and our study indicates that viral production will likely do the same. We show variable infection dynamics in response to temperature for the different virus-host strain systems examined, which complicates the assessment of the environmental relevance of each isolate. However, we do like to advocate that virus (and host) isolation, characterization of virus-host dynamics, and responses to changing ecologically relevant environmental factors are fundamentally essential to understanding the role of algal viruses in (Arctic) marine waters. The newly isolated viruses make it possible to comprehensively investigate the interactions of these unique virus-host combinations under climate change relevant environmental variables. Joli et al. [22] showed the importance of Arctic *Micromonas* viruses by metagenome sequencing. It would be interesting to investigate the ecological relevance of the strains tested in our study using molecular approaches. In the natural environment, selective effects of temperature may drive (intra)species diversity, potentially affecting the ability of *Micromonas* to respond to the changing conditions of the vulnerable Arctic. Modeling studies could help to comprehend (and predict) the extent to which the Arctic phytoplankton community would be influenced by changes in infection dynamics associated with temperature changes.

Supplementary Materials: The following are available online at www.mdpi.com/1999-4915/9/6/134/s1, Table S1: phytoplankton genera used for MpoV infection assays, Table S2: *p*-values belonging to Figure 6B, Table S3: Top 3 BLASTN hits of the isolates against KEGG environmental metagenomes. Figure S1: Phylogeny *Micromonas polaris* TX-01, Figure S2: Cytograms of MpoVs, Figure S3: Virus growth cycle of MpoV-44T infecting different *Micromonas* host strains, Figure S4: PFGE photograph for genome size estimation.

Acknowledgments: This work was part of the VIRPOL and VIRARCT projects (grants 851.40.010 and 866.12.404 awarded to C.P.D.B.) which were supported by the Earth and Life Sciences Foundation (ALW), with financial aid from the Netherlands Organisation for Scientific Research (NWO). B.E.D. was supported by NWO Vidi grant 864.14.004. We especially thank Anna Noordeloos, Kirsten Kooijman, Harry Witte, Astrid van Hoogstraten, and Henk van Ven for their technical support. We furthermore thank the following interns for their lab assistance: Dennis de Waard, Wessel Jellema, Merel Collenteur, Gizem Yikilmaz, Ryan Sewbaransingh, Sander Fokkes, Daniel Zorg, and Alex Janse. We thank the anonymous reviewers for their constructive comments.

Author Contributions: C.P.D.B. has initiated the research; C.E. and D.S.M. have isolated the MpoVs; D.S.M. isolated *M. polaris* TX-01; D.S.M., T.B., C.E., and C.P.D.B. have performed the virus characterization; D.S.M. and T.B. have designed, performed the experiments, and analyzed the data; N.N.W. was responsible for TEM analysis, J.D.L.v.B. for phylogenetic analysis, and B.E.D. for metagenomic analysis. All authors contributed to the writing of the paper (D.S.M., T.B., and C.P.D.B. are the lead authors).

Conflicts of Interest: The authors declare no conflict of interest.

References

1. Mayer, J.A.; Taylor, F.J.R. A virus which lyses the marine nanoflagellate *Micromonas pusilla*. *Nature* **1979**, *281*, 299–301. [CrossRef]

2. Waters, R.E.; Chan, A.T. *Micromonas pusilla* virus: The virus growth cycle and associated physiological events within the host cells; host range mutation. *J. Gen. Virol.* **1982**, *63*, 199–206. [CrossRef]

3. Cottrell, M.T.; Suttle, C.A. Wide-spread occurrence and clonal variation in viruses which cause lysis of a cosmopolitan, eukaryotic marine phytoplankter, *Micromonas pusilla*. *Mar. Ecol. Prog. Ser.* **1991**, *78*, 1–9. [CrossRef]

4. Cottrell, M.T.; Suttle, C.A. Dynamics of lytic virus infecting the photosynthetic marine picoflagellate *Micromonas pusilla*. *Limnol. Oceanogr.* **1995**, *40*, 730–739. [CrossRef]

5. Cottrell, M.T.; Suttle, C.A. Genetic Diversity of Algal Viruses Which Lyse the Photosynthetic Picoflagellate *Micromonas pusilla* (Prasinophyceae). *Appl. Environ. Microbiol.* **1995**, *61*, 3088–3091. [PubMed]

6. Worden, A.Z.; Lee, J.-H.; Mock, T.; Rouzé, P.; Simmons, M.P.; Aerts, A.L.; Allen, A.E.; Cuvelier, M.L.; Derelle, E.; Everett, M.V.; et al. Green evolution and dynamic adaptations revealed by genomes of the marine picoeukaryotes *Micromonas*. *Science* **2009**, *324*, 268–272. [CrossRef] [PubMed]

7. Not, F.; Latasa, M.; Marie, D.; Cariou, T.; Vaulot, D.; Simon, N. A single species, *Micromonas pusilla* (Prasinophyceae), dominates the eukaryotic picoplankton in the Western English Channel. *Appl. Environ. Microbiol.* **2004**, *70*, 4064–4072. [CrossRef] [PubMed]

8. Foulon, E.; Not, F.; Jalabert, F.; Cariou, T.; Massana, R.; Simon, N. Ecological niche partitioning in the picoplanktonic green alga *Micromonas pusilla*: Evidence from environmental surveys using phylogenetic probes. *Environ. Microbiol.* **2008**, *10*, 2433–2443. [CrossRef] [PubMed]

9. Martínez Martínez, J.; Boere, A.; Gilg, I.; van Lent, J.W.M.; Witte, H.J.; van Bleijswijk, J.D.L.; Brussaard, C.P.D. New lipid envelope-containing dsDNA virus isolates infecting *Micromonas pusilla* reveal a separate phylogenetic group. *Aquat. Microb. Ecol.* **2015**, *74*, 17–28. [CrossRef]

10. Brussaard, C.P.D.; Noordeloos, A.A.M.; Sandaa, R.-A.; Heldal, M.; Bratbak, G. Discovery of a dsRNA virus infecting the marine photosynthetic protist *Micromonas pusilla*. *Virology* **2004**, *319*, 280–291. [CrossRef] [PubMed]

11. Attoui, H.; Jaafar, F.M.; Belhouchet, M.; de Micco, P.; de Lamballerie, X.; Brussaard, C.P.D. *Micromonas pusilla* reovirus: A new member of the family Reoviridae assigned to a novel proposed genus (Mimoreovirus). *J. Gen. Virol.* **2006**, *87*, 1375–1383. [CrossRef] [PubMed]

12. Hingamp, P.; Grimsley, N.; Acinas, S.G.; Clerissi, C.; Subirana, L.; Poulain, J.; Ferrera, I.; Sarmento, H.; Villar, E.; Lima-Mendez, G.; et al. Exploring nucleo-cytoplasmic large DNA viruses in Tara Oceans microbial metagenomes. *ISME J.* **2013**, *7*, 1678–1695. [CrossRef] [PubMed]

13. Zingone, A.; Sarno, D.; Forlani, G. Seasonal dynamics in the abundance of *Micromonas pusilla* (Prasinophyceae) and its viruses in the Gulf of Naples (Mediterranean Sea). *J. Plankton Res.* **1999**, *21*, 2143–2159. [CrossRef]

14. Baudoux, A.C.; Lebredonchel, H.; Dehmer, H.; Latimier, M.; Edern, R.; Rigaut-Jalabert, F.; Ge, P.; Guillou, L.; Foulon, E.; Bozec, Y.; et al. Interplay between the genetic clades of *Micromonas* and their viruses in the Western English Channel. *Environ. Microbiol. Rep.* **2015**, *7*, 765–773. [CrossRef] [PubMed]

15. Not, F.; Massana, R.; Latasa, M.; Marie, D.; Colson, C.; Eikrem, W.; Pedrós-Alió, C.; Vaulot, D.; Simon, N. Late summer community composition and abundance of photosynthetic picoeukaryotes in Norwegian and Barents Seas. *Limnol. Oceanogr.* **2005**, *50*, 1677–1686. [CrossRef]

16. Lovejoy, C.; Vincent, W.F.; Bonilla, S.; Roy, S.; Martineau, M.J.; Terrado, R.; Potvin, M.; Massana, R.; Pedrós-Alió, C. Distribution, phylogeny, and growth of cold-adapted picoprasinophytes in Arctic Seas. *J. Phycol.* **2007**, *43*, 78–89. [CrossRef]

17. Balzano, S.; Marie, D.; Gourvil, P.; Vaulot, D. Composition of the summer photosynthetic pico and nanoplankton communities in the Beaufort Sea assessed by T-RFLP and sequences of the 18S rRNA gene from flow cytometry sorted samples. *ISME J.* **2012**, *6*, 1480–1498. [CrossRef] [PubMed]

18. Kilias, E.; Wolf, C.; Nöthig, E.M.; Peeken, I.; Metfies, K. Protist distribution in the Western Fram Strait in summer 2010 based on 454-pyrosequencing of 18S rDNA. *J. Phycol.* **2013**, *49*, 996–1010. [CrossRef] [PubMed]

19. Kilias, E.S.; Nöthig, E.M.; Wolf, C.; Metfies, K. Picoeukaryote plankton composition off west Spitsbergen at the entrance to the Arctic Ocean. *J. Euk. Microbiol.* **2014**, *61*, 569–579. [CrossRef] [PubMed]

20. Metfies, K.; von Appen, W.J.; Kilias, E.; Nicolaus, A.; Nöthig, E.M. Biogeography and photosynthetic biomass of arctic marine pico-eukaroytes during summer of the record sea ice minimum 2012. *PLoS ONE* **2016**, *11*, e0148512. [CrossRef] [PubMed]

21. Balzano, S.; Gourvil, P.; Siano, R.; Chanoine, M.; Marie, D.; Lessard, S.; Sarno, D.; Vaulot, D. Diversity of cultured photosynthetic flagellates in the northeast Pacific and Arctic Oceans in summer. *Biogeosciences* **2012**, *9*, 4553–4571. [CrossRef]

22. Joli, N.; Monier, A.; Logares, R.; Lovejoy, C. Seasonal patterns in Arctic prasinophytes and inferred ecology of *Bathycoccus* unveiled in an Arctic winter metagenome. *ISME J.* **2017**, *11*, 1372–1385. [CrossRef] [PubMed]
23. Richter-Menge, J.; Mathis, J. The Arctic. In *State of the Climate in 2015*; Blunden, J., Arndt, J.S., Eds.; American Meteorological Society: Boston, MA, USA, 2015; Volume 97, pp. S131–S153.
24. ACIA. Chapter 9: Marine Systems. In *Arctic Climate Impact Assessment*; Cambridge University Press: Cambridge, UK, 2005; pp. 453–538. 1042 p.
25. Timmermans, M.-L. Sea Surface Temperature. In Arctic Report Card. 2016. Available online: http://www.webcitation.org/6ogrhzGOI (accessed on 3 March 2017).
26. Simon, N.; Foulon, E.; Grulois, D.; Six, C.; Latimier, M.; Desdevises, Y.; Latimier, M.; Le Gall, F.; Tragin, M.; Houdan, A.; et al. Revision of the genus *Micromonas* (Manton et Parke) (Chlorophyta, Mamiellophyceae), of the type species *M. pusilla* (Butcher) Manton & Parke and of the species *M. commoda* (van Baren, Bacry and Worden) and description of two new species based on the genetic and phenotypic characterization of cultured isolates. *Protist* **2017**, in review.
27. Li, W.K.; McLaughlin, F.A.; Lovejoy, C.; Carmack, E.C. Smallest algae thrive as the Arctic Ocean freshens. *Science* **2009**, *326*, 539. [CrossRef] [PubMed]
28. Li, W.K.; Carmack, E.C.; McLaughlin, F.A.; Nelson, R.J.; Williams, W.J. Space-for-time substitution in predicting the state of picoplankton and nanoplankton in a changing Arctic Ocean. *J. Geophys. Res. Oceans* **2013**, *118*, 5750–5759. [CrossRef]
29. Coello-Camba, A.; Agustí, S.; Vaqué, D.; Holding, J.; Arrieta, J.M.; Wassmann, P.; Duarte, C.M. Experimental assessment of temperature thresholds for Arctic phytoplankton communities. *Estuar. Coast* **2015**, *38*, 873–885. [CrossRef]
30. Brussaard, C.P.D.; Noordeloos, A.A.M.; Witte, H.; Collenteur, M.C.J.; Schulz, K.G.; Ludwig, A.; Riebesell, U. Arctic microbial community dynamics influenced by elevated CO_2 levels. *Biogeosciences* **2013**, *10*, 719–731. [CrossRef]
31. Lara, E.; Arrieta, J.M.; Garcia-Zarandona, I.; Boras, J.A.; Duarte, C.M.; Agustí, S.; Wassmann, P.F.; Vaqué, D. Experimental evaluation of the warming effect on viral, bacterial and protistan communities in two contrasting Arctic systems. *Aquat. Microb. Ecol.* **2013**, *70*, 17–32. [CrossRef]
32. Payet, J.P.; Suttle, C.A. Viral infection of bacteria and phytoplankton in the Arctic Ocean as viewed through the lens of fingerprint analysis. *Aquat. Microb. Ecol.* **2014**, *72*, 47–61. [CrossRef]
33. Mojica, K.D.A.; Brussaard, C.P.D. Factors affecting virus dynamics and microbial host–virus interactions in marine environments. *FEMS Microbiol. Ecol.* **2014**, *89*, 495–515. [CrossRef] [PubMed]
34. Olsen, R.H. Isolation and growth of psychrophilic bacteriophage. *Appl. Microbiol.* **1967**, *15*, 198. [PubMed]
35. Borriss, M.; Helmke, E.; Hanschke, R.; Schweder, T. Isolation and characterization of marine psychrophilic phage-host systems from Arctic sea ice. *Extremophiles* **2003**, *7*, 377–384. [CrossRef] [PubMed]
36. D'amico, S.; Collins, T.; Marx, J.C.; Feller, G.; Gerday, C. Psychrophilic microorganisms: Challenges for life. *EMBO Rep.* **2006**, *7*, 385–389. [CrossRef] [PubMed]
37. Wells, L.E. Cold-active viruses. In *Psychrophiles: From Biodiversity to Biotechnology*; Margesin, R., Schinner, F., Marx, J.-C., Gerday, C., Eds.; Springer: Berlin/Heidelberg, Germany, 2008; pp. 157–173.
38. Nagasaki, K.; Yamaguchi, M. Effect of temperature on the algicidal activity and the stability of HaV (*Heterosigma akashiwo* virus). *Aquat. Microb. Ecol.* **1998**, *15*, 211–216. [CrossRef]
39. Toseland, A.D.S.J.; Daines, S.J.; Clark, J.R.; Kirkham, A.; Strauss, J.; Uhlig, C.; Lenton, T.M.; Valentin, K.; Pearson, G.A.; Moulton, V.; et al. The impact of temperature on marine phytoplankton resource allocation and metabolism. *Nat. Clim. Chang.* **2013**, *3*, 979–984. [CrossRef]
40. Zachary, A. An ecological study of bacteriophages of *Vibrio natriegens*. *Can. J. Microbiol.* **1978**, *24*, 321–324. [CrossRef] [PubMed]
41. Demory, D.; Arsenieff, L.; Simon, N.; Six, C.; Rigaut-Jalabert, F.; Marie, D.; Ge, P.; Bigeard, E.; Jacquet, S.; Sciandra, A.; et al. Temperature is a key factor in *Micromonas*–virus interactions. *ISME J.* **2017**, *11*, 601–612. [CrossRef] [PubMed]
42. Guillard, R.R.L.; Ryther, J.H. Studies of marine planktonic diatoms: I. *Cyclotella Nana* Hustedt, and *Detonula Confervacea* (CLEVE) Gran. *Can. J. Microbiol.* **1962**, *8*, 229–239. [CrossRef] [PubMed]
43. Šlapeta, J.; López-García, P.; Moreira, D. Global dispersal and ancient cryptic species in the smallest marine eukaryotes. *Mol. Biol. Evol.* **2006**, *23*, 23–29. [CrossRef] [PubMed]

44. Stamatakis, A. RAxML version 8: A tool for phylogenetic analysis and post-analysis of large phylogenies. *Bioinformatics* **2014**, *30*, 1312–1313. [CrossRef] [PubMed]

45. Ludwig, W.; Strunk, O.; Westram, R.; Richter, L.; Meier, H.; Buchner, A.; Lai, T.; Steppi, S.; Jobb, G.; Förster, W.; et al. ARB: A software environment for sequence data. *Nucleic Acids Res.* **2004**, *32*, 1363–1371. [CrossRef] [PubMed]

46. Romari, K.; Vaulot, D. Composition and temporal variability of picoeukaryote communities at a coastal site of the English Channel from 18S rDNA sequences. *Limnol. Oceanogr.* **2004**, *49*, 784–798. [CrossRef]

47. van Baren, M.J.; Bachy, C.; Reistetter, E.N.; Purvine, S.O.; Grimwood, J.; Sudek, S.; Yu, H.; Poirier, C.; Deerinck, T.J.; Kuo, A.; et al. Evidence-based green algal genomics reveals marine diversity and ancestral characteristics of land plants. *BMC Genomics* **2016**, *17*, 267. [CrossRef] [PubMed]

48. Harrison, P.J.; Waters, R.E.; Taylor, F.J.R. A broad-spectrum artificial seawater medium for coastal and open ocean phytoplankton. *J. Phycol.* **1980**, *16*, 28–35. [CrossRef]

49. Marie, D.; Brussaard, C.P.D.; Thyrhaug, R.; Bratbak, G.; Vaulot, D. Enumeration of marine viruses in culture and natural samples by flow cytometry. *Appl. Environ. Microbiol.* **1999**, *65*, 45–52. [PubMed]

50. Brussaard, C.P.D. Optimization of procedures for counting viruses by flow cytometry. *Appl. Environ. Microbiol.* **2004**, *70*, 1506–1513. [CrossRef] [PubMed]

51. Feldman, H.A.; Wang, S.S. Sensitivity of various viruses to chloroform. *Exp. Biol. Med.* **1961**, *106*, 736–738. [CrossRef]

52. Olsen, R.H.; Siak, J.-S.; Gray, R.H. Characteristics of PRD1, a plasmid-dependent broad host range DNA bacteriophage. *J. Virol.* **1974**, *14*, 689–699. [CrossRef] [PubMed]

53. Baudoux, A.-C.; Brussaard, C.P.D. Characterization of different viruses infecting the marine harmful algal bloom species *Phaeocystis globosa*. *Virology* **2005**, *341*, 80–90. [CrossRef] [PubMed]

54. Chen, F.; Suttle, C.A. Evolutionary relationships among large double-stranded DNA viruses that infect microalgae and other organisms as inferred from DNA polymerase genes. *Virology* **1996**, *219*, 170–178. [CrossRef] [PubMed]

55. Mustapha, S.B.; Larouche, P.; Dubois, J.-M. Spatial and temporal variability of sea-surface temperature fronts in the coastal Beaufort Sea. *Cont. Shelf Res.* **2016**, *124*, 134–141. [CrossRef]

56. Hop, H.; Falk-Petersen, S.; Svendsen, H.; Kwasniewski, S.; Pavlov, V.; Pavlova, O.; Søreide, J.E. Physical and biological characteristics of the pelagic system across Fram Strait to Kongsfjorden. *Prog. Oceanogr.* **2006**, *71*, 182–231. [CrossRef]

57. Passmore, R.; Hsu, J.; Liu, R.X.; Tam, E.; Cai, Y.; Su, W.; Frasca, J.; Brigden, S.M.; Comeau, A.M.; Ortmann, A.C. 2000. MPN Assay Analyzer. Available online: http://www.webcitation.org/6ogxAqLbE (accessed on 3 March 2017).

58. Sunagawa, S.; Coelho, L.P.; Chaffron, S.; Kultima, J.R.; Labadie, K.; Salazar, G.; Djahanschiri, B.; Zeller, G.; Mende, D.R.; Alberti, A.; et al. Structure and function of the global ocean microbiome. *Science* **2015**, *348*, 1261359. [CrossRef] [PubMed]

59. Altschul, S.F.; Gish, W.; Miller, W.; Myers, E.W.; Lipman, D.J. Basic local alignment search tool. *J. Mol. Biol.* **1990**, *215*, 403–410. [CrossRef]

60. Sievers, F.; Wilm, A.; Dineen, D.; Gibson, T.J.; Karplus, K.; Li, W.; Lopez, R.; McWilliam, H.; Remmert, M.; Söding, J.; et al. Fast, scalable generation of high-quality protein multiple sequence alignments using Clustal Omega. *Mol. Syst. Biol.* **2011**, *7*, 539. [CrossRef] [PubMed]

61. Guindon, S.; Dufayard, J.F.; Lefort, V.; Anisimova, M.; Hordijk, W.; Gascuel, O. New algorithms and methods to estimate maximum-likelihood phylogenies: Assessing the performance of PhyML 3.0. *Syst. Biol.* **2010**, *59*, 307–321. [CrossRef] [PubMed]

62. Li, H.; Durbin, R. Fast and accurate short read alignment with Burrows-Wheeler transform. *Bioinformatics* **2009**, *25*, 1754–1760. [CrossRef] [PubMed]

63. Baudoux, A.-C.; Brussaard, C.P.D. Influence of irradiance on virus-algal host interactions. *J. Phycol.* **2008**, *44*, 902–908. [CrossRef] [PubMed]

64. Maat, D.S.; de Blok, R.; Brussaard, C.P.D. Combined phosphorus limitation and light stress prevent viral proliferation in the phytoplankton species *Phaeocystis globosa*, but not in *Micromonas pusilla*. *Front. Mar. Sci.* **2016**, *3*, 160. [CrossRef]

65. Wilson, W.H.; van Etten, J.L.; Allen, M.J. The Phycodnaviridae: The story of how tiny giants rule the world. *Curr. Top. Microbiol. Immunol.* **2009**, *328*, 1–42. [CrossRef] [PubMed]

66. Clerissi, C.; Grimsley, N.; Ogata, H.; Hingamp, P.; Poulain, J.; Desdevises, Y. Unveiling of the diversity of Prasinoviruses (Phycodnaviridae) in marine samples by using high-throughput sequencing analyses of PCR-amplified DNA polymerase and major capsid protein genes. *Appl. Environ. Microbiol.* **2014**, *80*, 3150–3160. [CrossRef] [PubMed]

67. Sahlsten, E. Seasonal abundance in Skagerrak-Kattegat coastal waters and host specificity of viruses infecting the marine photosynthetic flagellate *Micromonas pusilla. Aquat. Microb. Ecol.* **1998**, *16*, 103–108. [CrossRef]

68. Pagarete, A.; Chow, C.E.; Johannessen, T.; Fuhrman, J.A.; Thingstad, T.F.; Sandaa, R.A. Strong seasonality and interannual recurrence in marine myovirus communities. *Appl. Environ. Microbiol.* **2000**, *79*, 6253–6259. [CrossRef] [PubMed]

69. McKie-Krisberg, Z.M.; Sanders, R.W. Phagotrophy by the picoeukaryotic green alga Micromonas: Implications for Arctic Oceans. *ISME J.* **2014**, *8*, 1953–1961. [CrossRef] [PubMed]

70. Hop, H.; Pearson, T.; Hegseth, E.N.; Kovacs, K.M.; Wiencke, C.; Kwasniewski, S.; Eiane, K.; Mehlum, F.; Gulliksen, B.; Wlodarska-Kowalczuk, M.; et al. The marine ecosystem of Kongsfjorden, Svalbard. *Polar Res.* **2002**, *21*, 167–208. [CrossRef]

71. Tverberg, V.; Nilsen, F.; Goszczko, I.; Cottier, F.; Svendsen, H.; Gerland, S. The warm winter temperatures of 2006 and 2007 in the Kongsfjorden water masses compared to historical data. In *8th Ny-Ålesund (NySMAC) Science Managers Committee*; Polarnet Technical Report; 2008; pp. 40–43.

72. Suttle, C.A. Marine viruses—Major players in the global ecosystem. *Nat. Rev. Microbiol.* **2007**, *5*, 801–812. [CrossRef] [PubMed]

73. Wells, L.E.; Deming, J.W. Modelled and measured dynamics of viruses in Arctic winter sea-ice brines. *Environ. Microbiol.* **2006**, *8*, 1115–1121. [CrossRef] [PubMed]

74. Long, A.M.; Short, S.M. Seasonal determinations of algal virus decay rates reveal overwintering in a temperate freshwater pond. *ISME J.* **2016**, *10*, 1602–1612. [CrossRef] [PubMed]

75. Tomaru, Y.; Katanozaka, N.; Nishida, K.; Shirai, Y.; Tarutani, K.; Yamaguchi, M.; Nagasaki, K. Isolation and characterization of two distinct types of HcRNAV, a single-stranded RNA virus infecting the bivalve-killing microalga *Heterocapsa circularisquama. Aquat. Microb. Ecol.* **2004**, *34*, 207–218. [CrossRef]

76. Tomaru, Y.; Tanabe, H.; Yamanaka, S.; Nagasaki, K. Effects of temperature and light on stability of microalgal viruses, HaV, HcV and HcRNAV. *Plankton Biol. Ecol.* **2005**, *52*, 1–6.

77. Nagasaki, K.; Shirai, Y.; Tomaru, Y.; Nishida, K.; Pietrokovski, S. Algal viruses with distinct intraspecies host specificities include identical intein elements. *Appl. Environ. Microbiol.* **2005**, *71*, 3599–3607. [CrossRef] [PubMed]

78. Wells, L.E.; Deming, J.W. Characterization of a cold-active bacteriophage on two psychrophilic marine hosts. *Aquat. Microb. Ecol.* **2006**, *45*, 15–29. [CrossRef]

79. Luhtanen, A.M.; Eronen-Rasimus, E.; Kaartokallio, H.; Rintala, J.M.; Autio, R.; Roine, E. Isolation and characterization of phage–host systems from the Baltic Sea ice. *Extremophiles* **2014**, *18*, 121–130. [CrossRef] [PubMed]

80. Tarutani, K.; Nagasaki, K.; Yamaguchi, M. Viral impacts on total abundance and clonal composition of the harmful bloom-forming phytoplankton *Heterosigma akashiwo. Appl. Environ. Microbiol.* **2000**, *66*, 4916–4920. [CrossRef] [PubMed]

![viruses logo] *viruses*

MDPI

Article

A Pelagic Microbiome (Viruses to Protists) from a Small Cup of Seawater

Flavia Flaviani [1,2], Declan C. Schroeder [2,*], Cecilia Balestreri [2], Joanna L. Schroeder [2], Karen Moore [3], Konrad Paszkiewicz [3], Maya C. Pfaff [4] and Edward P. Rybicki [1,*]

[1] Department of Molecular and Cell Biology, University of Cape Town, Private Bag X3, Rondebosch 7701, South Africa; flafla@mba.ac.uk
[2] Marine Biological Association of the UK, Citadel Hill, Plymouth PL1 2PB, UK; cecilia.balestreri@gmail.com (C.B.); joanna.schroeder.uk@gmail.com (J.L.S.)
[3] University of Exeter Sequencing Service, Biosciences, Stocker Rd., University of Exeter, Exeter EX4 4QD, UK; K.A.Moore@exeter.ac.uk (K.M.); K.H.Paszkiewicz@exeter.ac.uk (K.P.)
[4] Department of Environmental Affairs, Oceans and Coasts, P.O. Box 52126, Victoria and Alfred Waterfront, Cape Town 8000, South Africa; maya.pfaff@gmail.com
*   Correspondence: dsch@mba.ac.uk (D.C.S.); ed.rybicki@uct.ac.za (E.P.R.); Tel.: +44-1752-426-484 (D.C.S.); +27-21-650-3265 (E.P.R.)

Academic Editors: Mathias Middelboe and Corina Brussaard
Received: 13 January 2017; Accepted: 13 March 2017; Published: 17 March 2017

Abstract: The aquatic microbiome is composed of a multi-phylotype community of microbes, ranging from the numerically dominant viruses to the phylogenetically diverse unicellular phytoplankton. They influence key biogeochemical processes and form the base of marine food webs, becoming food for secondary consumers. Due to recent advances in next-generation sequencing, this previously overlooked component of our hydrosphere is starting to reveal its true diversity and biological complexity. We report here that 250 mL of seawater is sufficient to provide a comprehensive description of the microbial diversity in an oceanic environment. We found that there was a dominance of the order *Caudovirales* (59%), with the family *Myoviridae* being the most prevalent. The families *Phycodnaviridae* and *Mimiviridae* made up the remainder of pelagic double-stranded DNA (dsDNA) virome. Consistent with this analysis, the Cyanobacteria dominate (52%) the prokaryotic diversity. While the dinoflagellates and their endosymbionts, the superphylum Alveolata dominates (92%) the microbial eukaryotic diversity. A total of 834 prokaryotic, 346 eukaryotic and 254 unique virus phylotypes were recorded in this relatively small sample of water. We also provide evidence, through a metagenomic-barcoding comparative analysis, that viruses are the likely source of microbial environmental DNA (meDNA). This study opens the door to a more integrated approach to oceanographic sampling and data analysis.

Keywords: microbiome; viruses; prokaryote; eukaryote; NGS; diversity; phylotypes; eDNA; meDNA

1. Introduction

The paradigm of "everything is everywhere, but the environment selects" [1] suggests that all microbial taxa have the potential to be found everywhere. This largely holds true for the main marine bacteriophage taxa, with the presence of cyanophage-like sequences of the order *Caudovirales* dominating all ocean viromes, including the recently sampled Indian Ocean [2–5]. The order *Caudovirales* is comprised of three families: *Myoviridae* (contractile tails), *Siphoviridae* (non-contractile tails) and *Podoviridae* (short tails) [6]. During the Global Ocean Sampling (GOS) expedition [5], myovirus-associated sequences were ubiquitously distributed among sampling sites with the highest prevalence in tropical oligotrophic locations. Podo- and siphoviruses showed site-specific distributions, with the highest abundances recorded in temperate mesotrophic waters and hypersaline lagoons, respectively. Within the Indian Ocean, 32% of the viral fraction (VF) was attributed to known viruses, with 95% of the known viruses identified as belonging to the order *Caudovirales* (*Myoviridae*, 54.3%; *Podoviridae*, 27.6%; *Siphoviridae*, 17%) [4]. The nucleo-cytoplasmic large DNA viruses (NCLDVs) were often the next major lineage present, with the family *Phycodnaviridae* representing 83.9% of this group, followed by *Iridoviridae* at 8.5% and *Mimiviridae* at 7.3%.

Most of the virome-based studies carried out so far do not report on the diversity of the likely hosts that the viruses infect, making it unclear as to whether the viruses present in the water column are the result of active or past infections. An exception is the Tara Oceans expedition, where eukaryotic and prokaryotic diversity [7,8] was reported in conjunction with viral diversity [9–11]. Global surveys, which include the southwest Indian Ocean, indicate that the α-Proteobacteria dominate the prokaryotic communities in both surface waters and at the deep chlorophyll maxima. The second most represented group are either the Cyanobacteria or γ-Proteobacteria, depending on location [8]. For the eukaryotic fraction, samples collected during the Tara Ocean expedition showed that the pico- and nano-plankton was dominated by photosynthetic dinoflagellates (of the family Dinophyceae). Parasites of the superphylum Alveolata, specifically the marine alveolates (MALV)-I and MALV-II clusters, routinely infect members of the family Dinophyceae and can account for up to 88% of the eukaryote fraction in some locations. These two MALV clusters have recently been renamed Syndiniales groups I and II, respectively [12]. Specifically for the southwest Indian Ocean, the eukaryotic fraction was dominated by alveolates including the Dinophyceae and their Syndiniales parasites [7].

Studies on microbial diversity in aquatic environments rely on sample volumes ranging from tens of litres to as much as a thousand litres of water [2,13,14]. Sampling of large volumes was thought to be a necessity for early sequencing technologies, which required considerable quantities (micrograms) of DNA. Newer technologies, such as linear amplification deep sequencing with Illumina, require smaller quantities (nanograms) of DNA [15]. Additionally, various sample concentration methods have been developed in order to collect the greatest quantities of DNA possible from water samples [16]. Standard viral filtration methods involve the use of filters with a pore size of 0.2 μm to remove bacteria from the sample and collect only the virus fraction. However, this 0.2 μm size fraction results in underreporting of giant viruses [17,18], as the giant virus particles can have diameters varying from ~0.2 to 1.5 μm, with *Pithovirus sibericum* being the largest known member of this group [19]. In addition, the <0.2 μm size fraction also contains large amounts of dissolved DNA. Jiang and Paul concluded that, in this size fraction, viral particles makes up only a small component of the filterable DNA, the majority being dissolved DNA of bacterial and eukaryotic origin [20].

Dissolved DNA forms part of environmental DNA (eDNA), derived from cellular debris produced from biota living in that environment [21]. Therefore, eDNA is being used as a tool to determine whether an invasion has taken place [22] or to track an endangered species [23]. The size fraction used to describe eDNA is the size fraction that removes larger eukaryotes (passing through a 0.5 mm mesh) but retains microbes (>0.45 μm filter). Therefore, the eDNA concept excludes the microbial community as they are retained in this size range. To our knowledge no study has yet addressed the question of whether the <0.45 μm size fraction, the microbial environmental DNA (meDNA) fraction, can be used as a proxy to describe the complete biota in any given environment.

In this study, we tested the hypothesis that the volume equivalent to a cup of seawater (250 mL) is sufficient to describe the most abundant microbial taxa (from viruses to protists) in the marine environment. Serendipitously, our study site is within 548 nautical miles of station 64, previously sampled by the *Tara* Oceans expedition (−29.5333, 37.9117), thereby allowing for a semi-qualitative comparison to be made. Our protocol differed from previous studies, including that of *Tara* Oceans, as it contained no concentration steps. In addition, only 50 mL of the 0.45 μm 250 mL permeate was used to describe the combined dissolved DNA and viral fraction (meDNA). The 0.45 μm size fraction was chosen because we wanted to limit the removal of giant viruses. Here we report how a relatively small water sample can be used to capture the dominant microbial taxa within any given aquatic system.

2. Materials and Methods

2.1. Sample Collection

The water sample analysed in this study was collected during the second transect of the Great Southern Coccolithophore Belt expedition (GSCB-cruise RR1202) in the southwest Indian Ocean in February 2012 [24]. The location of the sampling station S1 (-38.314983, 40.958083, water temperature 20.83 °C, pH 8.08) was mapped using RgoogleMaps_1.2.0.7 [25] under R version 3.3.0 (accessed on 3 May 2016) (Figure 1a).

One litre of water was gathered from the conductivity, temperature, and depth (CTD) rosette sampler from the chlorophyll maximum layer (5 m). Of this, an aliquot of 250 mL of seawater was filtered through a 0.45-μm polycarbonate filter and the filter was used for the DNA extraction onboard the R/V Roger Revelle using Qiagen DNeasy Blood and Tissue protocol (Qiagen, Valencia, CA, USA). The DNA was stored at -21 °C and subsequently transferred to Plymouth, UK, for further processing. Fifty millilitres of filtered water were set aside, wrapped in tin foil and stored in a fridge. This too was returned to Plymouth, UK, for further processing.

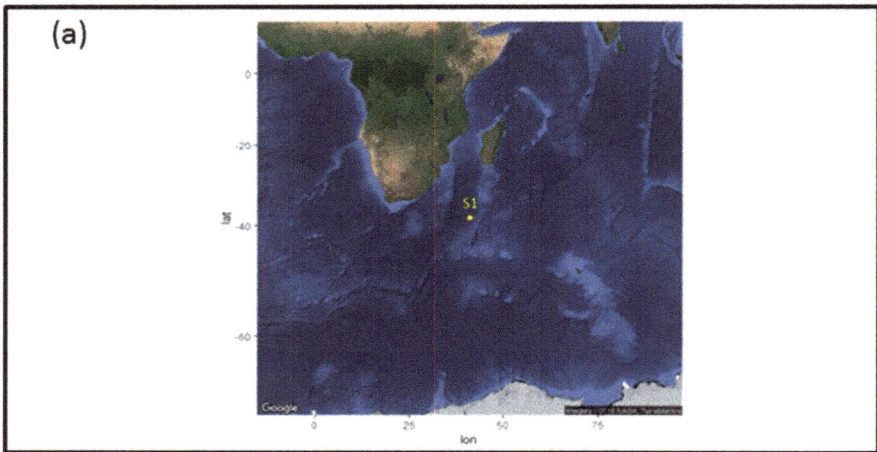

Figure 1. *Cont.*

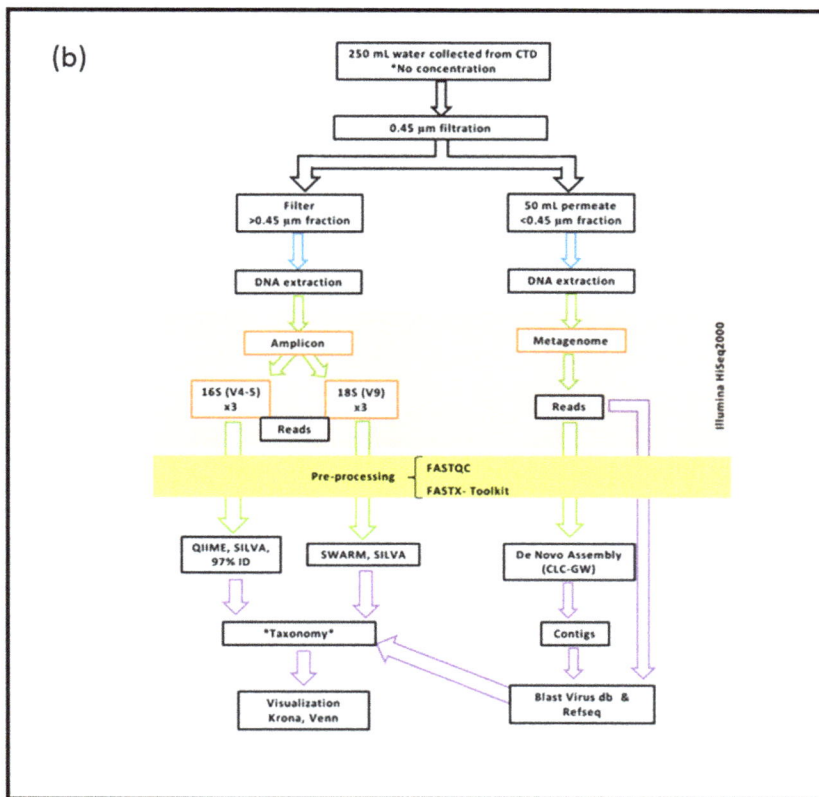

Figure 1. (**a**) Map showing the location of sample collection; (**b**) schematics of the bioinformatics pipeline.

2.2. DNA Extraction, Preparation and Sequencing of the >0.45 μm Fraction

The V4 region, along the prokaryotic 16S ribosomal RNA gene was amplified using the universal primer pair 515F and Illumina tagged primer 806R7, 806R10 and 806R15 (Illumina, San Diego, CA, USA) [26]. For eukaryotic 18S ribosomal RNA gene, we used the primer pair 1391F and Illumina tagged EukB6, EukB16 and EukB23 to amplify the V9 region [27]. For all polymerase chain reaction (PCRs), we added 1–5 μL of the eDNA (concentration range from 1.47 to 32.51 ng/μL), to 5X Colourless GoTaq Flexi Buffer (Promega, Madison, WI, USA), 1.5 μL MgCl$_2$ Solution 25 mM, 2.5 μL dNTPs (10 mM final concentration), 1 μL Evagreen Dye 20X (Biotium, Fremont, CA, USA), 0.1 μL GoTaq DNA Polymerase (5u/μL) and 12.9 μL of sterile water for a final volume of 25 μL for each reaction. This was done to determine the mid-exponential threshold of each reaction, ran on a Corbette Rotor-Gene™ 6000 (Qiagen). The real-time PCR proceeded with an initial denaturation at 94 °C for 3 min, followed by 40 cycles of a three-step PCR: 94 °C for 45 s and 50 °C for 60 s and 72 °C for 90 s. The fluorescence was acquired at the end of each annealing/extension step on the green channel. The cycle threshold of the amplification in the exponential phase was recorded for amplification.

A second standard PCR amplification was carried out in triplicate and run at the same conditions, excluding the addition of the Evagreen Dye. The sample was removed from the machine when it reached the cycle threshold, as previously determined. Products were run on a 1.4% agarose gel to confirm the success of the amplification and the product size of the amplification. The bands were cut out and purified using the Zymoclean Gel DNA Recovery Kit (Zymo Research,

Irvine, CA, USA). Quantity and quality was verified on the NanoDrop 1000 (Thermo Scientific, Wilmington, DE, USA) and QuantiFluor E6090 (Promega). V4-16S and V9-18S were prepared mixing an equimolar concentration of each amplicon triplicate into the pool for which concentration was checked on the Bioanalyser (Agilent Technologies, Santa Clara, CA, USA). The final pooled samples were denatured and diluted to 6 pM and mixed with 1 pM PhiX control (Illumina), read 1 sequencing primer was diluted in HT1, before the flowcell was clustered on the cBOT (Illumina). Multiplexing sequencing primers and read two sequencing primers were mixed with Illumina HP8 and HP7 sequencing primers, respectively. The flowcell was sequenced (100 PE) on HiSeq 2000 using sequencing by synthesis (SBS) reagents (Version 3.0). The raw sequences are available at the European Nucleotide Archive (ENA) under accession number PRJEB16346 and PRJEB16674.

2.3. DNA Extraction, Preparation and Sequencing of the <0.45 µm Fraction

The whole 50 mL permeate was used in the nucleic acid extraction procedure. We added 100 µL of proteinase K (10 mg/mL; Sigma-Aldrich, St. Louis, MO, USA) and 200 µL of 10% sodium dodecyl sulfate (SDS) (Sigma-Aldrich) to the permeate and incubated the solution for two hours with constant rotation at 55 °C. The lysate was then collected through multiple centrifugations on a Qiagen DNeasy Blood and Tissue column (Qiagen). The standard Qiagen protocol was followed with 20 µL nuclease-free water (Sigma-Aldrich) used as the elution agent. Quantity and quality was determined using the NanoDrop 1000 (Thermo Scientific) and QuantiFluor E6090 (Promega). Two hundred microliters of DNA (<40 ng) were fragmented using a Bioruptor (Diagenode, Seraing (Ougrée), Belgium) on medium for 15 bursts of 30 s with a 30 s pause and concentrated to 30 µL on a Minelute column (Qiagen). Fragments were made into libraries using the Nextflex ChipSeq library preparation kit (BIOO scientific, Austin, TX, USA) without size selection and with 18 cycles of PCR amplification. Bioanalyser (Agilent Technologies) analysis indicated the final library contained insert between 30 basepairs (bp) to 870 bp. The library was multiplexed with other samples and sequenced (100 paired end) on a HiSeq 2000 (Illumina) using RTA1.9 and CASAVA1.8.

2.4. Bioinformatics Pipeline for the Prokaryotic (16S) and Eukaryotic (18S) Amplicon

The complete bioinformatics pipeline is illustrated in Figure 1b. The read quality was first assessed using Fast-QC [28]. FASTX-Toolkit [29] was utilised for the trimming and filtering steps; the first and last 10 bases were trimmed in order to remove low quality nucleotides. Reads were then filtered in order to retain only reads with more than 95% of nucleotide positions called with a quality score of 20. Trimmed and cleaned reads from each of the triplicate V4-16S and V9-18S PCRs were pooled in order to assign OTUs using Qiime [30] with 97% similarities for clustering and Swarm analysis [31], respectively. A taxonomy was assigned using BLASTn implemented in Qiime and Swarm using SILVA Version 119 [32] with a minimum e-value of 1×10^{-5}.

2.5. Bioinformatics Pipeline of the <0.45 µm Fraction (Metagenome)

As for the amplicon dataset, the quality of the reads was first assessed using Fast-QC [28]. The FASTX-Toolkit [29] was used to trim the first last bases to remove low quality nucleotides, and subsequently to filter out reads with fewer than 95% of nucleotide positions called with a quality score of 20. The forward read (R1) of the 100 bp pair-end HiSeq reads have been subjected to random library size normalization using Qiime script subsample_fasta.py; reverse reads (R2) had poor quality and were therefore discarded. The reads were used in a BLASTX [33] analysis against a Virus database (db; courtesy of Pascal Hingamp) with e-values less than 1×10^{-5}. The Virus database consisted of Refseq curated viral genomes, together with additional new genomes [11], and 20% of R1 Refseq whole organism db [34]. In addition, the pair-end reads were assembled into contigs using a de Bruijn de novo assembly program in CLC Genomic Workbench (Version 7.1.5; CLCbio, Cambridge, MA, USA) using global alignment with automatics bubble and word size, minimum contigs length of 250, mismatch cost of 2, insertion and deletion cost of 3, length fraction of 0.5 and similarity threshold of 0.8. The contigs were annotated with the BLASTX as described for the R1 normalised reads. Blast analyses were performed by using the University of Cape Town's HPC hex cluster.

The top hits from all the blast searches were selected through the use of a parser Perl script (http://www.bioinformatics-made-simple.com), and then a customised R script was developed to assign taxonomy. A complete viral taxonomy was assigned through a manually curated implementation of the International Committee on Taxonomy of Viruses (ICTV) database 2013 v1 with the National Center for Biotechnology Information (NCBI) taxonomy database.

2.6. Visualization of Community Diversity

Krona tools [35] were used to visualize community diversity as characterized by the Silva (v119), Refseq and Virus db genes taxonomy assignments. Venn diagrams were created using the R package VennDiagram_1.6.17 on R (Version 3.3.0; 2016-05-03).

2.7. Filters Applied to Annotated Datasets

We performed independent analyses on three independent PCR replicates (V4-16S and V9-18S) and assigned a taxonomy using Silva [36]. By using replication, we removed the level of noise in the sample introduced by PCR and sequencing artefacts, while retaining rare organisms. Therefore, we considered four levels of stringency at the phylotype level: (1) T0, all phylotypes present across the three replicates; (2) T1, removing singletons from each replicate; (3) T10, a minimum of 10 copies per phylotype had to be present in any one of the replicates, (4) T10-R1, a minimum of 10 copies per phylotype present in any two replicates and (5) T10-R2, a minimum of 10 copies per phylotype present in all three replicates.

3. Results

3.1. Microbiota in the >0.45 µm Fraction

After pre-processing, which included a specific subsampling to an equal read length of 125 bases, we retained an average 0.9 million reads for the prokaryotic and around 270 thousand for the eukaryotic dataset (Table 1). These reads clustered (T0 applied to combination of the three replicates) into around 46 thousand unique Operational Taxonomic Units (OTUs) for the prokaryotes, which clustered into 1409 phylotypes. For the eukaryotes 6836 OTUs clustered into 477 phylotypes (Table 1). Four different filters were applied which resulted in an increase in selection stringency (T0 to T10-R2) without the removal of significant numbers of reads from the prokaryotes (Figure 2a) and eukaryotes (Figure 3a) datasets, independent of sequence depth. However, the greatest change observed due to the application of the filters, was seen in the number of phylotypes observed (Figures 2b and 3b). A total number of 1886 phylotypes was observed in the 250 mL of southwest Indian Ocean, made up of 1409 prokaryotic and 477 eukaryotic phylotypes. When the singletons were removed (T1), the number of prokaryotic phylotypes dropped by nearly a half to 834 (59.19%, phylotypes retained) (Figure 2b); this was also observable in the OTUs (Table 1) moving from 45,826 to 23,081. Similarly, the number of eukaryotic phylotypes dropped by a third to 346 phylotypes (72.54% phylotypes retained) (Figure 3b), whilst OTUs dropped from 6836 to 2930 (Table 1). When a further filter, that a minimum of at least 10 reads per phylotype must be present in any of the replicates (T10), was applied, the diversity dropped by an additional 36% (compared to T0) to just under 77% for prokaryotes—retaining only 23% (Figure 2b), and 24% to 51% in eukaryotes—retaining only 49% (Figure 3b), leaving a total number of phylotypes as 554.

The phylotypes removed after applying the singleton filter (T1) (Supplementary Table S1) included *Cicer arietinum* (chickpeas), *Sesamum indicum* (sesame) and *Nicotiana sylvestris* (tobacco), which were not expected to be present in the marine environment. The application of the T10 filter resulted in the removal of a few marine species instead, such as *Noctiluca scintillans*, *Amphidinium mootonorum* and *Pandorina morum*. The additional application of replication filters, present in greater than 10 copies in at least any two (T10-R1) and all three (T10-R2) replicates, revealed a further but minimal reduction in the overall phylotype content (Figures 2b and 3b): both the prokaryotes and eukaryotes dropped to 17% and 38% (from T10 to T10-R1, Figure 2b) and 13% and 34% (from T10 to T10-R2, Figure 3b), respectively. We could identify a core of 184 phylotypes for the prokaryotes (Figure 2c) and 163 for the eukaryotes (Figure 3c) which were common across all filters. If no filter was applied, 575 prokaryotes (41%) and 131 eukaryotes (27%) unique or rare were observed, however, irrespective of which filter is applied no phylotype unique to their stringency were observed (Figures 2c and 3c).

In summary, we have identified a total of 1886 phylotypes of prokaryotes and eukaryotes without the application of any filter (T0), which was reduced to 1,180 after singletons were removed (T1). A further decrease in phylotype composition to 554, 423 and 347 was identified after application of T10, T10-R1 and T10-R2 filters.

We then considered the three replicates independently in order to understand how phylotypes differ across the three PCR replicates (Figures 2d and 3d). Prokaryotic diversity ranged from 767 phylotypes in replicate 3 to 1077 in replicate 2 (Figure 2d), corresponding to the sequence depth (Figure 2a). This was however not observed for the eukaryotes (Figure 3d), ranging from 339 of replicate 1 to 353 of replicate 2 (Figure 3d), irrespective of the sequence depth (Figure 3a). When applying the T1 filter, the number of phylotypes retained were on average 65% (from 882 to 561 in replicate 1, from 1077 to 697 replicate 2 and from 767 to 505 in replicate 3) and 79% (from 339 to 267 in replicate 1, from 353 to 279 in replicate 2 and from 346 to 278 in replicate 3) of the prokaryotes and eukaryotes, respectively (Figures 2d and 3d). Applying stringency filter T10 reduced the prokaryotic diversity in replicate 1 to 28%, in replicate 2 to 27% and replicate 3 to 26% (Figure 2d), whilst for the eukaryotes across replicates 1, 2 and 3 to 57%, 55% and 58%, respectively (Figure 3d).

Phylotype composition at T0 had 36% prokaryotic and 50% eukaryotic phylotypes in common across all replicates (Figures 2e and 3e). Between 9% and 22% of prokaryotes and 10% and 22% of eukaryotes were unique to each replicate. When singletons (T1) were removed and the T10 filter applied, the phylotypes common across all replicates increased to 45% and 58% for prokaryotes (Figure 2e), whilst for the eukaryotes, increased to 61% and 70% (Figure 3e). This coincided with the reduction in unique phylotypes retained per replicate. Replicate 1, 2 and 3 changing from 164 to 22, 309 to 55 and 124 to 2 unique prokaryotic phylotypes (Figure 2e). Similarly, replicate 1, 2 and 3 changed from 48 to 16, 57 to 12 and 49 to 16 unique eukaryotic phylotypes (Figure 3e).

Viruses **2017**, 9, 47

Table 1. Description of sequences generated in this study.

Fraction	Reads/Contigs	Sequence Length	OTUs and Phylotypes after Applying Filter				
			T0		T1		
			OTUs	Phylotypes	OTU		Phylotypes
Prokaryote	741,033	125	20,381	882	11,341		561
	1,117,576	125	30,642 [1] 45,826	1077 [1] 1409	16,593 [1] 23,081		697 [1] 834
	841,639	125	24,756	767	13,416		505
Eukaryote	223,814	125	2972	339	1714		267
	275,201	125	3271 [1] 6836	353 [1] 477	1780 [1] 2930		279 [1] 346
	308,208	125	3470	346	1836		278
Metagenome	4,962	Average 78.9 min: 240 max: 74,442 x̄: 1045		254 virus			

[1] sum from the three replicates with duplicates removed.

115

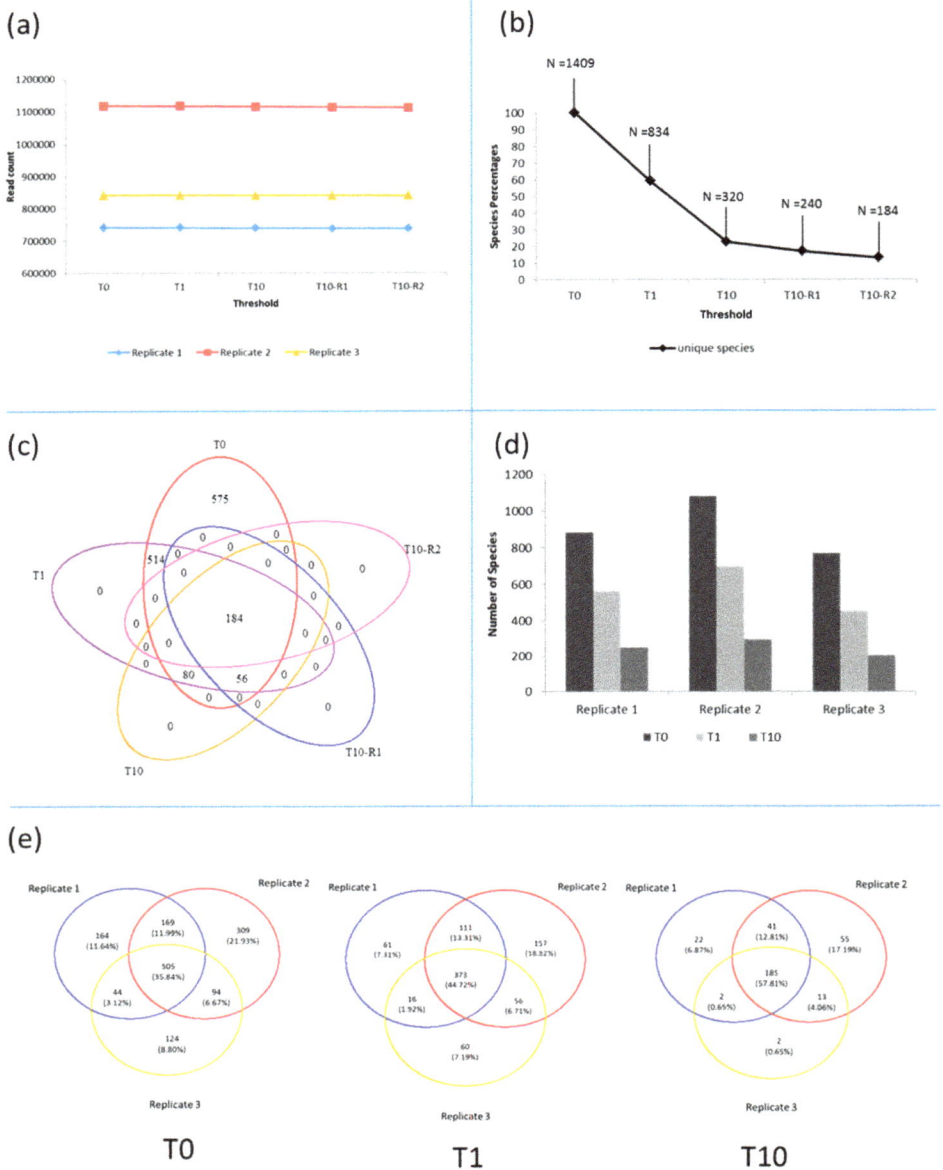

Figure 2. Analyses of the prokaryotic fraction. (**a**) Reduction in number of reads when filters are applied; (**b**) percentage and phylotype count when filter are applied; (**c**) presence–absence analyses at phylotype level before and after application of the filters; (**d**) number of phylotype analyses by replicate; (**e**) presence–absence analyses at phylotype level when filters are applied to each replicate.

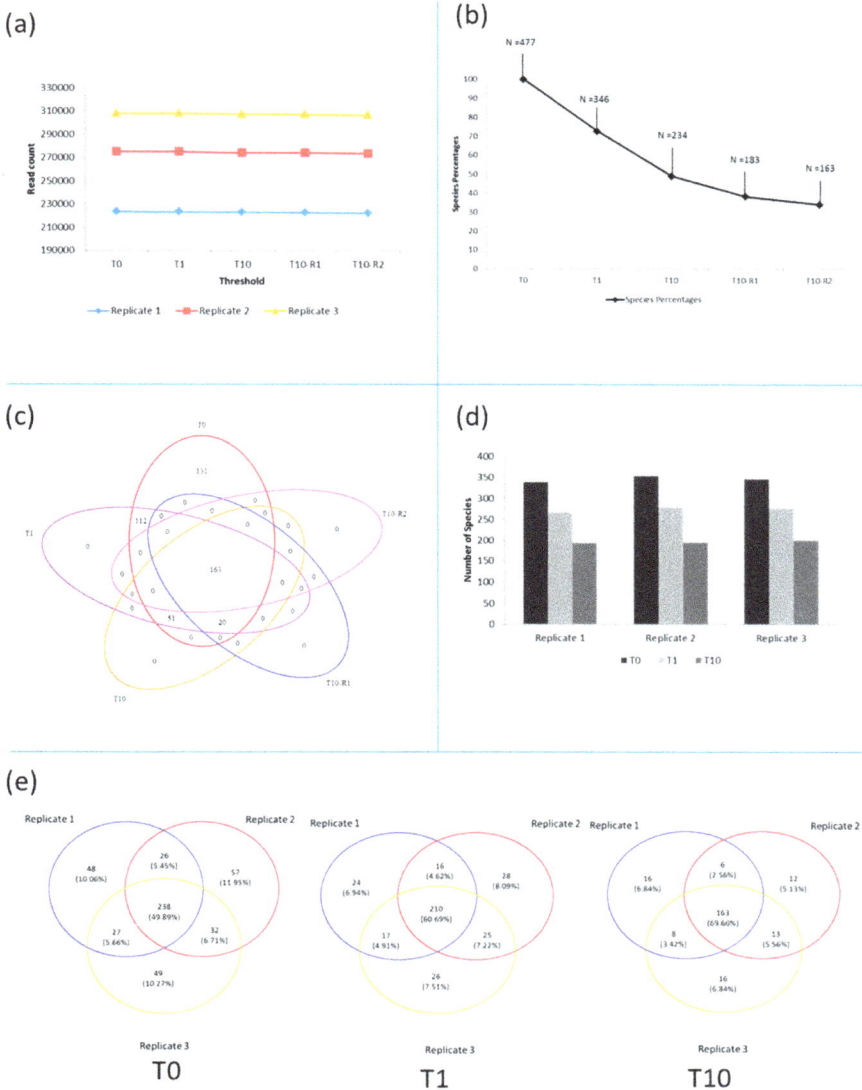

Figure 3. Analyses of the eukaryotic fraction. (**a**) Reduction in number of reads when filters are applied; (**b**) percentage and phylotype count when filters are applied; (**c**) presence–absence analyses at phylotype level before and after application of the filter; (**d**) number of phylotypes analyses by replicate; (**e**) presence–absence analyses at phylotype level when filters are applied to each replicate.

3.2. Diversity and Community Structure of the >0.45 µm Fraction

Cyanobacteria made up 42% of the prokaryotic community diversity; their composition was dominated by the genera *Synechococcus* (30%) and *Prochlorococcus* (9%) (Figure S1). The V4-16S universal primers also amplified the eukaryote plastid ribosomal genes, making up 2.68% of the total sequences. The second most diverse bacterial group were the Proteobacteria (32%), comprising the orders α-Proteobacteria (20%), γ-Proteobacteria (8%) and δ-proteobacteria (3%). The α-Proteobacteria comprised the orders Rhodospirallales (5%), SAR11 clade (5%), Rickettsiales (5%), Rhodobacteriales (4%) and the OCS116 clade (0.4%). The γ-Proteobacteria comprised the orders Oceanospirallales (6%), Alteromonadales (0.8%), Marinicella (0.7%) and K189A clade (0.5%). The δ-Proteobacteria were assigned to the SAR324 clade (3%). Bacteroidetes and Actinobacteria represented 4% and 2% of the prokaryote diversity. Finally, a large component (20%) of the prokaryotic community could not be assigned to any known sequences (Figure S1).

The eukaryotic community was dominated (92%) by the superphylum Alveolata (Figure S2), comprising the Protoalveolata (44%), Dinoflagellata (40%), Ciliophora (3%) and FV18-2D11 (3%). Protoalveolata were dominated by Syndiniales (97%), subdivided as: Group II (57%), Group I (18%), Amoebophyra (17%), Duboscquella (4%) and Perkinsidae (3%). The group Dinoflagellata was formed by Peridiniphycidae (16%), Gymnodiniphycidae (14%), Dinophysiales (1%) and Prorocentrum (0.7%).

3.3. Diversity of the <0.45 μm Fraction

After pre-processing 10 million paired reads were assembled to contigs with an average contig length of 1045 bp (Table 1), and a subsample of 1.5 million reads from R1 were utilised for analyses at the level of reads. The majority of sequences and predicted genes based on BLASTX against a virus database could be annotated as "other than virus" (Figure 4a). This was independent of whether the reads (99%) or the assembled contigs (86%) were used for the annotation (Figure 4b). Using the Refseq database, the metagenome could be divided into 59.92% Bacteria, 39.32% unknown, 0.71% Eukaryota and 0.05% Viruses at the read level, whilst for the contigs the hits could be divided into Bacteria (86.85%), unknown (11.03%), Eukaryota (1.35%), Viruses (0.75%) and Archaea (0.02%) (Figure 4c,d).

Utilizing the output from the Refseq database we compared annotation based on reads versus contigs. We observed very low similarities between the phylotypes annotated in the reads compared to the contigs (Figure 5). Only 8.81% of phylotypes were common across the two methods when no filter was applied (T0; Figure 5a), whereas 13.35% were common when T10 was applied (Figure 5c). To account for the high level of randomness associated with the top hits from BLAST outputs especially from universal conserved genes, we repeated the analyses using a lower stringency annotation, i.e., the genus as lowest level of classification instead of the phylotypes. Common annotations between the analysis based on reads versus contigs increased to 17.93% at T0 (Figure 5b) and 37.48% at T10 (Figure 5d). Therefore, from here on we focused our attention on the annotation based on the contigs. The Refseq annotation (Figure S3) produced an output highly dominated by Actinobacteria (47%) and Proteobacteria (38%). Specifically, the order Microcroccales made up 41% of sequences with the genus *Microbacterium* being the most dominant (33% of all the bacteria). The Proteobacteria comprised the classes' α-Proteobacteria (37%), γ-Proteobacteria (1%) and β-Proteobacteria (0.5%). The class α-Proteobacteria was dominated by the order Sphingomonadales (33%), with the genus *Erythrobacter* representing 24% of all the contigs, for which one coding sequence matched a 16S gene (Figure S3). Eukaryotes were represented in 1.35% of the metagenomic fraction and were dominated by the family Phaeophyceae (87%). Metazoa constituted only 0.07% of the eukaryotes (Figure S4).

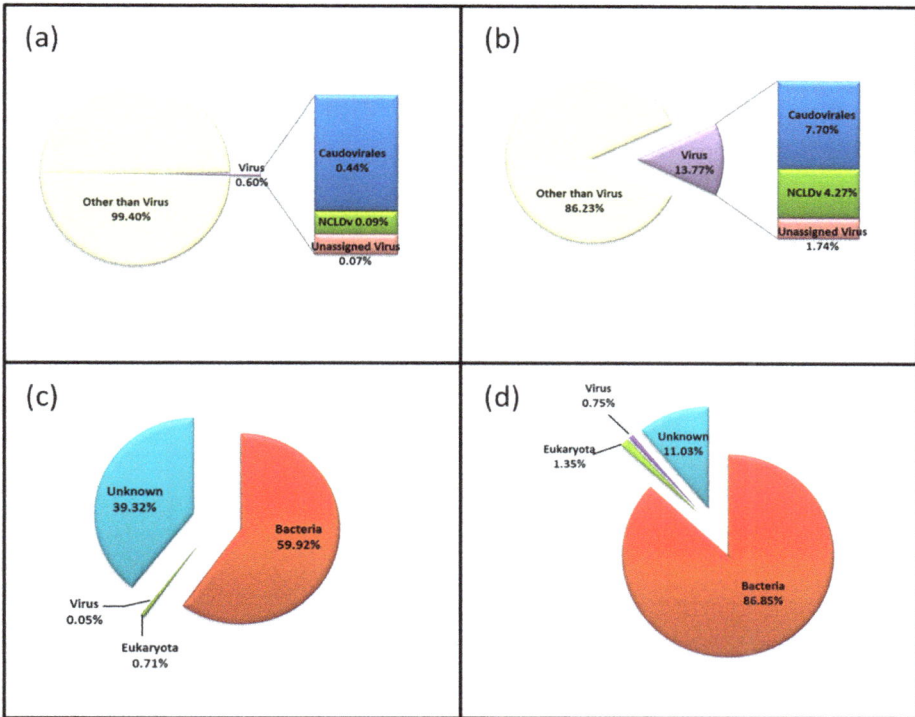

Figure 4. Taxonomic assignment based on reads (**a,c**) and contigs (**b,d**) analyses. Reads (R1) were annotated using (**a**) the Virus database and (**c**) the Refseq database; contigs were annotated using (**b**) the virus database and (**c**) the Refseq nr-protein database.

The viral contigs were further annotated using a curated Virus database (Figure 6). The virome was dominated by the order *Caudovirales* (59%) comprising the families *Myoviridae* (26%), *Siphoviridae* (22%) and *Podoviridae* (10%). The NCLDVs (28%) represented the second major order, with the families *Phycodnaviridae* (13%) and *Mimiviridae* (8%) as the main representatives.

3.4. Composition of Biota of the <0.45 µm versus the >0.45 µm Fraction

To understand if the prokaryotes and eukaryotes identified in the permeate (<0.45 µm) consisted of environmental DNA (debris or vesicles from extant Bacteria and Eukaryotes present in the water column), stable free DNA, or small Bacteria that passed through the filter, we ran presence–absence analyses comparing presence of microbiota in the <0.45 µm versus the >0.45 µm fraction for each filter (Figure 7). We also ran the analysis at the genus level or, when the genus annotation was not available, at the highest taxonomic level available. Very little overlap was observed across all levels of stringency (Figure 7). The genus *Phaeodactilum* (Table S1), shared between all datasets at T0, disappeared when singletons were removed (Figure 7b). Commonalities between eukaryotes and prokaryotes showed the presence of chloroplasts and mitochondria in the prokaryotic fraction with genera shared for 1.24% at T0, 0.83% at T1 and 0.45% at T10 (Figure 7). When the filter T1 was applied, it caused the removal of unusual genera such as *Cicer*, *Cucumis*, and *Porphyridium*, whilst genera such as *Chlorella*, *Chroomonas*, *Karlodium* and *Pedinomonas* disappeared with T10 filter (Table S2).

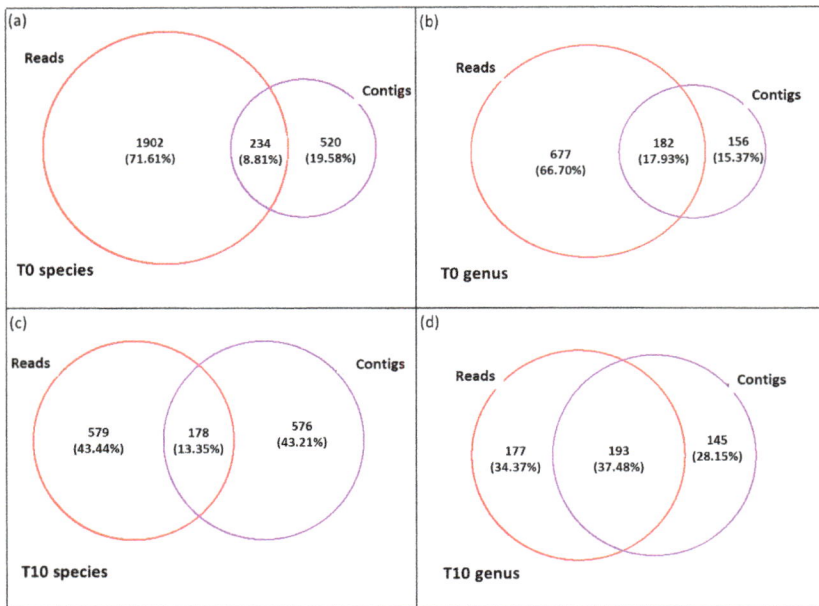

Figure 5. Presence–absence analyses of the <0.45 μm fraction. Comparison of phylotypes at the level of species (**a,c**) and genus (**b,d**) using a subsample of reads (R1) versus contigs at T0 (**a,b**) and T10 (**c,d**).

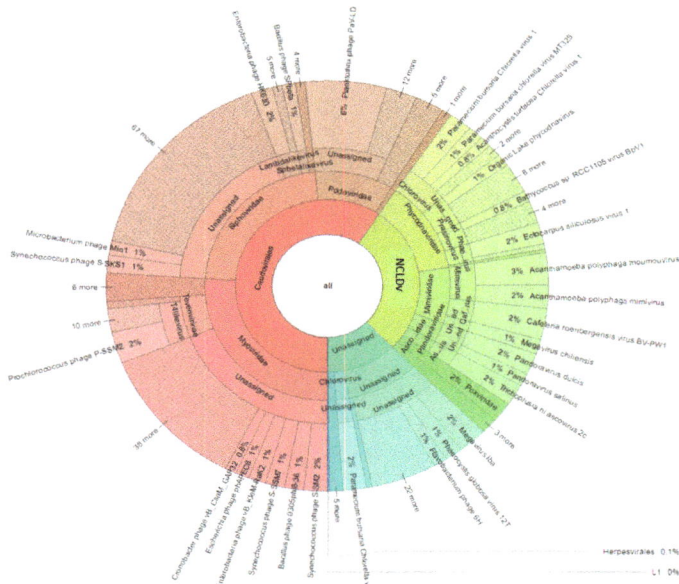

Figure 6. Krona chart of contigs annotation using the Virus db.

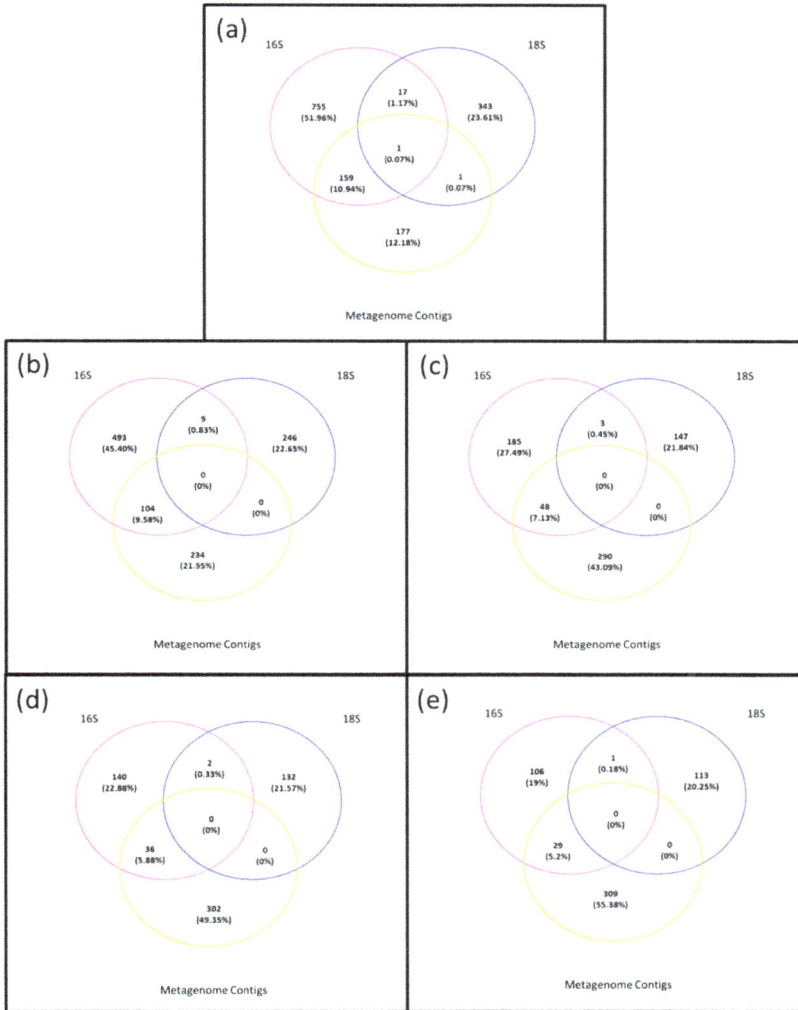

Figure 7. Presence–absence analyses between the >0.45 μm fraction (prokaryotes and eukaryotes) and the permeate (<0.45 μm). (**a**) T0: Metagenomic contigs, prokaryotes, eukaryotes; (**b**) T0: Metagenomic contigs, T1: prokaryotes, eukaryotes; (**c**) T0: Metagenomic contigs, T10: prokaryotes, eukaryotes; (**d**) T0: Metagenomic contigs, T10-R1: prokaryotes, eukaryotes; (**e**) T0: Metagenomic contigs, T10-R2: prokaryotes, eukaryotes.

4. Discussion

Microbes, from the smallest viruses to the largest unicellular protists, dominate our oceans, playing a central role in ocean food webs and as key drivers of biogeochemical processes [37]; yet the complex interactions and ecological significance of these relationships within and between biomes are largely unknown. The necessity of studying prokaryotes, eukaryotes and viruses together was highlighted in 2011 when it was estimated that only 11.2% and 2.2% of selected literature utilised two or three microbial groups, respectively [38]. For this reason, the more recent ocean expeditions sampling efforts include multiple trophic levels and ecosystem components in an attempt to better describe the complex microbial ecosystem structure and dynamics [39]. Describing and studying the hosts (prokaryotes and eukaryote assemblages) alongside their viruses can help improve our understanding of the roles of microbes in a more holistic way.

Given the patchiness of marine environments, changing rapidly both in time and space, the definition of a unique standard sample volume remains elusive [38]. Yet fingerprint profiles in the marine environment have shown the absence of significant difference in richness when utilizing from 10 to 1000 mL of seawater [40] as well as the low variability of the community structure when utilising more than 50 mL [41]. With this study, we used 250 mL of water, sampling the same seawater mass for all three microbial components (prokaryotes, eukaryotes and viruses). Here we demonstrate that the application of four levels of stringency allowed us to step-wise eliminate OTUs produced by sequencing errors and/or contamination. The removal of singletons resulted in the reduction of the overall phylotypes by around 700, while retaining over 99% of the reads. This step removed sequences of terrestrial origin (e.g., *Nicotiana* and *Cicer*), which are not expected to occupy the marine microbiome. Although singleton removal is a common practice, researchers do often retain these taxa under the label of "rare" microbiome. When singletons are removed in conjunction with replication of PCR runs a more stringent and precise description of the microbiota present in the environment can be obtained. This filtering step (T1 on the three replicates combined) allowed us to identify around 23,000 OTUs for the prokaryotic dataset and 3000 for the eukaryotic dataset grouping 834 and 346 as the lowest level of assigned taxa, respectively. Furthermore, the use of replication reduced the overall retained phylotypes when compared to individual replicates, because the duplicate values across the three replicates were removed, leaving only unique annotations, which constituted the dominate phylotypes of the sample. The further application of a more stringent filter, i.e., a phylotype was present with at least 10 reads in each PCR replicate, gave us the confidence that the rare microbiota were not included accidentally in the final dataset. However, this will invariably mean that genuine rare microbiota could be removed. This was the case of taxa such as *Chlorella*, *Pedinomonas*, *Marinobacter* and *Oceanicaulis*, which were removed by applying this filter level.

Bacterial composition at the location analysed by *Tara* Oceans expedition (station 64), based 548 nautical miles from ours, showed high abundance of α-Proteobacteria followed by Cyanobacteria (chloroplasts), γ-Proteobacteria and Bacteroidetes [8]. The microbial composition in our sample revealed the dominance of Cyanobacteria (*Synechococcus* and *Prochlorococcus*) followed by α-Proteobacteria, γ-Proteobacteria and Bacteroidetes. This Cyanobacteria dominance is more consistent with other viral abundance data (discussed further later on). Eukaryotes collected from *Tara* Oceans station 64 were dominated by the pico-nanoplankton, the Alveolata (Dinophyceae and Syndiniales clade MALV-I-II), followed in abundance by "other protists" [7]; our station was also dominated by Alveolata (Dinophyceae and Syndiniales). We hypothesise that the variation in composition from our station S1 and *Tara* Oceans' station 64 can be attributed to sampling different water masses as well as different sampling seasons: *Tara* Oceans' one was sampled in winter (July 2010), while our station S1 was collected in summer (February 2012). Given these differences, it is nonetheless remarkable how similar the microbial communities were, especially when considering the application of vastly different sampling protocols.

Analyses of the metagenomic fraction, 0.45 μm permeate, showed that annotations based on the assembled contigs lead to a more robust description of diversity. We found that the majority (86%) of our data did not match any viral genomes in our curated virus database. This was similar to what was reported by previous studies, i.e., 55% [42], 91.4% average [2], 88% [4] and 64.48% [43]. Marine viral metagenomics or metabarcoding studies currently apply various biomass or volume concentration methods before the extraction of DNA for sequencing. Such studies applied to our area of interest reported on the dominance of the order *Caudovirales*. Members of the family *Phycodnaviridae* were the second most abundant viral group, often followed by the family *Mimiviridae*. Our study demonstrated that a similar description of viral diversity is achievable from only 250 mL of seawater.

The high abundance of Prochlorococcus and Synechococcus phages was consistent with the observed dominance of their host cyanobacteria genera. Both *Prochlorococcus* and *Synechococcus* co-occurred and dominated the prokaryotic dataset with 30% and 9% of the sequences. It is curious to note that the *Tara* Oceans expedition [8] did not find any barcode sequences matching extant Cyanobacteria lineages despite the high abundance of both Prochlorococcus and Synechococcus phages in this locality. The reason for this anomaly might lie in the differing methodologies applied or indeed the difference in timing of sampling. Future side-by-side methodological comparative studies might resolve the reason behind these inconsistencies.

NCDLVs, such as *Phycodnaviridae* and *Mimiviridae,* surprisingly coincided with the presence of diatoms and dinoflagellates. These taxa, which constituted more than 90% of the eukaryotic dataset, are considered the most widespread protists on earth and are known to be routinely infected by RNA viruses [44]. Nevertheless, dinoflagellates are also infected by NCLDVs [44,45] and therefore our study suggests that further undescribed host-virus relationships can occur between dinoflagellates, diatoms and NCLDVs.

The 0.45 μm permeate or meDNA contains dissolved genetic material associated with cellular derived exudates (as part of the eDNA fraction [21]), viruses [20] or indeed small bacteria [46]. The comparative analyses of the two sampled size fractions revealed that bacteria and eukaryotes identified in this environment were not the source of the entire meDNA in our sample. On average 10% and 0% of the phylotypes was found in common between the meDNA and the bacterial and eukaryote permeate fractions, respectively. The likely explanation for the source of this DNA could be either the presence of viruses carrying host genes, since host genes have been identified in viral isolates or the presence of small bacteria (>0.45 μm). The latter included genera, identified in both datasets, such as *Pseudomonas, Flavobacterium, Serratia* and *Vibrio*, which are known to pass through 0.45 μm filters [46]. Nine coding sequences of the <0.45 μm fraction had hits with 16S proteins, six of which corresponded to *Microbacterium* (data not shown), and represented the main genera identified in this fraction. Furthermore it has been shown that, in adverse conditions, *Microbacterium* can present size reduction, which allowed it to pass through 0.45 μm filters [47,48]. Viruses often acquire host genes through horizontal gene transfer and since a large proportion of genetic material with unknown identity was also described, we hypothesise that viruses are the likely source of this meDNA. Whether this will be the case for all locations and situations remain to be determined. Our hypothesis contradicts Jiang and Paul [20], possibly because of their study locations being more productive. However, their study stopped short of confirming the species identified to actually being present in their water sample. Ultimately, eDNA, or, in our case, meDNA, do not appear to be a good proxy for describing the microbes present in this body of water.

5. Conclusions

To our knowledge, we report for the first time that 250 mL of seawater is sufficient to provide a comprehensive description of microbial diversity made up of 834 prokaryotic, 346 eukaryotic and 254 unique virus phylotypes. Moreover, given the paucity of fully curated marine viral genomes in searchable viral databases, we hypothesise that the meDNA fraction will be of viral origin. This in turn reinforces the potential of viruses to move host DNA around and even actively increase the repertoire of functional genes within any given individual, population, community or ecosystem. Finally, we do not recommend the use of meDNA as a proxy to describe the microbiome; however, this study needs to be replicated in other scenarios and locations.

Supplementary Materials: The following are available online at www.mdpi.com/1999-4915/9/3/47/s1. Figure S1: HTML link to the Prokaryotic Krona chart, Figure S2: HTML link to the Eukaryotic Krona chart, Table S1: 16S and 18S species lists in T0, Table S2: Common genera.

Acknowledgments: Computations were performed using facilities provided by the University of Cape Town's ICTS High Performance Computing team: http://hpc.uct.ac.za. Project was funded under National Research Foundation (NRF) grant to Ed Rybicki. (CPR20110717000020991). We would also like to thank Barney Balch for the opportunity to join the Coccolthophore Belt cruise. Declan Schroeder was funded by the FP7-OCEAN-2011 call, MicroB3 (grant number 287589) and the NERC eDNA award (grant number NE/N006151/1).

Author Contributions: Flavia Flaviani. and Declan Schroeder. wrote the manuscript; Cecilia Balestreri. collected the samples; Karen Moore and Konrad Pasckiewicz. prepared the DNA for Illumina sequencing and performed the sequencing; Flavia Flaviani and Joanna Schroeder performed the bioinformatics analysis; Maya Pfaff. reviewed the manuscript; Ed Rybicki and Declan Schroeder conceived the study.

Conflicts of Interest: The authors declare no conflict of interest.

References

1. De Wit, R.; Bouvier, T. 'Everything is everywhere, but, the environment selects'; what did baas becking and beijerinck really say? *Environ. Microbiol.* **2006**, *8*, 755–758. [CrossRef] [PubMed]
2. Angly, F.E.; Felts, B.; Breitbart, M.; Salamon, P.; Edwards, R.A.; Carlson, C.; Chan, A.M.; Haynes, M.; Kelley, S.; Liu, H.; et al. The marine viromes of four oceanic regions. *PLoS Biol.* **2006**, *4*, e368. [CrossRef] [PubMed]
3. Breitbart, M.; Rohwer, F. Here a virus, there a virus, everywhere the same virus? *Trends Microbiol.* **2005**, *13*, 278–284. [CrossRef] [PubMed]
4. Williamson, S.J.; Allen, L.Z.; Lorenzi, H.A.; Fadrosh, D.W.; Brami, D.; Thiagarajan, M.; McCrow, J.P.; Tovchigrechko, A.; Yooseph, S.; Venter, J.C. Metagenomic exploration of viruses throughout the indian ocean. *PLos ONE* **2012**, *7*, e42047. [CrossRef] [PubMed]
5. Williamson, S.J.; Rusch, D.B.; Yooseph, S.; Halpern, A.L.; Heidelberg, K.B.; Glass, J.I.; Andrews-Pfannkoch, C.; Fadrosh, D.; Miller, C.S.; Sutton, G.; et al. The sorcerer ii global ocean sampling expedition: Metagenomic characterization of viruses within aquatic microbial samples. *PLos ONE* **2008**, *3*, e1456. [CrossRef] [PubMed]
6. Ackermann, H.W. Bacteriophage observations and evolution. *Res. Microbiol.* **2003**, *154*, 245–251. [CrossRef]
7. de Vargas, C.; Audic, S.; Henry, N.; Decelle, J.; Mahé, F.; Logares, R.; Lara, E.; Berney, C.; Le Bescot, N.; Probert, I.; et al. Eukaryotic plankton diversity in the sunlit ocean. *Science* **2015**, *348*. [CrossRef]
8. Sunagawa, S.; Coelho, L.P.; Chaffron, S.; Kultima, J.R.; Labadie, K.; Salazar, G.; Djahanschiri, B.; Zeller, G.; Mende, D.R.; Alberti, A.; et al. Structure and function of the global ocean microbiome. *Science* **2015**, *348*. [CrossRef] [PubMed]
9. Brum, J.R.; Hurwitz, B.L.; Schofield, O.; Ducklow, H.W.; Sullivan, M.B. Seasonal time bombs: Dominant temperate viruses affect southern ocean microbial dynamics. *ISME J.* **2016**, *10*, 437–449. [CrossRef] [PubMed]
10. Brum, J.R.; Ignacio-Espinoza, J.C.; Roux, S.; Doulcier, G.; Acinas, S.G.; Alberti, A.; Chaffron, S.; Cruaud, C.; de Vargas, C.; Gasol, J.M.; et al. Patterns and ecological drivers of ocean viral communities. *Science* **2015**, *348*. [CrossRef] [PubMed]
11. Mihara, T.; Nishimura, Y.; Shimizu, Y.; Nishiyama, H.; Yoshikawa, G.; Uehara, H.; Hingamp, P.; Goto, S.; Ogata, H. Linking virus genomes with host taxonomy. *Viruses* **2016**, *8*, 66. [CrossRef] [PubMed]
12. Horiguchi, T. Diversity and phylogeny of marine parasitic dinoflagellates. In *Marine Protists: Diversity and Dynamics*; Ohtsuka, S., Suzaki, T., Horiguchi, T., Suzuki, N., Not, F., Eds.; Springer: Japan, Tokyo, 2015; pp. 397–419.
13. Hurwitz, B.L.; Sullivan, M.B. The pacific ocean virome (pov): A marine viral metagenomic dataset and associated protein clusters for quantitative viral ecology. *PLos ONE* **2013**, *8*, e57355.
14. Lima-Mendez, G.; Faust, K.; Henry, N.; Decelle, J.; Colin, S.; Carcillo, F.; Chaffron, S.; Ignacio-Espinosa, J.C.; Roux, S.; Vincent, F.; et al. Determinants of community structure in the global plankton interactome. *Science* **2015**, *348*. [CrossRef] [PubMed]
15. Hoeijmakers, W.A.M.; Bartfai, R.; Francoijs, K.-J.; Stunnenberg, H.G. Linear amplification for deep sequencing. *Nat. Protocols* **2011**, *6*, 1026–1036. [CrossRef] [PubMed]
16. John, S.G.; Mendez, C.B.; Deng, L.; Poulos, B.; Kauffman, A.K.M.; Kern, S.; Brum, J.; Polz, M.F.; Boyle, E.A.; Sullivan, M.B. A simple and efficient method for concentration of ocean viruses by chemical flocculation. *Environ. Microbiol. Rep.* **2011**, *3*, 195–202. [CrossRef] [PubMed]
17. Claverie, J.-M.; Ogata, H.; Audic, S.; Abergel, C.; Suhre, K.; Fournier, P.-E. Mimivirus and the emerging concept of "giant" virus. *Virus Res.* **2006**, *117*, 133–144. [CrossRef] [PubMed]
18. Martínez, J.M.; Swan, B.K.; Wilson, W.H. Marine viruses, a genetic reservoir revealed by targeted viromics. *ISME J.* **2014**, *8*, 1079–1088. [CrossRef] [PubMed]

19. Legendre, M.; Bartoli, J.; Shmakova, L.; Jeudy, S.; Labadie, K.; Adrait, A.; Lescot, M.; Poirot, O.; Bertaux, L.; Bruley, C.; et al. Thirty-thousand-year-old distant relative of giant icosahedral DNA viruses with a pandoravirus morphology. *Proc. Natl. Acad. Sci. USA* **2014**, *111*, 4274–4279. [CrossRef] [PubMed]

20. Jiang, S.C.; Paul, J.H. Viral contribution to dissolved DNA in the marine environment as determined by differential centrifugation and kingdom probing. *Appl. Environ. Microbiol.* **1995**, *61*, 317–325. [PubMed]

21. Creer, S.; Deiner, K.; Frey, S.; Porazinska, D.; Taberlet, P.; Thomas, W.K.; Potter, C.; Bik, H.M. The ecologist's field guide to sequence-based identification of biodiversity. *Methods Ecol. Evol.* **2016**, *7*, 1008–1018. [CrossRef]

22. Dejean, T.; Valentini, A.; Miquel, C.; Taberlet, P.; Bellemain, E.; Miaud, C. Improved detection of an alien invasive species through environmental DNA barcoding: The example of the american bullfrog lithobates catesbeianus. *J. Appl. Ecol.* **2012**, *49*, 953–959. [CrossRef]

23. Ikeda, K.; Doi, H.; Tanaka, K.; Kawai, T.; Negishi, J.N. Using environmental DNA to detect an endangered crayfish cambaroides japonicus in streams. *Conserv. Genet. Resour.* **2016**, *8*, 231–234. [CrossRef]

24. The great southern coccolithophore belt. Available online: http://www.Bco-dmo.Org/project/473206 (accessed on 13 March 2017).

25. Loecher, M.; Ropkins, K. Rgooglemaps and loa: Unleashing r graphics power on map tiles. *J. Stat. Softw.* **2015**, *63*, 1–18. [CrossRef]

26. Caporaso, J.G.; Paszkiewicz, K.; Field, D.; Knight, R.; Gilbert, J.A. The western english channel contains a persistent microbial seed bank. *ISME J.* **2012**, *6*, 1089–1093. [CrossRef] [PubMed]

27. Stoeck, T.; Bass, D.; Nebel, M.; Christen, R.; Jones, M.D.M.; Breiner, H.-W.; Richards, T.A. Multiple marker parallel tag environmental DNA sequencing reveals a highly complex eukaryotic community in marine anoxic water. *Mol. Ecol.* **2010**, *19*, 21–31. [CrossRef] [PubMed]

28. A quality control tool for high throughput sequence data. Available online: http://www.Bioinformatics.Babraham.Ac.Uk/projects/fastqc/ (accessed on 13 March 2017).

29. Fastx-toolkit. Available online: http://hannonlab.Cshl.Edu/fastx_toolkit/ (accessed on 13 March 2017).

30. Qiime-Quantitative Insights Into Microbial Ecology. Available online: http://qiime.Org (accessed on 13 March 2017).

31. Mahé, F.; Rognes, T.; Quince, C.; de Vargas, C.; Dunthorn, M. Swarm: Robust and fast clustering method for amplicon-based studies. *PeerJ* **2014**, *2*, e593. [CrossRef] [PubMed]

32. Silva rrna database project. Available online: https://www.Arb-silva.De (accessed on 13 March 2017).

33. Altschul, S.F.; Gish, W.; Miller, W.; Myers, E.W.; Lipman, D.J. Basic local alignment search tool. *J. Mol. Biol.* **1990**, *215*, 403–410. [CrossRef]

34. Tatusova, T.; Ciufo, S.; Fedorov, B.; O'Neill, K.; Tolstoy, I. Refseq microbial genomes database: New representation and annotation strategy. *Nucleic Acids Res.* **2014**, *42*, D553–D559. [CrossRef] [PubMed]

35. Ondov, B.D.; Bergman, N.H.; Phillippy, A.M. Interactive metagenomic visualization in a web browser. *BMC Bioinform.* **2011**, *12*, 385. [CrossRef] [PubMed]

36. Quast, C.; Pruesse, E.; Yilmaz, P.; Gerken, J.; Schweer, T.; Yarza, P.; Peplies, J.; Glöckner, F.O. The silva ribosomal rna gene database project: Improved data processing and web-based tools. *Nucleic Acids Res.* **2013**, *41*, D590–D596. [CrossRef] [PubMed]

37. Fuhrman, J.A. Microbial community structure and its functional implications. *Nature* **2009**, *459*, 193–199. [CrossRef] [PubMed]

38. Zinger, L.; Gobet, A.; Pommier, T. Two decades of describing the unseen majority of aquatic microbial diversity. *Mol. Ecol.* **2012**, *21*, 1878–1896. [CrossRef] [PubMed]

39. Marine ecology: Ocean survey finds huge diversity. *Nature* **2015**, *521*, 396.

40. Dorigo, U.; Fontvieille, D.; Humbert, J.-F. Spatial variability in the abundance and composition of the free-living bacterioplankton community in the pelagic zone of lake bourget (france). *FEMS Microbiol. Ecol.* **2006**, *58*, 109–119. [CrossRef] [PubMed]

41. Ghiglione, J.-F.; Larcher, M.; Lebaron, P. Spatial and temporal scales of variation in bacterioplankton community structure in the nw mediterranean sea. *Aquat. Microb. Ecol.* **2005**, *40*, 229–240. [CrossRef]

42. Brum, J.R.; Culley, A.I.; Steward, G.F. Assembly of a marine viral metagenome after physical fractionation. *PLos ONE* **2013**, *8*, e60604. [CrossRef] [PubMed]

43. Breitbart, M.; Salamon, P.; Andresen, B.; Mahaffy, J.M.; Segall, A.M.; Mead, D.; Azam, F.; Rohwer, F. Genomic analysis of uncultured marine viral communities. *Proc. Natl. Acad. Sci. USA* **2002**, *99*, 14250–14255. [CrossRef] [PubMed]

44. Nagasaki, K. Dinoflagellates, diatoms, and their viruses. *J. Microbiol.* **2008**, *46*, 235–243. [CrossRef] [PubMed]

45. Correa, A.M.S.; Welsh, R.M.; Vega Thurber, R.L. Unique nucleocytoplasmic dsdna and +ssrna viruses are associated with the dinoflagellate endosymbionts of corals. *ISME J.* **2013**, *7*, 13–27. [CrossRef] [PubMed]

46. Tabor, P.S.; Ohwada, K.; Colwell, R.R. Filterable marine bacteria found in the deep sea: Distribution, taxonomy, and response to starvation. *Microb. Ecol.* **1981**, *7*, 67–83. [CrossRef] [PubMed]

47. Chicote, E.; García, A.M.; Moreno, D.A.; Sarró, M.I.; Lorenzo, P.I.; Montero, F. Isolation and identification of bacteria from spent nuclear fuel pools. *J. Ind. Microbiol. Biotechnol.* **2005**, *32*, 155–162. [CrossRef] [PubMed]

48. Iizuka, T.; Yamanaka, S.; Nishiyama, T.; Hiraishi, A. Isolation and phylogenetic analysis of aerobic copiotrophic ultramicrobacteria from urban soil. *J. Gen. Appl. Microbiol.* **1998**, *44*, 75–84. [CrossRef] [PubMed]

viruses

MDPI

Article

Isolation and Characterization of a Double Stranded DNA Megavirus Infecting the Toxin-Producing Haptophyte *Prymnesium parvum*

Ben A. Wagstaff [1], Iulia C. Vladu [1], J. Elaine Barclay [1], Declan C. Schroeder [2], Gill Malin [3] and Robert A. Field [1,*]

[1] Department of Biological Chemistry, John Innes Centre, Norwich Research Park, Norwich NR4 7UH, UK; ben.wagstaff@live.co.uk (B.A.W.); vladu.iulia@yahoo.com (I.C.V.); elaine.barclay@jic.ac.uk (J.E.B.)
[2] Marine Biological Association of the UK, Plymouth PL1 2PB, UK; dsch@mba.ac.uk
[3] Centre for Ocean and Atmospheric Studies, School of Environmental Sciences, University of East Anglia, Norwich Research Park, Norwich NR4 7TJ, UK; g.malin@uea.ac.uk
* Correspondence: rob.field@jic.ac.uk; Tel.: +44-1603-450720

Academic Editors: Mathias Middelboe and Corina Brussaard
Received: 29 January 2017; Accepted: 27 February 2017; Published: 9 March 2017

Abstract: *Prymnesium parvum* is a toxin-producing haptophyte that causes harmful algal blooms globally, leading to large-scale fish kills that have severe ecological and economic implications. For the model haptophyte, *Emiliania huxleyi*, it has been shown that large dsDNA viruses play an important role in regulating blooms and therefore biogeochemical cycling, but much less work has been done looking at viruses that infect *P. parvum*, or the role that these viruses may play in regulating harmful algal blooms. In this study, we report the isolation and characterization of a lytic nucleo-cytoplasmic large DNA virus (NCLDV) collected from the site of a harmful *P. parvum* bloom. In subsequent experiments, this virus was shown to infect cultures of *Prymnesium* sp. and showed phylogenetic similarity to the extended *Megaviridae* family of algal viruses.

Keywords: *Prymnesium parvum*; haptophyte; algal bloom; algal virus; *Megaviridae*; NCLDV

1. Introduction

The last two decades have seen a boom in the study of marine viruses and the role that they play in regulating both bacterial and unicellular eukaryote bloom dynamics [1,2]. Although phages and the bacteria that they infect have been studied for many years, the more recently discovered *Acanthamoeba polyphaga mimivirus* (APMV) and its *Megaviridae* relatives have brought about a new age in photosynthetic protist virology. It has recently been shown that dsDNA viruses infecting algae do not form monophyletic lineages [3], with divergence occurring even within the host division. A good example of this evolutionary divergence can be found in viruses that infect the coccolithophore *Emiliania huxleyi* (EhV) [4,5] and the prymnesiophyte *Phaeocystis globosa* (PgV) [6], which along with other algal viruses have been proposed to form an extended branch of the *Megaviridae* [7]. It is widely accepted that these viruses not only play a crucial role in ecosystem dynamics [8,9], but also contribute significantly to biogeochemical cycles [10,11]. A lesser studied impact, however, lies in the role that such viruses may play in the termination of toxic eukaryotic algal blooms. Lytic viruses that infect the toxic raphidophyte *Heterosigma akashiwo* have been extensively studied [12–19] but, because of the elusive nature of *H. akashiwo* toxicity to fish, none of these studies sought to investigate the role of viral infection on levels of algal toxicity.

The toxin-producing haptophyte *Prymnesium parvum* forms dense blooms in marine, brackish and inland waters, devastating fish populations through the release of natural product toxins [20,21].

The haptophytes are a diverse division of microalgae that include the bloom-forming *Emiliania huxleyi* and *Phaeocystis globosa*, both of which play crucial roles in oceanic carbon and sulfur cycles [22,23]. Virus infection of these organisms has been studied in some detail, with the genome of the dsDNA *Phaeocystis globosa* virus (PgV-16T) being recently described [3]. From a metabolomics perspective, *Phaeocystis pouchetti* lysis by a strain-specific virus has been shown to cause substantial release of dimethyl sulphide and its major precursor dimethylsulphoniopropionate [24], an action that is believed to contribute significantly to the global sulfur cycle. Although much effort has gone into studying the relationship between *E. huxleyi* and its infecting viruses, viruses infecting toxin-producing algal species within the haptophyte family are much less well studied. These include the euryhaline species *Prymnesium* spp. and *Chrysochromulina* spp., whose blooms can often result in severe economic damage through loss of fish stocks [25,26]. Viruses that infect the non-toxic *P. kappa* have recently been described, but to date no viruses have been isolated and characterized that infect the toxin-producing *P. parvum* species, even though Schwierzke et al. have previously suggested a role for viruses in regulating natural *P. parvum* populations [27].

In this study, we isolated a novel lytic virus of *P. parvum* 946/6, *Prymnesium parvum* DNA virus BW1 (henceforth referred to as PpDNAV), from the site of a recent harmful bloom event of this species in Norfolk, England. We show that the virus has a typical narrow host range; using morphological characterisation and phylogenetics, we also show that the virus lies in the recently described clade of algal megaviruses.

2. Materials and Methods

2.1. Prymnesium parvum Culture Conditions

For choice of host cell, *P. parvum* 946/6 was obtained from the Culture Collection of Algae and Protozoa (CCAP—www.ccap.ac.uk). The additional 14 strains used for host range screening were obtained from the Marine Biological Association Culture Collection (https://www.mba.ac.uk/culture-collection/). Batch cultures were maintained at 22 °C on a 14:10 light cycle at 100 μmol·photons·m^{-2}·s^{-1}. Cultures were grown in f/2–Si medium at a salinity of 7–8 practical salinity unit (PSU). Under these conditions, cell densities of ~3 × 10^6 cells·mL^{-1} could be achieved after 12–16 days of growth.

2.2. Isolation of Lytic Virus Particles

PpDNAV was isolated from surface water samples taken at various locations on Hickling Broad, Norfolk, England on 9 February 2016. In brief, 4 × 100 mL water samples from various locations around the Broad were centrifuged at 3000× *g* and the supernatant subsequently filtered through 0.45 μm pore-size filters (Sartorius AG, Goettingen, Germany). The resulting solutions were then concentrated 100- to 200-fold using 100 kDa MW cut off spin filters (Amicon Ultra 15, Merck Millipore, Watford, UK) to give 0.5 to 1 mL of viral concentrate, which was stored at 4 °C in the dark until use. Small volumes (0.2 mL) of concentrate from each location were added to 1.8 mL of exponentially growing cultures of *P. parvum* 946/6. Blank culture medium was used as a control. Cultures were visually inspected for signs of cell lysis (culture clearing) after 7–10 days where the control cultures continued to grow. Culture clearing was then followed up by Transmission Electron Microscopy (TEM) analysis of the culture lysates. Clonal populations of PpDNAV were obtained by taking the supernatant of a lysed culture, and exhaustively diluting with media. These diluted samples (0.2 mL) were added to 1.8 mL of an exponentially growing culture of *P. parvum* 946/6. The highest dilution that still produced cell lysis after seven days was taken through to the next round. This was repeated at least three times and resulted in a population of PpDNAV free of morphologically different viruses, as judged by TEM.

2.3. Transmission Electron Microscopy

For TEM analysis of virus-like particles (VLPs) in culture supernatant, 2 mL of a virus-lysed culture was filtered through 0.45 μm filters and 10 μL of the filtrate was adsorbed onto a 400 mesh copper palladium grid with a carbon-coated pyroxylin support film before being negatively stained with 2% aqueous uranyl acetate [28]. The grids were viewed in a FEI Tecnai 20 transmission electron microscope (Eindhoven, The Netherlands) at 200 kV and digital TIFF images were taken with an AMT XR60B digital camera (Deben, Bury St Edmunds, UK).

For analysing intracellular VLPs, 1 mL of infected cultures of *P. parvum* 946/6 was taken at 24 and 48 h post-infection (p.i.). These were centrifuged at 3000× *g* to pellet algal cells and the supernatant was discarded. The pellet was washed twice with sterile medium. The pelleted cells were then resuspended in 2.5% (*v/v*) aqueous glutaraldehyde solution and left overnight. This suspension was then centrifuged at 3000× *g* to pellet the algal cells. Half the volume of the supernatant was then discarded and an equal volume of warm (60 °C) low gelling temperature agarose (Sigma Aldrich, Haverhill, UK) was added, before resuspension of the cells and placing on ice to solidify. The solidified samples were then put into 2.5% (*v/v*) glutaraldehyde with 0.05 M sodium cacodylate, pH 7.3 [29] and left overnight. Using a Leica EM TP machine (Leica Microsystems, Cambridge, UK), the samples were washed in 0.05 M sodium cacodylate and then post-fixed with 1% (*w/v*) OsO4 in 0.05 M sodium cacodylate for 60 min at room temperature. After washing and dehydration with ethanol, the samples were gradually infiltrated with LR White resin (London Resin Company, London, UK) according to the manufacturer's instructions. After polymerization, the resulting material was sectioned with a diamond knife using a Leica EM UC6 ultramicrotome (Leica Microsystems). Ultrathin sections of approximately 90 nm were picked up on 200 mesh gold grids that had been coated in pyroxylin and carbon. The grids were then contrast-stained with 2% (*w/v*) uranyl acetate for 1 h and 1% (*w/v*) lead citrate for 1 min, washed in distilled water and air-dried. The grids were then viewed with a FEI Tecnai 20 transmission electron microscope (Eindhoven, The Netherlands) at 200 kV and digital TIFF images were produced.

2.4. Host Specificity

Fifteen different strains of *Prymnesium* were tested in triplicate for signs of cell lysis by PpDNAV using the infection methodology described above. Cell lysis, as observed by culture clearing, was noted for five of the 15 strains tested (Table 1).

Table 1. Host range of PpDNAV. + lysed culture, − culture not lysed.

Genus/Species	Strain Code	Lysis with PpDNAV
Prymnesium parvum	946/6	+
Prymnesium parvum	94A	-
Prymnesium parvum	94C	+
Prymnesium parvum	579	-
Prymnesium patelliferum	527A	+
Prymnesium patelliferum	527C	+
Prymnesium patelliferum	527D	-
Prymnesium sp.	522	-
Prymnesium sp.	569	-
Prymnesium sp.	592	+
Prymnesium sp.	593	-
Prymnesium sp.	595	-
Prymnesium sp.	596	-
Prymnesium sp.	597	-
Prymnesium sp.	598	-

2.5. Infection Cycle

The virus–algae lytic cycle was investigated by accurately recording algal cell abundance during an infection cycle. A late-log phase culture of *P. parvum* 946/6 was infected with PpDNAV (0.1% *v/v*) and triplicate 2 mL aliquots were taken at various time points post infection (p.i.). These were diluted with 0.2 µm filtered seawater prior to counting using a Multisizer 3 Analyser (Beckman Coulter, High Wycombe, UK) fitted with a 100 µm aperture tube. The control culture continued to grow throughout the experiment, whilst the infected algal culture was lysed rapidly after 48 h.

2.6. Chloroform Sensitivity

To test the virus sensitivity to chloroform, an adaptation of the method of Martínez Martínez et al. was employed [30]. Briefly, 1 mL of 0.45 µm-filtered PpDNAV was added to an equivalent volume of chloroform and shaken vigorously for 5 min. The resulting mixture was then centrifuged at $4000\times g$ in a benchtop centrifuge for 5 min to separate the organic and polar layers. The aqueous phase was transferred by pipetting to a clean microcentrifuge tube and incubated at 37 °C for 1 h to remove residual chloroform. As a control, 1 mL of chloroform was added to 1 mL of f/2 medium. Chloroform-treated PpDNAV, chloroform-treated medium and untreated PpDNAV were added to *P. parvum* 946/6 as described above in the infectivity experiment protocol; signs of lysis, as judged by culture clearing, were recorded after one week.

2.7. Viral DNA Extraction, Sequencing, and Phylogenetic Analyses

For DNA extraction, 1 L of late log phase *P. parvum* 946/6 was infected with axenic PpDNAV (0.1% *v/v*). Lysis was allowed to occur over a period of five days, by which point almost all cells had been lysed. The culture was centrifuged at $6500\times g$ to pellet cell debris, before being filtered through 0.22 µm filters to remove remaining cell debris or contaminating bacteria. The filtrate was incubated for 72 h with 100 µg/mL carbenicillin before being concentrated to 30 mL using 100 kDa mw cut-off spin filters. Ultracentrifugation at $150,000\times g$ was used to pellet viral particles, and these were re-suspended in 2 mL of $\varrho = 1.4$ CsCl and layered onto a CsCl gradient which was resolved at $150,000\times g$ for 18 h. Fractions from $\varrho = 1.3$ to $\varrho = 1.4$ were pooled and DNA extracted using a PureLink Viral RNA/DNA Kit, according to the manufacturer's protocol.

An amount of 1 µg of purified viral DNA was then sent to The Earlham Institute, UK, for Illumina MiSeq sequencing (Illumina, Inc., San Diego, CA, USA) and assembly. The initial assembly was then analysed using GeneMarkS [31] which identified 332 protein-coding sequences. BLASTp analysis was then performed against the National Center for Biotechnology Information (NCBI) GenBank nonredundant (nr) protein sequence database [32] to identify major capsid protein and DNA Pol B candidates. Nucleic acid and amino acid sequences for the major capsid protein (MCP) and DNA Polymerase B (DNA polB) were submitted to Genbank with the accession codes KY509047 and KY509048, respectively.

Phylogenetic analysis was performed using the obtained sequences for MCP and DNA polB, as well as other related sequences from previously discovered algal viruses, identified using BLASTp. These sequences were aligned using the default settings of multiple sequence alignment software version 7 (MAFFT) [33], and trees were constructed from the neighbour-joining method [34] (midpoint-rooted) using Molecular Evolutionary Genetics Analysis version 7.0 (MEGA7) [35].

3. Results

3.1. Isolation of Lytic Virus Particles

PpDNAV isolation was conducted from water samples collected at Hickling Broad, Norfolk, England. Among four water samples from which viral lysates were prepared, lysis of *P. parvum* 946/6 occurred with three samples (Figure 1). Transmission electron micrographs of the viral lysates showed that icosahedral VLPs were present in all three samples, but samples 1 and 2 also contained significant levels of phage-like particles; we suspect that these were a result of infection with the low levels of bacteria that were present in the non-axenic *P. parvum* 946/6 cultures. To avoid further downstream separation of viruses, we chose to continue working with sample 4 only (sourced at—52°44′19.12″ N, Long—1°34′39.49″ E), which appeared by TEM to be free of phages. After a triplicate dilution series, the resulting monoclonal viral lysate still lysed the host cells and TEM of thin-sectioned cells confirmed the presence of VLPs (Figure 2A,B); thereby fulfilling Koch's postulates.

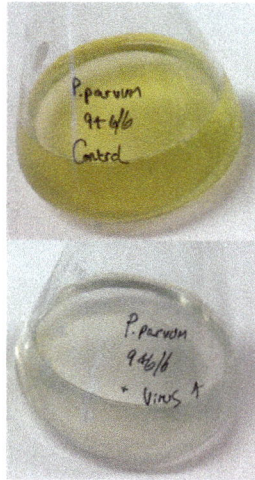

Figure 1. (**Top**) control culture; (**Bottom**) 'Cleared' culture 96 h post viral infection.

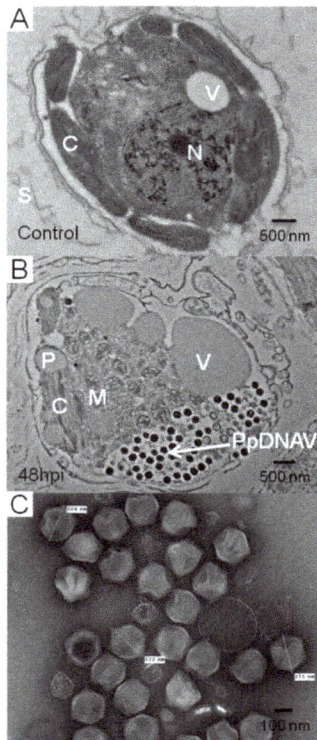

Figure 2. (**A**) Thin-sections of healthy *P. parvum* 946/6 cells; (**B**) Thin-sections of *P. parvum* 946/6 48 h post infection. (**C**) Free *Prymnesium parvum* DNA virus (PpDNAV) particles in culture supernatant 72 h post infection. C: chloroplast; V: contractile vacuole; N: nucleus; S: scales; M: mitochondria, P: pyrenoid.

3.2. Virus Morphology, Host Range, and Infectious Properties

Transmission electron microscopy of isolated and intracellular (thin sectioned) viruses revealed an icosahedral capsid with an average diameter of 221 nm ($n = 71$) (Figure 2). Although no external viral lipid membrane was evident, some viral particles showed an internal white 'halo' between the capsid and the DNA of PpDNAV, suggestive of the virus having an internal membrane. The presence of a viral factory or viroplasm [36] in the host cytoplasm, and in some cases an imperfect vertex or a tail-like structure were also observed (Figures S1 and S2). As seen in Figure 2B, establishment of a viral factory in the host cytoplasm also results in a loss of the nuclear envelope and therefore loss of the nucleus. This was observed in the majority of infected cells examined at 48 h p.i.

Fifteen different strains of *Prymnesium* were screened for sensitivity to PpDNAV (Table 1). PpDNAV was found to be sensitive to chloroform, whereby the chloroform-treated virus no longer caused lysis of *P. parvum* 946/6 (Figure S3). This supports the notion of a viral membrane in this system.

The lytic cycle of the virus was explored to determine both the incubation period and eclipse period (Figure 3). At 48 h p.i., the cells had clearly lost mobility and sedimented at the base of the culture flask. Re-suspension of the cells by shaking led to similar cell counts as seen at 24 h p.i., as determined by Coulter counting. The time before symptoms of viral infection, the incubation period, was therefore judged to be 24 h. The eclipse period reflects the time between infection and appearance of mature virus particles within the host; as new mature virions were first observed 48 h p.i., the eclipse period was judged to be 24–48 h. At 72 h p.i., the onset of cell lysis had occurred. PpDNAV appeared to lyse >95% of host cells by 120 h p.i., whilst uninfected control cultures continued to grow over the full course of the experiment.

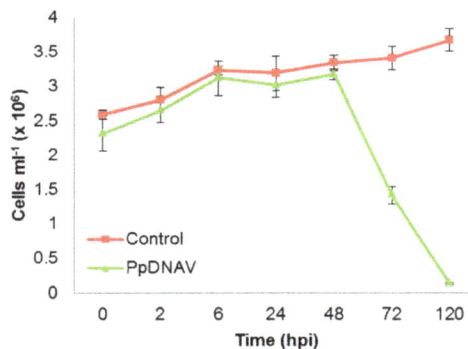

Figure 3. PpDNAV infection cycle propagated on *P. parvum* 946/6. Graph shows the average number of algal cells in control cultures (squares) and PpDNAV infected cultures (circles). Error bars represent the standard error for triplicate cultures.

3.3. Genome Sequencing and Phylogenetic Analysis

Predicted proteins from the initial genome assembly included the MCP1 protein (KY509047) and DNA polB (KY509048) which were used for phylogenetic analysis. The 525 aa sequence for MCP1 was found to have 91% sequence similarity to the major capsid protein 1 of *Phaeocystis globosa* virus (YP_008052475.1) and 84% similarity to MCP1 of Organic Lake Phycodnavirus 2 (ADX06358.1) with E-values of 0.0 in each case. This alignment allowed construction of a phylogenetic tree (Figure S4) that shows clustering with other megaviruses, including PgV-16T.

For DNA polB (KY509048), the 1281 aa sequence displayed 77% sequence similarity to DNA polB of PgV-16T (YP_008052566.1) and 64% similarity to DNA polB of Organic Lake phycodnavirus 2 (ADX06483.1). The resulting phylogenetic tree (Figure 4) shows a similar clustering of PpDNAV to the algal *Megaviridae* family, but also illustrates an obvious divergence between algal viruses that fall within the *Megaviridae* family and those that do not; with EhV-86 and *Heterosigma akashiwo* virus (HaV)-1 rightfully placed outside of the *Megaviridae* clade.

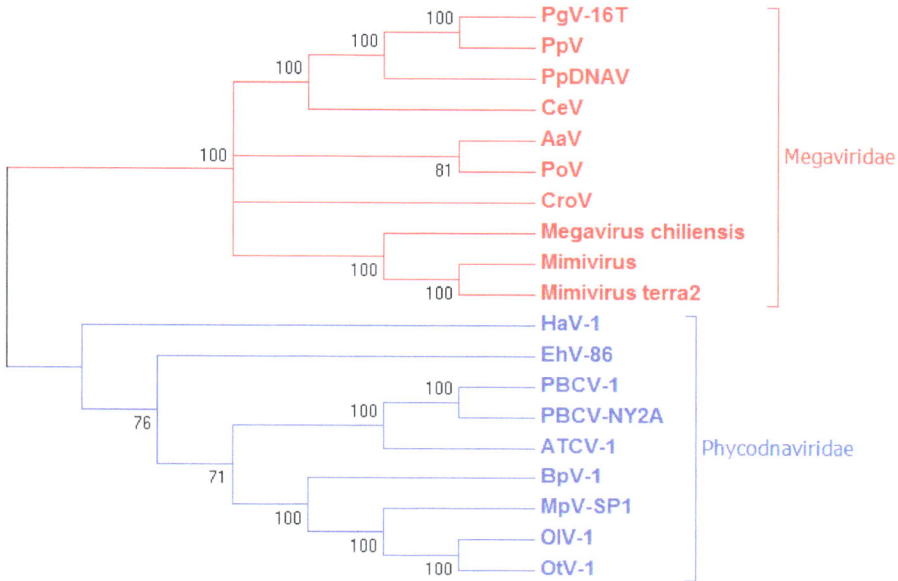

Figure 4. Phylogenetic clustering of PpDNAV with other large algal *Megaviridae*. Alignment was performed using the default settings of multiple sequence alignment software version 7 (MAFFT) [33], and the neighbour-joining method (midpoint-rooted) [34] was used to construct a tree from 19 viral DNA Polymerase Beta (polB) sequences using Molecular Evolutionary Genetics Analysis version 7.0 (MEGA7) [35]. The final tree was based on 630 ungapped positions, 500 resampling permutations, and was collapsed for bootstrap values <50. The tree shows that PpDNAV clusters with the well-defined clade of *Megaviridae* and the algal-infecting *Megaviridae* (red), and not with the *Phycodnaviridae* (blue).

4. Discussion

Haptophytes are abundant in marine waters but can also thrive in brackish inland waters. Whilst a significant amount of work has been done on the marine dwelling coccolithophore *Emiliania huxleyi* and its associated viruses, little work has looked at the toxin-producing members of the haptophytes. In the present study, we isolated and characterized a novel megavirus, PpDNAV, from brackish inland waters where harmful blooms of *Prymnesium parvum* frequently occur [37]. We showed that this lytic virus was able to infect *P. parvum* 946/6, later expanded to five out of 15 *Prymnesium* strains tested. Morphological and phylogenetic analysis of two core dsDNA virus conserved genes suggests that this virus belongs to the extended *Megaviridae* family of algal-infecting viruses.

Transmission electron microscopy of negatively stained virus particles from a lysed culture supernatant revealed icosahedral capsids with an average diameter of 221 nm. Many particles appeared to have one imperfect vertex, with some showing material protruding from what appeared to be a stargate [38] (Figure S1). These likely represent particles in an advanced stage of packing or unpacking genetic material [38], and suggested early on that PpDNAV lies in the extended *Megaviridae* branch of algal viruses. Thin sections of infected *P. parvum* 946/6 cells showed evidence for a viroplasm as the site of replication, where empty capsids could be seen closer to the centre of the viroplasm (Figure S2). This further supported the inclusion of PpDNAV in the extended *Megaviridae* family [39]. The infectivity of PpDNAV is chloroform sensitive (Figure S3), and the lack of an obvious external lipid membrane observed by TEM may suggest that internal membrane/s are present; although chloroform sensitivity cannot always be used to confirm lipid membrane presence [40].

New mature virions were first observed by electron microscopy at 48 h p.i., so the eclipse period of the virus in infected algal cells was estimated to be 24–48 h. At 48 h, the algal cell count as recorded by Coulter counting was still the same as at 24 h, but a complete sedimentation of cells had occurred, suggesting a loss in motility and a likely shutdown of important cellular processes. By 72 h, a rapid decline in cell abundance could be observed, showing that the loss of motility precedes the host lysis event, as is seen for some other flagellated algae [18]. In its natural environment, this may lead to accumulation of viral particles at the sediment surface rather than dispersed in the water column.

The algal host species specificity of PpDNAV was assessed against *P. parvum* 946/6, which had been kept in 7–8 PSU f/2 medium for two years, and 14 other *Prymnesium* strains which had been maintained in a full strength seawater medium. Initially, PpDNAV only infected *P. parvum* 946/6, but after ~6 months of sub-culturing of the other 14 strains in 7–8 PSU f/2 medium, the host range broadened to five out of the 15 strains. We speculate that the change in salinity contributed to the change in sensitivity to PpDNAV; recent work by Nedbalová et al. [41] suggests that a change in membrane lipid composition in different salinities may account for this situation. This somewhat less restricted host range is similar to that found for *Haptolina ericina* virus (HeV RF02), *Prymnesium kappa* virus (PkV RF01) and *Prymnesium kappa* virus (PkV RF02) [42]. Taken together, this suggested that PpDNAV was a member of the algal *Megavirus* family [43].

Phylogenetic analysis using sequences for MCP1 and DNA polB of PpDNAV confirmed morphological findings, showing that PpDNAV clusters amongst the algal viruses belonging to the *Megaviridae* family, such as PgV-16T [3,6], *Chrysochromulina ericina* virus (CeV) [44], *Pyramimonas orientalis* virus (PoV) [45] and the recently reassigned *Aureococcus anophageffens* virus (AaV) [46]. With the exception of *Emliania huxleyi* virus (EhV-86), which appears to branch independently, the close clustering of viruses infecting haptophytes, as well as the chlorella viruses clustering together, supports the notion that viruses co-evolve with their hosts [6,46,47].

Of the algal viruses compared in this study, only HaV-1 is known to infect a toxin producing host [12–14]. However, the toxic metabolites responsible for bloom toxicity are not established in *Heterosigma akashiwo*, making studies of viral impact on toxicity difficult. On the other hand, reports of toxic *P. parvum* metabolites are numerous and include fatty acids [48], glycerolipids [49] and very large ladder-frame polyether toxins, known collectively as the prymnesins [50–52]. Reports of cases of toxic and non-toxic blooms of *Prymnesium* and other harmful algal species [37] has led to speculation that an ecological trigger exists for toxicity. While efforts have been made to associate nutrients, pH and other conditions to bloom toxicity [21], the identity of the full spectrum of toxicity-causing agents remains to be establshed; there may conceivably be a role for viral infection in *Prymnesium* cell lysis and hence toxin release. We now have the opportunity to use this algae–virus system in clearing up some of these unanswered questions. Further studies into the effect of viral infection and host algal cell lysis on toxic bloom events need to be explored in order to fully understand the underlying mechanisms behind production and release of toxins from Prymnesium. In addition, as further sequences of algal viruses become available, new opportunities will open up for accurate monitoring of viral population fluctuations with respect their host. Furthermore, the increase in characterized viruses will provide more information when analysing metagenomic data sets such as those generated by the *Tara* Oceans expedition [53,54]. Hence the discovery and characterization of PpDNAV in this study will aid this burgeoning field of scientific endeavour.

Supplementary Materials: The following are available online at www.mdpi.com/1999-4915/9/3/40/s1. Figure S1: Electron micrographs showing PpDNAV particles with an imperfect vertex; Figure S2: Electron micrographs showing both mature and immature virions close to a viroplasm; Figure S3: Chloroform sensitivity as judged by culture clearing; Figure S4: Neighbor-joining phylogenetic tree of MCPs.

Acknowledgments: These studies were supported by the BBSRC Institute Strategic Programme on Understanding and Exploiting Metabolism (MET) [BB/j004561/1] and the John Innes Foundation. B.A.W. was supported by a BBSRC industrial CASE PhD studentship supported by Environment Agency. We thank Willie Wilson for helpful advice of viral morphology. We also thank Steve Lane, Andy Hindes, John Currie, Jenny Pratscher and Colin Murrell for their ongoing support.

Author Contributions: B.A.W. and R.A.F. conceived and designed the experiments; B.A.W. performed the experiments, with help from I.C.V. for chloroform sensitivity and culture maintenance, and J.E.B. for thin-section preparation and EM imaging; D.C.S. provided 14 *Prymnesium* strains for host range screening and reviewed the manuscript; G.M. also reviewed the manuscript; B.A.W., G.M. and R.A.F. analysed the data; B.A.W. and R.A.F. wrote the manuscript.

Conflicts of Interest: The authors declare no conflict of interest.

References

1. Thingstad, T.F. Elements of a theory for the mechanisms controlling abundance, diversity, and biogeochemical role of lytic bacterial viruses in aquatic systems. *Limnol. Oceanogr.* **2000**, *45*, 1320–1328. [CrossRef]

2. Fuhrman, J.A. Marine viruses and their biogeochemical and ecological effects. *Nature* **1999**, *399*, 541–548. [CrossRef] [PubMed]

3. Santini, S.; Jeudy, S.; Bartoli, J.; Poirot, O.; Lescot, M.; Abergel, C.; Barbe, V.; Wommack, K.E.; Noordeloos, A.A.; Brussaard, C.P.; et al. Genome of *Phaeocystis globosa* virus PgV-16T highlights the common ancestry of the largest known DNA viruses infecting eukaryotes. *Proc. Natl. Acad. Sci. USA* **2013**, *110*, 10800–10805. [CrossRef] [PubMed]

4. Wilson, W.H.; Tarran, G.A.; Schroeder, D.; Cox, M.; Oke, J.; Malin, G. Isolation of viruses responsible for the demise of an *Emiliania huxleyi* bloom in the English Channel. *J. Mar. Biol. Assoc. UK* **2002**, *82*, 369–377. [CrossRef]

5. Schroeder, D.C.; Oke, J.; Malin, G.; Wilson, W.H. Coccolithovirus (Phycodnaviridae): Characterisation of a new large dsDNA algal virus that infects *Emiliana huxleyi*. *Arch. Virol.* **2002**, *147*, 1685–1698. [CrossRef] [PubMed]

6. Brussaard, C.P.D.; Short, S.M.; Frederickson, C.M.; Suttle, C.A. Isolation and Phylogenetic Analysis of Novel Viruses Infecting the Phytoplankton *Phaeocystis globosa* (Prymnesiophyceae). *Appl. Environ. Microbiol.* **2004**, *70*, 3700–3705. [CrossRef] [PubMed]

7. Moniruzzaman, M.; Gann, E.R.; LeCleir, G.R.; Kang, Y.; Gobler, C.J.; Wilhelm, S.W. Diversity and dynamics of algal Megaviridae members during a harmful brown tide caused by the pelagophyte, *Aureococcus anophagefferens*. *FEMS Microbiol. Ecol.* **2016**, *92*, fiw058. [CrossRef] [PubMed]

8. Short, S.M. The ecology of viruses that infect eukaryotic algae. *Environ. Microbiol.* **2012**, *14*, 2253–2271. [CrossRef] [PubMed]

9. Brussaard, C.P.D. Viral Control of Phytoplankton Populations—A Review. *J. Eukaryot. Microbiol.* **2004**, *51*, 125–138. [CrossRef] [PubMed]

10. Suttle, C.A. Marine viruses—Major players in the global ecosystem. *Nat. Rev. Microl.* **2007**, *5*, 801–812. [CrossRef] [PubMed]

11. Wilhelm, S.W.; Suttle, C.A. Viruses and Nutrient Cycles in the Sea: Viruses play critical roles in the structure and function of aquatic food webs. *Bioscience* **1999**, *49*, 781–788. [CrossRef]

12. Nagasaki, K.; Yamaguchi, M. Isolation of a virus infectious to the harmful bloom causing microalga *Heterosigma akashiwo* (Raphidophyceae). *Aquat. Microb. Ecol.* **1997**, *13*, 135–140. [CrossRef]

13. Keizo, N.; Mineo, Y. Intra-species host specificity of HaV (*Heterosigma akashiwo* virus) clones. *Aquat. Microb. Ecol.* **1998**, *14*, 109–112. [CrossRef]

14. Keizo, N.; Mineo, Y. Effect of temperature on the algicidal activity and the stability of HaV (*Heterosigma akashiwo* virus). *Aquat. Microb. Ecol.* **1998**, *15*, 211–216. [CrossRef]

15. Lawrence, J.E.; Chan, A.M.; Suttle, C.A. A novel virus (HaNIV) causes lysis of the toxic bloom-forming alga *Heterosigma akashiwo* (Raphidophyceae). *J. Phycol.* **2001**, *37*, 216–222. [CrossRef]

16. Lawrence, J.E.; Chan, A.M.; Suttle, C.A. Viruses causing lysis of the toxic bloom-forming alga *Heterosigma akashiwo* (Raphidophyceae) are widespread in coastal sediments of British Columbia, Canada. *Limnol. Oceanogr.* **2002**, *47*, 545–550. [CrossRef]

17. Tai, V.; Lawrence, J.E.; Lang, A.S.; Chan, A.M.; Culley, A.I.; Suttle, C.A. Characterization of HaRNAV, a single-stranded RNA virus causing lysis of *Heterosigma akashiwo* (Raphidophyceae). *J. Phycol.* **2003**, *39*, 343–352. [CrossRef]

18. Janice, E.L.; Curtis, A.S. Effect of viral infection on sinking rates of *Heterosigma akashiwo* and its implications for bloom termination. *Aquat. Microb. Ecol.* **2004**, *37*, 1–7. [CrossRef]

19. Lawrence, J.E.; Brussaard, C.P.D.; Suttle, C.A. Virus-Specific Responses of *Heterosigma akashiwo* to Infection. *Appl. Environ. Microbiol.* **2006**, *72*, 7829–7834. [CrossRef] [PubMed]

20. Granéli, E.; Edvardsen, B.; Roelke, D.L.; Hagström, J.A. The ecophysiology and bloom dynamics of *Prymnesium* spp. *Harmful Algae* **2012**, *14*, 260–270. [CrossRef]

21. Manning, S.R.; La Claire, J.W. Prymnesins: Toxic Metabolites of the Golden Alga, *Prymnesium parvum* Carter (*Haptophyta*). *Mar. Drugs* **2010**, *8*, 678–704. [CrossRef] [PubMed]

22. Schoemann, V.; Becquevort, S.; Stefels, J.; Rousseau, V.; Lancelot, C. *Phaeocystis* blooms in the global ocean and their controlling mechanisms: A review. *J. Sea Res.* **2005**, *53*, 43–66. [CrossRef]

23. Leblanc, K.; Hare, C.E.; Feng, Y.; Berg, G.M.; DiTullio, G.R.; Neeley, A.; Benner, I.; Sprengel, C.; Beck, A.; Sanudo-Wilhelmy, S.A.; et al. Distribution of calcifying and silicifying phytoplankton in relation to environmental and biogeochemical parameters during the late stages of the 2005 North East Atlantic Spring Bloom. *Biogeosciences* **2009**, *6*, 2155–2179. [CrossRef]

24. Malin, G.; Wilson, W.H.; Bratbak, G.; Liss, P.S.; Mann, N.H. Elevated production of dimethylsulfide resulting from viral infection of cultures of *Phaeocystis pouchetii*. *Limnol. Oceanogr.* **1998**, *43*, 1389–1393. [CrossRef]

25. Edvardsen, B.; Paasche, E. *Bloom Dynamics and Physiology of Prymnesium and Chrysochromulina in Physiological Ecology of Harmful Algal Blooms*; Springer-Verlag: Berlin/Heidelberg, Germany, 1998.

26. Roelke, D.L.; Barkoh, A.; Brooks, B.W.; Grover, J.P.; Hambright, K.D.; LaClaire, J.W.; Moeller, P.D.R.; Patino, R. A chronicle of a killer alga in the west: Ecology, assessment, and management of *Prymnesium parvum* blooms. *Hydrobiologia* **2016**, *764*, 29–50. [CrossRef]

27. Schwierzke, L.; Roelke, D.L.; Brooks, B.W.; Grover, J.P.; Valenti, T.W.; Lahousse, M.; Miller, C.J.; Pinckney, J.L. *Prymnesium parvum* Population Dynamics During Bloom Development: A Role Assessment of Grazers and Virus. *J. Am. Water. Resour. Assoc.* **2010**, *46*, 63–75. [CrossRef]

28. Brenner, S.; Horne, R.W. A negative staining method for high resolution electron microscopy of viruses. *Biochim. Biophys. Acta* **1959**, *34*, 103–110. [CrossRef]

29. Gordon, G.B.; Miller, L.R.; Bensch, K.G. Fixation of tissue culture cells for ultrastructural cytochemistry. *Exp. Cell Res.* **1963**, *31*, 440–443. [CrossRef]

30. Martínez Martínez, J.; Boere, A.; Gilg, I.; van Lent, J.W.M.; Witte, H.J.; van Bleijswijk, J.D.L.; Brussaard, C.P.D. New lipid envelope-containing dsDNA virus isolates infecting *Micromonas pusilla* reveal a separate phylogenetic group. *Aquat. Microb. Ecol.* **2015**, *74*, 17–28. [CrossRef]

31. Besemer, J.; Lomsadze, A.; Borodovsky, M. GeneMarkS: A self-training method for prediction of gene starts in microbial genomes. Implications for finding sequence motifs in regulatory regions. *Nucleic Acids Res.* **2001**, *29*, 2607–2618. [CrossRef] [PubMed]

32. Altschul, S.F.; Gish, W.; Miller, W.; Myers, E.W.; Lipman, D.J. Basic local alignment search tool. *J. Mol. Biol.* **1990**, *215*, 403–410. [CrossRef]

33. Katoh, K.; Toh, H. Recent developments in the MAFFT multiple sequence alignment program. *Brief. Bioinform.* **2008**, *9*, 286–298. [CrossRef] [PubMed]

34. Saitou, N.; Nei, M. The neighbor-joining method: A new method for reconstructing phylogenetic trees. *Mol. Biol. Evol.* **1987**, *4*, 406–425. [CrossRef] [PubMed]

35. Kumar, S.; Stecher, G.; Tamura, K. MEGA7: Molecular Evolutionary Genetics Analysis version 7.0 for bigger datasets. *Mol. Biol. Evol.* **2016**, *33*, 1870–1874. [CrossRef] [PubMed]

36. Den Boon, J.A.; Diaz, A.; Ahlquist, P. Cytoplasmic Viral Replication Complexes. *Cell Host Microbe* **2010**, *8*, 77–85. [CrossRef] [PubMed]

37. Holdway, P.A.; Watson, R.A.; Moss, B. Aspects of the ecology of *Prymnesium parvum* (Haptophyta) and water chemistry in the Norfolk Broads, England. *Freshwat. Biol.* **1978**, *8*, 295–311. [CrossRef]

38. Zauberman, N.; Mutsafi, Y.; Halevy, D.B.; Shimoni, E.; Klein, E.; Xiao, C.; Sun, S.; Minsky, A. Distinct DNA Exit and Packaging Portals in the Virus *Acanthamoeba polyphaga mimivirus*. *PLoS Biol.* **2008**, *6*, e114. [CrossRef] [PubMed]

39. Mutsafi, Y.; Fridmann-Sirkis, Y.; Milrot, E.; Hevroni, L.; Minsky, A. Infection cycles of large DNA viruses: Emerging themes and underlying questions. *Virology* **2014**, *466–467*, 3–14. [CrossRef] [PubMed]

40. Feldman, H.A.; Wang, S.S. Sensitivity of various viruses to chloroform. *Proc. Soc. Exp. Biol. Med.* **1961**, *106*, 736–738. [CrossRef] [PubMed]

41. Nedbalová, L.; Střížek, A.; Sigler, K.; Řezanka, T. Effect of salinity on the fatty acid and triacylglycerol composition of five haptophyte algae from the genera Coccolithophora, Isochrysis and Prymnesium determined by LC-MS/APCI. *Phytochemistry* **2016**, *130*, 64–76. [CrossRef] [PubMed]

42. Johannessen, T.V.; Bratbak, G.; Larsen, A.; Ogata, H.; Egge, E.S.; Edvardsen, B.; Eikrem, W.; Sandaa, R.-A. Characterisation of three novel giant viruses reveals huge diversity among viruses infecting *Prymnesiales* (*Haptophyta*). *Virology* **2015**, *476*, 180–188. [CrossRef] [PubMed]

43. Wilhelm, S.W.; Coy, S.R.; Gann, E.R.; Moniruzzaman, M.; Stough, J.M.A. Standing on the Shoulders of Giant Viruses: Five Lessons Learned about Large Viruses Infecting Small Eukaryotes and the Opportunities They Create. *PLoS Pathog.* **2016**, *12*, e1005752. [CrossRef] [PubMed]

44. Gallot-Lavallée, L.; Pagarete, A.; Legendre, M.; Santini, S.; Sandaa, R.-A.; Himmelbauer, H.; Ogata, H.; Bratbak, G.; Claverie, J.-M. The 474-Kilobase-Pair Complete Genome Sequence of CeV-01B, a Virus Infecting *Haptolina* (*Chrysochromulina*) *ericina* (*Prymnesiophyceae*). *Genome Announc.* **2015**, *3*, e01413–e01415. [CrossRef]

45. Sandaa, R.-A.; Heldal, M.; Castberg, T.; Thyrhaug, R.; Bratbak, G. Isolation and Characterization of Two Viruses with Large Genome Size Infecting *Chrysochromulina ericina* (*Prymnesiophyceae*) and *Pyramimonas orientalis* (*Prasinophyceae*). *Virology* **2001**, *290*, 272–280. [CrossRef] [PubMed]

46. Moniruzzaman, M.; LeCleir, G.R.; Brown, C.M.; Gobler, C.J.; Bidle, K.D.; Wilson, W.H.; Wilhelm, S.W. Genome of brown tide virus (AaV), the little giant of the Megaviridae, elucidates NCLDV genome expansion and host-virus coevolution. *Virology* **2014**, *466–467*, 60–70. [CrossRef] [PubMed]

47. Mirza, S.F.; Staniewski, M.A.; Short, C.M.; Long, A.M.; Chaban, Y.V.; Short, S.M. Isolation and characterization of a virus infecting the freshwater algae *Chrysochromulina parva*. *Virology* **2015**, *486*, 105–115. [CrossRef] [PubMed]

48. Henrikson, J.C.; Gharfeh, M.S.; Easton, A.C.; Easton, J.D.; Glenn, K.L.; Shadfan, M.; Mooberry, S.L.; Hambright, K.D.; Cichewicz, R.H. Reassessing the ichthyotoxin profile of cultured *Prymnesium parvum* (golden algae) and comparing it to samples collected from recent freshwater bloom and fish kill events in North America. *Toxicon* **2010**, *55*, 1396–1404. [CrossRef] [PubMed]

49. Kozakai, H.; Oshima, Y.; Yasumoto, T. Isolation and Structural Elucidation of Hemolysin from the Phytoflagellate *Prymnesium parvum*. *Agric. Biol. Chem.* **1982**, *46*, 233–236. [CrossRef]

50. Igarashi, T.; Satake, M.; Yasumoto, T. Prymnesin-2: A Potent Ichthyotoxic and Hemolytic Glycoside Isolated from the Red Tide Alga *Prymnesium parvum*. *J. Am. Chem. Soc.* **1996**, *118*, 479–480. [CrossRef]

51. Igarashi, T.; Satake, M.; Yasumoto, T. Structures and Partial Stereochemical Assignments for Prymnesin-1 and Prymnesin-2: Potent Hemolytic and Ichthyotoxic Glycosides Isolated from the Red Tide Alga *Prymnesium parvum*. *J. Am. Chem. Soc.* **1999**, *121*, 8499–8511. [CrossRef]

52. Rasmussen, S.A.; Meier, S.; Andersen, N.G.; Blossom, H.E.; Duus, J.Ø.; Nielsen, K.F.; Hansen, P.J.; Larsen, T.O. Chemodiversity of Ladder-Frame Prymnesin Polyethers in *Prymnesium parvum*. *J. Nat. Prod.* **2016**, *79*, 2250–2256. [CrossRef] [PubMed]

53. Roux, S.; Brum, J.R.; Dutilh, B.E.; Sunagawa, S.; Duhaime, M.B.; Loy, A.; Poulos, B.T.; Solonenko, N.; Lara, E.; Poulain, J.; et al. Ecogenomics and potential biogeochemical impacts of globally abundant ocean viruses. *Nature* **2016**, *537*, 689–693. [CrossRef] [PubMed]

54. Karsenti, E.; Acinas, S.G.; Bork, P.; Bowler, C.; De Vargas, C.; Raes, J.; Sullivan, M.; Arendt, D.; Benzoni, F.; Claverie, J.-M.; et al. A Holistic Approach to Marine Eco-Systems Biology. *PLoS Biol.* **2011**, *9*, e1001177. [CrossRef] [PubMed]

Article

Isolation and Characterization of a *Shewanella* Phage–Host System from the Gut of the Tunicate, *Ciona intestinalis*

Brittany Leigh [1], Charlotte Karrer [2], John P. Cannon [2], Mya Breitbart [1] and Larry J. Dishaw [2,*]

[1] College of Marine Science, University of South Florida, St. Petersburg, FL 33701, USA;
    bleigh@mail.usf.edu (B.L.); mya@usf.edu (M.B.)
[2] Department of Pediatrics, University of South Florida, St. Petersburg, FL 33701, USA;
    ckarrer@mail.usf.edu (C.K.); jcannon@health.usf.edu (J.P.C.)
* Correspondence: LJDishaw@health.usf.edu; Tel.: +1-727-553-3608

Academic Editors: Mathias Middelboe and Corina P. D. Brussaard
Received: 29 January 2017; Accepted: 17 March 2017; Published: 22 March 2017

Abstract: Outnumbering all other biological entities on earth, bacteriophages (phages) play critical roles in structuring microbial communities through bacterial infection and subsequent lysis, as well as through horizontal gene transfer. While numerous studies have examined the effects of phages on free-living bacterial cells, much less is known regarding the role of phage infection in host-associated biofilms, which help to stabilize adherent microbial communities. Here we report the cultivation and characterization of a novel strain of *Shewanella fidelis* from the gut of the marine tunicate *Ciona intestinalis*, inducible prophages from the *S. fidelis* genome, and a strain-specific lytic phage recovered from surrounding seawater. In vitro biofilm assays demonstrated that lytic phage infection affects biofilm formation in a process likely influenced by the accumulation and integration of the extracellular DNA released during cell lysis, similar to the mechanism that has been previously shown for prophage induction.

Keywords: *Shewanella*; bacteriophage; biofilm; extracellular DNA

1. Introduction

A significant proportion of microbes in the marine environment, including both bacteria and bacteriophages (i.e., phages), are associated with eukaryotic hosts, where they form stable symbiotic relationships. These symbiotic relationships are often specific and necessary in maintaining animal health via carefully orchestrated exchanges (i.e., homeostasis). Although phages are the most abundant biological entities in the natural world [1,2], little is known about their role in structuring and maintaining host-associated microbial communities, or how they influence bacteria within a biofilm, a lifestyle many aquatic bacteria exhibit. Even less is known about how perturbation of these microbial communities influences the eukaryotic host. Many animals maintain a "core" assemblage of bacteria (i.e., a core microbiome) that likely provides advantages to the host [3–5]. Some of these bacteria are consistently found within the same environments (e.g., animal intestines) and across diverse animal hosts, where they are presumed to serve distinct functions for either the animal host and/or the surrounding microbes. One such bacterial genus is *Shewanella* [6–9].

Shewanella species from a wide range of environments are known for their highly versatile metabolic capabilities that utilize diverse electron acceptors including nitrate, nitrite, thiosulfate, elemental sulfur, iron oxide and manganese oxide [10–12]. *Shewanella* species shuttle electrons across their membranes during anaerobic respiration, resulting in electrical activity within their biofilms and the transformation of insoluble compounds to bioavailable ones. Interestingly, biofilms with electrical activity have been documented to influence host cellular responses [13]. These bacteria make stable biofilms and because they can respire almost any compound, they likely represent important symbionts of animals as well. Despite extensive genomic rearrangements within *Shewanella* genomes [14], members of the genus retain a core set of metabolic genes that facilitate their survival in diverse environments [15], including the gut of a number of organisms [6–9]. *Shewanella putrefaciens*, which is closely related to *S. fidelis*, has shown promise as a probiotic for aquaculture [16], further emphasizing important roles for *Shewanella* in aquatic animal-microbe relationships.

To date, a number of *Shewanella* phages (both lytic and temperate) have been described from marine and freshwater environments [17–21], and in *Shewanella oneidensis*, prophages have also been implicated as vital for biofilm formation through excision-mediated lysis [21]. Stably integrated prophage-like elements are common within the genomes of most marine bacterial species [22], and prophages are also thought to be important among bacteria that colonize the gut mucosa of animals [23], often forming biofilms [24,25]. These biofilms are thought to serve as physical structures that can enhance pathogen defense by contributing to physical barriers, and through the production of diverse antimicrobials [26,27].

To begin to understand the role of phages in shaping the microbiome of sessile, filter-feeding marine invertebrates, we isolated and characterized a core member of the gut microbiome in the tunicate, *Ciona intestinalis* [4]. This novel strain of *Shewanella fidelis* (3313) was sequenced, and its inducible prophages and a strain-specific lytic phage (SFCi1, which was isolated separately from seawater) were characterized. Previously, it has been shown that spontaneous prophage induction can augment biofilms in some strains of bacteria [28–31]; we demonstrate here that infection of *S. fidelis* 3313 by lytic phage SFCi1 also enhances biofilm formation in vitro in a similar DNA-dependent manner.

2. Materials and Methods

2.1. Bacterial Isolation from the Gut of Ciona Intestinalis

Ciona intestinalis specimens were collected from Mission Bay in San Diego (M-REP Animal Collection Services, San Diego, CA, USA) during the Spring of 2014. Animals were cleared for 48 h in seawater filtered through a 0.22 μm pore size filter (Millipore Sterivex, Merck, Darmstadt, Germany) (with water changes every several hours), before the entire gut (stomach, midgut, hindgut) of five animals was dissected and homogenized using a dounce homogenizer. The gut homogenate was filtered through a 0.45 μm pore size filter (Millipore Sterivex, Merck) to remove host tissue, and the bacteria were pelleted by centrifugation at 12,500 xg for 10 min and washed three times through resuspension and centrifugation in 1 mL of sterile (filtered through a 0.22 μm pore size filter and autoclaved) artificial seawater (Instant Ocean AS9519, Marine Depot, Garden Grove, CA, USA). Serial dilutions of the bacterial homogenate were plated on marine agar (MA) 2216 (Becton Dickinson Company, Franklin Lakes, NJ, USA). Colonies displaying distinct phenotypes were randomly chosen, purified by streaking, and grown separately in the corresponding liquid broth (marine broth (MB) 2216, pH 7.6) at 20 °C; subsequently, a 20% glycerol stock was made for each isolate and stored at −80 °C. DNA was isolated using the PowerSoil DNA Kit (MoBio Laboratories, Carlsbad, CA, USA) and the 16S rRNA gene amplified using universal primers 27F and 1492R [32] (polymerase chain reaction (PCR) conditions: denature at 95 °C for 5 min, cycle 35 times through 94 °C for 30 s, 56 °C for 30 s, 72 °C for 1 min 30 s, and end with a final extension at 72 °C for 10 min), sequenced via the Sanger platform and identified using BLAST against the NCBI non-redundant database [33].

2.2. Phage Isolation, Propagation, and Purification for Transmission Electron Microscopy

S. fidelis 3313 recovered from the *Ciona* gut homogenate was screened for lytic phages via standard plaque assays using seawater from which the animals were shipped (i.e., bag water) filtered through a 0.22 μm pore size filter. Approximately 500 mL of the filtered seawater was concentrated using Amicon Ultra-15 concentration units (molecular weight cut-off (MWCO) 100 kDa; EMD (Merck Millipore,

Darmstadt, Germany) by centrifugation to a final volume of ~15 mL. Lytic phages were isolated with the double agar method (0.5% low-melt top agar) [34] using the prepared seawater concentrate and the bacterial host grown to log phase (OD_{600} = 0.25) in MB. Each plaque was then cored, plaque-purified three times and resuspended in 500 μL of sterile modified sodium magnesium (MSM) buffer (450 mM NaCl, 102 mM $MgSO_4$, 50 mM Tris Base, pH 8). The purified phage was propagated on *S. fidelis* 3313 lawns on MA at room temperature. The resulting lysate was filtered through a 0.22 μm pore size filter and stored in MSM buffer at 4 °C.

To estimate phage–host growth dynamics including the latent period and burst size, a one-step infection curve was performed according to Hyman and Abedon [35], with slight modifications. Latent period is defined as the period between the adsorption time and the initial phage lysis of the bacterial culture, prior to any significant rise in phage particles [35]. For this procedure, a 10 min adsorption step at a multiplicity of infection (MOI) of 1 was followed by centrifugation at 13,000× *g* for 30 s to pellet the bacteria with adsorbed phages. The pellet was then resuspended in 1 mL of sterile MB. Triplicate samples were taken at 10 min intervals for up to 2 h and directly plated using the double agar method to determine phage titer. Additionally, burst size was measured as the ratio of final phage particles to the number of bacterial cells at the onset of phage exposure.

A portion of the lysate was further purified via cesium chloride (CsCl) gradient ultracentrifugation [36] for morphological analysis by transmission electron microscopy (TEM) using an Hitachi 7100 (Hitachi Ltd., Tokyo, Japan). The purified virus particles were prepared for imaging on a formvar grid (Electron Microscopy Sciences, Hatfield, PA, USA) using a negative stain with 2% uranyl acetate, as described previously [37]. Images were captured using an Orius SC600 bottom mount camera (Gatan Inc., Pleasanton, CA, USA) at 100 kV. A separate aliquot of this purified viral suspension was reserved for DNA extraction using the QIAmp MinElute Virus Spin Kit (Qiagen Inc., Valencia, CA, USA) and for sequencing, as described below.

2.3. DNA Extraction, Sequencing and Analysis

Shewanella fidelis 3313 was cultured in MB overnight at 20 °C with shaking at 90 RPM, and its lytic phage was propagated and purified as described above. Bacterial DNA was extracted using the PowerSoil DNA Kit (MoBio Laboratories, Carlsbad, CA, USA) as described above. All viral DNA was amplified using a GenomiPhi V2 DNA amplification kit (GE Healthcare Life Sciences, Pittsburgh, PA, USA) to generate adequate template for sequencing (~1 μg). To minimize bias introduced by the amplification process, three identical reactions were prepared and pooled.

Bacterial, phage and prophage DNA were sequenced with the Illumina MiSeq platform generating mate-pair (2 × 250) libraries (Operon, Eurofins MWG Operon LLC, Huntsville, AL, USA). The NuGen UltraLow DNA kit was used (Eurofins, Louisville, KY, USA) to prepare libraries, and DNA quality was assessed on a BioAnalyzer 2100 (Agilent Technologies, Santa Clara, CA, USA), with size selection (400–600 bp) conducted to remove outlier DNA fragments after sonication. The resulting mate-pair reads were assembled using approaches described in Deng et al. [38], first by a de Bruijn graph assembler (Velvet de novo assembler with a k-mer of 35 (phage) and 27 (bacteria) [39]), followed by the default consensus algorithm in Geneious 8.1.7 (Biomatters Ltd, Auckland, New Zealand) [40]. Viral genome open reading frames (ORFs) were identified using Glimmer3 [41] through Geneious 8.1.7; annotations were improved with the BLASTX algorithm against non-redundant protein databases in GenBank, Protein Data Bank (PDB), SwissProt, Protein Information Resource (PIR) and Protein Research Foundation (PRF) via Geneious 8.1.7. All resulting bacterial contigs were uploaded to the Rapid Annotation using Subsystem Technology (RAST) server [42,43] under sample ID mgs422948, with full annotations based on the SEED Database. All contigs passed Metagenomics (MG)-RAST quality control and all predicted proteins were annotated. The complete 16S rRNA gene of *S. fidelis* 3313 was compared via PHYML maximum likelihood trees to other described *Shewanella* species in the GenBank database (Figure S1). The closest species was determined to be *Shewanella fidelis* (ATCC BAA-318; GenBank ID: 17801). All assembled contigs greater than 1500 bp (a total of 24) were compared to this nearest neighbor to determine average nucleotide identity (ANI) of the entire genome using the ANI calculator [44]. Genomes with ANI values above 95% are considered to belong to the same species [45].

2.4. Prophage Induction and Identification

Prophage regions were identified and annotated by screening all bacterial contigs using the VirSorter pipeline [46]. To determine if *S. fidelis* 3313 possessed any inducible prophages, mitomycin C was introduced to an early log-phase culture (OD_{600} of 0.025) at a final concentration of 1 µg/mL and incubated for 24 h [47]. The resulting culture, along with an untreated control culture, was then stained with SYBR Gold nucleic acid stain and induced phage particles were enumerated using epifluorescence microscopy, as described previously in Patel et al. [48]. Induced phage particles were then CsCl-purified and treated with 2U of DNase I Turbo (Invitrogen, Carlsbad, CA, USA), before DNA was extracted using the QIAmp MinElute Virus Spin Kit (Qiagen Inc.). Mate-pair (2×250) libraries were produced using the NuGen UltraLow DNA kit, quality controlled with the BioAnalyzer as previously stated (Operon, Eurofins MWG Operon LLC), and sequenced on the Illumina MiSeq platform. Reads were then mapped back to the assembled *S. fidelis* 3313 genome utilizing the Geneious 8.1.7 software, with default parameters to determine which of the predicted prophages were induced. Additionally, TEM analysis of the CsCl-purified induced prophage fraction was performed as described above.

2.5. Biofilm Assays

Single colonies of *S. fidelis* 3313 were grown in MB at 20 °C overnight with shaking at 90 RPM. Concentration was estimated by optical density at OD_{600} based on previously-calibrated growth curves and colony forming units (data not shown). For biofilm assays, stationary cultures were diluted to a final concentration of 10^6 cells mL^{-1} (early log phase OD_{600} of 0.025), and phages were added at 10^6 plaque forming units (PFUs) mL^{-1} immediately before plating. All bacterial treatments and controls were plated on 12-well plates (Thermo Scientific, Waltham, MA, USA) in triplicate; bacteria were also plated in duplicate on 35 mm glass bottom dishes (No. 1.5, uncoated; MatTek Corporation, Ashland, MA, USA) for extracellular DNA detection using the TOTO-1 Iodide 514/533 stain (Molecular Probes, Invitrogen) and counterstained for live cells using SYTO60 red (Molecular Probes, Invitrogen) [49]. Additionally, purified salmon sperm DNA (Invitrogen) was added to separate cultures at a final volume of 300 ng mL^{-1}. All stationary culture dishes were incubated at 20 °C for up to two days to allow biofilm formation, before excess liquid and planktonic bacteria were removed by gentle pipetting. To quantify biofilm formation, culture dishes were allowed to dry, and then subsequently stained with a 0.1% crystal violet solution for 10 min, as per Merritt et al. [50]. Crystal violet was removed by decanting, and the dishes were washed twice with distilled water to remove excess stain and then allowed to dry completely. The dried crystal violet was resuspended in 30% acetic acid for ~10 min, and the OD_{590} was determined for each culture dish [50]. At each time point, biofilms in the 12-well dishes and one of the MatTek dishes at 24 h (when the difference was the most drastic) were treated with 2U of DNase I Turbo (Invitrogen), for 10 min at 37 °C. Both dishes for fluorescent microscopy staining were then washed once with $1\times$ PBS before TOTO-1 staining for 10 min. Excess dye was then removed and the biofilm washed twice in $1\times$ PBS before counterstaining with SYTO60 for 10 min. Excess dye was removed, washed once in $1\times$ PBS and held in $1\times$ PBS for imaging of 10 random fields using the Leica Application Suite (Leica Microsystems, Wetzlar, Germany) and the Metamorph version 7.5 software (Molecular Devices, LLC, Sunnyvale, CA, USA), using consistent exposure, aperture, and magnification settings. Images were exported as TIF files from Metamorph (Molecular Devices) and imported into ImageJ 1.48v [51]. Without any additional enhancements, the channels were separated, thresholds permanently set, signal intensity was averaged and standard deviations determined over the area selected.

3. Results

3.1. Bacterial Cultivation and Genome Sequencing

The 16S rRNA gene of *S. fidelis* 3313 is identical to the V3-V4 region of a core operational taxonomic unit (OTU) described previously from the *Ciona* gut [4]; members of the core microbiome likely are of functional relevance within the host, perhaps provisioning nutrients or essential elements, or contributing to the establishment of gut-associated bacterial communities. Next-generation sequencing of the *S. fidelis* 3313 genome resulted in 5,940,000 reads which assembled into 129 scaffolds (N_{50} = 937,903), with an average coverage of 527× and a mean guanine-cytosine (GC) content of 43.3%. Collectively, these scaffolds contained 3,536 predicted ORFs, all of which could be annotated through at least one of the four MG-RAST protein databases [42,43]. Pairwise sequence identity comparisons and nearest neighbor analyses suggest that isolate 3313 from the *Ciona* gut is related most closely to a partially-sequenced *S. fidelis* species previously isolated from sediments and seawater in the South China Sea [52]. The near-full-length 16S rRNA gene of the *S. fidelis* 3313 isolate is 99% identical to this previously-described *S. fidelis* (Figure S1), and was deposited separately into GenBank (ID: KY696838); however, the assembled genomic contigs reveal an average overall nucleotide identity of 97.35% (Figure S2), making isolate 3313 a novel strain of *Shewanella fidelis* [45,53].

3.2. Prophage Induction and Genome Sequencing

VirSorter screening of the *S. fidelis* 3313 assembled genome revealed the presence of at least three genetic loci with sequence similarities to previously-described prophages. To determine if *S. fidelis* 3313 possessed inducible prophages, cultures in early log phase (OD_{600} = 0.025) were treated overnight with mitomycin C, a commonly-used mutagen for prophage induction; the resulting virus-like particles (VLPs) were enumerated via epifluorescence microscopy (Figure 1). Mitomycin C induction resulted in an increase in VLPs (Figure 1A), concurrent with a decrease in culture turbidity as measured by optical density (Figure 1B).

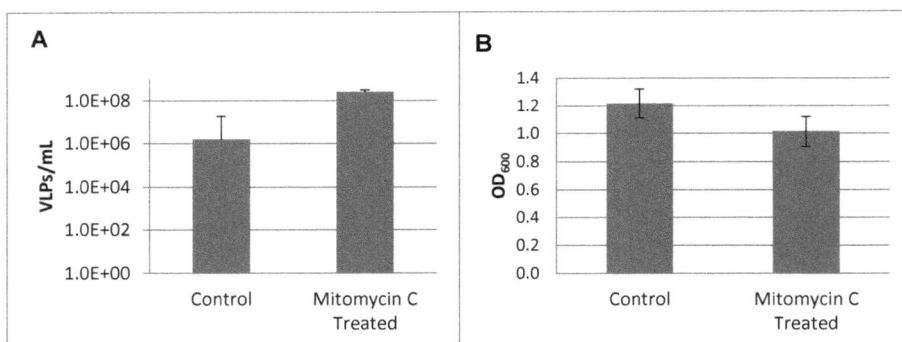

Figure 1. (**A**) Viral-like particles (VLPs) enumerated via epifluorescence microscopy after mitomycin C induction of *Shewanella fidelis* 3313. (**B**) Bacterial turbidity measured at OD_{600}. All measurements were recorded at 24 h after treatment.

Imaging of the same supernatant after cesium chloride (CsCl)-purification by transmission electron microscopy (TEM) confirmed two morphologies of intact phage particles (Figure 2). The DNA of these purified phage particles was then sequenced and mapped to the *S. fidelis* 3313 genome, verifying that they represent two intact and active prophage elements. The first prophage, detected in contig 6, spans 19,321 bp and is predicted to encode 31 genes. This prophage, hereby referred to as SFPat, encodes mostly hypothetical proteins. The second prophage, hereby referred to SFMu1, was detected in contig 7, spans 45,796 bp, and is predicted to encode 57 genes. Phage SFMu1 possesses several Mu-like gene elements described previously in a number of Gram-negative bacteria, including *Shewanella oneidensis* [54]. Annotations of predicted ORFs corresponding to each of these elements are outlined in Supplementary Materials Tables S1–S2.

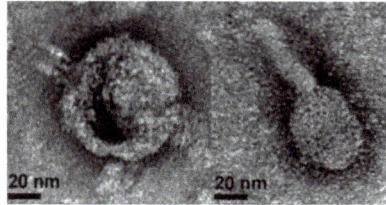

Figure 2. Representative transmission electron microscopy (TEM) images of mitomycin C-induced phages recovered from the supernatant of *S. fidelis* 3313.

3.3. Lytic Phage Isolation and Genome Sequencing

To identify phages capable of lytic infection of *S. fidelis* 3313, water from the collection site (also used in shipping live animals) was sequentially filtered, concentrated, and screened by standard plaque assays. A candidate lytic phage was identified, amplified, purified, and sequenced. The *S. fidelis* 3313 lytic phage, named SFCi1, has a circular, double-stranded DNA genome consisting of 42,279 bp (38,089× genome coverage), and a GC content of 59.1% (Figure 3).

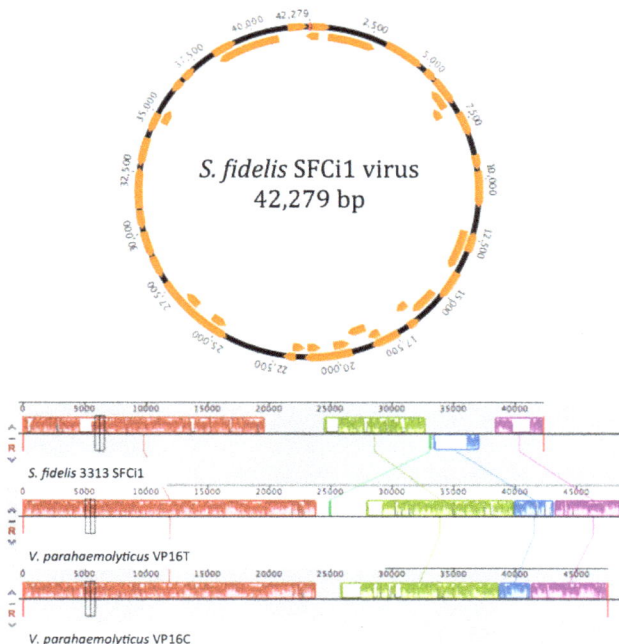

Figure 3. (**A**) Circular genome of lytic phage SFCi1 depicts 40 open reading frames (ORFs). (**B**) Mauve alignment of the SFCi1 genome against two most closely related phage genomes, VP16C and VP16T. Colored boxes indicate sequence blocks with shared sequence identity; regions lacking sequence homology are indicated in white.

The assembled genome is predicted to possess 40 ORFs, summarized in Supplementary Materials Table S3. The infection curve generated by the SFCi1 phage revealed a decline in optical density of the *S. fidelis* 3313 culture within one-hour post-infection (approximate latent period), with complete culture lysis by 4 h (Figure 4). Additionally, the average burst size was calculated to be 62.

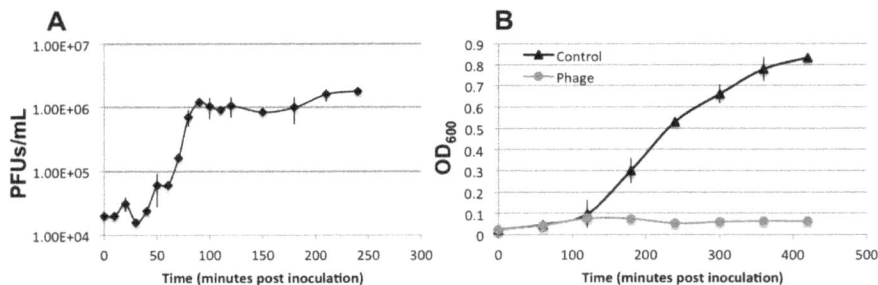

Figure 4. One-step growth curve of lytic phage SFCi1 with its host, *S. fidelis* 3313, as determined by plaque forming units (PFUs) and bacterial turbidity measurements (OD_{600}).

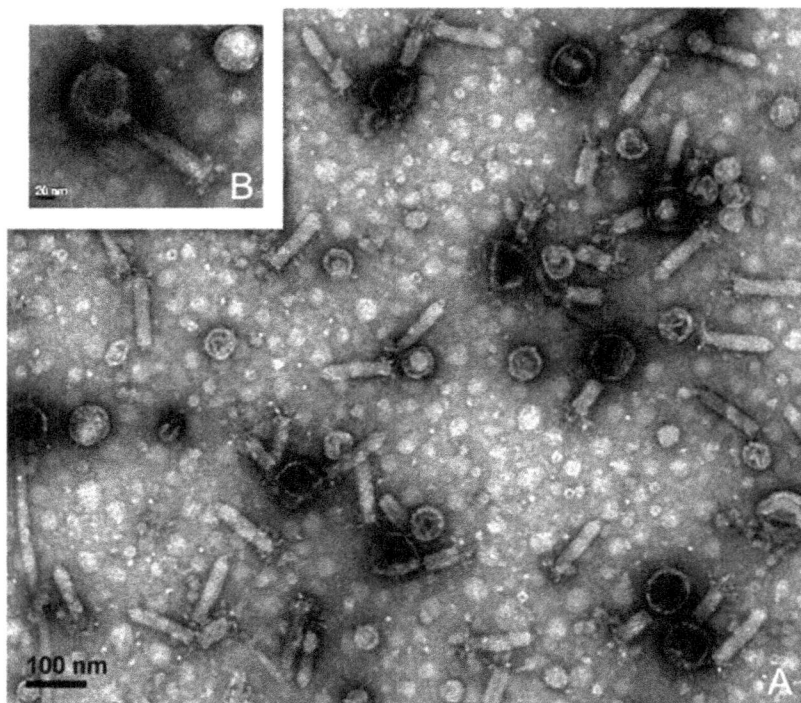

Figure 5. (**A**) TEM image of a pure culture of lytic phage, SFCi1, indicates that it is a myophage; (**B**) Magnified image of a single viral particle outlining the icosahedral head and details of the contractile tail and baseplate.

Morphological analysis via TEM suggests that SFCi1 belongs to the *Myoviridae* family [55], with a capsid diameter of approximately 70 nm (StDev: 7.4 nm), and a tail length of approximately 120 nm

(StDev: 5.5 nm) (Figure 5). Members of this family possess a contractile tail through which DNA is inserted into the bacterial cell after degradation of surface structures by lysozyme, an enzyme also encoded in the SFCi1 genome. Phage SFCi1 shares the greatest sequence similarity with two siphophages (VP16C and VP16T) isolated from San Diego that infect *Vibrio parahaemolyticus* [56]. However, SFCi1 is unable to infect the *Vibrio parahaemolyticus* host of VP16C/T (data not shown). Figure 3, which depicts the regions of nucleotide identity between the three phage genomes via Mauve alignment [57], reveals substantial syntenic regions. It is notable that the genes encoding the tail of SFCi1 are more similar to that of *Vibrio* phage H188, a recently discovered myophage from the Yellow Sea [58]. The tail proteins characteristic of myophages generally allow for a broader host range than that seen for siphophages [59]. The SFCi1 phage genome was also placed onto the phage proteomic tree [60], which approximated the closest viral relatives as VP16T/C (Figure S3).

3.4. Similarity to Vibrio Phages

Because phages utilize host machinery for protein translation during replication, their codon usage tends to evolve towards that of the host genome [61]. The codon adaptation index (CAI), which is expected to increase upon phage–host coevolution, was higher between *S. fidelis* 3313 and the three phages examined (SFCi1 (0.753), VP16C (0.747), and VP16T (0.758)), than between *Vibrio parahaemolyticus* (RIMD 2210633) and those phages (SFCi1 (0.624), VP16C (0.607), and VP16T (0.604)). No tRNAs were detected in the SFCi1 phage genome that could independently influence the codon usage bias [62].

3.5. Biofilm Development

In vitro, the addition of lytic phage SFCi1 into pure static cultures of *S. fidelis* 3313 caused an increase in biofilm formation, beginning at 8 h post addition, and led to a more robust biofilm by 24 h, as detected by crystal violet staining of adherent biofilm structures (Figure 6).

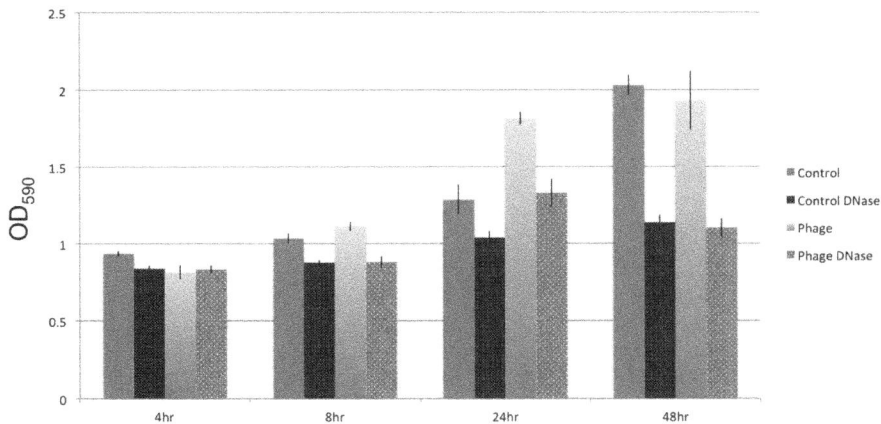

Figure 6. Development of a biofilm by *S. fidelis* over 48 h in stationary culture, as measured by crystal violet staining of the biofilm, in the presence (Phage) and absence (Control) of lytic phage SFCi1 and/or DNase I treatment.

Similar effects on biofilm formation could be achieved with the inclusion of foreign DNA, such as salmon sperm DNA, into the cultures (data not shown). When DNase I was added to static cultures, whether or not exposed to lytic phages, biofilm density decreased. This suggests extracellular DNA is an important structural component of the *S. fidelis* 3313 biofilms, seemingly as a result of cell lysis. The extracellular DNA was quantified by measuring fluorescent signal (TOTO-1 Iodide 514/533 stain, Molecular Probes), and determining the percent area coverage represented by the labeled extracellular DNA signal. Fluorescent microscopy images depicted an increase in free extracellular DNA (80.96% ± 7.9% pixel area coverage) within the biofilm-rich cultures that include phage SFCi1 (Figure 7). Control cultures also revealed the presence of extracellular DNA (30.23% ± 5.1% pixel area coverage), suggesting that lysis by spontaneous prophage induction as shown in controls in Figure 1) may play a role in natural biofilm formation by this strain. Treatment with DNase I resulted in the depletion of extracellular DNA from both the control (0.754% ± 0.01% area) and phage SFCi1 (0.772% ± 0.017% area) cultures. However, DNase I alone was not capable of completely eliminating detectable biofilms, suggesting that extracellular DNA is not the only requirement for biofilm formation.

3.6. GenBank Accession Numbers

Shewanella fidelis 3313 SFCi1 virus was submitted to GenBank under the Accession ID number KX196154. The *Shewanella fidelis* 3313 16S rDNA gene has been submitted to GenBank under the Accession ID number KY696838. The *Shewanella fidelis* 3313 draft genome is available in MG-RAST under sample ID mgs422948.

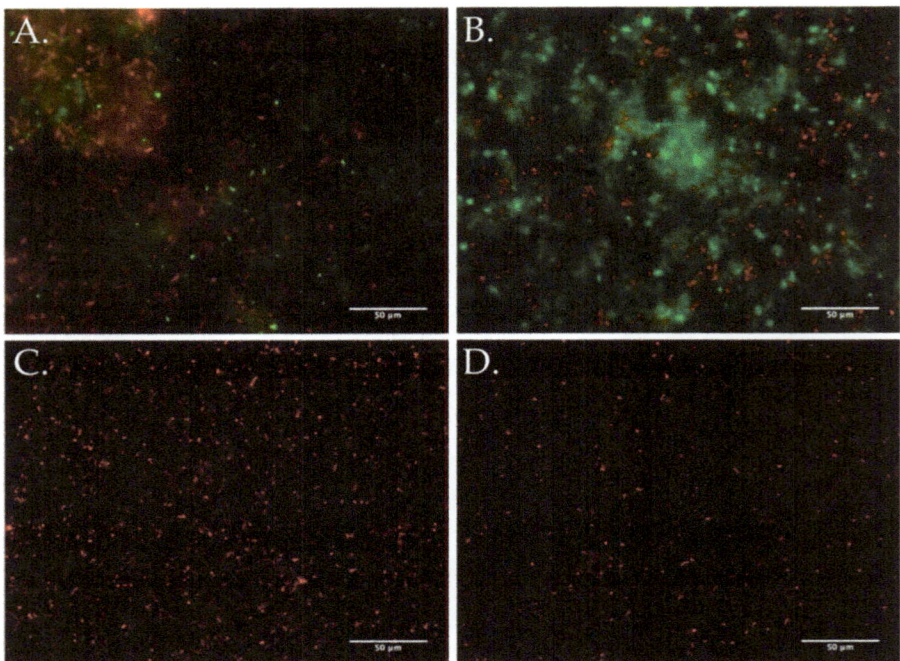

Figure 7. TOTO-1 Iodide 514/533 stain and SYTO60 red counterstain reveals extracellular DNA as a major component of the stationary culture biofilm. (**A**) Untreated *S. fidelis* 3313 control culture (30.23% ± 5.1% pixel area coverage); (**B**) *S. fidelis* 3313 exposed to SFCi1 lytic phage (80.96% ± 7.9% area); (**C**) control culture treated with DNase I (0.754% ± 0.01% area); and (**D**) lytic-phage treated culture co-treated with DNase I (0.772% ± 0.017% area).

4. Discussion

Despite being surrounded by abundant and diverse microbes, filter-feeding sessile aquatic invertebrates maintain stable and often species-specific resident microbial communities (i.e., core microbiomes) [3–5]. A stable microbiome can contribute gene products of functional relevance, influencing metabolic pathways and nutrient acquisition [63]. *Ciona intestinalis* is a marine protochordate with a fully sequenced genome that is being developed for gut microbiome studies utilizing germ-free mariculture [64]; recent evidence also indicates the presence of a core microbiome [4]. Culturing members of the *Ciona* gut microbiome, along with corresponding lytic phages, is a first step in designing experimental approaches to study the processes governing bacterial colonization of mucosal surfaces. Mono-association studies in *Ciona* (e.g., with *S. fidelis* 3313) could provide key insights into host-bacterial interactions at the onset of colonization in naïve tissue surfaces. Furthermore, investigations of mixed community colonization could reveal patterns of succession that are influenced by the activity of phages, along with other external and host-derived factors [65–67].

The establishment and maintenance of homeostasis between a host and its gut microbiome is an exceedingly complex, multifaceted process, likely to be influenced by many parameters, including external factors (e.g., nutrient availability, other microbes and microbial products), viruses that infect these microbes (i.e., phages), and host factors (e.g., mucus, immunity). The abundance, ubiquity, and high genetic diversity of phages, together with their roles in shaping bacterial communities and their influences on horizontal gene transfer [2,68–70], suggest that phages are highly influential members of complex microbial communities [71]. Filter-feeding invertebrates come into contact with a continuous assortment of microbes and free phages, which have the potential to profoundly alter community dynamics in established microbiomes. In addition to the free phages in seawater ($\sim10^7$ phages per milliliter [2]), most marine bacterial species have been documented to possess stable and active prophages (i.e., phages integrated into bacterial genomes) [22]. While an increasing number of studies are examining the diversity and variability of phages within microbiomes [72–75], little is known about how these phages influence microbiomes and their role(s) in homeostasis.

In the *S. fidelis* 3313 draft genome assembly, a total of three prophage regions were identified by VirSorter. Activity of two of these prophages was confirmed experimentally through the identification and sequencing of intact viral particles after induction of live cultures with the mutagen mitomycin C. Prophage SFPat (from contig 6) consists largely of hypothetical proteins. However, prophage SFMu1 (from contig 7) possesses genes more similar to those in public databases, facilitating annotation. For example, the Mu-like phages replicate via DNA transposition, and have evolved genes such as a *cis*-acting transposition enhancer and a centrally-located strong gyrase binding site [76], genes that also are present in SFMu1.

From the animal's surrounding seawater, this study also identified a lytic phage, SFCi1, which infects *S. fidelis* 3313. Placement of the SFCi1 genome on the phage proteomic tree identified the *Vibrio* phages VP16T and VP16C as the closest relatives. Genome comparisons of SFCi1 to all other known *Shewanella* phages indicated little to no similarity (data not shown), and with less than 95% ANI to any other phage, SFCi1 represents a new phage species. Codon usage assessments suggest that SFCi1 may have infected *Shewanella* long before it encountered and/or infected *Vibrio parahaemolyticus*, which is consistent with previous observations that VP16C/T only recently infected *V. parahaemolyticus* [56]. Both *Shewanella* and *Vibrio* species are abundant in the marine water column, and horizontal gene transfer between the two bacterial families has been described [77]. Genetic exchange between the VP16C/T phages and phage SFCi1 could represent similar examples.

Previous studies have suggested that phages may influences bacterial competition in the environment [78], a phenomenon that likely also affects gut ecosystem dynamics. Activated prophages are known to influence biofilms by facilitating the transient liberation of extracellular DNA, which can then become a component of the biofilm matrix [21,31]. Biofilms can shelter microbial inhabitants from both physical and mechanical stressors, and consist mostly of microbially-derived polysaccharides, nucleic acids and lipids. Extracellular DNA has been shown to be an important component of biofilms in several species [79]. While the extracellular DNA is often composed of chromosomal DNA from the biofilm-producing bacteria [80], foreign extracellular DNA can similarly influence the growth and maturation of biofilms, suggesting that the origin of the DNA is less important than the structural support it lends [81]. Several sources of extracellular DNA within biofilms have been identified, including release in response to quorum sensing [82–84], autolysis [85], and prophage induction [21,30,86]. However, additional methods of extracellular DNA release by external factors such as lytic phages and host products (e.g., immune molecules, mucus) have yet to be elucidated in the context of biofilm formation.

Previously, *Vibrio anguillarum* phages have been shown to have different influences on biofilm formation over long-term cultures, due mostly to varied microcolony morphologies with one strain having flat single layers (BA35), and the other forming complex 3D structures (PF430-3) [87]. This structural distinction resulted in differential penetrability of the biofilm structure by phages during the initial stages of biofilm formation, and ultimately impeded the formation of the BA35 single layer biofilm compared to the control. Phage infection of BA35 resulted in a large percentage of resistant mutants (~70% of isolated cells); however, PF430-3 formed more complex 3D structures in the presence of its lytic phage, resulting in fewer resistant mutants. This result suggested that aggregation within these structures could be reducing phage adsorption [87]. The phage particles appeared to be trapped within the aggregate matrix, a complex consortia of extracellular polymeric substances of unknown composition that seemingly provided physical protection against phage infection; the mechanism for this differential aggregation phenotype has yet to be elucidated.

Lytic phage infection has also been shown to influence biofilm formation in *Pseudomonas aeruginosa*, *Salmonella enterica* and *Staphylococcus aureus*, with different outcomes in each bacteria-phage system; *S. aureus* was the only system where biofilm formation was enhanced [88], with *S. aureus* thought to transition to the biofilm phenotype as a means of physiological protection from phage infection. Enhanced formation of biofilms took place at early time points, with treated cultures returning to control levels over time, likely as a result of early biofilm dispersal. Although not investigated, the authors suggested that extracellular DNA released by lytic phage infection likely played a role in biofilm development [88].

The current study adds to this growing body of evidence regarding phage influence on biofilm formation, by demonstrating that lytic phage SFCi1 infection can similarly enhance biofilm formation in vitro by *S. fidelis* 3313. Fluorescence imaging shows that extracellular DNA is much more prominent in SFCi1 phage-exposed, biofilm-rich, cultures of *S. fidelis* 3313. The presence of some extracellular DNA in control cultures is consistent with spontaneous prophage induction. Treatment of phage-infected cultures with DNase I reduces biofilms, indicating that extracellular DNA is likely a major structural component of these biofilms. As seen for prophage induction, lytic phage dynamics can influence biofilm formation through the release of extracellular DNA. Additional mechanisms of biofilm enhancement have been described and include the generation of phage resistant mutants [89], quorum sensing [90], and the "wall effect", whereby two subpopulations of the same strain co-exist with one protecting the other [91]. Whether any or all of these mechanisms are also involved in the phage-related biofilm increase described in *S. fidelis* 3313 remains to be demonstrated.

Among mucus-rich tissues, it has also been suggested that lytic phages can directly associate with mucus, a process that appears to influence selection of microbiota and/or help protect against certain pathogens [92]. However, the role of these phages at the gut mucus interface in vivo, and the ability of phages to infiltrate established biofilms and modify community structure, is not yet clear. Biofilms derived from distinct bacterial species, and sometimes composed from mixed communities, can demonstrate unique physical and chemical compositions of exopolysaccharides that exert distinct influences on phages [24,93]. Once phages become integrated into biofilms, some can be rapidly incorporated into the bacterial genome as prophages and lie dormant, making the bacterium a reservoir of future phage particles [94]. A variety of effectors can influence the induction of prophages. For example, viral excision from bacterial genomes in response to environmental stress has been implicated in biofilm restructuring and dispersal [31,86,95]. Since a large number of bacteria within the gut contain prophages, their induction likely imposes a significant restructuring of the associated microbial communities in response to host and environmental factors. However, natural triggers for prophage induction among members of mucus-associated microbiomes remain to be defined.

Phage manipulation of host mucosal epithelium-associated biofilms contributes to an already complex relationship between the host and its associated microbiome [96,97]. Characterizing members of the gut microbial community and their potential interactions is essential to understanding their influence on host health, and the establishment and maintenance of homeostasis. Deciphering these complex interactions is aided by carefully-designed in vitro and in vivo studies that enable careful control and observation of the dialogue between host immune factors and various members of the microbiome (including phages). Because bacteria in a biofilm often are recalcitrant to interventional therapies, dissection of the processes modulating the development and dispersal of biofilms in vivo within a natural host may aid in the development of new therapeutic approaches.

Supplementary Materials: The following are available online at www.mdpi.com/1999-4915/9/3/60/s1, Figure S1: A PHYML-based tree (100 bootstrap replicates; 70% stringency) of *Shewanella* sp. 16S rRNA gene comparisons. Figure S2: Average nucleotide identity (ANI) comparison between the *S. fidelis* ATCC genome (GI:655363240) and *S. fidelis* 3313 as determined by the ANI Calculator (http://enve-omics.ce.gatech.edu/ani/); Figure S3: *S. fidelis* SFCi1 phage placement on phage proteomic tree. Table S1: ORF table for SFPat prophage; Table S2: ORF table for SFMu1 prophage; Table S3: ORF table for SFCi1 lytic phage.

Acknowledgments: These studies were supported by grant IOS-1456301 from the National Science Foundation to L.J.D. and M.B. and by a National Science Foundation Graduate Research Fellowship (Award No. 1144244) to B.A.L. The authors thank Tony Greco, University of South Florida College of Marine Science Electron Microscopy Suite, for assistance in sample imaging and Lauren McDaniel for lending her knowledge and expertise, assistance with flow cytometry, and providing the *V. parahaemolyticus* strain. The authors would also like to thank Terry F.F. Ng for bioinformatics assistance and Robert Edwards for SFCi1 assessment on the phage proteomic tree. The authors would also like to thank Gary W. Litman for constructive feedback on previous versions of the manuscript.

Author Contributions: B.L., J.P.C., M.B. and L.D. conceived and designed the experiments; B.L. and C.K. performed the experiments; B.L., J.P.C., M.B. and L.J.D. analyzed the data; M.B. and L.D. contributed reagents/materials/analysis tools; B.L., M.B. and L.J.D. wrote the paper. All authors read and approved the final manuscript.

Conflicts of Interest: The authors declare no conflict of interest.

References

1. Brussow, H.; Hendrix, R.W. Phage genomics: Small is beautiful. *Cell* **2002**, *108*, 13–16. [CrossRef]
2. Suttle, C.A. Viruses in the sea. *Nature* **2005**, *437*, 356–361. [CrossRef] [PubMed]
3. Schmitt, S.; Tsai, P.; Bell, J.; Fromont, J.; Ilan, M.; Lindquist, N.; Perez, T.; Rodrigo, A.; Schupp, P.J.; Vacelet, J.; et al. Assessing the complex sponge microbiota: Core, variable and species-specific bacterial communities in marine sponges. *ISME J.* **2012**, *6*, 564–576. [CrossRef] [PubMed]
4. Dishaw, L.J.; Flores-Torres, J.; Lax, S.; Gemayel, K.; Leigh, B.; Melillo, D.; Mueller, M.G.; Natale, L.; Zucchetti, I.; De Santis, R.; et al. The gut of geographically disparate *Ciona intestinalis* harbors a core microbiota. *PLoS ONE* **2014**, *9*, e93386. [CrossRef] [PubMed]
5. Franzenburg, S.; Walter, J.; Kunzel, S.; Wang, J.; Baines, J.F.; Bosch, T.C.; Fraune, S. Distinct antimicrobial peptide expression determines host species-specific bacterial associations. *Proc. Natl. Acad. Sci. USA* **2013**, *110*, E3730–E3738. [CrossRef] [PubMed]
6. Kim, D.; Baik, K.S.; Kim, M.S.; Jung, B.M.; Shin, T.S.; Chung, G.H.; Rhee, M.S.; Seong, C.N. *Shewanella haliotis* sp. Nov., isolated from the gut microflora of abalone, *Haliotis discus hannai*. *Int. J. Syst. Evol. Microbiol.* **2007**, *57*, 2926–2931. [CrossRef] [PubMed]
7. Franchini, P.; Fruciano, C.; Frickey, T.; Jones, J.C.; Meyer, A. The gut microbial community of midas cichlid fish in repeatedly evolved limnetic-benthic species pairs. *PLoS ONE* **2014**, *9*, e95027. [CrossRef] [PubMed]
8. Navarrete, P.; Espejo, R.T.; Romero, J. Molecular analysis of microbiota along the digestive tract of juvenile Atlantic salmon (*Salmo salar* L.). *Microbial Ecol.* **2009**, *57*, 550–561. [CrossRef] [PubMed]
9. Li, M.; Yang, H.; Gu, J.D. Phylogenetic diversity and axial distribution of microbes in the intestinal tract of the polychaete *Neanthes glandicincta*. *Microbial Ecol.* **2009**, *58*, 892–902. [CrossRef] [PubMed]

10. Fredrickson, J.K.; Romine, M.F.; Beliaev, A.S.; Auchtung, J.M.; Driscoll, M.E.; Gardner, T.S.; Nealson, K.H.; Osterman, A.L.; Pinchuk, G.; Reed, J.L.; et al. Towards environmental systems biology of *Shewanella*. *Nat. Rev. Microbiol.* **2008**, *6*, 592–603. [CrossRef] [PubMed]

11. Hau, H.H.; Gralnick, J.A. Ecology and biotechnology of the genus *Shewanella*. *Ann. Rev. Microbiol.* **2007**, *61*, 237–258. [CrossRef] [PubMed]

12. Karpinets, T.V.; Romine, M.F.; Schmoyer, D.D.; Kora, G.H.; Syed, M.H.; Leuze, M.R.; Serres, M.H.; Park, B.H.; Samatova, N.F.; Uberbacher, E.C. *Shewanella* knowledgebase: Integration of the experimental data and computational predictions suggests a biological role for transcription of intergenic regions. *Database* **2010**, baq012. [CrossRef] [PubMed]

13. Ericsson, A.C.; Davis, D.J.; Franklin, C.L.; Hagan, C.E. Exoelectrogenic capacity of host microbiota predicts lymphocyte recruitment to the gut. *Physiol. Genomics* **2015**, *47*, 243–252. [CrossRef] [PubMed]

14. Caro-Quintero, A.; Deng, J.; Auchtung, J.; Brettar, I.; Hofle, M.G.; Klappenbach, J.; Konstantinidis, K.T. Unprecedented levels of horizontal gene transfer among spatially co-occurring *Shewanella* bacteria from the Baltic Sea. *ISME J.* **2011**, *5*, 131–140. [CrossRef] [PubMed]

15. Ong, W.K.; Vu, T.T.; Lovendahl, K.N.; Llull, J.M.; Serres, M.H.; Romine, M.F.; Reed, J.L. Comparisons of *Shewanella* strains based on genome annotations, modeling, and experiments. *BMC Syst. Biol.* **2014**, *8*, 31. [CrossRef] [PubMed]

16. Lobo, C.; Moreno-Ventas, X.; Tapia-Paniagua, S.; Rodriguez, C.; Morinigo, M.A.; de La Banda, I.G. Dietary probiotic supplementation (*Shewanella putrefaciens* Pdp11) modulates gut microbiota and promotes growth and condition in *Senegalese sole* larviculture. *Fish Physiol. Biochem.* **2014**, *40*, 295–309. [CrossRef] [PubMed]

17. Luhtanen, A.M.; Eronen-Rasimus, E.; Kaartokallio, H.; Rintala, J.M.; Autio, R.; Roine, E. Isolation and characterization of phage-host systems from the Baltic Sea ice. *Extremophiles* **2014**, *18*, 121–130. [CrossRef] [PubMed]

18. Jian, H.; Xiao, X.; Wang, F. Role of filamentous phage SW1 in regulating the lateral flagella of *Shewanella piezotolerans* strain WP3 at low temperatures. *Appl. Environ. Microbiol.* **2013**, *79*, 7101–7109. [CrossRef] [PubMed]

19. Han, F.; Li, M.; Lin, H.; Wang, J.; Cao, L.; Khan, M.N. The novel *Shewanella putrefaciens*-infecting bacteriophage Spp001: Genome sequence and lytic enzymes. *J. Ind. Microbiol. Biotechnol.* **2014**, *41*, 1017–1026. [CrossRef] [PubMed]

20. Wang, F.; Wang, F.; Li, Q.; Xiao, X. A novel filamentous phage from the deep-sea bacterium *Shewanella piezotolerans* WP3 is induced at low temperature. *J. Bacteriol.* **2007**, *189*, 7151–7153. [CrossRef] [PubMed]

21. Godeke, J.; Paul, K.; Lassak, J.; Thormann, K.M. Phage-induced lysis enhances biofilm formation in *Shewanella oneidensis* MR-1. *ISME J.* **2011**, *5*, 613–626. [CrossRef] [PubMed]

22. Paul, J.H. Prophages in marine bacteria: Dangerous molecular time bombs or the key to survival in the seas? *ISME J.* **2008**, *2*, 579–589. [CrossRef] [PubMed]

23. Waller, A.S.; Yamada, T.; Kristensen, D.M.; Kultima, J.R.; Sunagawa, S.; Koonin, E.V.; Bork, P. Classification and quantification of bacteriophage taxa in human gut metagenomes. *ISME J.* **2014**, *8*, 1391–1402. [CrossRef] [PubMed]

24. Flemming, H.C.; Wingender, J. The biofilm matrix. *Nat. Rev. Microbiol.* **2010**, *8*, 623–633. [CrossRef] [PubMed]

25. Jefferson, K.K. What drives bacteria to produce a biofilm? *FEMS Microbiol. Lett.* **2004**, *236*, 163–173. [CrossRef] [PubMed]

26. Tlaskalova-Hogenova, H.; Stepankova, R.; Kozakova, H.; Hudcovic, T.; Vannucci, L.; Tuckova, L.; Rossmann, P.; Hrncir, T.; Kverka, M.; Zakostelska, Z.; et al. The role of gut microbiota (commensal bacteria) and the mucosal barrier in the pathogenesis of inflammatory and autoimmune diseases and cancer: Contribution of germ-free and gnotobiotic animal models of human diseases. *Cell Mol. Immunol.* **2011**, *8*, 110–120. [CrossRef] [PubMed]

27. Sommer, F.; Backhed, F. The gut microbiota—Masters of host development and physiology. *Nat. Rev. Microbiol.* **2013**, *11*, 227–238. [CrossRef] [PubMed]

28. Rice, S.A.; Tan, C.H.; Mikkelsen, P.J.; Kung, V.; Woo, J.; Tay, M.; Hauser, A.; McDougald, D.; Webb, J.S.; Kjelleberg, S. The biofilm life cycle and virulence of *Pseudomonas aeruginosa* are dependent on a filamentous prophage. *ISME J.* **2009**, *3*, 271–282. [CrossRef] [PubMed]

29. Wang, X.; Kim, Y.; Ma, Q.; Hong, S.H.; Pokusaeva, K.; Sturino, J.M.; Wood, T.K. Cryptic prophages help bacteria cope with adverse environments. *Nat. Commun.* **2010**, *1*, 147. [CrossRef] [PubMed]

30. Carrolo, M.; Frias, M.J.; Pinto, F.R.; Melo-Cristino, J.; Ramirez, M. Prophage spontaneous activation promotes DNA release enhancing biofilm formation in *Streptococcus pneumoniae*. *PLoS ONE* **2010**, *5*, e15678. [CrossRef] [PubMed]

31. Binnenkade, L.; Teichmann, L.; Thormann, K.M. Iron triggers lambdaSo prophage induction and release of extracellular DNA in *Shewanella oneidensis* MR-1 biofilms. *Appl. Environ. Microbiol.* **2014**, *80*, 5304–5316. [CrossRef] [PubMed]

32. Weisburg, W.G.; Barns, S.M.; Pelletier, D.A.; Lane, D.J. 16s ribosomal DNA amplification for phylogenetic study. *J. Bacteriol.* **1991**, *173*, 697–703. [CrossRef] [PubMed]

33. Altschul, S.F.; Madden, T.L.; Schäffer, A.A.; Zhang, J.; Zhang, Z.; Miller, W.; Lipman, D.J. Gapped BLAST and PSI-BLAST: A new generation of protein database search programs. *Nucleic Acids Res.* **1997**, *25*, 3389–3402. [CrossRef]

34. Adams, M.H. *Bacteriophages*; Interscience Publishers: New York, NY, USA, 1959; p. 592.

35. Hyman, P.; Abedon, S.T. Practical methods for determining phage growth parameters. *Methods Mol. Biol.* **2009**, *501*, 175–202. [PubMed]

36. Thurber, R.V.; Haynes, M.; Breitbart, M.; Wegley, L.; Rohwer, F. Laboratory procedures to generate viral metagenomes. *Nat. Protoc.* **2009**, *4*, 470–483. [CrossRef] [PubMed]

37. Ackermann, H.-W.; Heldal, M. Basic electron microscopy of aquatic viruses. *Man Aquatic Viral Ecol.* **2010**, 182–192. [CrossRef]

38. Deng, X.; Naccache, S.N.; Ng, T.; Federman, S.; Li, L.; Chiu, C.Y.; Delwart, E.L. An ensemble strategy that significantly improves *de novo* assembly of microbial genomes from metagenomic next-generation sequencing data. *Nucleic Acids Res.* **2015**, *43*, e46. [CrossRef] [PubMed]

39. Zerbino, D.R.; Birney, E. Velvet: Algorithms for *de novo* short read assembly using de Bruijn graphs. *Genome Res.* **2008**, *18*, 821–829. [CrossRef] [PubMed]

40. Kearse, M.; Moir, R.; Wilson, A.; Stones-Havas, S.; Cheung, M.; Sturrock, S.; Buxton, S.; Cooper, A.; Markowitz, S.; Duran, C.; et al. Geneious basic: An integrated and extendable desktop software platform for the organization and analysis of sequence data. *Bioinformatics* **2012**, *28*, 1647–1649. [CrossRef] [PubMed]

41. Delcher, A.L.; Bratke, K.A.; Powers, E.C.; Salzberg, S.L. Identifying bacterial genes and endosymbiont DNA with Glimmer. *Bioinformatics* **2007**, *23*, 673–679. [CrossRef] [PubMed]

42. Aziz, R.K.; Bartels, D.; Best, A.A.; DeJongh, M.; Disz, T.; Edwards, R.A.; Formsma, K.; Gerdes, S.; Glass, E.M.; Kubal, M.; et al. The RAST server: Rapid annotations using subsystems technology. *BMC Genomics* **2008**, *9*, 75. [CrossRef] [PubMed]

43. Overbeek, R.; Olson, R.; Pusch, G.D.; Olsen, G.J.; Davis, J.J.; Disz, T.; Edwards, R.A.; Gerdes, S.; Parrello, B.; Shukla, M.; et al. The SEED and the rapid annotation of microbial genomes using subsystems technology (RAST). *Nucleic Acids Res.* **2014**, *42*, D206–D214. [CrossRef] [PubMed]

44. Goris, J.; Konstantinidis, K.T.; Klappenbach, J.A.; Coenye, T.; Vandamme, P.; Tiedje, J.M. DNA-DNA hybridization values and their relationship to whole-genome sequence similarities. *Int. J. Syst. Evol. Microbiol.* **2007**, *57*, 81–91, ANI Average Nucleotide Identity. Available online: http://enve-omics.ce.gatech.edu/ani/ (accessed on 12 May 2016). [CrossRef] [PubMed]

45. Rodriguez-R, L.M.; Konstantinos, T. Bypassing cultivation to identify bacterial species. *Microbe* **2014**, *9*, 111–118. [CrossRef]

46. Roux, S.; Enault, F.; Hurwitz, B.L.; Sullivan, M.B. VirSorter: Mining viral signal from microbial genomic data. *PeerJ* **2015**, *3*, e985. [CrossRef] [PubMed]

47. Jiang, S.C.; Paul, J.H. Significance of lysogeny in the marine environment: Studies with isolates and a model of lysogenic phage production. *Microb. Ecol.* **1998**, *35*, 235–243. [CrossRef] [PubMed]

48. Patel, A.; Noble, R.T.; Steele, J.A.; Schwalbach, M.S.; Hewson, I.; Fuhrman, J.A. Virus and prokaryote enumeration from planktonic aquatic environments by epifluorescence microscopy with SYBR green I. *Nat. Protoc.* **2007**, *2*, 269–276. [CrossRef] [PubMed]

49. Okshevsky, M.; Meyer, R.L. Evaluation of fluorescent stains for visualizing extracellular DNA in biofilms. *J. Microbiol. Methods* **2014**, *105*, 102–104. [CrossRef] [PubMed]

50. Merritt, J.H.; Kadouri, D.E.; O'Toole, G.A. Growing and analyzing static biofilms. *Curr. Protoc. Microbiol.* **2005**, 1B-1.

51. Schneider, C.A.; Rasband, W.S.; Eliceiri, K.W. NIH image to ImageJ: 25 Years of image analysis. *Nat. Methods* **2012**, *9*, 671–675. [CrossRef] [PubMed]

52. Ivanova, E.P.; Sawabe, T.; Hayashi, K.; Gorshkova, N.M.; Zhukova, N.V.; Nedashkovskaya, O.I.; Mikhailov, V.V.; Nicolau, D.V.; Christen, R. *Shewanella fidelis* sp. Nov., isolated from sediments and sea water. *Int. J. Syst. Evol. Microbiol.* **2003**, *53*, 577–582. [CrossRef] [PubMed]

53. Zhang, W.; Du, P.; Zheng, H.; Yu, W.; Wan, L.; Chen, C. Whole-genome sequence comparison as a method for improving bacterial species definition. *J. Gen. Appl. Microbiol.* **2014**, *60*, 75–78. [CrossRef] [PubMed]

54. Heidelberg, J.F.; Paulsen, I.T.; Nelson, K.E.; Gaidos, E.J.; Nelson, W.C.; Read, T.D.; Eisen, J.A.; Seshadri, R.; Ward, N.; Methe, B.; et al. Genome sequence of the dissimilatory metal ion-reducing bacterium *Shewanella oneidensis*. *Nat. Biotechnol.* **2002**, *20*, 1118–1123. [CrossRef] [PubMed]

55. Fauquet, C.M.; Maniloff, J.; Desselberger, U.; Ball, A. *Virus Taxonomy: Viiith Report of the International Committee on Taxonomy of Viruses*; Elsevier Academic Press: New York, NY, USA, 2005.

56. Seguritan, V.; Feng, I.W.; Rohwer, F.; Swift, M.; Segall, A.M. Genome sequences of two closely related *Vibrio parahaemolyticus* phages, VP16T and VP16C. *J. Bacteriol.* **2003**, *185*, 6434–6447. [CrossRef] [PubMed]

57. Darling, A.E.; Mau, B.; Perna, N.T. Progressive mauve: Multiple genome alignment with gene gain, loss and rearrangement. *PLoS ONE* **2010**, *5*, e11147. [CrossRef] [PubMed]

58. Li, Y.; Wang, M.; Liu, Q.; Song, X.; Wang, D.; Ma, Y.; Shao, H.; Jiang, Y. Complete genomic sequence of bacteriophage H188: A novel *Vibrio kanaloae* phage isolated from Yellow Sea. *Curr. Microbiol.* **2016**, *72*, 628–633. [CrossRef] [PubMed]

59. Sime-Ngando, T. Environmental bacteriophages: Viruses of microbes in aquatic ecosystems. *Front. Microbiol.* **2014**, *5*, 355. [CrossRef] [PubMed]

60. Rohwer, F.; Edwards, R. The phage proteomic tree: A genome-based taxonomy for phage. *J. Bacteriol.* **2002**, *184*, 4529–4535. [CrossRef] [PubMed]

61. Lucks, J.B.; Nelson, D.R.; Kudla, G.R.; Plotkin, J.B. Genome landscapes and bacteriophage codon usage. *PLoS Comput. Biol.* **2008**, *4*, e1000001. [CrossRef] [PubMed]

62. Chithambaram, S.; Prabhakaran, R.; Xia, X. Differential codon adaptation between dsDNA and ssDNA phages in *Escherichia coli*. *Mol. Biol. Evol.* **2014**, *31*, 1606–1617. [CrossRef] [PubMed]

63. Nicholson, J.K.; Holmes, E.; Kinross, J.; Burcelin, R.; Gibson, G.; Jia, W.; Pettersson, S. Host-gut microbiota metabolic interactions. *Science* **2012**, *336*, 1262–1267. [CrossRef] [PubMed]

64. Leigh, B.A.; Liberti, A.; Dishaw, L.J. Generation of germ-free *Ciona intestinalis* for studies of gut-microbe interactions. *Front. Microbiol.* **2016**, *7*, 2092. [CrossRef] [PubMed]

65. Hooper, L.V.; Macpherson, A.J. Immune adaptations that maintain homeostasis with the intestinal microbiota. *Nat. Rev. Immunol.* **2010**, *10*, 159–169. [CrossRef] [PubMed]

66. Mazmanian, S.K.; Liu, C.H.; Tzianabos, A.O.; Kasper, D.L. An immunomodulatory molecule of symbiotic bacteria directs maturation of the host immune system. *Cell* **2005**, *122*, 107–118. [CrossRef] [PubMed]

67. Fujimura, K.E.; Slusher, N.A.; Cabana, M.D.; Lynch, S.V. Role of the gut microbiota in defining human health. *Expert Rev. Anti Infect Ther.* **2010**, *8*, 435–454. [CrossRef] [PubMed]

68. Breitbart, M. Marine viruses: Truth or dare. *Ann. Rev. Mar. Sci.* **2012**, *4*, 425–448. [CrossRef] [PubMed]

69. Weitz, J.S.; Stock, C.A.; Wilhelm, S.W.; Bourouiba, L.; Coleman, M.L.; Buchan, A.; Follows, M.J.; Fuhrman, J.A.; Jover, L.F.; Lennon, J.T.; et al. A multitrophic model to quantify the effects of marine viruses on microbial food webs and ecosystem processes. *ISME J.* **2015**, *9*, 1352–1364. [CrossRef] [PubMed]

70. Suttle, C.A. Marine viruses—Major players in the global ecosystem. *Nat. Rev. Microbiol.* **2007**, *5*, 801–812. [CrossRef] [PubMed]

71. Ogilvie, L.A.; Jones, B.V. The human gut virome: A multifaceted majority. *Front. Microbiol.* **2015**, *6*, 918. [CrossRef] [PubMed]

72. Stern, A.; Mick, E.; Tirosh, I.; Sagy, O.; Sorek, R. CRISPR targeting reveals a reservoir of common phages associated with the human gut microbiome. *Genome Res.* **2012**, *22*, 1985–1994. [CrossRef] [PubMed]

73. Lepage, P.; Leclerc, M.C.; Joossens, M.; Mondot, S.; Blottiere, H.M.; Raes, J.; Ehrlich, D.; Dore, J. A metagenomic insight into our gut's microbiome. *Gut* **2013**, *62*, 146–158. [CrossRef] [PubMed]

74. Reyes, A.; Semenkovich, N.P.; Whiteson, K.; Rohwer, F.; Gordon, J.I. Going viral: Next-generation sequencing applied to phage populations in the human gut. *Nat. Rev. Microbiol.* **2012**, *10*, 607–617. [CrossRef] [PubMed]

75. Modi, S.R.; Lee, H.H.; Spina, C.S.; Collins, J.J. Antibiotic treatment expands the resistance reservoir and ecological network of the phage metagenome. *Nature* **2013**, *499*, 219–222. [CrossRef] [PubMed]

76. Watson, M.A.; Chaconas, G. Three-site synapsis during Mu DNA transposition: A critical intermediate preceding engagement of the active site. *Cell* **1996**, *85*, 435–445. [CrossRef]

77. Erauso, G.; Lakhal, F.; Bidault-Toffin, A.; Le Chevalier, P.; Bouloc, P.; Paillard, C.; Jacq, A. Evidence for the role of horizontal transfer in generating Pvt1, a large mosaic conjugative plasmid from the clam pathogen, *Vibrio tapetis*. *PLoS ONE* **2011**, *6*, e16759. [CrossRef] [PubMed]

78. Duerkop, B.A.; Clements, C.V.; Rollins, D.; Rodrigues, J.L.; Hooper, L.V. A composite bacteriophage alters colonization by an intestinal commensal bacterium. *Proc. Nat. Acad. Sci. USA* **2012**, *109*, 17621–17626. [CrossRef] [PubMed]

79. Okshevsky, M.; Meyer, R.L. The role of extracellular DNA in the establishment, maintenance and perpetuation of bacterial biofilms. *Crit. Rev. Microbiol.* **2015**, *41*, 341–352. [CrossRef] [PubMed]

80. Steinberger, R.E.; Holden, P.A. Extracellular DNA in single- and multiple-species unsaturated biofilms. *Appl. Environ. Microbiol.* **2005**, *71*, 5404–5410. [CrossRef] [PubMed]

81. Liu, H.; Xie, Z.; Shen, P. Involvement of DNA in biofilm formation II: From bacterial adhesion to biofilm formation. *J. Nat. Sci. Wuhan* **2012**, *17*, 162–168. [CrossRef]

82. Allesen-Holm, M.; Barken, K.B.; Yang, L.; Klausen, M.; Webb, J.S.; Kjelleberg, S.; Molin, S.; Givskov, M.; Tolker-Nielsen, T. A characterization of DNA release in *Pseudomonas aeruginosa* cultures and biofilms. *Mol. Microbiol.* **2006**, *59*, 1114–1128. [CrossRef] [PubMed]

83. Chandramohan, L.; Ahn, J.S.; Weaver, K.E.; Bayles, K.W. An overlap between the control of programmed cell death in *Bacillus anthracis* and sporulation. *J. Bacteriol.* **2009**, *191*, 4103–4110. [CrossRef] [PubMed]

84. Rice, K.C.; Bayles, K.W. Death's toolbox: Examining the molecular components of bacterial programmed cell death. *Mol. Microbiol.* **2003**, *50*, 729–738. [CrossRef] [PubMed]

85. Webb, J.S.; Thompson, L.S.; James, S.; Charlton, T.; Tolker-Nielsen, T.; Koch, B.; Givskov, M.; Kjelleberg, S. Cell death in *Pseudomonas aeruginosa* biofilm development. *J. Bacteriol.* **2003**, *185*, 4585–4592. [CrossRef] [PubMed]

86. Nanda, A.M.; Thormann, K.; Frunzke, J. Impact of spontaneous prophage induction on the fitness of bacterial populations and host-microbe interactions. *J. Bacteriol.* **2015**, *197*, 410–419. [CrossRef] [PubMed]

87. Tan, D.; Dahl, A.; Middelboe, M. Vibriophages differentially influence biofilm formation by *Vibrio anguillarum* strains. *Appl. Environ. Microbiol.* **2015**, *81*, 4489–4497. [CrossRef] [PubMed]

88. Hosseinidoust, Z.; Tufenkji, N.; van de Ven, T.G. Formation of biofilms under phage predation: Considerations concerning a biofilm increase. *Biofouling* **2013**, *29*, 457–468. [PubMed]

89. Lacqua, A.; Wanner, O.; Colangelo, T.; Martinotti, M.G.; Landini, P. Emergence of biofilm-forming subpopulations upon exposure of *Escherichia coli* to environmental bacteriophages. *Appl. Environ. Microbiol.* **2006**, *72*, 956–959. [CrossRef] [PubMed]

90. Spoering, A.L.; Gilmore, M.S. Quorum sensing and DNA release in bacterial biofilms. *Curr. Opin. Microbiol.* **2006**, *9*, 133–137. [CrossRef] [PubMed]

91. Chao, L.; Ramsdell, G. The effects of wall populations on coexistence of bacteria in the liquid phase of chemostat cultures. *J. Gen. Microbiol.* **1985**, *131*, 1229–1236. [CrossRef] [PubMed]

92. Barr, J.J.; Auro, R.; Furlan, M.; Whiteson, K.L.; Erb, M.L.; Pogliano, J.; Stotland, A.; Wolkowicz, R.; Cutting, A.S.; Doran, K.S.; et al. Bacteriophage adhering to mucus provide a non-host-derived immunity. *Proc. Natl. Acad. Sci. USA* **2013**, *110*, 10771–10776. [CrossRef] [PubMed]

93. Sutherland, I. Biofilm exopolysaccharides: A strong and sticky framework. *Microbiology* **2001**, *147*, 3–9. [CrossRef] [PubMed]

94. Moons, P.; Faster, D.; Aertsen, A. Lysogenic conversion and phage resistance development in phage exposed *Escherichia coli* biofilms. *Viruses* **2013**, *5*, 150–161. [CrossRef] [PubMed]

95. Bayles, K.W. Are the molecular strategies that control apoptosis conserved in bacteria? *Trends Microbiol.* **2003**, *11*, 306–311. [CrossRef]

96. Backhed, F.; Ley, R.E.; Sonnenburg, J.L.; Peterson, D.A.; Gordon, J.I. Host-bacterial mutualism in the human intestine. *Science* **2005**, *307*, 1915–1920. [CrossRef] [PubMed]

97. Thomas, A.D.; Parker, W. Cultivation of epithelial-associated microbiota by the immune system. *Future Microbiol.* **2010**, *5*, 1483–1492. [CrossRef] [PubMed]

viruses

MDPI

Article

Stumbling across the Same Phage: Comparative Genomics of Widespread Temperate Phages Infecting the Fish Pathogen *Vibrio anguillarum*

Panos G. Kalatzis [1,2,†], Nanna Rørbo [1,†], Daniel Castillo [1], Jesper Juel Mauritzen [1], Jóhanna Jørgensen [1], Constantina Kokkari [2], Faxing Zhang [3], Pantelis Katharios [2] and Mathias Middelboe [1,*]

[1] Marine Biological Section, University of Copenhagen, DK-3000 Helsingør, Denmark; panos.kalatzis@bio.ku.dk (P.G.K.); nanna.rorbo@bio.ku.dk (N.R.); danielcastillobq@gmail.com (D.C.); kcq442@alumni.ku.dk (J.J.M.) ; johannajorgensen88@gmail.com (J.J.);
[2] Institute of Marine Biology, Biotechnology and Aquaculture, Hellenic Centre for Marine Research, Crete 71500, Greece; dkok@hcmr.gr (C.K.); katharios@hcmr.gr (P.K.)
[3] Beijing Genomics Institute (BGI) Park, No.21 Hongan 3rd Street, Building NO. 7, Yantian District, Shenzhen 518083, China; jason.zhang@genomics.cn
*   Correspondence: mmiddelboe@bio.ku.dk; Tel.: +45-3532-1991
†   The two first co-authors contributed equally to this work.

Academic Editor: Eric O. Freed
Received: 1 March 2017; Accepted: 12 May 2017; Published: 20 May 2017

Abstract: Nineteen *Vibrio anguillarum*-specific temperate bacteriophages isolated across Europe and Chile from aquaculture and environmental sites were genome sequenced and analyzed for host range, morphology and life cycle characteristics. The phages were classified as Siphoviridae with genome sizes between 46,006 and 54,201 bp. All 19 phages showed high genetic similarity, and 13 phages were genetically identical. Apart from sporadically distributed single nucleotide polymorphisms (SNPs), genetic diversifications were located in three variable regions (VR1, VR2 and VR3) in six of the phage genomes. Identification of specific genes, such as N6-adenine methyltransferase and lambda like repressor, as well as the presence of a tRNA^{Arg}, suggested a both mutualistic and parasitic interaction between phages and hosts. During short term phage exposure experiments, 28% of a *V. anguillarum* host population was lysogenized by the temperate phages and a genomic analysis of a collection of 31 virulent *V. anguillarum* showed that the isolated phages were present as prophages in >50% of the strains covering large geographical distances. Further, phage sequences were widely distributed among CRISPR-Cas arrays of publicly available sequenced *Vibrio*s. The observed distribution of these specific temperate Vibriophages across large geographical scales may be explained by efficient dispersal of phages and bacteria in the marine environment combined with a mutualistic interaction between temperate phages and their hosts which selects for co-existence rather than arms race dynamics.

Keywords: bacteriophages; temperate; *Vibrio anguillarum*; Siphovirus; genetic similarity; omnipresent; lysogenic conversion.

1. Introduction

Vibrios are a genetically and metabolically diverse group of marine heterotrophic bacteria that plays significant roles in marine biogeochemical cycling [1]. The genus is globally distributed in marine and brackish environments and occurs as both free-living and surface-associated communities ranging from coastal areas to the open sea [1–3]. The group contains several pathogens, including *Vibrio anguillarum*, which infects more than 50 species of fish, mollusks and crustaceans, causing vibriosis [4], a devastating disease affecting the aquaculture industry worldwide. *V. anguillarum* consists of both pathogenic and non-pathogenic strains [5,6], and the mechanism of pathogenesis of *V. anguillarum* is not completely understood yet. Recent advancements in genome sequencing have started to shed light in putative bacterial virulence and fitness factors including exotoxins, adherence/colonization proteins, invasion, capsule and cell surface components as well as iron uptake system [7,8].

Virulence in *Vibrios* can also be associated with the expression of prophage-encoded genes through the process of lysogenic conversion [9–11]. A well-known example of prophage generated virulence is the human pathogen *V. cholerae*, in which the key toxins (CTX and Zot) are encoded by specific prophages [12,13]. More recently, these prophage-associated toxins have also been identified in environmental *Vibrios* (e.g., *V. coralliilyticus*), indicating that prophage-encoded genes are disseminated between environmental *Vibrio* populations [14]. In addition, in the fish pathogen *V. harveyi*, the prophage VHML conferred virulence by lysogenic conversion [15,16] and recent bioinformatic analyses have identified numerous phage genes potentially encoding virulence and fitness factors in environmental *Vibrio* strains [14,17]. Therefore, there is increasing evidence that lysogenic conversion plays a significant role in shaping environmental *Vibrio* populations contributing to their functional and genetic diversity. In addition to lysogenic conversion, prophages can also affect host properties by serving as anchor points for genomic rearrangements [18], by disrupting critical genes or operons during integration and excision [19] and by preventing infection by similar phages with superinfection exclusion mechanisms [20]. Further, phage-derived bacterial immunity can be obtained by phage-encoded interactions with restriction-modification (RM) system, such as expression of methyltransferases [21,22].

The presence of prophages in marine bacteria has also been connected with more subtle benefits for the bacterial host. Marine prophages can suppress unneeded metabolic activities of their host under unfavorable environmental conditions, serving as energy saving and survival enhancing tool [23]. Thus, *Vibrio* prophages and temperate phages constitute potentially a major reservoir of functional traits in the marine environment, including virulence. The mobilization of these phage-encoded properties may be a key driving force for dissemination of such properties in environmental marine *Vibrio* communities.

Bacteriophages can display biogeographical patterns following their hosts' geographical patterns [24] and both cosmopolitan phages with a worldwide distribution [25], and phages constrained to a specific environment have been observed [9,26]. Previous studies have found indications that populations of genetically related vibriophages are widely distributed over large geographic distances in the oceans [27]. However, little is known about the geographical distribution of specific vibriophages. The diversity of bacteriophages can be high when examined locally but when a more global approach is attempted, diversity can be relatively limited due to the phages ability move among different biomes [27–29].

Studies on *V. anguillarum*-specific bacteriophages have so far mainly focused on lytic phages and their potential use in phage therapy application [30–32] while very little is known about the role of temperate phage-encoded virulence and other fitness factors as drivers of functional dynamics in *V. anguillarum*. Improving our understanding of how phages may influence the genomic and phenotypic characteristics of *V. anguillarum* thus requires more detailed knowledge of the genetic properties and dispersal of temperate phages infecting this pathogen as well as their distribution as prophages in the host community.

In this study, we report the complete nucleotide sequences, annotations and genome comparison of 19 *V. anguillarum*-specific temperate bacteriophages isolated from Denmark, Norway, Greece and Chile using both aquaculture and environmental sites. Further, we analyze the distribution of the temperate phages as prophages in a collection of 31 virulent *V. anguillarum* strains and test their ability to lysogenize *V. anguillarum*. The bacteriophage sequences were mapped against the available CRISPR-Cas arrays of the *Vibrio* genus in order to assess past events of phage-bacterial interactions leading to spacer acquisition. The geographical distribution and genomic characterization, along with the phage morphology, host range and kinetic parameters of selected phages unraveled a novel insight into a group of temperate phages, designated H20-like phages and their potential for lysogenic conversion and interactions with *V. anguillarum* across large spatial scales.

2. Materials and Methods

2.1. Bacterial Strains and Growth Conditions

The *V. anguillarum* strains A023, T265, BA35 and VaKef were used as hosts for bacteriophage isolation. The host strains are part of a *V. anguillarum* collection, which consists of 31 well characterized and whole genome sequenced strains [33]. All bacterial strains were stored at −80 °C in Luria-Bertani (LB) medium (Invitrogen, Carlsbad, CA, USA) supplied with 15% glycerol. For proliferation, the strains were inoculated in LB medium and incubated at 25 °C with constant agitation.

2.2. Isolation, Purification and Propagation of Bacteriophages

More than 100 aquaculture and environmental marine water samples have been collected from Denmark, Norway, Greece and Chile and isolation of phages was performed using standard enrichment methodology [34] with minor modifications. The bacterial strains used for phage isolation were added separately to the water samples, supplemented with 10% LB medium. Samples were incubated overnight at 25 °C with constant agitation, following centrifugation at $6000 \times g$ for 10 min. Supernatants were filtered through 0.22 μm syringe filters (Whatman, Maidstone, UK) and the presence of phages was examined with plaque assay using the double-layer agar method [35]. Briefly, 100 μL filtrate was added to 300 μL liquid cultures of each of the bacterial strain that were used for phage isolation. The mixture was then added to 4 mL melted soft agar (0.4% agar and 1% sea salts (Sigma-Aldrich, Saint Louis, MO, USA)) and poured onto LB 1.5% agar plates. Following overnight incubation at 25 °C, the plates were inspected for plaque forming units (pfu) on the host bacterial lawn. Single phage plaques were picked and purified by 5× re-plating. In this way, 19 clonal phage isolates were obtained from individual water samples (Table 1). Proliferation of bacteriophage isolates was conducted in 50 mL LB medium, by mixing the bacterial host strain at its early exponential phase with the corresponding bacteriophages at a multiplicity of infection (MOI) of 10, and incubating the mixture overnight at 25 °C with constant agitation. Following centrifugation and filtration through 0.22 μm syringe filters, the 19 final bacteriophage stocks were titered and stored at 4 °C.

Water samplings did not require any special permission.

2.3. Morphology, Life Cycle and Host Range of Bacteriophages

Virion morphology of two of the H20-like bacteriophages, φVaK and pVa-7, was determined under Transmission Electron Microscopy (TEM) observation at the facilities of University of Crete, Greece. Samples were prepared on collodium copper grids, negatively contrasted with 2% uranyl acetate, and examined by JEOL JEM2100 electron microscope at 80 kV, applying instrumental magnification of 120,000. The morphology of the phages (H20 and H8) was determined previously [36].

One-step growth experiments were performed with φH20, pVa-3, φCLA and φP3 by adding the phages to 1 mL of their bacterial host *V. anguillarum* strain BA35 (MOI:0.001) at early exponential phase in LB medium and incubating at 25 °C for 15 min. Following centrifugation at $6000 \times g$ for 10 min, supernatants containing free, unattached phages were discarded, while phages which had attached to the bacterial hosts were pelleted. The bacterial cells were then resuspended in 20 mL medium containing 1% Tryptone (Difco, Livonia, MI, USA), 0.5% Yeast extract (Difco), and 2% sea salts (Sigma-Aldrich). Setting that time point as t = 0, 20 μL drops of serial dilutions were spotted on bacterial lawns prepared on LB agar plates every 10 min for total duration of 130 min. Plaque forming units in the spotted areas were counted after overnight incubation at 25 °C. One-step curves were plotted from 3 individual experiments performed for each of the 4 bacteriophages assayed.

Phage adsorption was evaluated for φH20, pVa-3, φCla and φP3, using 6-well plates each containing 3 mL of *V. anguillarum* strain BA35 liquid culture at early exponential phase of growth in LB medium. Each bacteriophage tested was added to 3 wells (triplicates) applying a MOI of 0.001. Samples of 200 μL were collected every 5 min, followed by centrifugation at 10,000× *g* for 2 min. Drops of 20 μL containing un-adsorbed phages were taken from the supernatant and spotted on the host bacterial lawn as above. The total duration of the experiment was 30 min and plaque forming units were counted after overnight incubation at 25 °C. The phage adsorption constant was calculated from the decrease in the free, unattached phages over the incubation period [37,38].

The lytic spectrum (host range) of the isolated bacteriophages was assessed by spotting them on the bacterial lawns of the 31 *V. anguillarum* pathogenic strains collection. Twenty-microliter drops of each phage concentrate were spotted in triplicate on bacterial lawns followed by overnight incubation at 25 °C.

2.4. DNA Extraction and Sequencing

DNA was extracted from all 19 isolated bacteriophages using high titer samples (10^{10} pfu·mL^{-1}). The viral particles were concentrated by adding poly-ethylene glycol to the 4 °C phage stocks, followed by centrifugation (20,000× *g*, 1 h) and resuspension of the phage pellet in 100 μL of NaCl solution (150 mM). QIAamp DNA Blood Mini Kit (QIAGEN, Hilden, Germany) was used for the bacteriophage genome extraction, according to the manufacturer's protocol. The concentration of phage DNA was quantified by NanoDrop (Thermo Fisher Scientific, Waltham, MA, USA) and the DNA quality was assessed in an agarose gel.

Genome sequencing for all isolated bacteriophages was conducted using Illumina Hi Seq 2000 sequencer (Illumina, San Diego, CA, USA) at Beijing Genomic Institute (Shenzhen, Guangdong, China) using pair-end read sizes of 100 bp. Sequencing process, library construction and trimming of contaminated reads and primers, were performed in accordance with the manufacturer's protocols. De novo assembly of the produced reads was done by Geneious algorithm, with Geneious software bioinformatics platform R9 version [39], resulting in single contigs for all phages. Short and low-coverage contigs were discarded.

2.5. Annotation, Comparative Genomics, and Phylogeny

The genes of the sequenced bacteriophages were predicted by Glimmer 3 [40] and the open reading frames (ORFs) were generated. Annotations of the genes were performed automatically by Rapid Annotation Subsystem Technology (R.A.S.T.) [41,42]. Both annotated ORFs and hypothetical proteins ORFs that represented coding sequences (CDSs), were crosschecked manually by protein Basic Local Alignment Tool (BLASTP) and by Protein Fold Recognition Server, Phyre^2 [43,44], defining the gene functions both by genetic similarity and by protein structure. The presence of tRNA was assessed by the online tool ARAGORN [45]. Alignment of whole viral genomes was conducted by progressive MAUVE algorithm [46] using the Geneious software bioinformatics platform [39]. The direct terminal repeats (DTR) that were detected at the right and left ends of each phage contig marked the physical ends of the phage genomes (151 bp for φVaK and 143 bp for the rest phages). Additionally, attL and attR were characterized by the sequence TCGTGATTCCTTGC(T)CACCG CCACATCCAAGCCTCTTG(A)GGTATCAAGAGGCTTTATTTTATCTGACAGACCCCGCAAT(T) and direction of all sequences is attL → attR. Phylogenetic analyses of both the isolated phages and the *V. anguillarum* prophages included in the study, were conducted by Geneious Tree Builder, using Neighbor-Joining method and Tamura-Nei genetic distance model (1000 bootstraps).

2.6. CRISPR Arrays and Prophage Detection in Vibrio

All publicly available *Vibrio* genomes and downloaded from the National Center for Biotechnology Information (NCBI) GenBank [47]. The presence of CRISPR arrays in the *Vibrio* genomes was evaluated in 1185 genomes representing 64 different *Vibrio* species, using CRISPR Finder online tool [48].

Further, potentially inducible intact prophages were detected in the genomes of our *V. anguillarum* collection [34] using the online prophage finder tool, PHAST [49].

2.7. Integration of Phages in V. anguillarum *Strain BA35*

The ability of the isolated temperate phages to integrate into the bacterial host's genome was evaluated by coculturing *V. anguillarum* strain BA35 with ϕH20. Triplicate bottles with 48 mL LB medium were inoculated with the addition of host strain BA35 and ϕH20 (MOI: 50), incubated at 25 °C for 12 h with agitation. Bacterial growth was monitored by measuring optical density at 600 nm (OD_{600}) every hour using Novaspec Plus Visible Spectrophotometer (Amersham Biosciences, Uppsala, Sweden). Exponential growth was observed after 10–12 h of incubation and triplicates of 1 mL were sampled and serially diluted in SM buffer (Sodium Magnesium; 100 mM NaCl, 8 mM $MgSO_4 \cdot 7H_2O$, 50 mM Tris-Cl, 0.01% gelatin, pH: 7.5). Dilutions were poured onto LB agar plates and the following day, 50 bacterial colonies were isolated and recultured at 25 °C. The ability of the bacteriophage ϕH20 to lysogenize its host was assessed by Polymerase Chain Reaction (PCR) with 3 sets of specific primers (Supplementary Table S1). They were able to target and propagate 3 specific phage genes: a structural tail protein gene, a terminase large subunit gene and a hypothetical protein gene. All 50 isolated bacterial strains were washed 5 times with SM buffer in order to eliminate the possibility of a false positive reaction caused by any remnant phage particles. A triple positive PCR would confirm the lysogenization.

2.8. GenBank Accession Numbers

The GenBank accession numbers for the new sequenced bacteriophages are: KX581090, KX581091, KX581092, KX581093, KX581094, KX581095, KX581096, KX581097, KX581098, KX581099, KX581100, KY658673, KY658674, KY658675, KY658676, KY658677, KY658678, KY658679, and KY658680. Since all 19 phages were genetically similar, they were designated as H20-like phages named after ϕH20, which was isolated and described as the first [36].

3. Results

3.1. Isolation and Characterization of Bacteriophages

Sixteen bacteriophages were isolated in the current study using two *V. anguillarum* strains as hosts: strain BA35 (ϕP2, ϕP3, pVa-1, pVa-2, pVa-3, pVa-4, pVa-5, pVa-6, pVa-7, pVa-8, ϕCLA, ϕHer, ϕLen, ϕPel and ϕStrym) and strain VaKef (ϕVaK). Further, three phages were isolated in a previous study [36] using three different hosts: BA35 (ϕH20), T265 (ϕH8) and A023 (ϕH2). None of the used host strains contained H20-like prophages, implying that the isolated phages originated from the water samples. All phages produce similar 1 mm plaques on their host bacterial lawns. They were isolated from four different countries (Denmark, Norway, Greece and Chile), both from aquaculture and environmental marine water samples. Their genome sizes ranged between 46,006 and 54,201 base pair (bp) with a similar GC content of 43%–43.1% and with a number of ORFs between 76 and 94 (Table 1).

Table 1. Genomic information about the 19 sequenced bacteriophages of the current study. The bacteriophages were isolated against four different clinical *Vibrio anguillarum* strains (A023, T265, BA35, and VaKef). DK: Denmark, N: Norway, GR: Greece, CL: Chile, Aq: Aquaculture. M: Marine environment.

Bacteriophages	Origin	Environment	Host Strain	Genome Size (bp)	GC Content (%)	ORFs	GenBank Accession Number
φH2	DK	Aq	A023	46,149	43.1	76	KY658673
φH8	DK	Aq	T265	46,157	43.1	76	KY658674
φH20	DK	Aq	BA35	53,224	43	91	KY658675
φP2	N	Aq	BA35	46,149	43.1	76	KY658676
φP3	N	Aq	BA35	53,242	43	91	KY658677
pVa-1	GR	M	BA35	53,227	43.1	91	KX581095
pVa-2	GR	M	BA35	53,286	43	91	KX581094
pVa-3	GR	M	BA35	54,344	43.1	94	KY658678
pVa-4	GR	M	BA35	54,295	43.1	93	KY658679
pVa-5	GR	M	BA35	53,227	43	91	KX581096
pVa-6	GR	M	BA35	53,2274	43	91	KX581097
pVa-7	GR	M	BA35	54,268	43.1	93	KX581110
pVa-8	GR	M	BA35	53,227	43	91	KY658680
φCLA	CL	M	BA35	53,226	43	91	KX581091
φHer	GR	M	BA35	53,226	43	91	KX581090
φLen	GR	M	BA35	53,226	43	91	KX581092
φPel	GR	M	BA35	53,227	43	91	KX581093
φStrym	GR	M	BA35	53,226	43	91	KX581099
φVaK	GR	Aq	VaKef	53,216	43	92	KX581098

In addition to the genomic characterization, a selection of the H20-like phages group was further analyzed for virion morphology and life cycle parameters. The observed morphology of the virions under TEM classified the bacteriophages to the Siphoviridae family. Virions had a long, non-contractile tail of about 150 nm and a head of approximately 50 nm in diameter (Figure 1).

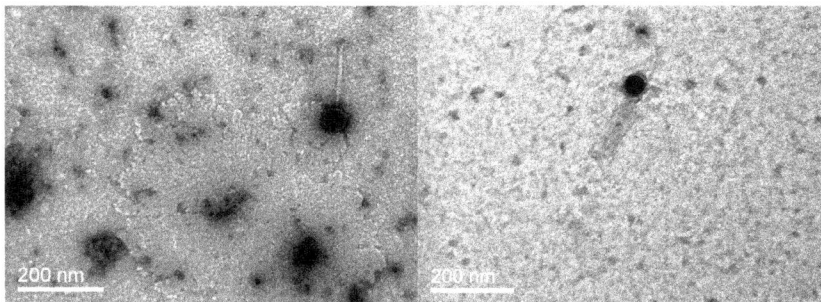

Figure 1. Transmission electron microscopy micrographs of bacteriophages: φVaK (**left**); and pVa-7 (**right**) classifying them to Siphoviridae family.

One-step growth and phage adsorption curves have been generated to determine the life cycle parameters of the bacteriophages (Supplementary Figure S1A,B). Latency time, burst size and adsorption constant ranged from 50 to 70 min, 100 to 145 virions per cell and 8.6×10^{-8} to 1.2×10^{-7}, respectively (Table 2).

Table 2. Kinetic parameters (latency time, burst size and adsorption constant) of the bacteriophages φH20, pVa-3, φCLA and φP3. All values are means ± standard deviation of three independent experiments.

Bacteriophage	Latency Time (min)	Burst Size (Virions/Cell)	Adsorption Constant K_{30}
φH20	60	112 ± 9	$1.18 \times 10^{-7} \pm 1.28 \times 10^{-8}$
pVa-3	60	100 ± 13	$8.63 \times 10^{-8} \pm 2.46 \times 10^{-9}$
φP3	70	101 ± 16	$6.40 \times 10^{-8} \pm 4.39 \times 10^{-9}$
φCla	50	145 ± 24	$2.11 \times 10^{-7} \pm 2.68 \times 10^{-8}$

3.2. Host Range Analysis and Phylogeny

Thirty-one pathogenic strains of *V. anguillarum* were used to determine the lytic spectrum of the 19 bacteriophage isolates (Figure 2). Seventeen out of 19 phages had almost identical host range with only one minor difference in the case of φH8. Two phages showed a different host range: φH2 only infected strain A023, whereas bacteriophage φVaK had a broader host range and infected eight out of 31 bacterial strains tested. According to the phylogenetic analysis of the 19 isolated bacteriophages, φP3, φH8, φP2, φH2 and φVaK had a higher number of substitutions per site. The combined host range table and the phylogenetic tree of the bacteriophages (Figure 2) indicated that the differences in the lytic spectrum are accompanied by differences at the genetic level. Especially in the case of φVaK, where most genetic differences are observed, the lytic spectrum is much broader than the others.

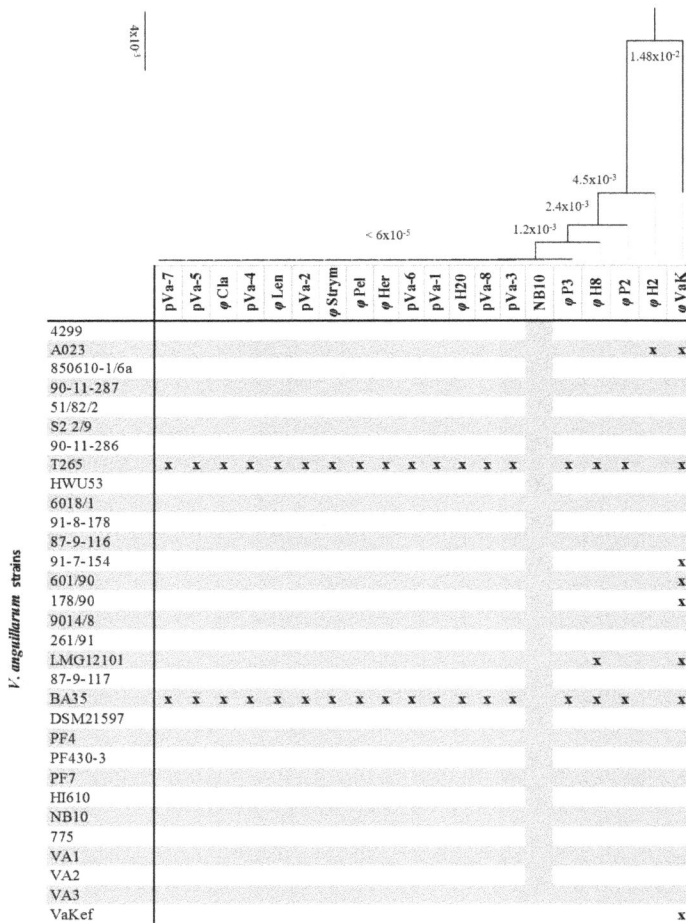

Figure 2. Host range of the 19 bacteriophage isolates against 31 pathogenic *V. anguillarum* strains. Blocks marked with x, indicate the infection of the bacteria by the corresponding bacteriophage isolate. The phylogenetic tree based on the whole genomes of the 19 sequenced phages of the current study (Neighbor-Joining method). The prophage from *V. anguillarum* strain NB10 was included as a reference phage. The branch labels indicate the substitutions per site.

3.3. Comparative and Functional Phage Genomics

The genomes of all 19 bacteriophages closely resemble a prophage genome which is present in the *Vibrio anguillarum* strain NB10 (GenBank No. LK021130) [50]. Additionally, H20-like bacteriophages were also found to be a part of several *V. anguillarum* genomes that have been recently released in the NCBI GenBank [34]. However, genomic comparisons of the phages have revealed some differences at the genetic level, which are depicted in an aligned sequence graph (Figure 3).

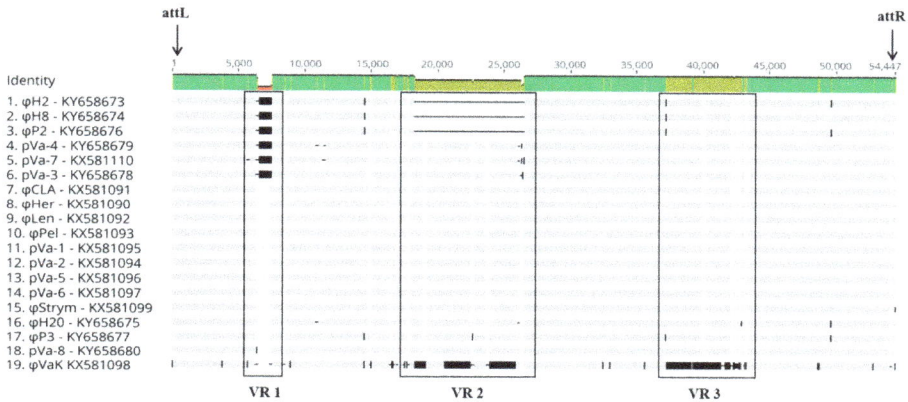

Figure 3. Genomic alignment of the 19 newly sequenced temperate bacteriophages. The black boxes indicate the variable regions (VR 1, VR 2 and VR 3). Single nucleotide polymorphisms (SNPs) and genetic differences are found sporadically in the genomes as black vertical lines either thin or thick depending on the length of the differences. Genome gaps are depicted with the black horizontal lines. The physical ends of the genomes are indicated by direct terminal repeats (DTRs) following the direction of attL → attR.

SNPs are sporadically distributed through the 19 phage genomes. However, there are three variable regions, named VR1, VR2 and VR3, which show systematic modifications among phages (Figure 3). In phages φP2, φH8 and φH2, the biggest part of VR2 is completely missing, whereas VR1 is only present in phages φP2, φH8, φH2, pVa-3, pVa-4 and pVa-7. As a result of these genomic modifications, the H20-like bacteriophages demonstrate a range of genome sizes, between 46-kb and 54-kb (Table 1).

Apart from the presence/absence of VR1 and VR2, most of the phages are genetically homogenous with the exception φVaK, where both VR1 and VR3 are very diverse compared to all other phage genomes, while VR1 is missing. SNPs in the genome of φVaK are also more frequent compared to the other bacteriophages. These genomic differences seem to be reflected in the phage's phenotype, since the host range of this phage is much broader than the others (Figure 2).

For most of the genes in the H20-like phages, the function remains unknown. However, taking into consideration both the genetic similarity and the protein folding structure of the viral gene products, it was possible to define the predicted function of approximately one third of the viral genes. The viral genes were predicted and then annotated based on their genetic similarity with the NCBI GenBank database, mainly referring to the prophage of strain NB10 annotation. Additional information on gene function that came from their protein folding pattern, determined by amino acid composition, contributed significantly in complementing the annotation of the bacteriophages. Thus, 15 gene functions and protein families were attributed to the one third of the viral genes, with structural and DNA binding proteins being mostly represented. Seventy-four genes compose the core-gene content of all 19 bacteriophages and this is the common basis on which accessory genes are added in some of the phage genomes (Figure 4). These 74 genes encode all the structural proteins, most of the DNA binding proteins and hydrolases and all the proteins related to regulation of gene expression, biosynthesis and metabolism, peptidase and lysozyme activity, ribonuclease activity, HNH endonucleases, RNA binding and DNA repairing. The presence of an integrase gene suggests that the H20-like bacteriophages are temperate phages able to be integrated in their host's genome, as it has happened in the case of the *V. anguillarum* NB10 prophage. Repressor, integrase, methyltransferase and tRNA^{Arg}, are also part of the core gene content present in all 19 bacteriophages. The core genome thus supports the functionality of the H20-like bacteriophages. However, 21 different accessory genes are present in some of the bacteriophage genomes. VR1 and VR2 are the main genomic parts of the accessory genome since they possess two and 17 genes, respectively. VR1 has a transposase and a putative 5-methycytocin specific restriction enzyme and VR2 has 13 genes of unknown function, two hydrolase, one DNA binding and one lysozyme activity proteins (Figure 4). The presence of accessory genes was not associated with any of the measured phenotypic properties.

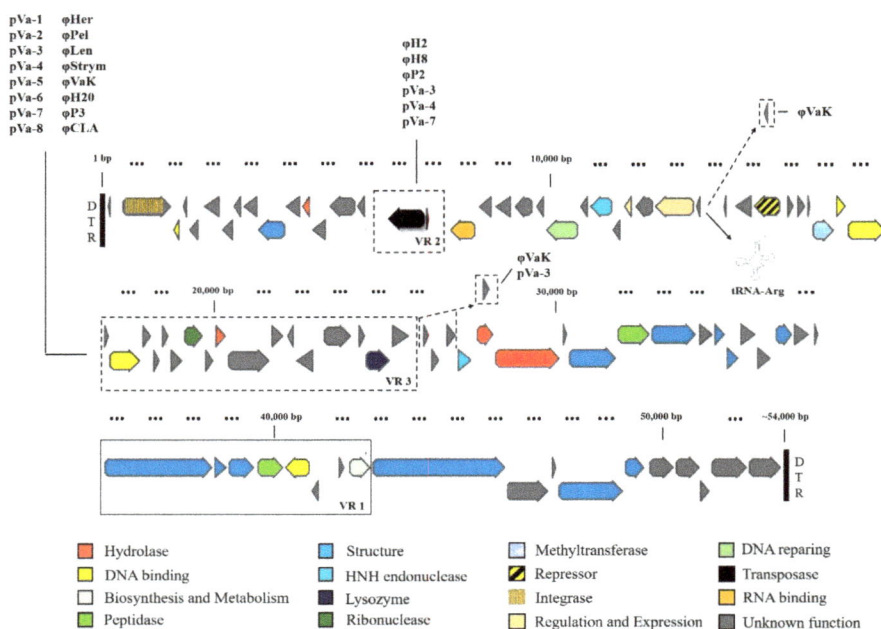

Figure 4. The core- and accessory- gene content of the H20-like bacteriophages. Core-gene content is composed by 74 genes shared by all 19 bacteriophages. Accessory-gene content is indicated by dashed lines, with VR1 and VR2 representing the main components. The bacteriophages that contain the accessory gene blocks VR1 and/or VR2 are correspondingly indicated in columns beside. Two more hypothetical proteins complement the accessory gene content and their corresponding carrier phages (pVa-3 and φVaK) are also indicated. DTRs mark the physical ends of the genome (bold vertical lines).

In the case of φVaK, VR2 and VR3 are the most diverse compared to the other homogenous phage genomes. However, the vast majority of the changes are silent and do not translate into any differences in the predicted gene function at the amino acid level.

The detailed annotations for the 19 temperate bacteriophages are listed in the Supplementary Table S2.

3.4. Vibrio *CRISPR Spacers in H20-Like Phages*

Although H20-like bacteriophages were isolated from distant and different locations, their genetic differences are minor. In order to assess the interaction of H20-like bacteriophages with *Vibrio* hosts in general, a matrix containing the CRISPR arrays from all the published *Vibrio* genomes was composed. Keeping the cut-off value of >80%, 10 CRISPR spacers from three different *Vibrio* species were found to match with the consensus genome of the H20-like phages (Supplementary Table S3), implying a *Vibrio*–H20-like phage contact at some point through evolutionary time. Furthermore, the mapping of H20-like phage genomes against the two publicly available CRISPR arrays of *V. anguillarum* strain PF7, led to 24 spacers matching with >80% similarity. Eighteen of them were >95% similar, whereas eight of them were 100% identical to genomic parts of the H20-like phages. The spacers matched against several regions in the phage, and were randomly scattered in the genome. Additionally, in the case of *V. navarrensis* (ATCC 51183), the spacer 21 matches 100% with its corresponding part in the H20-like phage genome, whereas 9 more spacers mapped with >80% similarity. In order to assess the presence of H20-like prophages in *V. anguillarum,* the genomes of the bacterial collection of *V. anguillarum* strains were analyzed. The sequences of 17 H20-like prophages were bioinformatically detected as integrated in the 31 *V. anguillarum* genomes (Table 3).

Table 3. Bioinformatic analysis indicates the presence or absence of H20-like prophages in the genomes of 31 pathogenic *V. anguillarum* strains originated from several different aquaculture producing countries. n/a: not available.

Stain Code	Origin	Phylogenetic Group	H20-Like Prophage	Reference
DSM21597	Norway	I	-	[51]
4299	Norway	II	-	[52]
HI610	Norway	II	-	[53]
S2 2/9	Denmark	III	-	[54]
90-11-286	Denmark	III	-	[54,55]
PF430-3	Chile	IV	-	[36]
PF4	Chile	IV	-	[56]
PF7	Chile	IV	-	[56]
A023	Spain	V	X	[57]
90-11-287	Denmark	V	X	[55]
51/82/2	Germany	V	X	[57]
T265	UK	V	-	[57]
HWU53	Denmark	V	X	[57]
6018/1	Denmark	V	X	[57]
91-8-178	Norway	V	X	[57]
87-9-116	Finland	V	X	[54,55]
91-7-154	Denmark	V	X	[54,55]
601/90	Italy	V	X	[54]
178/90	Italy	V	X	[54]
9014/8	Denmark	V	X	[54]
261/91	Italy	V	X	[57]
LMG12010	n/a	V	X	[57]
87-9-117	Finland	V	X	[54,55]
BA35	USA	V	-	[57]
775	USA	V	-	[58,59]
NB10	North Baltic	V	X	[49,60]
VA1	Denmark	V	X	[36]
850610-1/6a	Denmark	n/a	X	[57]
VA2	Denmark	n/a	-	[36]
VA3	Denmark	n/a	-	[36]
VaKef	Greece	n/a	X	Clinical strain—Unpublished

3.5. H20-Like Prophages Presence in V. anguillarum

According to the phylogenetic tree that was constructed from the identified prophages and isolated temperate phages, five out of 17 prophages and three out of 19 free-living phages are forming two different small monophyletic groups. However, 11 out of 17 prophages and 15 out of 19 free-living temperate phages are clustered together in a genetically and statistically robust and homogenous taxon, independent of their origin. Bacteriophage ϕVaK is clustered together with the VA1 prophage forming a separate monophyletic taxon (Figure 5).

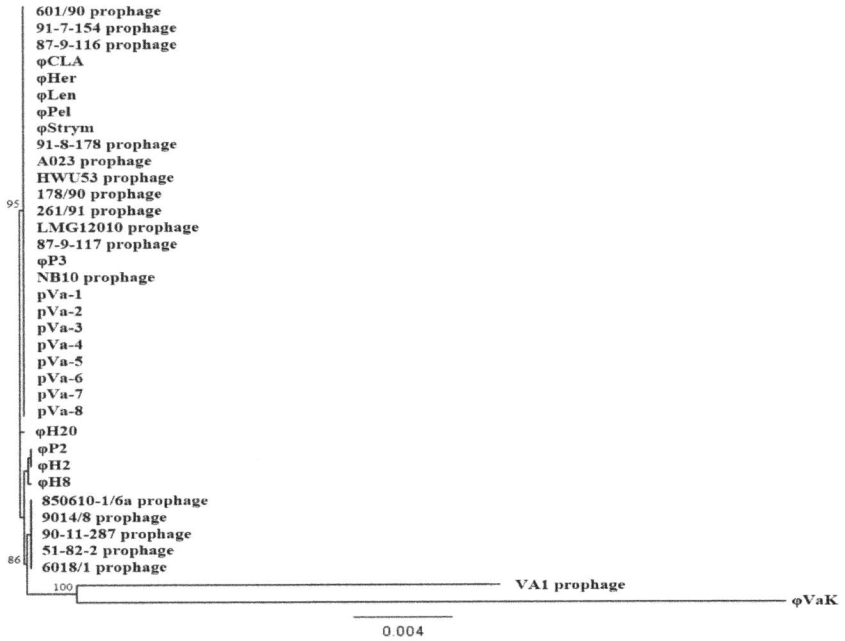

601/90 prophage
91-7-154 prophage
87-9-116 prophage
φCLA
φHer
φLen
φPel
φStrym
91-8-178 prophage
A023 prophage
HWU53 prophage
178/90 prophage
261/91 prophage
LMG12010 prophage
87-9-117 prophage
φP3
NB10 prophage
pVa-1
pVa-2
pVa-3
pVa-4
pVa-5
pVa-6
pVa-7
pVa-8
φH20
φP2
φH2
φH8
850610-1/6a prophage
9014/8 prophage
90-11-287 prophage
51-82-2 prophage
6018/1 prophage
VA1 prophage
φVaK

0.004

Figure 5. Phylogenetic tree (Neighbor-Joining method) based on the whole genomes of the 19 temperate bacteriophage isolates of the current study and 17 prophages which were bioinformatically detected in the genome of their *V. anguillarum* hosts. The branch labels indicate the percentage of statistical support.

The in vitro cell lysis experiment (Supplementary Figure S2) confirmed that the temperate bacteriophage H20 isolated as a free-living phage can be integrated in the host's bacterial genome. The PCR with specific primers performed on the isolated bacterial strains was positive in amplifying the three selected gene parts of the H20-like phages. In 28% of the bacterial host strains (14 out of 50), the bacteriophage φH20 was successfully integrated after 10 h of co-culturing.

3.6. Distribution of H20-Like Phage on a Global Scale

Free-living H20-like bacteriophages have been isolated from Denmark, Norway, Greece and Chile. Combining these data with geographical distribution of the *Vibrio* CRISPR spacers matches and the functional prophages contained in the *V. anguillarum* genomes, the H20-like bacteriophages seem to show a global scale distribution (Figure 6).

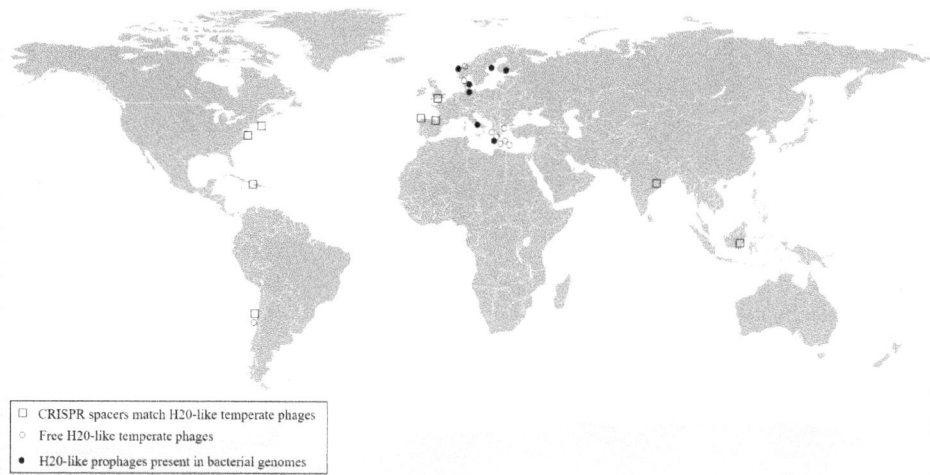

Figure 6. Distribution of H20-like phages on a global scale. The presence of the phage in the different locations was confirmed by: (a) isolating them as free phages from the environment (○); (b) bioinformatic detection of their presence as prophages in the bacterial host genomes (●); and (c) high similarity matches of the temperate phages with CRISPR spacers in the *Vibrio* genus (□).

4. Discussion

Several lytic Vibriophages have previously been sequenced and characterized in connection with their potential use as a strategy against vibriosis in aquaculture [61]. However, there are very few reports on temperate phages against non-cholera *Vibrio* strains [15,62] and their potential importance for lateral transfer of virulence and metabolic capacities among other *Vibrio* pathogens [11,20,63].

The H20-like phage genomes showed no similarity to available phage sequences in GenBank. However, all the phages displayed high genetic similarity with a sequence in in *V. anguillarum* strain NB10, which was then identified as a previously undetected prophage in genome of NB10 [49], and subsequently, in 17 other *V. anguillarum* strains. The most prominent differences between NB10 prophage and H20-like viruses were observed in the variable regions VR1, VR2 and VR3.

Among the H20-like phages cluster, φVaK showed the largest genetic differences. Most of the genetic differences were synonymous, leaving the encoded genes unaltered. However, the variations of φVaK in the VR1 may explain the differences in its host range, since even small changes in structural proteins such as minor tail proteins and tail fiber proteins may affect host range [64]. The bacterial host strain VaKef that was used for the isolation of φVaK was also different from the closely related BA35 and T265 and A023 that were used for isolating the rest of the H20-like Siphoviruses.

Phage tail length tape measure protein has a crucial role in phages' genome injection into the bacterial host [65]. Interestingly, the structure of this protein in H20-like phages resembled the structure of the channel forming toxin colicin Ia [66], suggesting a dual role for the phage tail length tape measure gene. Since only one molecule of colicin Ia is needed to kill a bacterial cell (single hit kinetics) [67], these pore forming toxins are lethal for the bacteria [68,69]. A temperate phage encoded colicin Ia could therefore potentially confer a fitness advantage to its host against competing bacterial strains but could also be lethal for its own host. Lytic activity of phage tail length tape measure protein has previously been reported in *Staphylococcus aureus* specific Siphovirus vB_SauS-phiIPLA35, where a lysozyme-like domain of the tape measure protein had muramidase activity able to lyse *S. aureus* cells [70].

The detection of N6-adenine methyltransferase gene in the H20-like phages may contribute to explaining their wide distribution, as it potentially interferes with the Restriction-Modification (RM) bacterial phage defense system [71,72] thus improving the infection efficiency of the H20-like phages. RM systems generally consist of a restriction endonuclease which cleaves invading phage dsDNA and a methyltransferase, which catalyzes the methylation of specific bacterial dsDNA sequences, protecting the hosts own DNA from enzymatic cleavage. The presence of methyltransferase in the H20-like bacteriophages indicated that these phages can methylate their own DNA to avoid degradation by the host's restriction enzymes. This finding is in accordance with other studies showing that the presence of virus-encoded N6-adenine methyltransferase can provide protection against restriction endonuclease activity [21,73]. However, the role of N6-adenine methyltransferase does not seem to operate solely as a counter defense mechanism to bacterial RM systems. Methyltransferases can also affect bacterial virulence [74] or function as transcriptional regulators by either activating or repressing bacterial genes [75,76]. N6-adenine methyltransferase was previously found in the temperate Vibriophage VHML where it was linked with the virulence of *V. harveyi* host strain upon integration [77]. Epigenetic control of gene expression through methylation is a powerful tool in prokaryotes emphasizing the potential influence of the H20-like phages in host functional properties. The presence of N6-adenine methyltransferase gene in H20-like phages genomes seems to be an important mutual benefit in the host-phage interaction which may promote coexistence. In general, methyltransferase-encoding genes are found in about 20% of the currently annotated bacteriophage genomes [78], supporting an important role of the gene for the phage–host interaction.

Examples of co-evolution of temperate bacteriophages and their bacterial hosts also includes phage-encoded prevention of infection by other phages by superinfection exclusion (Sie) [79]. The H20-like phages carry a repressor gene which genetically and structurally resembles the lambda temperate phage repressor. Such repressor proteins could potentially protect the lysogenized bacteria from similar superinfecting bacteriophages by blocking the expression of the lytic pathway genes [80,81] and this mechanism possibly confers repressor-mediated immunity to other H20-like phages in their *V. anguillarum* host.

Prophages have previously been shown to potentially affect fitness and metabolic properties in fish pathogenic *Vibrios*. For example, the temperate Vibriophage VHML decreased the nutrient uptake in lysogenized *V. harveyi* cells, through a generalized suppression of metabolic activity as a potential energy-saving mechanism under nutrient-limited conditions [82]. Further, evidence of phage-encoded hemagglutinin, which is potentially involved in virulence of *V. pelagius* [63,83] and experimental verification of prophage-mediated virulence in *V. harveyi* [16] support that lysogenic conversion in *Vibrios* represents an important mechanism of adaptation to changing environmental conditions. The presence of functional genes in the H20 like-phages suggests that these phages may also represent a significant contribution to the phenotypic properties of *V. anguillarum* upon integration.

The tRNA^{Arg} that was found in the genome of H20-like phages has also been reported in other Siphoviruses [84] and encodes the amino acid codon AGA which is considered a rare arginine tRNA [85]. Rare codons are generally responsible for encoding transcriptional and translational properties that are distinct from those encoded by the prevalent arginine codons and therefore affecting the expression of regulatory genes [86]. In accordance with this, all four arginine amino acids which are encoded by repressor gene in the H20-like bacteriophages are translated by the rare codon AGA. This suggests that the tRNA in the H20-bacteriophages participates in the regulation of the expression of the repressor, and thus in the decision of lytic or temperate life cycle, implying also that there might be other reasons for phages to carry tRNAs beyond rare codon usage.

The isolation of phages and prophages belonging to the H20 group across large geographical scales and in marine environments with and without aquaculture activities suggests that these *V. anguillarum* phages are common and widespread in the marine environment. Further, the high levels of identity with spacers of the CRISPR systems detected in both *V. anguillarum* and in other *Vibrio* bacteria, indicated long term interactions between *Vibrios* and H20- like phages. This is in line with recent observations of susceptibility to specific phages in 36 isolates of the cosmopolitan Roseobacter-clade species, *Rugeria mobilis* obtained across the world's oceans covering large ranges in temperature, oxygen concentration and habitat (free-living, particle attached, sediment) [25]. These results suggest the co-existence of specific phages and bacteria on a global scale in groups of ubiquitous marine bacteria such as *Roseobacter* and *Vibrio*. This seems inconsistent with the perception of phage–host interactions as a driver of phage and bacterial co-evolution and diversification, as this mechanism would be expected to promote local diversification in response to selection for phage resistance in bacterial populations. The ability of phages to move across biomes have shown to result in a high viral diversity on a local scale, but relatively low diversity when examined globally [28,87]. In addition, different types of phages seem to show different distribution patterns as phages belonging to Myoviridae and Podoviridae have demonstrated specific geographical distribution, whereas Siphoviruses displayed a global distribution [88].

The distribution patterns of specific phages and hosts and the implications of phage–host interactions on evolution and diversification thus seems to vary across spatial scales and between groups of bacteria and phages. While the co-evolutionary dynamics of phages and hosts have traditionally been characterized in terms of selection pressures on host defense and phage infectivity, the mutualistic nature of the interaction between temperate phages and their hosts may select for co-existence rather than arms race dynamics [20]. The efficient lysogenization of susceptible *V. anguillarum* by temperate phage H20 and thus transfer of phage-encoded genes between bacterial strains demonstrated in the current study emphasizes the potential of H20-like phages for integration in their host genome and thus the dispersal of the phage genes in *V. anguillarum*. Consequently, efficient dispersal of phages and bacteria across large spatial scales in the marine environment and a selective advantage of the phage–host interaction by lysogenic conversion of the host may select for phage–host co-existence in the global ocean and thus contribute to explaining the currently observed large scale distribution of H20-like temperate phages and prophages.

Supplementary Materials: The following are available online at www.mdpi.com/1999-4915/9/5/122/s1, Table S1: Three sets of specific primers designed for picking a hypothetical protein, a structural protein gene and a terminase protein gene, respectively. Table S2: Detailed annotations for the 19 temperate bacteriophages. Table S3: Spacers. Figure S1: One step growth curves and phage adsorption curves, for bacteriophages ϕH20, pVa-3, ϕCLA and ϕP3. Figure S2: In vitro cell lysis experiment of bacteriophage ϕH20 against its bacterial host V. anguillarum strain BA35. Arrow indicates the sampling point.

Acknowledgments: The study was supported by the Danish Council for Strategic Research (ProAqua project 12-132390), the Danish Agency for Science, Technology and Innovation (INP Grant# 5132-00014B) and by the Greek National Strategic Reference Framework 2007–2013 (co-funded by European Social Fund and Greek National Funds) FISHPHAGE project 131, www.gsrt.gr. We would like to thank the staff of Venøsund and Sogndal aquaculture units for providing us water samples and George Chalepakis and the staff of the Laboratory of Electron Microscopy of the University of Crete for the TEM work.

Author Contributions: Conceived and designed the experiments: P.G.K., N.R., P.K., M.M. Performed the experiments: P.G.K., N.R., D.C., J.J., J.J.M., C.K. Analyzed the data: P.G.K., N.R., D.C., M.M. Contributed reagents/materials/analysis tools: P.K., F.Z., M.M. Wrote the paper: P.G.K., N.R., M.M.

Conflicts of Interest: The authors declare no conflict of interest.

References

1. Takemura, A.F.; Chien, D.M.; Polz, M.F. Associations and dynamics of vibrionaceae in the environment, from the genus to the population level. *Front. Microbiol.* **2014**, *5*, 1–26. [CrossRef] [PubMed]

2. Thompson, F.L.; Iida, T.; Swings, J. Biodiversity of Vibrios. *Microbiol. Mol. Biol. Rev.* **2004**, 403–431. [CrossRef] [PubMed]

3. Thompson, J.; Polz, M. Dynamics of Vibrio Populations and Their Role in Environmental Nutrient Cycling. In *The Biology of Vibrios*; Thompson, F., Austin, B., Swings, J., Eds.; ASM Press: Washington, DC, USA, 2006; pp. 190–203.

4. Frans, I.; Michiels, C.W.; Bossier, P.; Willems, K.A.; Lievens, B.; Rediers, H. Vibrio anguillarum as a fish pathogen: Virulence factors, diagnosis and prevention. *J. Fish Dis.* **2011**, *34*, 643–661. [CrossRef] [PubMed]

5. Frans, I.; Dierckens, K.; Crauwels, S.; Van Assche, A.; Leisner, J.; Larsen, M.H.; Michiels, C.W.; Willems, K.A.; Lievens, B.; Bossier, P.; et al. Does Virulence Assessment of Vibrio anguillarum Using Sea Bass (*Dicentrarchus labrax*) Larvae Correspond with Genotypic and Phenotypic Characterization? *PLoS ONE* **2013**, *8*, 2–10. [CrossRef]

6. Rønneseth, A.; Castillo, D.; D'Alvise, P.; Tønnesen, Ø.; Haugland, G.; Grotkjaer, T.; Engell-Sørensen, K.; Nørremark, L.; Bergh, Ø.; Wergeland, H.I.; et al. Comparative assessment of *Vibrio* virulence in marine fish larvae. *J. Fish Dis.* **2017**. [CrossRef] [PubMed]

7. Rodkhum, C.; Hirono, I.; Stork, M.; Di Lorenzo, M.; Crosa, J.H.; Aoki, T. Putative virulence-related genes in *Vibrio anguillarum* identified by random genome sequencing. *J. Fish Dis.* **2006**, *29*, 157–166. [CrossRef] [PubMed]

8. Naka, H.; Crosa, J.H. Genetic Determinants of Virulence in the Marine Fish Pathogen *Vibrio anguillarum*. *Fish Pathol.* **2011**, *46*, 1–10. [CrossRef] [PubMed]

9. Clokie, M.R.; Millard, A.D.; Letarov, A.V.; Heaphy, S. Phages in nature. *Bacteriophage* **2011**, *1*, 31–45. [CrossRef] [PubMed]

10. Wagner, P.L.; Waldor, M.K. Bacteriophage Control of Bacterial Virulence. *Infect. Immnunity* **2002**, *70*, 3985–3993. [CrossRef]

11. Davies, E.V.; Winstanley, C.; Fothergill, J.L.; James, C.E. The role of temperate bacteriophages in bacterial infection. *FEMS Microbiol. Lett.* **2016**, *363*, 1–10. [CrossRef] [PubMed]

12. Waldor, M.K.; Mekalanos, J.J. Lysogenic Conversion by a Filamentous Phage Encoding Cholera Toxin. *Science* **1996**, *272*, 1910–1914. [CrossRef] [PubMed]

13. Fasano, A; Baudry, B.; Pumplin, D.W.; Wasserman, S.S.; Tall, B.D.; Ketley, J.M.; Kaper, J.B. Vibrio cholerae produces a second enterotoxin, which affects intestinal tight junctions. *Proc. Natl. Acad. Sci. USA* **1991**, *88*, 5242–5246. [CrossRef] [PubMed]

14. Weynberg, K.D.; Voolstra, C.R.; Neave, M.J.; Buerger, P.; van Oppen, M.J.H. From cholera to corals: Viruses as drivers of virulence in a major coral bacterial pathogen. *Sci. Rep.* **2015**, *5*, 17889. [CrossRef] [PubMed]

15. Oakey, H.J.; Owens, L. A new bacteriophage, VHML, isolated from a toxin-producing strain of Vibrio harveyi in tropical Australia. *J. Appl. Microbiol.* **2000**, *89*, 702–709. [CrossRef] [PubMed]

16. Munro, J.; Oakey, J.; Bromage, E.; Owens, L. Experimental bacteriophage-mediated virulence in strains of Vibrio harveyi. *Dis. Aquat. Organ.* **2003**, *54*, 187–194. [CrossRef] [PubMed]

17. Hasan, N.A.; Grim, C.J.; Lipp, E.K.; Rivera, I.N.G.; Chun, J.; Haley, B.J.; Taviani, E.; Choi, S.Y.; Hoq, M.; Munk, A.C.; et al. Deep-sea hydrothermal vent bacteria related to human pathogenic Vibrio species. *Proc. Natl. Acad. Sci. USA* **2015**, *112*, E2813–E2819. [CrossRef] [PubMed]

18. Brüssow, H.; Canchaya, C.; Hardt, W.; Bru, H. Phages and the Evolution of Bacterial Pathogens: From Genomic Rearrangements to Lysogenic Conversion. *Microbiol. Mol. Biol. Rev.* **2004**, *68*, 560–602. [CrossRef] [PubMed]

19. Feiner, R.; Argov, T.; Rabinovich, L.; Sigal, N.; Borovok, I.; Herskovits, A.A. A new perspective on lysogeny: prophages as active regulatory switches of bacteria. *Nat. Rev. Microbiol.* **2015**, *13*, 641–650. [CrossRef] [PubMed]

20. Obeng, N.; Pratama, A.A.; van Elsas, J.D. The Significance of Mutualistic Phages for Bacterial Ecology and Evolution. *Trends Microbiol.* **2016**, *24*, 440–449. [CrossRef] [PubMed]

21. Murphy, J.; Klumpp, J.; Mahony, J.; O'Connell-Motherway, M.; Nauta, A.; van Sinderen, D. Methyltransferases acquired by lactococcal 936-type phage provide protection against restriction endonuclease activity. *BMC Genom.* **2014**, *15*, 831. [CrossRef] [PubMed]

22. Bochow, S.; Elliman, J.; Owens, L. Bacteriophage adenine methyltransferase: A life cycle regulator? Modelled using Vibrio harveyi myovirus like. *J. Appl. Microbiol.* **2012**, *113*, 1001–1013. [CrossRef] [PubMed]

23. Paul, J.H. Prophages in marine bacteria: dangerous molecular time bombs or the key to survival in the seas? *ISME J.* **2008**, *2*, 579–589. [CrossRef] [PubMed]

24. Spencer, R.G.M. Indigenous marine bacteriophages. *J. Bacteriol.* **1960**, 614.

25. Sonnenschein, E.C.; Nielsen, K.F.; D'Alvise, P.; Porsby, C.H.; Melchiorsen, J.; Heilmann, J.; Kalatzis, P.G.; Lopez-Perez, M.; Bunk, B.; Sproer, C.; et al. Global occurrence and heterogeneity of the Roseobacter-clade species Ruegeria mobilis. *ISME J.* **2016**, *11*, 569–583. [CrossRef] [PubMed]

26. Thurber, R.V. Current insights into phage biodiversity and biogeography. *Curr. Opin. Microbiol.* **2009**, *12*, 582–587. [CrossRef] [PubMed]

27. Kellogg, C.A.; Rose, J.B.; Jaing, S.C.; Paul, J.H. Genetic Diversity of Related Vibriophages Isolated from Marine Environments Around Florida and Hawaii, USA. *Mar. Sci. Fac. Pub.* **1995**, *120*, 89–98. [CrossRef]

28. Breitbart, M.; Rohwer, F. Here a virus, there a virus, everywhere the same virus? *Trends Microbiol.* **2005**, *13*, 243–293. [CrossRef] [PubMed]

29. Comeau, A.M.; Suttle, C.A. Distribution, genetic richness and phage sensitivity of Vibrio spp. from coastal British Columbia. *Environ. Microbiol.* **2007**, *9*, 1790–1800. [CrossRef] [PubMed]

30. Romero, J.; Higuera, G.; Gajardo, F.; Castillo, D.; Middelboe, M.; Garcia, K.; Ramirez, C.; Espejo, R.T. Complete Genome Sequence of Vibrio anguillarum Phage CHOED Successfully Used for Phage Therapy in Aquaculture. *Genome Announc.* **2014**, *2*, 2013–2014. [CrossRef] [PubMed]

31. Higuera, G.; Bastías, R.; Tsertsvadze, G.; Romero, J.; Espejo, R.T. Recently discovered Vibrio anguillarum phages can protect against experimentally induced vibriosis in Atlantic salmon, Salmo salar. *Aquaculture* **2013**, *392–395*, 128–133. [CrossRef]

32. Mateus, L.; Costa, L.; Silva, Y.J.; Pereira, C.; Cunha, A.; Almeida, A. Efficiency of phage cocktails in the inactivation of Vibrio in aquaculture. **2014**, *424–425*, 167–173. [CrossRef]

33. Castillo, D.; Alvise, P.D.; Xu, R.; Zhang, F.; Middelboe, M.; Gram, L. Comparative genome analyses of *Vibrio anguillarum* strains reveal a link with pathogenicity traits. *mSystems* **2017**, *2*. [CrossRef] [PubMed]

34. Comeau, A.M.; Chan, A.M.; Suttle, C.A. Genetic richness of vibriophages isolated in a coastal environment. *Environ. Microbiol.* **2006**, *8*, 1164–1176. [CrossRef] [PubMed]

35. Stenholm, A.R.; Dalsgaard, I.; Middelboe, M. Isolation and characterization of bacteriophages infecting the fish pathogen *Flavobacterium psychrophilum*. *Appl. Environ. Microbiol.* **2008**, *74*, 4070–4078. [CrossRef] [PubMed]

36. Tan, D.; Gram, L.; Middelboe, M. Vibriophages and their interactions with the fish pathogen Vibrio anguillarum. *Appl. Environ. Microbiol.* **2014**, *80*, 3128–3140. [CrossRef] [PubMed]

37. Hyman, P.; Abedon, S.T. Practical methods for determining phage growth paramenters. In *Bacteriophages: Methods and Protocols, Volume 1: Isolation, Characterization and Interactions*; Clokie, M., Kropinski, A., Eds.; Humana Press: New York, NY, USA, 2009; pp. 175–202.

38. Castillo, D.; Christiansen, R.H.; Dalsgaard, I.; Madsen, L.; Middelboe, M. Bacteriophage resistance mechanisms in the fish pathogen *Flavobacterium psychrophilum*: Linking genomic mutations to changes in bacterial virulence factos. *Appl. Environ. Microbiol.* **2015**, *81*, 1157–1167. [CrossRef] [PubMed]

39. Kearse, M.; Moir, R.; Wilson, A.; Stones-Havas, S.; Cheung, M.; Sturrock, S.; Buxton, S.; Cooper, A.; Markowitz, S.; Duran, C.; et al. Geneious Basic: An integrated and extendable desktop software platform for the organization and analysis of sequence data. *Bioinformatics* **2012**, *28*, 1647–1649. [CrossRef] [PubMed]

40. Delcher, A.L.; Bratke, K.A.; Powers, E.C.; Salzberg, S.L. Identifying bacterial genes and endosymbiong DNA with Glimmer. *Bioinformatics* **2007**, *23*, 673–679. [CrossRef] [PubMed]

41. Aziz, R.K.; Bartels, D.; Best, A.A.; DeJongh, M.; Disz, T.; Edwards, R.A.; Formsma, K.; Gerdes, S.; Glass, E.M.; Kubal, M.; et al. The RAST Server: Rapid Annotations using Subsystems Technology. *BMC Genomics* **2008**, *9*, 75. [CrossRef] [PubMed]

42. Overbeek, R.; Olson, R.; Pusch, G.D.; Olsen, G.J.; Davis, J.J.; Disz, T.; Edwards, R.A.; Gerdes, S.; Parrello, B.; Shukla, M.; Vonstein, V.; Wattam, A.R.; Xia, F.; Stevens, R. The SEED and the Rapid Annotation of microbial genomes using Subsystems Technology (RAST). *Nucleic Acids Res.* **2014**, *42*, 206–214. [CrossRef] [PubMed]

43. Kelly, L.A.; Mezulis, S.; Yates, C.; Wass, M.; Sternberg, M. The Phyre2 web portal for protein modelling, prediction, and analysis. *Nat. Protoc.* **2015**, *10*, 845–858. [CrossRef] [PubMed]

44. Phyre[2]. Available online: http://www.sbg.bio.ic.ac.uk/phyre2/html/page.cgi?id=index (accessed on 1 March 2017).

45. Laslett, D.; Canback, B. ARAGORN, a program to detect tRNA genes and tmRNA genes in nucleotide sequences. *Nucleic Acids Res.* **2004**, *32*, 11–16. [CrossRef] [PubMed]

46. Darling, A.C.E.; Mau, B.; Blattner, F.R.; Perna, N.T. Mauve: Multiple alignment of conserved genomic sequence with rearrangements. *Genome Res.* **2004**, 1394–1404. [CrossRef] [PubMed]

47. Agarwala, R.; Barrett, T.; Beck, J.; Benson, D.A.; Bollin, C.; Bolton, E.; Bourexis, D.; Brister, J.R.; Bryant, S.H.; Canese, K.; et al. Database resources of the National Center for Biotechnology Information. *Nucleic Acids Res.* **2016**, *44*, D7–D19.

48. Grissa, I.; Vergnaud, G.; Pourcel, C. CRISPRFinder: a web tool to identify clustered regularly interspaced short palindromic repeats. *Nucleic Acids Res.* **2007**, *35*, 52–57. [CrossRef] [PubMed]

49. Zhou, Y.; Liang, Y.; Lynch, K.H.; Dennis, J.J.; Wishart, D.S. PHAST: A Fast Phage Search Tool. *Nucleic Acids Res.* **2011**, *39*, 1–6. [CrossRef] [PubMed]

50. Holm, K.O.; Nilsson, K.; Hjerde, E.; Willassen, N.-P.; Milton, D.L. Complete genome sequence of Vibrio anguillarum strain NB10, a virulent isolate from the Gulf of Bothnia. *Stand. Genomic Sci.* **2015**, *10*, 60. [CrossRef] [PubMed]

51. MacDonell, M.T.; Colwell, R.R. Phylogeny of the *Vibrionaceae*, and Recommendation for Two New Genera, Listonella and Shewanella. *Syst. Appl. Microbiol.* **1985**, *6*, 171–182. [CrossRef]

52. Mikkelsen, H.; Schrøder, M.B.; Lund, V. Vibriosis and atypical furunculosis vaccines; efficacy, specificity and side effects in Atlantic cod, Gadus morhua L. *Aquaculture* **2004**, *242*, 81–91. [CrossRef]

53. Samuelsen, O.B.; Bergh, Ø. Efficacy of orally administered florfenicol and oxolinic acid for the treatment of vibriosis in cod (*Gadus morhua*). *Aquaculture* **2004**, *235*, 27–35. [CrossRef]

54. Pedersen, K.; Gram, L.; Austin, D.A.; Austin, B. Pathogenicity of Vibrio anguillarum serogroup O1 strains compared to plasmids, outer membrane protein profiles and siderophore production. *J. Appl. Microbiol.* **1997**, *82*, 365–371. [CrossRef] [PubMed]

55. Skov, M.N.; Pedersen, K.; Larsen, J.L. Comparison of Pulsed-Field Gel Electrophoresis, Ribotyping, and Plasmid Profiling for Typing of Vibrio anguillarum Serovar O1. *Appl Environ Microbiol.* **1995**, *61*, 1540–1545. [PubMed]

56. Silva-Rubio, A.; Avendaño-Herrera, R.; Jaureguiberry, B.; Toranzo, A.E.; Magariños, B. First description of serotype O3 in *Vibrio anguillarum* strains isolated from salmonids in Chile. *J. Fish Dis.* **2008**, *31*, 235–239. [CrossRef] [PubMed]

57. Austin, B.; Alsina, M.; Austin, D.A.; Blanch, A.R.; Grimont, F.; Grimont, P.A.D.; Jofre, J.; Koblavi, S.; Larsen, J.L.; Pedersen, K.; Tiainen, T.; Verdonck, L.; Swings, J. Identification and Typing of *Vibrio anguillarum*: A Comparison of Different Methods. *Syst. Appl. Microbiol.* **1995**, *18*, 285–302. [CrossRef]

58. Actis, L.A.; Fish, W.; Crosa, J.H.; Kellerman, K.; Ellenberger, S.R.; Hauser, F.M.; Sanders-Loehr, J. Characterization of anguibactin, a novel siderophore from Vibrio anguillarum 775(pJM1). *J. Bacteriol.* **1986**, *167*, 57–65. [CrossRef] [PubMed]

59. Kosters, W.L.; Actisp, L.A.; Waldbeserp, L.S.; Tolmaskyp, M.E.; Crosapfl, J.H. Molecular characterization of the iron transport system mediated by the pJM1 plasmid in Vibrio anguillarum 775. *J. Biol. Chem.* **1991**, 23829–23833.

60. Rehnstam, A.S.; Norqvist, A.; Wolf-Watz, H.; Hagstrom, A. Identification of Vibrio anguillarum in fish by using partial 16S rRNA sequences and a specific 16S rRNA oligonucleotide probe. *Appl. Envir. Microbiol.* **1989**, *55*, 1907–1910.

61. Oliveira, J.; Castilho, F.; Cunha, A.; Pereira, M.J. Bacteriophage therapy as a bacterial control strategy in aquaculture. *Aquac. Int.* **2012**, *20*, 879–910. [CrossRef]

62. Williamson, S.J.; McLaughlin, M.R.; Paul, J.H. Interaction of the ΦHSIC virus with its host: lysogeny or pseudolysogeny? *Appl. Environ. Microbiol.* **2001**, *67*, 1682–1688. [CrossRef] [PubMed]

63. Thompson, F.L.; Klose, K.E.; Group, A. Vibrio2005: The First International Conference on the Biology of Vibrios. *J. Bacteriol.* **2006**, *188*, 4592–4596. [CrossRef] [PubMed]

64. Jacobs-Sera, D.; Marinelli, L.; Bowman, C.; Broussard, G.; Guerrero, C.; Boyle, M.; Petrova, Z.; Dedrick, R.; Pope, W.; SEA-PHAGES; et al. On the nature of mycobacteriophage diversity and host preference. *Virology* **2012**, *434*, 187–201. [CrossRef] [PubMed]

65. Cumby, N.; Reimer, K.; Mengin-Lecreulx, D.; Davidson, A.R.; Maxwell, K.L. The phage tail tape measure protein, an inner membrane protein and a periplasmic chaperone play connected roles in the genome injection process of E.coli phage HK97. *Mol. Microbiol.* **2015**, *96*, 437–447. [CrossRef] [PubMed]

66. Wiener, M.; Freymann, D.; Ghosh, P.; Stroud, R.M. Crystal structure of colicin Ia. *Nature* **1997**, *385*, 461–464. [CrossRef] [PubMed]

67. Kim, Y.C.; Tarr, A.W.; Penfold, C.N. Colicin import into *E. coli* cells: A model system for insights into the import mechanisms of bacteriocins. *Biochim. Biophys. Acta - Mol. Cell Res.* **2014**, *1843*, 1717–1731. [CrossRef] [PubMed]

68. Alonso, G. How bacteria protect themselves against channel-forming colicins. *Int. Microbiol.* **2000**, *3*, 81–88. [PubMed]

69. Penfold, C.N.; Walker, D.; Kleanthous, C. How bugs kill bugs: progress and challenges in bacteriocin research. *Biochem. Soc. Trans.* **2012**, *40*, 1433–1437. [CrossRef] [PubMed]

70. Rodríguez-Rubio, L.; Gutiérrez, D.; Martínez, B.; Rodríguez, A.; Götz, F.; García, P. The tape measure protein of the *Staphylococcus aureus* bacteriophage vB_SauS-phiIPLA35 has an active muramidase domain. *Appl. Environ. Microbiol.* **2012**, *78*, 6369–6371. [CrossRef] [PubMed]

71. Luria, S.E. Host-induced modifications of viruses. *Cold Spring Harb. Symp. Quant. Biol.* **1953**, *18*, 237–244. [CrossRef] [PubMed]

72. Nathans, D.; Smith, H.O. Restriction endonucleases in the analysis and restructuring of dna molecules. *Annu. Rev. Biochem.* **1975**, *44*, 273–293. [CrossRef] [PubMed]

73. Park, Y.; Kim, G. Do; Choi, T.J. Molecular cloning and characterization of the DNA adenine methyltransferase gene in Feldmannia sp. virus. *Virus Genes* **2007**, *34*, 177–183. [CrossRef] [PubMed]

74. Wion, D.; Casadesús, J. N6-methyl-adenine: an epigenetic signal for DNA—Protein interactions. *Nat. Rev. Microbiol.* **2006**, *4*, 183–192. [CrossRef] [PubMed]

75. Low, D.A.; Weyand, N.J.; Mahan, M.J. Roles of DNA adenine methylation in regulating bacterial gene expression and virulence. *Infect. Immun.* **2001**, *69*, 7197–7204. [CrossRef] [PubMed]

76. Portillo, F.G.-D.; Pucciarelli, M.G.; Casadesus, J. DNA adenine methylase mutants of *Salmonella typhimurium* show defects in protein secretion, cell invasion, and M cell cytotoxicity. *Proc. Natl. Acad. Sci. USA* **1999**, *96*, 11578–11583. [CrossRef]

77. Oakey, H.J.; Cullen, B.R.; Owens, L. The complete nucleotide sequence of the *Vibrio harveyi* bacteriophage VHML. *J. Appl. Microbiol.* **2002**, *93*, 1089–1098. [CrossRef] [PubMed]

78. Murphy, J.; Mahony, J.; Ainsworth, S.; Nauta, A.; van Sinderen, D. Bacteriophage orphan DNA methyltransferases: Insights from their bacterial origin, function, and occurrence. *Appl. Environ. Microbiol.* **2013**, *79*, 7547–7555. [CrossRef] [PubMed]

79. Van Houte, S.; Buckling, A.; Westra, E.R. Evolutionary Ecology of Prokaryotic Immune Mechanisms. *Microbiol. Mol. Biol. Rev.* **2016**, *80*, 745–763. [CrossRef] [PubMed]

80. Pope, W.H.; Jacobs-Sera, D.; Russel, D.A.; Peebles, C.L.; Al-Atrache, Z.; Alcoser, T.A.; Alexander, L.M.; Alfano, M.B.; Alford, S.T.; Amy, N.E.; et al. Expanding the diversity of mycobacteriophages: Insights into genome architecture and evolution. *PLoS One* **2011**, *6*, e16329. [CrossRef] [PubMed]

81. Berngruber, T.W.; Weissing, F.J.; Gandon, S. Inhibition of superinfection and the evolution of viral latency. *J. Virol.* **2010**, *84*, 10200–10208. [CrossRef] [PubMed]

82. Vidgen, M.; Carson, J.; Higgins, M.; Owens, L. Changes to the phenotypic profile of *Vibrio harveyi* when infected with the Vibrio harveyi myovirus-like (VHML) bacteriophage. *J. Appl. Microbiol.* **2006**, *100*, 481–487. [CrossRef] [PubMed]

83. Williamson, S.J.; Paul, J.H. Environmental factors that influence the transition from lysogenic to lytic existence in the HSIC/*Listonella pelagia* marine phage-host system. *Microb. Ecol.* **2006**, *52*, 217–225. [CrossRef] [PubMed]

84. Wang, J.; Jiang, Y.; Vincent, M.; Sun, Y.; Yu, H.; Wang, J.; Bao, Q.; Kong, H.; Hu, S. Complete genome sequence of bacteriophage t5. *Virology* **2005**, *332*, 45–65. [CrossRef] [PubMed]

85. Zahn, K.; Landy, A. Modulation of lambda integrase synthesis by rare arginine tRNA. *Mol. Microbiol.* **1996**, *21*, 69–76. [CrossRef] [PubMed]

86. Napolitano, M.G.; Landon, M.; Gregg, C.J.; Lajoie, M.J.; Govindarajan, L.; Mosberg, J.A.; Kuznetsov, G.; Goodman, D.B.; Vargas-Rodriguez, O.; Isaacs, F.J.; et al. Emergent rules for codon choice elucidated by editing rare arginine codons in *Escherichia coli*. *Proc. Natl. Acad. Sci. U.S.A.* **2016**, *113*, E5588–E5597. [CrossRef] [PubMed]

87. Sano, E.; Carlson, S.; Wegley, L.; Rohwer, F. Movement of viruses between biomes. *Appl. Environ. Microbiol.* **2004**, *70*, 5842–5846. [CrossRef] [PubMed]

88. Williamson, S.J.; Rusch, D.B.; Yooseph, S.; Halpern, A.L.; Heidelberg, K.B.; Glass, J.I.; Pfannkoch, C.A.; Fadrosh, D.; Miller, C.S.; Sutton, G.; Frazler, M.; Venter, J.C. The sorcerer II global ocean sampling expedition: Metagenomic characterization of viruses within aquatic microbial samples. *PLoS ONE* **2008**, *3*, e1456. [CrossRef] [PubMed]

![viruses logo] *viruses*

MDPI

Article

A Novel Roseosiphophage Isolated from the Oligotrophic South China Sea

Yunlan Yang [1,†], Lanlan Cai [1,†], Ruijie Ma [1], Yongle Xu [1], Yigang Tong [2], Yong Huang [2], Nianzhi Jiao [1,*] and Rui Zhang [1,*]

[1] State Key Laboratory of Marine Environmental Science, Institute of Marine Microbes and Ecospheres, Xiamen University (Xiang'an), Xiamen 361102, Fujian, China; yangyunlan@stu.xmu.edu.cn (Y.Y.); cailanlan@stu.xmu.edu.cn (L.C.); maruijie@stu.xmu.edu.cn (R.M.); xuyongle3884778@xmu.edu.cn (Y.X.)

[2] Beijing Institute of Microbiology and Epidemiology, State Key Laboratory of Pathogen and Biosecurity, Beijing 100071, China; tong.yigang@gmail.com (Y.T.); presidenthuang@126.com (Y.H.)

* Correspondence: jiao@xmu.edu.cn (N.J.); ruizhang@xmu.edu.cn (R.Z.); Tel.: +86-592-288-0199 (N.J.); +86-592-288-0152 (R.Z.)

† These authors contributed equally to this work.

Academic Editors: Mathias Middelboe and Corina P. D. Brussaard

Received: 28 February 2017; Accepted: 10 May 2017; Published: 15 May 2017

Abstract: The *Roseobacter* clade is abundant and widespread in marine environments and plays an important role in oceanic biogeochemical cycling. In this present study, a lytic siphophage (labeled vB_DshS-R5C) infecting the strain type of *Dinoroseobacter shibae* named DFL12^T, which is part of the *Roseobacter* clade, was isolated from the oligotrophic South China Sea. Phage R5C showed a narrow host range, short latent period and low burst size. The genome length of phage R5C was 77, 874 bp with a G+C content of 61.5%. Genomic comparisons detected no genome matches in the GenBank database and phylogenetic analysis based on DNA polymerase I revealed phylogenetic features that were distinct to other phages, suggesting the novelty of R5C. Several auxiliary metabolic genes (e.g., *phoH* gene, heat shock protein and queuosine biosynthesis genes) were identified in the R5C genome that may be beneficial to the host and/or offer a competitive advantage for the phage. Among siphophages infecting the *Roseobacter* clade (roseosiphophages), four gene transfer agent-like genes were commonly located with close proximity to structural genes, suggesting that their function may be related to the tail of siphoviruses. The isolation and characterization of R5C demonstrated the high genomic and physiological diversity of roseophages as well as improved our understanding of host–phage interactions and the ecology of the marine *Roseobacter*.

Keywords: roseophage; genome; phylogenetic analysis; environmental distribution

1. Introduction

As the most abundant biological entities, viruses play an important role in nutrient cycles and energy flow in marine environments through viral lysis [1]. Viruses are also one of the major contributors to horizontal gene transfer and evolution of their hosts, with approximately 10^{23} infections occurring every second in seawater [2]. Recent investigations of viruses using metagenomics fundamentally changed our estimation of their diversity and community structure as well as our understanding of their interaction with their hosts [3]. Despite the tremendous amount of genetic information provided by virome studies, most are considered "dark material" owing to the lack of similarity to known sequences. It is proposed and demonstrated that this problem can be partially solved by the isolation and genetic characterization of viruses, especially those that infect dominant bacterial groups, such as *Synechococcus* and *Vibrio* in coastal areas, in addition to SAR 11 and *Prochlorococcus* in open ocean [4–8]. In addition, the physiological and ecological characterization of these isolated viruses, as well as their interaction with hosts, has improved our understanding of their ecological and biogeochemical roles in real environments [9,10].

In recent years, increasing attention has been paid to *Roseobacter* and their phages because of their worldwide distribution, high abundance, and possible ability to be cultured [11–16]. It is estimated that the *Roseobacter* clade contributes up to 15–20% of the total bacterioplankton in typical coastal areas and open oceans [17]. They are detected from a variety of marine habitats ranging from marine snow, micro and macro algae, microbial mats, sediments, sea ice, and hydrothermal vents [11]. Several major biogeochemical processes, such as the transformation of organic and inorganic sulfur compounds, carbon monoxide oxidation, and the degradation of high molecular weight organic matter, are mediated by *Roseobacter* [11,17]. In addition, they are easily isolated and fast-growing, which is advantageous for phage studies compared with other numerically dominant marine bacteria, such as SAR11 and *Prochlorococcus*.

The genus *Dinoroseobacter* is one of the most well-studied groups of marine bacteria [11,18]. *Dinoroseobacter shibae*, the type species of *Dinoroseobacter*, was isolated from the dinoflagellate *Prorocentrum lima* and lives in a symbiotic relationship with marine algae [19]. *D. shibae* is an aerobic anoxygenic phototrophic bacterium and is competitive in unpredictable, changing environments, because of its ability to perform light-driven ATP synthesis and its novel acylated homoserine lacton compounds [20–23]. The strain type *D. shibae* DFL12^T has been completely sequenced and exhibits complex viral defense systems (i.e., clustered regularly interspaced small palindromic repeats, CRISPRs) encoded in its genome [19,24].

To date, 17 phages and prophages infecting *Roseobacter* have been reported [12–14,16,25–32]. It has been shown that roseophages contain a considerable degree of genomic variability [29,33]. However, compared with the biogeographical, physiological and ecological diversity of *Roseobacter*, the range of roseophages studied so far remains relatively narrow. For example, most roseophages were isolated from coastal areas, but *Roseobacter* was also shown to be widely distributed in open ocean [13,29]. Viral metagenomic recruitments based on available roseophage genome sequences indicated the presence of roseophages in open ocean, but only one such organism has been isolated from this environment [34]. In addition, most of the roseophages belong to the *Podoviridae* family, but previous studies of dominating marine bacteria groups (e.g., *Synechococcus* and *Prochlorococcus*) have suggested that they are also infected by *Siphoviridae* or *Myoviridae* family phages. Phages adapted to different environments and with different phylogenetic properties are driving factors for ecological distribution and behavior of their hosts. Therefore, the aim of this study was to isolate and characterize more phages for *Roseobacter*, particularly *D. shibae*, from open ocean areas, such as the oligotrophic South China Sea. We expect that the phages isolated from distinct geographical environments and phylogenetic families will provide novel information regarding their genetics, diversity and distribution, which will expand our knowledge on the viral ecology of the marine *Roseobacter*.

2. Materials and Methods

2.1. Isolation and Purification of Phages

D. shibae DFL12^T was used as the host in this study and was maintained on RO medium (yeast extract 1 g·L^{-1}, peptone 1 g·L^{-1}, sodium acetate 1 g·L^{-1}, artificial seawater 1 L, pH 7.4–7.8) at room temperature. To increase the probability of phage isolation, viruses in the surface water of the South China Sea, which was characterized as an oligotrophic environment, were enriched by tangential flow filtration with a 30 kDa polysulfone cartridge (Labscale, Millipore, CA, USA). Double-layer agar was used for the isolation and purification of phages.

After purifying five times, phages were cultivated for expansion in liquid RO medium. The culture was centrifuged at 12,000× *g* at 4 °C for 10 min to obtain the phage-containing supernatant. Following this, phages were concentrated by precipitation with polyethylene glycol 8000 (final concentration: 100 g·L^{-1}) and centrifuged at 10,000× *g* for 60 min at 4 °C. The precipitate was resuspended in SM buffer (100 mM

NaCl, 8 mM $MgSO_4$, 50 mM Tris-HCl at pH 7.5) and purified by CsCl equilibrium gradient centrifugation (200,000× g, 4 °C, 24 h). The pellet was dialyzed through 30 kDa super-filters (UFC503096, Millipore) and the filtrate was collected for morphologic observation and DNA extraction.

2.2. Transmission Electron Microscopy (TEM)

The purified and desalted phages were diluted to the appropriate concentration with SM buffer and adsorbed onto 200-mesh carbon-coated coppers for 10–30 min in the dark. After being stained with 1% phosphotungstic acid and dried for 30 min, samples were viewed at 80 kV voltage using a JEM-2100 transmission electron microscope (JEOL, Tokyo, Japan). Images were collected using the CCD image transmission system (Gatan Inc., Pleasanton, CA, USA).

2.3. Host Range

Besides *D. shibae* DFL12^T, bacterial strains used in the host range test included *Citromicrobium* sp. WPS32, *Citromicrobium* sp. JL2201, *Citromicrobium* sp. JL1351, *Citromicrobium* sp. JL354, *Citromicrobium* sp. JL1363, *Dinoroseobacter* sp. JL1447, *Erythrobacter litoralis* DMS 8509, *Erythrobacter longus* DMS 6997, *Erythrobacter* sp. JL475, *Hoeflea phototrophica* DFL-43, *Paenibacillus* sp. JL1210, *Roseobacter denitrificans* OCh114, *Roseomonas* sp. JL2290, *Roseomonas* sp. JL2293, *Roseovarius* sp. JL2434, *Ruegeria* sp. JL126, *Spingobium* sp. JL1088 and *Silicibacter pomeroyi* DSS3. Exponentially growing cultures of these bacteria strains were incubated with phages, which were diluted to 10^2, 10^4, 10^6, and 10^8 PFU mL^{-1}, for 20 min and then plated using a plaque assay. Replication was conducted for each bacterium strain.

2.4. One-Step Growth Curve

To analyze the infectivity and replication ability of phages, a one-step growth curve test was performed. Phages were added to 1 mL of log-phase *D. shibae* DFL12^T with a multiplicity of infection of 0.01, before being incubated for 25 min at room temperature in the dark. The culture was centrifuged at 10,000× g at 4 °C for 5 min and resuspended in 1 mL of RO medium. Centrifugation was repeated twice. Thereafter, the pellet was transferred to 50 mL of RO medium and incubated over 7 h at 28 °C with continuous shaking. Samples were collected every 30 min and viral abundance was quantified using the double agar overlay plaque assay. The latent period was followed by a single burst of phages. The burst size was the average number of phages released per infected host cell and calculated as the ratio between the number of phages before and after the burst [35].

2.5. Lipid in the Viral Capsid

To investigate the presence of lipid in the viral capsid, 1 mL of phages were mixed by vibrating with 0 μL, 20 μL and 200 μL of chloroform, respectively, for 1 min and then kept at room temperature for 30 min. After centrifuging at a slow speed, phages remaining in the supernatant were dropped onto a *D. shibae* DFL12^T plate. The result was determined by the emergence of plaques.

2.6. DNA Extraction

Phage DNA was extracted using the phenol–chloroform extraction method. Briefly, purified phages were lysed by the addition of proteinase K (100 mg·mL^{-1}), SDS (10%, wt/vol) and EDTA (0.5 mol·L^{-1}, pH 8.0) and incubated at 55 °C in water for 3 h. The digested sample was then added to an equal volume of phenol/chloroform/isoamyl alcohol (25:24:1) and centrifuged at 12,000× g and 4 °C for 5 min to remove any debris. This step was repeated twice. The supernatant was sequentially purified by adding chloroform/isoamyl alcohol (24:1) and centrifuging at 12,000× g and 4 °C for 10 min. Following this, the supernatant was mixed with isoamyl alcohol and kept at −20 °C overnight. The precipitate was allowed to air-dry after slowly flushing with cold 70% ethanol. Samples were resuspended in 100 μL TE buffer (10 mM Tris-HCl, 1 mM EDTA, pH 8.0) and stored at 4 °C before analysis.

2.7. Genome Sequencing and Analysis, and Phylogenetic Analyses

The genomic DNA of R5C was sequenced on the Illumina Hiseq 2500 platform using the TruSeq PE Cluster Kit (Illumina, San Diego, CA, USA). The sequences were assembled using Velvet software (v1.2.03) (4699× coverage) [36]. A termini analysis was used to identify the phage's termini and genome packaging [37]. The raw reads were aligned to the genome sequence using a CLC Genomics Workbench (version 3.6.1). The GeneMarkS online server and ORF Finder were used to predict open reading frames (ORFs), while tRNAscan-SE was used to identify tRNA sequences [38–40]. Translated ORFs were analyzed and annotated by the algorithms of a BLASTP search against the NCBI database with E-value $\leq 10^{-3}$ [41,42]. Gene maps were created based on the genome annotations using the Java Operon. DNA polymerase I was used to investigate the phage R5C phylogeny. A maximum-likelihood method in the MEGA 6.0 software package was used to construct the phylogenetic tree.

The complete genome sequence was submitted to the GenBank database under the accession number KY606587.

2.8. Recruitment of Metagenomic Data

For a better understanding of the geographical distribution of R5C, the phage genes were used as queries to search against metagenomic databases of the Pacific Ocean Virome (POV) and Global Ocean Survey (GOS) [43,44]. Samples in POV were collected from various seasons (spring, summer, fall, and winter), depths (10 m to 4300 m), and proximities to land (coastal to open ocean). Samples in GOS were taken from a wide variety of aquatic habitats collected over 8000 km. Only sequences with a threshold E-value of 10^{-5} and a minimum amino acid length of 30 and a bit score greater than 40 were extracted from the database. In addition, we searched for homologs of the R5C genome in the Integrated Microbial Genomes/Virus (IMG/VR) database and the Environmental Viral Genomes databases (EVGs) generated by Nishimura et al. (2017) [45,46]. To test any spacers of CRISPR array within the phage sequence, the genome of phage was searched against viral spacer database of IMG/VR, as well as in CRISPRs loci of bacterial isolates used in the present study [47].

3. Results and Discussion

3.1. Isolation and Characterization of Phage R5C

Phages infecting *D. shibae* DFL12^T were isolated from seawater from the oligotrophic South China Sea and designated as vB_DshS-R5C based on nomenclature suggested by Kropinski et al. (2009) [48]. Plaques resulting from R5C lysis appeared small and blurry, with a diameter of 1–2 mm after 1–2 days incubation (Figure 1a). TEM showed that the phage R5C had a long hexagon head that measured about 114 ± 2 nm in length and 70 ± 2 nm for the greatest width in addition to a flexional long tail with a length of 142 ± 2 nm (Figure 1b). Based on its morphology, R5C belongs to the *Siphoviridae* family from the order *Caudovirales*.

Host infectivity was analyzed for 19 bacterium strains. The host infectivity test demonstrated that the phage R5C possesses a narrow host range and is only able to infect *D. shibae* DFL12^T (Table 1). Generally, myoviruses display the broadest host range among the three families of tailed dsDNA viruses, while podoviruses display the narrowest [1]. Interestingly, a narrow host range was observed for siphoviruses DSS3Φ8 and RDJLΦ1, while a broad host range was observed for podoviruses DS-1410Ws-06, RD-1410W1-01 and RD-1410Ws-07 (Table 1). This confirms the specificity and complexity of phage–host interactions and the rich diversity of roseophages. The phage R5C was resistant to all three concentrations of chloroform. These results indicated the absence of lipids in the capsid or the surrounding lipid layer, which has commonly been observed among all of the roseophages isolated so far.

Table 1. Host range of roseophages (+: infected; −: uninfected).

Strains	R5C	DSS3Φ8	RDJLΦ1	RLP1	RPP1	DSS3Φ2	ESS36Φ1	DS-1410Ws-06	RD-1410W1-01	RD-1410Ws-07
D. shibae DFL12^T	+							+	+	+
Acinetobacter sp. JL1404			−							
Alcanivorax sp. JL1378			−							
Alteromonas sp. JL1357			−							
Antarctobacter sp. JL351			−							
Citromicrobium sp. JL1363	−		−							
Citromicrobium sp. JL1351	−									
Citromicrobium sp. JL2201	−									
Citromicrobium sp. JL354	−		−							
Citromicrobium sp. WPS32	−									
Cytophaga sp. JL1362	−		−							
Dinoroseobacter sp. JL1447	−		−							
Erythrobacter litoralis DMS 8509	−		−							
Erythrobacter longus DSM 6997			−							
Erythrobacter sp. JL1350	−		−							
Erythrobacter sp. JL359			−							
Erythrobacter sp. JL475.	−		−							
Furnibacter sp. JL1383			−							
Hoeflea phototrophica DFL-43^T	−									
Leisingera methylohalidivorans MB2				−	−					
Marinovum algicola ATCC 51440^T				−	−					
Micrococcus sp. JL1389	−		−							
Nocardioides sp. JL1369										
Paenibacillus sp. JL1210	−		−							
Phaeobacter sp. 27-4										
Pseudoalteromonas sp. JL1391				−	−	−	−			
Rhodobacteraceae sp. 176				−	−					
Roseobacter denitrificans ATCC 33942	−	−	+					+	+	+
Roseobacter denitrificans OCh 114						−	−			
Roseobacter litoralis ATCC 49566			−			−	−	+		+
Roseobacter litoralis DMS 6996			−							
Roseobacter litoralis Och149								−	+	+
Roseobacter sp. JL1336			−							
Roseobacter sp. TM1038						−	−			
Roseobacter sp. TM1039						−	−			

Table 1. *Cont.*

Strains	R5C	DSS3Φ8	RDJLΦ1	RLP1	RPP1	DSS3Φ2	ESS36Φ1	DS-1410Ws-06	RD-1410W1-01	RD-1410Ws-07
Roseomonas sp. JL2290	−									
Roseomonas sp. JL2293	−									
Roseovarius sp.TM1042										
Roseovarius nubinhibens ISM^T		−		−		−	−	−	−	−
Roseovarius sp. 217				+	+					
Roseovarius sp. JL2434	−			+	−					
Roseovarius crassostreae CV919–312^T				−	−					
Roseovarius mucosus DFL–24^T				−	−					
Ruegeria atlantica AMA–03				−	−					
Ruegeria sp. 198				−	−					
Ruegeria sp. JL126	−									
Sagittula stellata E–37				−	−					
Silicibacter pomeroyi DSS3	−	+		−	−	+		+	+	+
Silicibacter sp. TM1040	−	−				−				
Sphingobium sp. JL1088	−		−							
Stappia sp. JL1358										
Sulfitobacter sp. 1921						−				
Sulfitobacter pseudomitzschiae H3							+			
Sulfitobacter sp. EE–36		−	−					−	−	−
Sulfitobacter sp. JL1353			−				−	−	−	−
Sulfitobacter sp CBB406						−	−			
Sulfitobacter sp. CHSB4						−				
Sulfitobacter sp. JL351						−	−			
Sulfitobacter sp. T11						−				

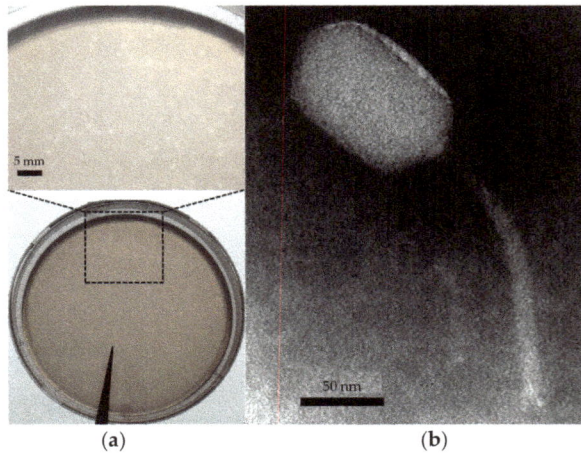

Figure 1. (a) Plaques and (b) transmission electron microscopy image of phage R5C.

3.2. Life Cycle

The results from the one-step growth curve are shown in Figure 2. The latent period for the phage R5C lasted about 1.5–2 h and a growth plateau was reached in 4 h (Table S1). R5C exhibited a small burst size of 65 PFU cell^{-1}. The burst size of roseophages appeared to be highly variable, ranging from 10 cell^{-1} to 1500 cell^{-1}. It should be noted that different methods were used for the enumeration of viral abundance. Generally, burst size is thought to be influenced by a number of factors such as bacterial/viral size, the metabolic activity of the host as well as the characteristic of the phage and host. A correlation between burst size and the trophic status of the environment has also been proposed [49,50]. However, this needs to be further verified because some phages (e.g., RPP1 and RD-1410Ws-01), which are isolated from eutrophic environments, also exhibit low burst sizes. Furthermore, even though R5C and DS-1410Ws-06 were isolated from the same host and cultured in a similar nutrient-rich medium, the burst sizes were different [32]. Similar to the host's physiological and ecological characteristics, features affecting phage burst size should not be ignored.

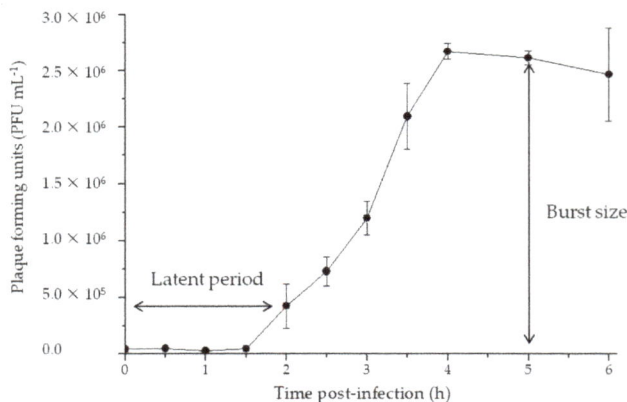

Figure 2. One-step growth curve of the phage R5C.

3.3. Genome Features

The terminal analysis revealed that no protruding cohesive end was found in the complete genome and suggested that phage R5C has a circular, double-stranded DNA genome according to Zhang et al. [37]. The genome size of R5C is 77,874 bp (Figure 3), which is the second largest among the published genomes of roseophages. It is suggested that the likelihood of phage interference with host cellular activities increases with genome size. R5C has a G+C content of 61.5%, which is the highest among all roseophages (Table S1). Generally, the G+C content is lower in phages than that in their hosts, while temperate phages have smaller biases towards G+C content [51]. For example, the average G+C values of the temperate phages ΦCB2047-A (58.8%) and ΦCB2047-C (59.0%) are close to that of their host *Sulfitobacter* sp. strain 2047 (60.3%). Interestingly, a small G+C deviation is also observed between R5C and its host (66.0%), which suggests that R5C may follow a temperate phage strategy. No tRNA sequences were detected in the R5C genome using the tRNAscan-SE program. The lack of tRNA was also found in other roseophages such as SIO1, P12053L, ΦCB2047-A, ΦCB2047-C, RDJLΦ1, RDJLΦ2, RD-1410W1-01, RD-1410Ws-07 and DS-1410Ws-06. Among the four roseosiphophages, DSS3Φ8 has the longest genome containing 24 tRNAs. In the literature, tRNA has been associated with longer genome length, higher codon usage bias and higher virulence [52].

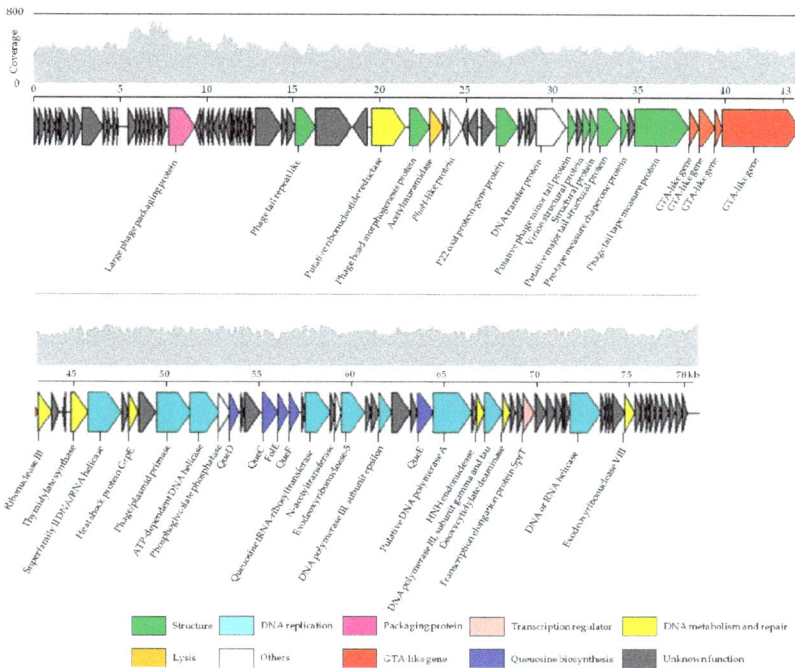

Figure 3. Genome map of the roseophage R5C. ORFs are depicted by leftward or rightward oriented arrows according to the direction of transcription. Gene features and genome modules (structure, lysis and DNA replication) are color-coded according to the legend below the figure.

In total, 123 ORFs were identified in the R5C genome using GeneMarkS and ORF Finder software (Table S2). A total of 66 gene products had homologous sequences in the NCBI non-redundant protein database and 41 of these could be assigned a recognizable function. At the amino acid level, genes homologous to that from other phages showed less than 71% similarity. About 66.7% of the

ORFs (82 ORFs, about 40% of the phage genome length) had no annotated features, while 57 of these ORFs had no matches in the databases. Single gene analysis showed R5C to be weakly similar to the known *Siphoviridae*. However, little or no nucleotide similarity was detected with these phages and protein homology was also detected with a few loci, with only one or two signature phage genes being shared between phages. Fifteen ORFs of R5C were homologous to that of both RDJLΦ1 and RDJLΦ2, showing similarly low identity levels (ranging from 24 to 72% and 25 to 74%, respectively). Furthermore, 19 ORFs with low identity (22–51%) were detected to be similar between R5C and DSS3Φ8. This suggested that R5C sequences presented high levels of divergence from known phage genomes and that proteins encoded by siphoviruses are under-represented in the database.

Among the 41 ORFs with recognizable functions, 10 were related to the structure and assembly of virions, such as a coat protein, a head-to-tail connecting protein, a tail fiber protein and the large subunit of terminase. Sixteen ORFs were predicted to encode proteins involved in DNA replication, metabolism and repair, while one conserved lysis ORF, acetylmuramidase, was predicted in the R5C genome. This was the first time that the DNA transfer protein, which is transcribed in the pre-early stage of infection in T5, had been detected in roseophages. Interestingly, four gene transfer agent (GTA) homologous genes and five queuosine biosynthesis genes were found in the R5C genome. Additionally, integrase and repressor genes, which indicate a potential for a lysogenic cycle, were not found in the R5C genome.

We compared the genomes of four roseosiphophages that possess gene transfer agent genes and found only seven conserved shared genes, including ribonucleotide reductase, DNA helicase, deoxycytidylate deaminase and GTA-like genes, with 22–50% identity at the amino acid level (Figure 4). This demonstrated the extremely high level of genetic divergence of roseosiphophages. The ribonucleotide reductase gene in R5C shares high amino acid identity with that of roseophages RDJLΦ1 (44%) and RDJLΦ2 (44%). As a key enzyme involved in DNA synthesis, ribonucleotide reductases are found in all organisms and convert nucleotides into deoxynucleotides [53]. In the phosphorus-limited marine environment, obtaining sufficient free nucleotides is critical for DNA synthesis [54,55]. DNA helicases are motor proteins that use the energy from NTP hydrolysis to separate transiently energetically-stable duplex DNA into single strands [56]. The ubiquity of helicases in prokaryotes, eukaryotes, and viruses indicates their fundamental importance in DNA metabolism [57]. Deoxycytidylate deaminases catalyze the deamination of dCMP to dUMP and thus provide the nucleotide substrates for thymidylate synthase [58]. All roseosiphophages isolated have highly conserved GTA-like genes, whereas all podophages infecting the *Roseobacter* clade roseopodophages lack similar genes. The four GTA-like genes (gp12–gp15) are close to genes encoding structural proteins, such as the tail tape measure protein of R5C, and the same structural phenomenon is also observed in other GTA-harboring phage genomes. These observations suggested that the function of gp12–gp15 may be related to the specific structure of siphophages, such as the tail. Further protein analyses are needed to verify this assumption.

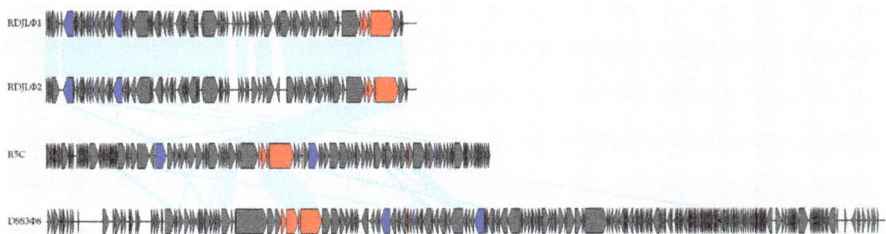

Figure 4. Comparison of four roseophages that possess GTA genes. Four GTA genes are indicated in red and other three conserved shared genes are indicated in blue. Genes sharing amino acid similarity between two phages are connected by light blue lines (E-value < 10^{-3}).

Like many other phages, the R5C genome contains a variety of auxiliary metabolic genes (AMGs). Currently, DNA metabolism and nucleotide synthesis genes are the most prevalent AMGs in roseophage. In R5C, we found AMGs frequently appeared in marine phages, such as *phoH* (ORF 47) and those firstly identified in roseophages (e.g., heat shock protein (ORF 74) and queuosine biosynthesis genes (ORF 79, ORF 82–84 and ORF 95)). A greater number of AMGs may broaden the role that phage play in their hosts' fitness during infection.

The *phoH* gene has been detected in phages infecting both heterotrophic and autotrophic bacteria, such as *Prochlorococcus* phage P-SSM2, *Synechococcus* phage Syn9, SAR11 phage HTVC008M, and *Vibrio* phage KVP40 [6,54,59,60]. Roseophage SIO1 and DSS3Φ8 also possess the *phoH* gene [13,26]. Phage-encoded *phoH* genes have previously been described as apparent parts of a multi-gene family with divergent functions and have played a part in phospholipid metabolism, RNA modification, and fatty acid beta-oxidation [54,61,62]. It is suggested that the *phoH* gene in the phages aids host regulation of phosphate uptake and metabolism under low-phosphate conditions, which is consistent with the environment from which R5C was isolated, namely the oligotrophic South China Sea.

Heat shock proteins are postulated to protect organisms from the toxic effects of heat and other forms of stress. These proteins exist in every organism studied from archaebacteria to eubacteria and from plants to animals [63]. Cellular heat-shock responses occur during the replication of many viruses, such as adenovirus and human cytomegalovirus [64,65]. This is the first report of a heat shock protein in roseophages. The *grpE* gene alone encodes a 24-kDa heat shock protein. The GrpE heat shock protein is important for bacteriophage λ DNA replication at all temperatures and for bacterial survival under certain conditions [66].

As a hypermodified nucleoside derivative of guanosine, queuosine occupies the wobble position (position 34) of the tRNAs coding for Asp, Asn, His or Tyr. The hypomodification of queuosine-modified tRNA plays an important role in cellular proliferation and metabolism [67]. The mechanisms of action of the queuosine biosynthesis genes in viruses remain unclear, even though similar gene clusters have been found in *Streptococcus* phage Dp-1, *Escherichia coli* phage 9g and other viruses [68–70]. The queuosine biosynthesis genes were detected for the first time in the genome of a roseophage in this study.

3.4. Phylogenetic Analyses

DNA polymerases, which play essential roles in viral replication, are found in many tailed bacteriophages, with three conserved motifs (motifs A, B, C) being present in all virioplankton metagenomic DNA polymerases [71]. In motif B, a link has been reported between leucine or tyrosine substitution in the site corresponding to phenylalanine and the phage lifestyle [72]. For example, all of the cultured phages with the tyrosine substitution were lytic, whereas lysogenic phages carrying the *pol*A gene possessed the leucine substitution. However, no evidence was found to link DNA polymerase I with the biological requirements for a lysogenic or lytic life cycle in R5C. Phylogenetic analyses based on DNA polymerase I showed that R5C was distantly related to other roseophages, most of which fell into the "N4-like" cluster. On the DNA polymerase I phylogenetic tree, R5C was most closely related to ctg DTF polA 1086, which was an environmental DNA polymerase sequence from Dry Tortugas surface water (Figure 5). Based on the currently available data, it is difficult to determine the taxonomic classification of R5C.

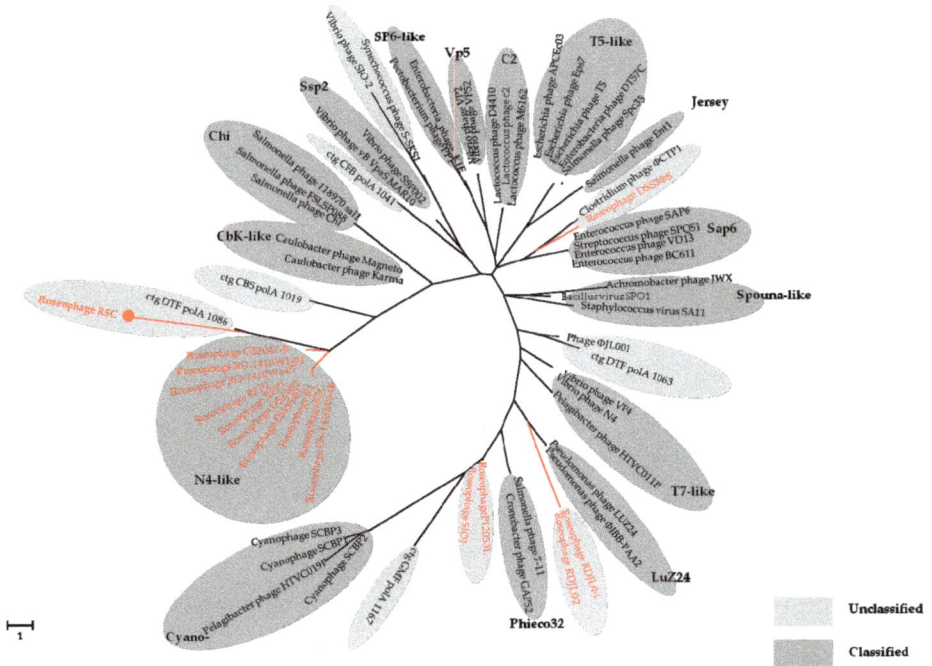

Figure 5. Unrooted maximum likelihood phylogenetic tree of DNA polymerase I of bacteriophages. Red color represents roseophages.

3.5. Environmental Distribution

To assess whether phage R5C is common in the environments, the R5C genome was searched against the IMG/VR and EVGs databases, which are assembled from ecologically diverse metagenomic samples [45,46]. However, no contig and genome with similarity to phage R5C was detected from the IMG/VR and EVGs databases. In addition, the genome of R5C was searched against spacers within CRISPRs of its host DFL12^T, the isolates used in this study, and the viral spacer database of IMG/VR. None of these analyses showed any match between R5C and viral spacer sequences within CRISPRs. When the R5C genes were searched against the POV and GOS databases, we found that their homologs were widespread from coastal regions to open oceans, similar to the previously shown genomic recruitment of N4-like roseophages [29]. A more detailed analysis showed that similar levels of identity (coastal, 22.2–90.0%; intermediate, 22.0–81.8%; open ocean, 22.6–83.3%) were found in different kinds of environment in the POV. Surprisingly, the highest frequency of counts was observed in samples from coastal areas in both the POV and GOS databases, despite the fact that R5C was isolated from open water (Figure 6). This distribution pattern was consistent with that found in previously published studies for roseophages DSS3Φ2 and EE36Φ1, but not RPP1 [16,29]. RPP1 was isolated from a coastal area but showed a higher distribution in open oceans [29]. These contrasting distribution patterns to isolation environments indicated that the interaction between roseophages and their hosts may be more complex and dynamic than previously thought, with further studies needed to reveal the global pattern of roseophages and their ecological significance.

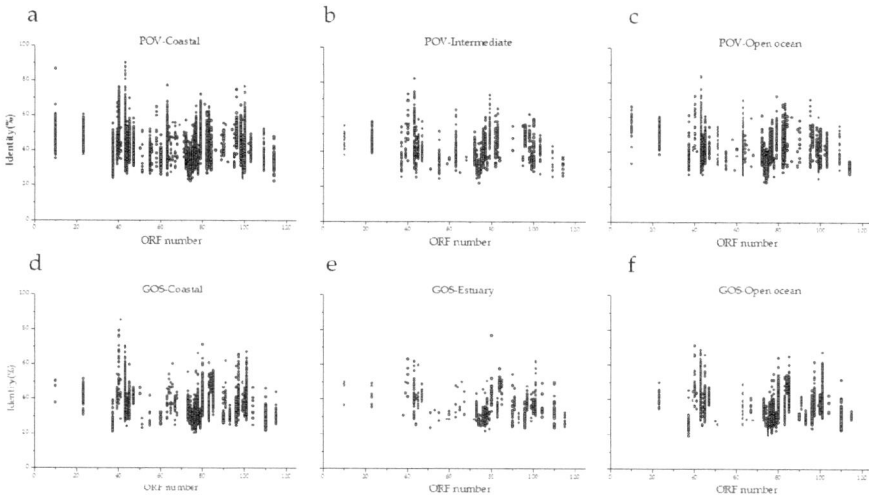

Figure 6. Prevalence of R5C-like ORFs in environmental viral metagenomic data. (**a**) Coastal metagenomes of POV; (**b**) Intermediate metagenomes of POV; (**c**) Open ocean metagenomes of POV; (**d**) Coastal metagenomes of GOS; (**e**) Estuary metagenomes of GOS; (**f**) Open ocean metagenomes of GOS.

4. Conclusions

Taken together, physiological and genomic characterization suggested that the phage R5C is a novel lytic *Siphoviridae* roseophage. Our study demonstrated that oligotrophic open ocean is also a source of roseophages, with novel roseophages possibly existing in this environment. The data for R5C provide valuable insight into our understanding of roseosiphophages, which have so far been under-investigated. However, these findings have raised several questions worth exploring in future studies: (1) What is the distribution pattern of roseophages and their major groups? (2) Do environmental conditions, such as nutrient levels, impact the ecological behavior of roseophages? and (3) What is the evolutionary forcing driving diversity of roseophage? With an increasing number of roseophages being isolated and considering the ecological significance of *Roseobacter*, *Roseobacter*–phages may serve as a model system for studying the interaction between marine bacteria and viruses. Tackling the questions above will expand our knowledge in this field.

Supplementary Materials: The following are available online at www.mdpi.com/1999-4915/9/5/109/s1, Table S1: Roseophages for which genome sequences are published; Table S2: Genome annotations of phage vB_DshS-R5C.

Acknowledgments: We thank Longfei Lu, Ziqiang Li, and Bu Xu of the State Key Laboratory of Marine Environmental Science for their useful suggestions and help. This work was supported by the National Programme on Global Change and Air-Sea Interaction (GASI-03-01-02-05) and National Natural Science Foundation of China (41522603, 31570172, 41376132, 91428308).

Author Contributions: Rui Zhang and Nianzhi Jiao supervised the project and revised manuscript; Yunlan Yang, Lanlan Cai, Yongle Xu, Yigang Tong, Yong Huang, and Rui Zhang analyzed data and wrote the manuscripts. Yunlan Yang, Lanlan Cai, and Ruijie Ma performed the experiments.

Conflicts of Interest: The authors declare no conflict of interest.

References

1. Suttle, C.A. Viruses in the sea. *Nature* **2005**, *437*, 356–361. [CrossRef] [PubMed]
2. Suttle, C.A. Marine viruses—Major players in the global ecosystem. *Nat. Rev. Microbiol.* **2007**, *5*, 801–812. [CrossRef] [PubMed]

3. Brum, J.R.; Ignacio-Espinoza, J.C.; Roux, S.; Doulcier, G.; Acinas, S.G.; Alberti, A.; Chaffron, S.; Cruaud, C.; Vargas, C.D.; Gasol, J.M.; et al. Patterns and ecological drivers of ocean viral communities. *Science* **2015**, *348*, 1261498. [CrossRef] [PubMed]

4. Labrie, S.J.; Frois-Moniz, K.; Osburne, M.S.; Kelly, L.; Roggensack, S.E.; Sullivan, M.B.; Gearin, G.; Zeng, Q.; Fitzgerald, M.; Henn, M.R.; et al. Genomes of marine cyanopodoviruses reveal multiple origins of diversity. *Environ. Microbiol.* **2013**, *15*, 1356–1376. [CrossRef] [PubMed]

5. Baudoux, A.C.; Hendrix, R.W.; Lander, G.C.; Bailly, X.; Podell, S.; Paillard, C.; Johnson, J.E.; Potter, C.S.; Carragher, B.; Azam, F. Genomic and functional analysis of *Vibrio* phage SIO-2 reveals novel insights into ecology and evolution of marine siphoviruses. *Environ. Microbiol.* **2012**, *14*, 2071–2086. [CrossRef] [PubMed]

6. Zhao, Y.; Temperton, B.; Thrash, J.C.; Schwalbach, M.S.; Vergin, K.L.; Landry, Z.C.; Ellisman, M.; Deerinck, T.; Sullivan, M.B.; Giovannoni, S.J. Abundant SAR11 viruses in the ocean. *Nature* **2013**, *494*, 357–360. [CrossRef] [PubMed]

7. Sullivan, M.B.; Coleman, M.L.; Quinlivan, V.; Rosenkrantz, J.E.; Defrancesco, A.S.; Tan, G.; Fu, R.; Lee, J.A.; Waterbury, J.B.; Bielawski, J.P. Portal protein diversity and phage ecology. *Environ. Microbiol.* **2008**, *10*, 2810–2823. [CrossRef] [PubMed]

8. Mann, N.H.; Clokie, M.R.; Millard, A.; Cook, A.; Wilson, W.H.; Wheatley, P.J.; Letarov, A.; Krisch, H.M. The genome of S-PM2, a "photosynthetic" T4-type bacteriophage that infects marine *Synechococcus* strains. *J. Bacteriol.* **2005**, *187*, 3188–3200. [CrossRef] [PubMed]

9. Stoddard, L.I.; Martiny, J.B.; Marston, M.F. Selection and characterization of cyanophage resistance in marine *Synechococcus* strains. *Appl. Environ. Microbiol.* **2007**, *73*, 5516–5522. [CrossRef] [PubMed]

10. Avrani, S.; Wurtzel, O.; Sharon, I.; Sorek, R.; Lindell, D. Genomic island variability facilitates *Prochlorococcus*-virus coexistence. *Nature* **2011**, *474*, 604–608. [CrossRef] [PubMed]

11. Brinkhoff, T.; Giebel, H.A.; Simon, M. Diversity, ecology, and genomics of the *Roseobacter* clade: A short overview. *Arch. Microbiol.* **2008**, *189*, 531–539. [CrossRef] [PubMed]

12. Huang, S.; Zhang, Y.; Chen, F.; Jiao, N. Complete genome sequence of a marine roseophage provides evidence into the evolution of gene transfer agents in *alphaproteobacteria*. *J. Virol.* **2011**, *8*, 124. [CrossRef] [PubMed]

13. Zhan, Y.; Huang, S.; Voget, S.; Simon, M.; Chen, F. A novel *roseobacter* phage possesses features of podoviruses, siphoviruses, prophages and gene transfer agents. *Sci. Rep.* **2016**, *6*, 30372. [CrossRef] [PubMed]

14. Cai, L.; Yang, Y.; Jiao, N.; Zhang, R. Complete genome sequence of vB_DshP-R2C, a N4-like lytic roseophage. *Mar. Genom.* **2015**, *22*, 15–17. [CrossRef] [PubMed]

15. Budinof, C.R. Diversity and Activity of Roseobacters and Roseophage. Ph.D. Thesis, University of Tennessee, Knoxville, TS, USA, 2012. Available online: http://trace.tennessee.edu/utk_graddiss/1276 (accessed on 18 April 2017).

16. Zhao, Y.; Wang, K.; Jiao, N.; Chen, F. Genome sequences of two novel phages infecting marine roseobacters. *Environ. Microbiol.* **2009**, *11*, 2055–2064. [CrossRef] [PubMed]

17. Buchan, A.; Gonzalez, J.M.; Moran, M.A. Overview of the marine *roseobacter* lineage. *Appl. Environ. Microbiol.* **2005**, *71*, 5665–5677. [CrossRef] [PubMed]

18. Christie-Oleza, J.A.; Armengaud, J. Proteomics of the *Roseobacter* clade, a window to the marine microbiology landscape. *Proteomics* **2015**, *15*, 3928–3942. [CrossRef] [PubMed]

19. Biebl, H.; Allgaier, M.; Tindall, B.J.; Koblizek, M.; Lunsdorf, H.; Pukall, R.; Wagner-Dobler, I. *Dinoroseobacter shibae* gen. nov., sp. nov., a new aerobic phototrophic bacterium isolated from dinoflagellates. *Int. J. Syst. Evol. Microbiol.* **2005**, *55*, 1089–1096. [CrossRef] [PubMed]

20. Jiao, N.; Zhang, Y.; Zeng, Y.; Hong, N.; Liu, R.; Chen, F.; Wang, P. Distinct distribution pattern of abundance and diversity of aerobic anoxygenic phototrophic bacteria in the global ocean. *Environ. Microbiol.* **2007**, *9*, 3091–3099. [CrossRef] [PubMed]

21. Kolber, Z.S.; Plumley, F.G.; Lang, A.S.; Beatty, J.T.; Blankenship, R.E.; VanDover, C.L.; Vetriani, C.; Koblizek, M.; Rathgeber, C.; Falkowski, P.G. Contribution of aerobic photoheterotrophic bacteria to the carbon cycle in the ocean. *Science* **2001**, *292*, 2492–2495. [CrossRef] [PubMed]

22. Wagner-Dobler, I.; Biebl, H. Environmental biology of the marine *Roseobacter* lineage. *Annu. Rev. Microbiol.* **2006**, *60*, 255–280. [CrossRef] [PubMed]

23. Allgaier, M.; Uphoff, H.; Felske, A.; Wagner-Dobler, I. Aerobic anoxygenic photosynthesis in *Roseobacter* clade bacteria from diverse marine habitats. *Appl. Environ. Microbiol.* **2003**, *69*, 5051–5059. [CrossRef] [PubMed]

24. Wagner-Dobler, I.; Ballhausen, B.; Berger, M.; Brinkhoff, T.; Buchholz, I.; Bunk, B.; Cypionka, H.; Daniel, R.; Drepper, T.; Gerdts, G.; et al. The complete genome sequence of the algal symbiont *Dinoroseobacter shibae*: A hitchhiker's guide to life in the sea. *ISME J.* **2010**, *4*, 61–77. [CrossRef] [PubMed]

25. Kang, I.; Jang, H.; Oh, H.-M.; Cho, J.-C. Complete genome sequence of *Celeribacter* bacteriophage P12053L. *J. Virol.* **2012**, *86*, 8339–8340. [CrossRef] [PubMed]

26. Rohwer, F.; Segall, A.; Steward, G.; Seguritan, V.; Breitbart, M.; Wolven, F.; Azam, F. The complete genomic sequence of the marine phage Roseophage SIO1 shares homology with nonmarine phages. *Limnol. Oceanogy* **2000**, *45*, 408–418. [CrossRef]

27. Ankrah, N.Y.D.; Budinoff, C.R.; Wilson, W.H.; Wilhelm, S.W.; Buchan, A. Genome sequence of the *Sulfitobacter* sp. strain 2047-infecting lytic phage ΦCB2047-B. *Genome Announc.* **2014**, *2*, e00945-13. [CrossRef] [PubMed]

28. Ankrah, N.Y.D.; Budinoff, C.R.; Wilson, W.H.; Wilhelm, S.W.; Buchana, A. Genome sequences of two temperate phages, ΦCB2047-A and ΦCB2047-C, infecting Sulfitobacter sp strain 2047. *Genome Announc.* **2014**, *2*, e00108-14. [CrossRef] [PubMed]

29. Chan, J.Z.; Millard, A.D.; Mann, N.H.; Schafer, H. Comparative genomics defines the core genome of the growing N4-like phage genus and identifies N4-like Roseophage specific genes. *Front. Microbiol.* **2014**, *5*, 506. [CrossRef] [PubMed]

30. Ji, J.; Zhang, R.; Jiao, N. Complete genome sequence of roseophage vB_DshP-R1, which infects *Dinoroseobacter shibae* DFL12. *Stand. Genomic. Sci.* **2015**, *10*, 6. [CrossRef] [PubMed]

31. Liang, Y.; Zhang, Y.; Zhou, C.; Chen, Z.; Yang, S.; Yan, C.; Jiao, N. Complete genome sequence of the siphovirus roseophage RDJLΦ2 infecting *Roseobacter denitrificans* OCh114. *Mar. Genom.* **2016**, *25*, 17–19. [CrossRef] [PubMed]

32. Li, B.; Zhang, S.; Long, L.; Huang, S. Characterization and complete genome sequences of three N4-like roseobacter phages isolated from the South China Sea. *Curr. Microbiol.* **2016**, *73*, 409–418. [CrossRef] [PubMed]

33. Angly, F.; Youle, M.; Nosrat, B.; Srinagesh, S.; Rodriguez-Brito, B.; McNairnie, P.; Deyanat-Yazdi, G.; Breitbart, M.; Rohwer, F. Genomic analysis of multiple Roseophage SIO1 strains. *Environ. Microbiol.* **2009**, *11*, 2863–2873. [CrossRef] [PubMed]

34. Zhang, Y.; Jiao, N. Roseophage RDJLΦ1, infecting the aerobic anoxygenic phototrophic bacterium *Roseobacter denitrificans* OCh114. *Appl. Environ. Microbiol.* **2009**, *75*, 1745–1749. [CrossRef] [PubMed]

35. Middelboe, M.; Chan, A.M.; Bertelsen, S.K. Isolation and life cycle characterization of lytic viruses infecting heterotrophic bacteria and cyanobacteria. In *Manual of Aquatic Viral Ecology*; Wilhelm, S.W., Weinbauer, M.G., Suttle, C.A., Eds.; American Society of Limnology and Oceanography: Waco, TX, USA, 2010; pp. 118–133. [CrossRef]

36. Zerbino, D.R.; Birney, E. Velvet: Algorithms for de novo short read assembly using de Bruijn graphs. *Genome Res.* **2008**, *18*, 821–829. [CrossRef] [PubMed]

37. Zhang, X.; Wang, Y.; Li, S.; An, X.; Pei, G.; Huang, Y.; Fan, H.; Mi, Z.; Zhang, Z.; Wang, W.; et al. A novel termini analysis theory using HTS data alone for the identification of Enterococcus phage EF4-like genome termini. *BMC Genom.* **2015**, *16*, 414. [CrossRef] [PubMed]

38. Besemer, J.; Borodovsky, M. GeneMark: Web software for gene finding in prokaryotes, eukaryotes and viruses. *Nucleic Acids Res.* **2005**, *33*, W451–W454. [CrossRef] [PubMed]

39. Rombel, I.T.; Sykes, K.F.; Rayner, S.; Johnston, S.A. ORF-FINDER: A vector for high-throughput gene identificatio. *Gene* **2002**, *282*, 33–41. [CrossRef]

40. Lowe, T.M.; Eddy, S.R. tRNAscan-SE: A program for improved detection of transfer RNA genes in genomic sequence. *Nucleic Acids Res.* **1997**, *25*, 955–964. [CrossRef] [PubMed]

41. Pruitt, K.D.; Tatusova, T.; Maglott, D.R. NCBI reference sequences (RefSeq): A curated non-redundant sequence database of genomes, transcripts and proteins. *Nucleic Acids Res.* **2007**, *35*, D61–D65. [CrossRef] [PubMed]

42. NCBI Resource Coordinators. Database resources of the national center for biotechnology information. *Nucleic Acids Res.* **2014**, *42*, D7–D17. [CrossRef]

43. Hurwitz, B.L.; Sullivan, M.B. The Pacific Ocean Virome (POV): A marine viral metagenomic dataset and associated protein clusters for quantitative viral ecology. *PLoS ONE* **2013**, *8*, e57355. [CrossRef] [PubMed]

44. Rusch, D.B.; Halpern, A.L.; Sutton, G.; Heidelberg, K.B.; Williamson, S.; Yooseph, S.; Wu, D.; Eisen, J.A.; Hoffman, J.M.; Remington, K.; et al. The Sorcerer II Global Ocean Sampling expedition: Northwest Atlantic through eastern tropical Pacific. *PLoS Biol.* **2007**, *5*, e77. [CrossRef] [PubMed]

45. Paez-Espino, D.; Chen, I.A.; Palaniappan, K.; Ratner, A.; Chu, K.; Szeto, E.; Pillay, M.; Huang, J.; Markowitz, V.M.; Nielsen, T.; et al. IMG/VR: A database of cultured and uncultured DNA viruses and retroviruses. *Nucleic Acids Res.* **2017**, *45*, D457–D465. [CrossRef] [PubMed]

46. Nishimura, Y.; Watai, H.; Honda, T.; Mihara, T.; Omae, K.; Roux, S.; Blanc-Mathieu, R.; Yamamoto, K.; Hingamp, P.; Sako, Y.; et al. Environmental viral genomes shed new light on virus-host interactions in the ocean. *mSphere* **2017**, *2*, e00359-16. [CrossRef] [PubMed]

47. Deveau, H.; Barrangou, R.; Garneau, J.E.; Labonte, J.; Fremaux, C.; Boyaval, P.; Romero, D.A.; Horvath, P.; Moineau, S. Phage response to CRISPR-encoded resistance in Streptococcus thermophilus. *J. Bacteriol.* **2008**, *190*, 1390–1400. [CrossRef] [PubMed]

48. Kropinski, A.M.; Prangishvili, D.; Lavigne, R. Position paper: The creation of a rational scheme for the nomenclature of viruses of *Bacteria* and *Archaea*. *Environ. Microbiol.* **2009**, *11*, 2775–2777. [CrossRef] [PubMed]

49. Parada, V.; Herndl, G.J.; Weinbauer, M.G. Viral burst size of heterotrophic prokaryotes in aquatic systems. *J. Mar. Biol. Assos. UK* **2006**, *86*, 613–621. [CrossRef]

50. Hwang, C.Y.; Cho, B.C. Virus-infected bacteria in oligotrophic open waters of the East Sea, Korea. *Aquat. Microb. Ecol.* **2002**, *30*, 1–9. [CrossRef]

51. Rocha, P.C.E.; Danchin, A. Base composition bias might result from competition for metabolic resources. *Trends Genet.* **2002**, *18*, 291–294. [CrossRef]

52. Bailly-Bechet, M.; Vergassola, M.; Rocha, E. Causes for the intriguing presence of tRNAs in phages. *Genome Res.* **2007**, *17*, 1486–1495. [CrossRef] [PubMed]

53. Stubbe, J. Ribonucleotide reductases: The link between an RNA and a DNA world? *Curr. Opin. Struct. Biol.* **2000**, *10*, 731–736. [CrossRef]

54. Sullivan, M.B.; Coleman, M.L.; Weigele, P.; Rohwer, F.; Chisholm, S.W. Three *Prochlorococcus* cyanophage genomes: Signature features and ecological interpretations. *PLoS Biol.* **2005**, *3*, e144. [CrossRef] [PubMed]

55. Chen, F.; Lu, J. Genomic sequence and evolution of marine cyanophage P60: A new insight on lytic and lysogenic phages. *Appl. Environ. Microbiol.* **2002**, *68*, 2589–2594. [CrossRef] [PubMed]

56. Matson, S.W.; Bean, D.W.; George, J.W. DNA helicases: Enzymes with essential roles in all aspects of DNA metabolism. *BioEssays* **1993**, *16*, 13–22. [CrossRef] [PubMed]

57. Tuteja, N.; Tuteja, R. Unraveling DNA helicases. Motif, structure, mechanism and function. *Eur. J. Biochem.* **2004**, *271*, 1849–1863. [CrossRef] [PubMed]

58. Moore, J.T.; Silversmith, R.E.; Maley, G.F.; Maley, F. T4-phage deoxycytidylate deaminase is a metalloprotein containing two zinc atoms per subunit. *J. Biol. Chem.* **1993**, *268*, 2288–2291. [PubMed]

59. Weigele, P.R.; Pope, W.H.; Pedulla, M.L.; Houtz, J.M.; Smith, A.L.; Conway, J.F.; King, J.; Hatfull, G.F.; Lawrence, J.G.; Hendrix, R.W. Genomic and structural analysis of Syn9, a cyanophage infecting marine *Prochlorococcus* and *Synechococcus*. *Environ. Microbiol.* **2007**, *9*, 1675–1695. [CrossRef] [PubMed]

60. Miller, E.S.; Heidelberg, J.F.; Eisen, J.A.; Nelson, W.C.; Durkin, A.S.; Ciecko, A.; Feldblyum, T.V.; White, O.; Paulsen, I.T.; Nierman, W.C.; et al. Complete genome sequence of the broad-host-range vibriophage KVP40: Comparative genomics of a T4-related bacteriophage. *J. Bacteriol.* **2003**, *185*, 5220–5233. [CrossRef] [PubMed]

61. Sullivan, M.B.; Huang, K.H.; Ignacio-Espinoza, J.C.; Berlin, A.M.; Kelly, L.; Weigele, P.R.; DeFrancesco, A.S.; Kern, S.E.; Thompson, L.R.; Young, S.; et al. Genomic analysis of oceanic cyanobacterial myoviruses compared with T4-like myoviruses from diverse hosts and environments. *Environ. Microbiol.* **2010**, *12*, 3035–3056. [CrossRef] [PubMed]

62. Kazakov, A.E.; Vassieva, O.; Gelfand, M.S.; Osterman, A.; Overbeek, R. Bioinformatics classification and functional analysis of phoH homologs. *In Silico Biol.* **2003**, *3*, 3–15. [PubMed]

63. Lindquist, S.; Craig, E.A. The heat-shock proteins. *Annu. Rev. Genet.* **1988**, *22*, 631–677. [CrossRef] [PubMed]

64. Glotzer, J.B.; Saltik, M.; Chiaocca, S.; Michou, A.-I.; Moseley, P.; Cotten, M. Activation of heat-shock response by an adenovirus is essential for virus replication. *Nature* **2000**, *407*, 207–211. [CrossRef] [PubMed]

65. Straus, D.; Walter, W.; Gross, C.A. DnaK, DnaJ, and GrpE heat shock proteins negatively regulate heat shock gene expression by controlling the synthesis and stability of sigma 32. *Genes Dev.* **1990**, *4*, 2202–2209. [CrossRef] [PubMed]

66. Liberek, K.; Marszalek, J.; Ang, D.; Georgopoulos, C.; Zylicz, M. Escherichia coli DnaJ and GrpE heat shock proteins jointly stimulate ATPase activity of DnaK. *Proc. Natl. Acad. Sci. USA* **1991**, *88*, 2874–2878. [CrossRef] [PubMed]

67. Vinayak, M.; Pathak, C. Queuosine modification of tRNA: Its divergent role in cellular machinery. *Biosci. Rep.* **2009**, *30*, 135–148. [CrossRef] [PubMed]

68. Sabri, M.; Hauser, R.; Ouellette, M.; Liu, J.; Dehbi, M.; Moeck, G.; Garcia, E.; Titz, B.; Uetz, P.; Moineau, S. Genome annotation and intraviral interactome for the *Streptococcus pneumoniae* virulent phage Dp-1. *J. Bacteriol.* **2011**, *193*, 551–562. [CrossRef] [PubMed]

69. Kulikov, E.E.; Golomidova, A.K.; Letarova, M.A.; Kostryukova, E.S.; Zelenin, A.S.; Prokhorov, N.S.; Letarov, A.V. Genomic sequencing and biological characteristics of a novel *Escherichia coli* bacteriophage 9g, a putative representative of a new *Siphoviridae* genus. *Viruses* **2014**, *6*, 5077–5092. [CrossRef] [PubMed]

70. Roux, S.; Enault, F.; Ravet, V.; Pereira, O.; Sullivan, M.B. Genomic characteristics and environmental distributions of the uncultivated Far-T4 phages. *Front. Microbiol.* **2015**, *6*, 199. [CrossRef] [PubMed]

71. Loh, E.; Loeb, L.A. Mutability of DNA polymerase I: Implications for the creation of mutant DNA polymerases. *DNA Repair* **2005**, *4*, 1390–1398. [CrossRef] [PubMed]

72. Schmidt, H.F.; Sakowski, E.G.; Williamson, S.J.; Polson, S.W.; Wommack, K.E. Shotgun metagenomics indicates novel family A DNA polymerases predominate within marine virioplankton. *ISME J.* **2014**, *8*, 103–114. [CrossRef] [PubMed]

viruses

MDPI

Article

Coccolithoviruses: A Review of Cross-Kingdom Genomic Thievery and Metabolic Thuggery

Jozef I. Nissimov [1,2], António Pagarete [3], Fangrui Ma [4], Sean Cody [4], David D. Dunigan [4], Susan A. Kimmance [1] and Michael J. Allen [1,*]

[1]   Plymouth Marine Laboratory, Prospect Place, The Hoe, Plymouth PL1 3DH, UK;
      jnissimov@marine.rutgers.edu (J.I.N.); sukim@pml.ac.uk (S.A.K.)
[2]   Department of Marine and Coastal Sciences, Rutgers University, New Brunswick, NJ 08901, USA
[3]   Department of Biology, University of Bergen, Bergen 7803, Norway; antonio.pagarete@uib.no
[4]   Nebraska Center for Virology, University of Nebraska, Lincoln, NE 68583, USA; fma2@unl.edu (F.M.);
      wfr1995@gmail.com (S.C.); ddunigan2@unl.edu (D.D.D.)
*   Correspondence: mija@pml.ac.uk; Tel.: +44-175-263-3100

Academic Editors: Mathias Middelboe and Corina Brussaard
Received: 18 January 2017; Accepted: 14 March 2017; Published: 18 March 2017

Abstract: Coccolithoviruses (*Phycodnaviridae*) infect and lyse the most ubiquitous and successful coccolithophorid in modern oceans, *Emiliania huxleyi*. So far, the genomes of 13 of these giant lytic viruses (i.e., *Emiliania huxleyi* viruses—EhVs) have been sequenced, assembled, and annotated. Here, we performed an in-depth comparison of their genomes to try and contextualize the ecological and evolutionary traits of these viruses. The genomes of these EhVs have from 444 to 548 coding sequences (CDSs). Presence/absence analysis of CDSs identified putative genes with particular ecological significance, namely sialidase, phosphate permease, and sphingolipid biosynthesis. The viruses clustered into distinct clades, based on their DNA polymerase gene as well as full genome comparisons. We discuss the use of such clustering and suggest that a gene-by-gene investigation approach may be more useful when the goal is to reveal differences related to functionally important genes. A multi domain "Best BLAST hit" analysis revealed that 84% of the EhV genes have closer similarities to the domain Eukarya. However, 16% of the EhV CDSs were very similar to bacterial genes, contributing to the idea that a significant portion of the gene flow in the planktonic world inter-crosses the domains of life.

Keywords: *E. huxleyi*; coccolithovirus; genome comparison; horizontal gene transfer; domains of life

1. Introduction

It has been more than 10 years since the complete genome sequencing of a giant double-stranded DNA-containing virus infecting the ubiquitous bloom-forming coccolithophorid species *Emiliania huxleyi* [1]. *Emiliania huxleyi* virus strain number 86 (EhV-86) was isolated in 1999 from the English Channel, and a subsequent analysis of its major capsid protein (MCP) and DNA polymerase genes placed it in a new separate genus, *Coccolithovirus*, within the family *Phycodnaviridae* [2–4]. The *Phycodnaviridae* comprises other giant viruses that infect algae, such as *Chlorella* sp. (e.g., PBCV-1), *Ectocarpus siliculousus* (e.g., EsV-1), *Micromonas pusilla* (e.g., MpV-SP1), *Chrysochromulina brevifilum* (e.g., CbV-PW1), and *Heterosigma akashiwo* (e.g., HaV01) [5,6]. Because of *E. huxleyi*'s global impact on biogeochemical cycles, the study of this host-virus system is particularly relevant. Interactions between *E. huxleyi* and EhVs have been investigated both in vivo, through large-scale semi-natural mesocosm experiments and natural *E. huxleyi* blooms [7–11], and in vitro [12–16], in laboratory experiments designed to elucidate specific aspects of the infection cycle, the infection dynamics and the host cellular response to infection [17].

As a result of these studies, considerable insight has been gained regarding (1) the EhV intein [18] and the expression profile of EhV genes during infection [19–21]; (2) the EhV life cycle, including its utilization of lipid rafts for budding from the host cells [22,23]; (3) host cellular processes in response to infection such as autophagy and the induction of programmed cell death (PCD) pathways [17,24–27]; (4) the manipulation of fatty acid and lipid metabolism within infected cells, leading to the production of virus-induced lipids crucial for the progression of the infection [10,28–31]; (5) vector transmission of EhVs in the natural environment via aerosols and zooplankton faecal pellets [32,33]; and (6) the co-occurring diversity of EhVs and their viruses in a range of habitats, including the Atlantic Ocean, Norwegian Fjords and coastal regions of the Black Sea [8,34–38]. Perhaps the most astonishing finding to date was the identification of a de novo sphingolipid biosynthesis pathway in the EhV genome [1,9,39], encoding virus-derived glycosphingolipids (vGSLs) that are crucial regulators of infection [28]. Thus, important parts of the metabolic activity in the planktonic realm (in this case lipid production) are viral-driven [40]. It has been proposed that this pathway is a result of horizontal gene transfer (HGT) between *E. huxleyi* and their coccolithoviruses [1,9,39]. Similar HGT events with far-reaching ecological relevance and consequences have been reported between other marine viruses and their hosts. These include photosynthesis-related genes such as photosystem II core reaction center protein D1 and a high-light-inducible protein in *Prochlorococcus* phages [41], as well as phosphate transporter genes (e.g., *pho4* and *phoH*) in viruses infecting *Bathycoccus*, *Micromonas* and *Ostreococcus* [42].

Since the sequencing of the EhV-86 genome, an additional 12 EhV strains have been isolated and their genomes sequenced [43–49]. This makes EhVs one of the largest collections of *Phycodnaviridae* viruses in culture with complete or near-complete genome sequences available. In this study, we conduct an in-depth comparison of these EhV genomes for the first time and consider the functional and ecological significance of variation observed between strains. Further, we investigate potential evolutionary links between these viruses and the other domains of life, in particular bacteria which co-occur with EhVs in the marine environment. Bacteria are important constituents of microbial food webs in marine habitats [50,51] and may themselves act as pathogens. As an example of the latter, certain bacteria have been found to induce caspase-like activity and cell death in *E. huxleyi* [52]. Diseases of macroalgae in coastal habitats have also been attributed to bacteria [53], which likely coexist with a diverse viral community [54]. Interestingly, bacterial-like genes have been reported in several *Phycodnaviruses*. For example, 48 to 57 bacterial-like genes were identified in three viruses of *Chlorella*, and 81% of the total bacterial-like genes identified in *Mimivirus* have no homologs in even distantly related eukaryotes [55,56]. The possibility that EhVs may also have acquired genes from marine microbes (other than their host) via HGT will be explored.

2. Methods

2.1. Coccolithoviruses and Their Phylogeny

The genomes of 13 EhV strains were retrieved from National Center for Biotechnology Information (NCBI) (Table 1). One genome (EhV-86) is complete. The remaining twelve are near-complete draft genomes which contain gaps between some of the contigs (see Section 3.3). A 14th genome (EhV-163) was not included in this analysis to avoid biases due to its incomplete sequence (80%) [43]. CLUSTALW multiple sequence alignment of two EhV genes, DNA polymerase and serine palmitoyltransferase (2921 and 2604 bp long, respectively), and subsequent phylogenetic analysis was conducted using MEGA6 [57]. The evolutionary history of these sequences was inferred using two independent methods: Neighbor-Joining and Maximum Likelihood [58].

Table 1. A brief description of the coccolithoviruses used in this study, their geographical origin, year of isolation and sea water depth from which they were obtained.

Strain #	Isolate Location	Isolation Date	Lat/Long	Depth (m)	Date Sequenced	NCBI Accession #	Reference *
EhV-84	E.C.	1999	50°15′ N, 04°13′ W	15	2011	JF974290	Schroeder et al., 2002 [2]; Nissimov et al., 2011 [45]
EhV-86	E.C.	1999	50°30′ N, 04°20′ W	surface	2005	AJ890364	Schroeder et al., 2002 [2]; Wilson et al., 2005 [1]
EhV-88	E.C.	1999	50°15′ N, 04°13′ W	5	2011	JF974310	Schroeder et al., 2002 [2]; Nissimov et al., 2012 [46]
EhV-201	E.C.	2001	49°56′ N, 04°19′ W	2	2011	JF974311	Schroeder et al., 2002 [2]; Nissimov et al., 2012 [46]
EhV-202	E.C.	2001	50°00′ N, 04°18′ W	15	2011	HQ634145	Schroeder et al., 2002 [2]; Nissimov et al., 2012 [47]
EhV-203	E.C.	2001	50°00′ N, 04°18′ W	15	2011	JF974291	Schroeder et al., 2002 [2]; Nissimov et al., 2011 [48]
EhV-207	E.C.	2001	50°15′ N, 04°13′ W	5	2011	JF974317	Schroeder et al., 2002 [2]; Nissimov et al., 2012 [46]
EhV-208	E.C.	2001	50°15′ N, 04°13′ W	5	2011	JF974318	Schroeder et al., 2002 [2]; Nissimov et al., 2012 [46]
EhV-99B1	N.F.	1999	60°20′ N, 05°20′ E	surface	2013	FN429076	Pagarete et al., 2013 [44]
EhV-18	E.C.	2008	50°15′ N, 04°13′ W	surface	2013	KF481685	Nissimov et al., 2014 [49]
EhV-145	Loss.	2008	57°72′ N, 03°29′W	surface	2013	KF481686	Nissimov et al., 2014 [49]
EhV-156	E.C.	2009	50°15′ N, 04°13′ W	surface	2013	KF481687	Nissimov et al., 2014 [49]
EhV-164	SSF	2009	56°26′ N, 02°63′ W	surface	2013	KF481688	Nissimov et al., 2014 [49]

* The references refer to literature that first presented information on each isolate or its genome. E.C.: English Channel; N.F.: Norwegian Fjord; Loss: Lossiemouth (off the UK coast); SSF: Scottish Shore of Fife (UK).

2.2. Whole Genome Reconstruction, Alignment and Comparison

Annotated EhV genomes [1,44–49] were used to build genomic alignments. Prior to analysis, the gaps between the assembled annotated contigs of each draft genome were made the same length by adding a known length of ambiguous nucleotides (i.e., a series of 10 Ns). The Artemis Comparison Tool (ACT) was then used for BLASTn pairwise alignment and visualization of the draft genome sequences against the fully assembled non-gapped reference genome of EhV-86 [59]. This enabled identification of the most likely orientation and placement of contigs (note that inversions are common in virus genomes and it is possible that inversions represent a real difference between the genomes). Misplaced contigs were placed in their appropriate location and converted into the correct orientation manually. Subsequently, MAUVE 2.0, a program for the identification of conserved genomic DNA regions and rearrangements, was used to elucidate the variable regions of each genome [60] in relation to EhV-86. Additional gene prediction analysis and functional annotations were performed within the publically available online Integrated Microbial Genomes—Expert Review platform (IMG/ER; [61]). Pairwise average nucleotide identity between each pair of genomes was also performed in this platform using Average Nucleotide Identity (ANI) analysis [62,63]. Briefly, ANI was used to compare all protein coding genes within each EhV genome to the non-gapped reference genome of EhV-86 (Table 2). Only bidirectional best hits (BBHs) with >70% sequence identity over >70% of the length of the shorter sequence in each BBH pair were retained.

Table 2. Average Nucleotide Identity (ANI) analysis of EhV genomes against EhV-86. The analysis included 12 draft EhV genomes, where a higher ANI score indicates greater genome similarity.

Reference Genome	Draft Genome	ANI Score	Total BBH *	Clade
	EhV-164	99.95	443	A1
	EhV-145	99.93	456	A1
	EhV-84	99.07	434	A1
	EhV-88	98.96	442	A1
	EhV-99B1	98.23	421	A3
EhV-86	EhV-208	96.78	399	A2
	EhV-207	96.67	411	A2
	EhV-201	96.6	399	A2
	EhV-203	96.6	402	A2
	EhV-18	79.52	307	B
	EhV-202	79.42	312	B
	EhV-156	79.4	308	B

* BBH: bidirectional best hit.

2.3. "Best BLAST Hit" Analysis of Genomes

A "Best BLAST hit" analysis was performed on the predicted CDSs of the EhV genomes using a BLASTp algorithm [64] with default parameters against the publically available NCBI non-redundant protein database, excluding viruses. The "Best BLAST hit" analyses used a comprehensive list of viral genes (in this case all the predicted CDSs of 13 EhV genomes) as a query against the NCBI non-redundant protein sequences and captured the top-hit taxonomic data if available (e.g., Domain, Phylum, Class, Order, Family, Genus, Species), associated with the subject homolog (Target), as well as sequence match scoring parameters. An initial search pulled all hits. These were clustered according to "Domain" and evaluated for alignment and Bit scores, which were normalized to the scoring system. While we recognize that there is no perfect cut-off, we have chosen to limit the dataset to Bit scores above 50 for "Domain", and above 100 (E-value $< 1 \times 10^{-19}$) for the subsequent taxon categories. These analyses fall within the suggested boundaries [65].

Specific BLASTp analysis of each EhV genome against the pan genome of *E. huxleyi* CCMP1516 [66] and against other EhVs was conducted within the IMG/ER platform using a cut-off E-value of $<1 \times 10^{-5}$ and identity of >50%. Additional BLASTp analysis of EhV genes highly similar to *E. huxleyi*

CCMP1516 host genes and those similar to other putative genes in other taxa discovered during the "Best BLAST hit" analysis (see Sections 3.5 and 3.6) were performed within NCBI against the non-redundant protein sequence database that includes viruses (Tables S1–S10).

3. Results and Discussion

3.1. Phylogenetic Relationships of Coccolithoviruses

Phylogenetic analysis of available DNA polymerase gene sequences, one of the commonly used phylogenetic markers for the dsDNA viral kingdom, indicated that the EhVs cluster into two major groups: A and B (Figure 1). The division into these groups was primarily due to a longer DNA polymerase gene present in EhV-18, EhV-156 and EhV-202 (caused by an insert of 12 bp, which results in the addition of glycine, threonine and two prolines at the end of the coding region). Group A was divided into three sub-clusters, where cluster A1 represents strains isolated primarily from the English Channel in 1999, A2 represents strains isolated primarily from the English Channel in 2001, and A3 represents a single strain isolated in 1999 from a Norwegian fjord. As such, these sub-clusters appear to be correlated with both year of isolation and location of origin. However, several exceptions were identified: EhV-145 and EhV-164 also clustered with A1 but were isolated after 1999. EhV-202, isolated in 2001 from the English Channel, clustered in B, with EhV-156 and EhV-18 (due to the 12 bp insertion mentioned previously), rather than A2. To further test these EhV groupings, phylogenetic analysis was also performed using the less conserved serine palmitoyltransferase (SPT) gene. This revealed similar groupings, with the exception of EhV-99B1 which clustered with A1.

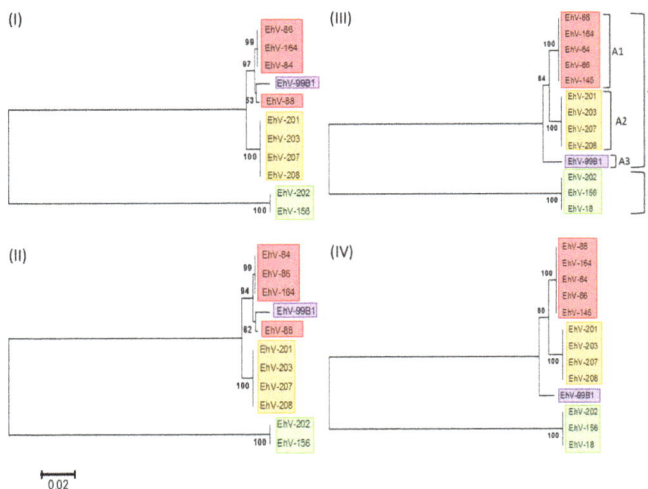

Figure 1. Phylogenetic analysis of coccolithoviruses based on their DNA polymerase and serine palmitoyltransferase (SPT) genes. The evolutionary history of 13 EhV strains was inferred based on the 2604 bp long SPT (I and II) and 2921 bp long DNA polymerase (III and IV) genes, using the Neighbor-Joining (I and III) and Maximum Likelihood (II and IV) methods. Note that EhV-18 and EhV-145 are absent from the serine palmitoyltransferase tree due to the full length SPT protein being split over two separate genes in their respective genomes. Based on the DNA polymerase phylogeny, the EhVs cluster into two main clades: A and B (green). Clade A is further divided into sub-clusters A1 (red), A2 (yellow), and A3 (purple). The percentage of replicate trees in which the associated taxa clustered together in the bootstrap test (1000 replicates) are shown next to the branches. The evolutionary distances were computed using the Tamura-Nei method and are in the units of the number of base substitutions per site.

The phylogenetic clades were also confirmed by the ANI analysis, in which the closest relative to EhV-86 based on the bidirectional pairwise alignment of the total number of genes was EhV-164 (placed in sub-clade A1), whereas those the furthest away were EhVs in clade B (Table 2).

Interestingly, significant differences have been observed in the infection dynamics of representative EhVs belonging to each phylogenetic clade identified in this study (Figure 1), both in respect to their lytic period [13] and their ability to utilize their host for the production of virally encoded glycosphingolipids, which are crucial for the successful demise of host cells [67].

3.2. Homology, Heterology and Genome Structure

Whole genome alignment revealed that the EhV genomes were syntenic with the exception of a ~80,000 bp section located in the middle of each genome (Figure 2) enriched with early transcription promoter elements [1,20,68]. As expected [37], this hyper-variable region was characterized by numerous inversions, rearrangements and base pair substitutions. One possibility is that the observed variation in this region is a result of the gaps in the draft genomes between the different contigs (Figure 2). Although this may be the case for some of the draft genomes (i.e., EhV-18, EhV-156, EhV-202, and EhV-208), where the highest number of gaps are in that region, it is not for others. For example, EhV-201 and EhV-99B1 sequences only have 2–3 gaps in that region, yet still exhibit variability. Further, many gaps are located in areas outside this region and show little variation; the draft genome of EhV-145 has 10 gaps in the section that ranges from ~300,000 bp to ~397,000 bp, yet variation in this section is minimal. It is therefore likely that the early transcriptional profile of these viruses can differ considerably.

The misalignment of different sections within the hyper-variable region (evident by numerous crossing lines between the co-linear blocks in Figure 2) becomes more prominent the further a genome is from those in cluster A1 (Figure 2). This is consistent with the DNA polymerase based phylogeny (Figure 1), in which viruses belonging to group B are more distant from those in group A1. Considered together, the genome alignments (Figure 2), and ANI analysis (Table 2) support the clustering of EhVs into the aforementioned groups and is consistent with DNA polymerase and SPT based phylogenies.

3.3. Genome Size and Putative Gene Differences

Previous pulsed-field gel electrophoresis (PFGE) of coccolithovirus genomes indicates that they have an average size of 410 kb [2]. In our study, the estimated length of the EhV genomes ranged from 376,759 to 421,891 bp (with a GC content of 39.94% to 40.49%), suggesting considerable variability in EhV genome size. For example, EhV-99B1 appears significantly shorter than the rest of the genomes (Figure 2). The genomes also differed in the number of predicted protein coding sequences (CDSs). They contained between 444 and 548 CDSs, from which an average of 90 had functional prediction, including three to six transfer RNA (tRNA) genes (Table 3). Due to the highly repetitive nature of the hyper-variable region, some of the variations observed could be attributed to sequencing errors, misalignments introduced during the original genome assembly, and the numerous gaps between the different contigs. However, these variations may also reflect real differences, as it is unlikely that regions as big as 45,132 bp long (the difference between the longest and shortest genome) were incorrectly sequenced or not included in the assemblies. These variations may have real consequences on the virus life cycle, affecting the time for viral genome synthesis within infected cells and, to an extent, the size of the capsid into which the DNA will be packed. As viruses exploit the resources of the host cell to synthesize new, infectious viral particles, these inherent differences may ultimately affect the number of viable viruses produced during infection.

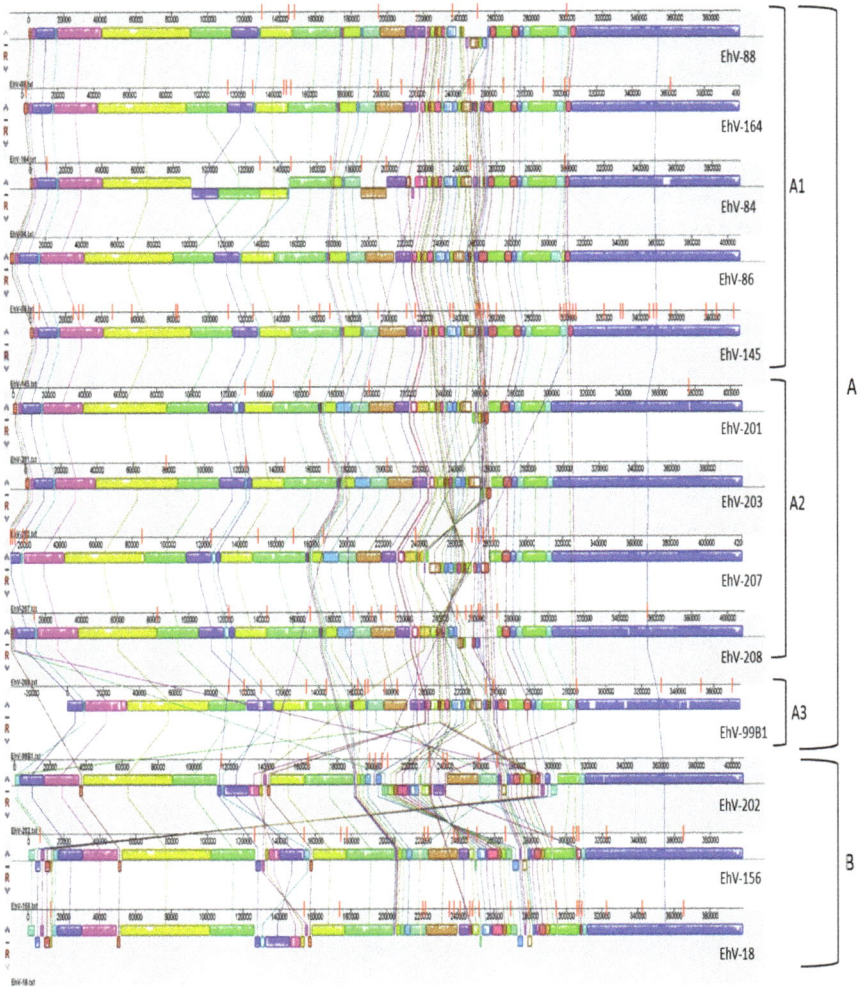

Figure 2. Whole genome alignment of sequenced coccolithovirus genomes. The genomes were aligned using MAUVE, in relation to the non-gapped backbone genome of EhV-86. Syntenous blocks are indicated in the same colours and the lines that connect them indicate the position of each block in relation to the same block of genes in the genome of EhV-86. The small red lines on each genome represent the exact positions of the gaps that separate the different contigs within each draft genome. The genomes are ordered based on their DNA polymerase phylogeny (Figure 1), based on the ANI analysis of this study (Table 2), and based on previously published microarray data that puts them into the aforementioned groups and sub-clades [20].

Table 3. Predicted genomic characteristics of sequenced coccolithoviruses. The statistics for each genome were obtained from the annotated ordered genomes uploaded into the Integrated Microbial Genomes—Expert Review (IMG/ER) online genome analysis pipeline [66]. Note that the numbers of genes, bases, coding sequences (CDSs), coding bases and transfer RNAs (tRNAs) here are underestimates (except for EhV-86) due to incomplete genome sequences.

Genome Name	Genes	Total Bases	CDS	Coding Bases	Genes with Function Prediction	tRNAs	GC (%)	Number of Gaps in Genome
E. huxleyi virus 84	486	396620	482	334463	85	4	40.17	8
E. huxleyi virus 86	478	407339	472	369157	90	5	40.18	0
E. huxleyi virus 88	480	397298	475	357803	90	5	40.18	7
E. huxleyi virus 201	457	407301	451	363714	89	6	40.46	6
E. huxleyi virus 202	488	407516	485	352215	93	3	40.3	11
E. huxleyi virus 203	470	400520	464	364178	91	6	40.12	5
E. huxleyi virus 207	479	421891	473	371313	93	6	40.49	15
E. huxleyi virus 208	461	411003	455	348386	90	6	40.42	16
E. huxleyi virus 99B1	451	376759	444	333400	90	6	40.04	16
E. huxleyi virus 18	508	399651	503	346161	91	5	40.49	21
E. huxleyi virus 145	552	397508	548	350414	103	4	39.94	41
E. huxleyi virus 156	498	399344	493	351083	88	5	40.47	19
E. huxleyi virus 164	514	400675	510	354290	95	4	40.11	17

Although strain-specific differences in the number and composition of transfer RNAs (tRNAs) encoded within the genomes were identified, the tRNAs for Arg, Asn and Gln were common to all (Table 4). Similarly, Pagarete et al. [44] report differences in the type of tRNAs encoded by EhV-86 and EhV-99B1. However, after an in-depth study of the different codon frequencies of the genomes of EhV-86 and EhV-99B1, they concluded that the difference in tRNAs does not correspond to the respective codon frequencies in each genome and that it is more likely that strain-specific tRNA differences represent adaptations to different host strains. If so, viruses with a larger number of tRNAs may have an advantage in terms of host range and translation within different host genotypes. Allen et al. [20] observe (in the absence of statistically robust data) that out of 23 different host strains tested, EhV-207 (which codes for six tRNAs) infects 10, whereas EhV-86 (which codes for five tRNAs) infects eight. Moreover, EhV-207 outcompetes EhV-86 when placed in direct competition over a single host genotype and has a higher rate of virus particle production [13]. The extent to which tRNAs are responsible for these observations has yet to be determined.

Most predicted CDSs were shared between all EhV genomes but a number of strain-specific CDSs were detected (Table 5). EhV-84 and EhV-202 contain the largest number of potentially "unique" genes (27 and 18, respectively). This is intriguing, as the other genomes in clade A1 and B (to which EhV-84 and EhV-202 belong respectively) do not exhibit a large number of "unique" genes (Table 5).

Table 4. Genes predicted to encode tRNAs in the genomes of 13 coccolithoviruses. Their presence (+) in each genome is indicated, and in grey shaded cells are those tRNAs common to all genomes. The phylogenetic clade of each genome (based on their DNA polymerase gene) is indicated above the column headers (see Figure 1). It is important to note that some of these tRNAs may still be present in the genomes of some EhVs in the unsequenced parts between the different contigs.

Phylogenetic Group		A											B	
		A1			A2						A3			
tRNA	EhV-84	EhV-86	EhV-88	EhV-164	EhV-145	EhV-201	EhV-203	EhV-207	EhV-208	EhV-99B1	EhV-18	EhV-156	EhV-202
Arg	+	+	+	+	+	+	+	+	+	+	+	+	+
Asn	+	+	+	+	+	+	+	+	+	+	+	+	+
Gln	+	+	+	+	+	+	+	+	+	+	+	+	+
Glu	+		+	+	+				+	+			
Ile		+	+	+						+			
Leu		+		+		+	+	+	+	+	+	+	
Lys		+				+	+	+	+	+	+	+	

Table 5. Coccolithovirus strain-specific CDSs that are not shared among the different viruses and are "unique" to each strain. Analysis was done using the "build in BLASTp" algorithm in IMG/ER [61] using a maximum E-value of 1×10^{-5} and a minimum % identity of 30. The phylogenetic clades regrouping genomes are shown above the column headers (see Figure 1).

Phylogenetic Group	A											B	
	A1			A2						A3		B	
Predicted CDS	EhV-84	EhV-86	EhV-88	EhV-145	EhV-164	EhV-201	EhV-203	EhV-207	EhV-208	EhV-99B1	EhV-202	EhV-18	EhV-156
hypothetical protein	27	3	4	7	8	4	4	9	9	6	18	9	6
putative endonuclease										2			
putative membrane protein		2								4			
putative transposase										1			
putative DUF814 domain containing protein										1			
zinc finger protein							1						
putative ribonuclease												1	
glycosyltransferase family 29 (sialyltransferase)												1	
Total	27	5	4	8	9	4	5	9	9	15	18	11	8

The observed strain-specific genomic differences within members of the same clade (Tables 3–5) highlight the limitations of virus taxonomy using a single gene such as DNA polymerase, when one tries to infer upon the evolutionary connections of these large viruses, the origin of certain genes, and potentially their functional implications. These limitations also become apparent in large-scale phylogenetic studies that include numerous taxa across the domains of life. Almost two decades ago, Villarreal and DeFilippis [69] suggested an alternative hypothesis for the flow of genetic material, with an emphasis on the virus DNA polymerase. Based on a large number of hits of virus DNA polymerases to other taxa, they suggest that algal viruses did not acquire replication genes from their eukaryotic hosts but vice versa [69]. We performed a similar analysis using the DNA polymerase gene of EhV-86 against the non-redundant protein sequence database in NCBI. The closest hit was to a DNA polymerase of the Yellowstone lake phycodnavirus 1 (reference sequence: YP_009174732.1), with a query cover of 81%, E-value of 9×10^{-137} and identity of 35% (Table S1), other *Phycodnaviruses*, and other organisms such as *Klebsormidium laccidum* (which belongs to a genus of filamentous charophyte green algae).

3.4. Coccolithovirus Gene Similarities to the Emiliania huxleyi Host

The prevailing dogma is that many EhV genes were acquired from *Emiliania huxleyi* through HGT. BLASTp analysis of EhV genomes revealed that, on average, ~25 EhV protein coding sequences were highly similar to counterparts in the host genome of *E. huxleyi* CCMP1516 (consistent with previous analysis of this host genome [66]). Unfortunately, most of these genes have no assigned function yet. Those with a predicted function include a group of genes that encode a de novo sphingolipid biosynthesis pathway (i.e., serine palmitoyltransferase [45%], sterol desaturase [42%], transmembrane fatty acid elongation protein [56%], and lipid phosphate phosphatase [29%]). BLASTp analysis of the EhV serine palmitoyltransferase gene (which encodes the rate-limiting enzyme in this pathway) showed that the most similar genes in the non-redundant protein sequence database of NCBI are putative serine palmitoyltransferases in *E. huxleyi*, *Chrysochromulina* sp. and *Perkinsus marinus* (Table S2). These genes are essential for the progression of infection through the production of virus-derived glycosphingolipids [28] and induction of PCD pathways within infected cells [17], and are present in all EhV genomes analysed. As previously reported [49], the two protein subunits of SPT (LCB1 and LCB2) are encoded by two separate genes in the genomes of EhV-18 and EhV-145. This separation over two genes may reflect the ancestral form prior to the fusion of the two subunits into a single gene (as seen in all other EhVs). The functional implications of this during infection are so far unknown. Nevertheless, the acquisition of this near complete pathway represents a classical example of HGT of a virus with its host, and illustrates a finely tuned co-evolutionary relationship between a host and its virus. In this case, it is manifested by a control for sphingolipid biosynthesis [29].

With the exception of EhV-99B1, EhV-202, EhV-18 and EhV-156, the CDS with the highest identity (i.e., 86.4%) to a similar gene in *E. huxleyi* CCMP1516 is a putative phosphate permease transporter (denoted as *ehv117* in EhV-86). This transporter is encoded by all EhV isolates from the English Channel and the Scottish coast, but absent in the Norwegian isolate EhV-99B1 ([44]; and this study) and the partially sequenced genome of the Norwegian isolate EhV-163 (not included in this study) [43]. Instead, a 75-bp scar remnant of the transporter gene is still present at the 3′ end of this gene in both Norwegian isolates, and is replaced by a putative endonuclease [43]. This indicates that the Norwegian fjord EhVs once possessed the gene [44] and have since lost it. Such genes are common in *E. huxleyi*, which has six inorganic phosphate transporters and an alkaline phosphatase that enable it to thrive in low phosphorous conditions [66]. Phosphorus is also essential for successful viral replication in most host-virus systems [70–72]. However, we do not currently know if the *ehv117* gene product is used to enhance phosphate acquisition from the external environment during infection. Its replacement with an endonuclease encoding gene in EhV-99B1 is intriguing, as this may function to provide additional phosphate via utilization of internal cellular resources, rather than acquisition of external phosphate via transporter activity. This could reflect the selective pressure imposed by particular geographic locations that are characterized by specific physicochemical conditions. For instance, the frequency of phosphate uptake genes in metagenomic datasets is higher in locations where the average phosphate concentration is lower (e.g., North Atlantic Gyre) [42,73]. Like SPT, it is most similar to a homologue in *E. huxleyi* (Table S3) and was likely acquired by HGT with its host. Other high hits in NCBI include similar genes in *Ostreococcus lucimarinus* and *Ectocarpus siliculosus*. Similarly, a phosphate transporter gene (i.e., *pho4*) was identified in the viruses infecting *Ostreococcus* and it was proposed that this was also a result of HGT [42].

For EhVs where the *ehv117* transporter gene is absent, the CDS with the highest percentage similarity to a CDS in *E. huxleyi* CCMP1516 varies. In EhV-99B1, it is a predicted ribonucleoside-diphosphate reductase protein (RNR) (71.9% similar) whereas in EhV-202, EhV-18, and EhV-156 it is a predicted polyubiquitin protein (94.7% similar). While the former is an enzyme crucial for the conversion of ribonucleotides to deoxyribonucleotides for DNA synthesis, the latter is involved in protein degradation. We cannot confirm if these genes are also a result of HGT with their host, as there were many top hits in the NCBI non-redundant protein sequence database to other taxa, including the haptophyte *Chrysochromulina*, *Chitinophaga niabensis* bacteria, and the Gram-negative bacteria *Cnuella takakiae*, in the case of RNR (Table S4), and radiolarians such as *Collozoum inerme*, and protists such as *Vitrella brassicaformis*, in the case of polyubiquitin (Table S5).

3.5. Coccolithovirus Gene Similarities to Other Eukaryotes

As previously mentioned, the current consensus is that many EhV genes are a result of HGT between coccolithoviruses and their specific host [39]. If this is true, it demonstrates that these viruses are entwined within the molecular workings of their hosts and have managed to acquire genes that target and coerce core aspects of their cellular metabolism. Alternatively, many EhV genes (including some of those mentioned in Section 3.4) may have an alternative origin. Thirty-three percent of the total EhV gene hits in the "Best BLAST hit" analysis (which excluded viruses) matched Eukaryotes other than *Isochrysidales* (the order to which *E. huxleyi* belongs) (Figure 3). These genes included those involved in nucleotide transport and metabolism, replication, recombination and repair, and transcription (Table 6). Among these other Eukaryotes are phylogenetically distant organisms such as moulds, fungi, unicellular flagellated protozoa, centric diatoms, amoeba, and green algae. Most of these can be found in the planktonic realm where *E. huxleyi* also dwells, particularly diatoms whose development is often tightly linked to *E. huxleyi* growth in ecological successions [74]. In this study, some of the top EhV gene hits against other genera include DNA-directed RNA polymerase subunit-B (an enzyme that produces mainly RNA transcripts) and DNA ligase (an enzyme that facilitates the joining of DNA strands) (Table 6). The former had top hits against members of the genus *Dictyostelium*, a group of soil living amoeba, whereas the latter had top hits against *Micromonas*, a group of chlorophyte green algae (Tables S6 and S7).

It is also plausible that EhVs have, or have had in their evolutionary past, alternative hosts that have enabled them to facilitate HGT with other taxa. Some giant algal viruses have the capacity to alternate between hosts belonging to different genera. For example, viruses capable of infecting *Haptolina hirta* infect also *Prymnesium kappa* and vice versa; [75]). Alternatively, there may have been a retroviral stage during the evolution of the *E. huxleyi*-virus system. Several EhV genes thought to have originated via HGT with *E. huxleyi* (including those for the de novo sphingolipid biosynthesis) lack introns, unlike their counterparts in the host genome. This suggests the possible involvement of coccolithophore-infecting RNA viruses in mediating HGT, although such viruses have not been observed yet.

Table 6. "Best BLAST hit" analysis of EhVs against the Bacterial and Eukaryotic domains of life, as well as against the known host *Emiliania huxleyi*. The top EhV genes with function predictions are shown based on the "Best BLAST hit" analysis using a BitScore of >100 and are ordered based on their E-value. The COG (Clusters of Orthologous Groups of genes with similar functions across the domains of life) cluster for each gene was identified on IMG/ER, where the different letters represent metabolic pathways involved in nucleotide transport and metabolism [F]; carbohydrate transport and metabolism [G]; general function prediction only [R]; replication, recombination and repair [L]; lipid transport and metabolism [I]; transcription [K]; inorganic ion transport and metabolism [P]; intracellular trafficking, and secretion and vesicular transport [U].

Ehv Predicted Gene	COG ID	Identity (%)	E-Value	Bit-Score	Phylum	Class	Order	Genus
Major Hits against Eukarya								
DNA-directed RNA polymerase subunit B	K	40.39	0	825	NA	NA	Dictyosteliida	Dictyostelium
DNA ligase	L	51.29	0	612	Chlorophyta	Mamiellophyceae	Mamiellales	Micromonas
DNA topoisomerase	L	33	0	586	Microsporidia	NA	NA	Nematocida
DNA-dependent RNA polymerase II largest subunit	K	34.09	7×10^{-171}	555	Ascomycota	Sordariomycetes	Xylariales	Eutypa
thymidylate synthase	F	49.7	1×10^{-168}	497	NA	NA	Peronosporales	Phytophthora
DNA polymerase delta catalytic subunit	L	35.11	2×10^{-132}	436	Arthropoda	Malacostraca	Decapoda	Procambarus
DNA helicase	L	49.3	3×10^{-137}	419	Bacillariophyta	Coscinodiscophyceae	Thalassiosirales	Thalassiosira
Major Hits against Bacteria								
deoxycytidylate deaminase	F	62.96	8×10^{-64}	206	Firmicutes	Clostridia	Clostridiales	Clostridium
Sialidase	G	30.03	2×10^{-37}	148	Bacteroidetes	Sphingobacteriia	Sphingobacteriales	Sphingobacterium
DNA-binding protein	R	34.36	7×10^{-32}	129	Verrucomicrobia	NA	NA	NA
endonuclease	L	45.67	4×10^{-32}	121	Proteobacteria	Alphaproteobacteria	Rhizobiales	Pseudochrobactrum
fatty acid desaturase	I	34.76	2×10^{-28}	121	Proteobacteria	Alphaproteobacteria	Rhodospirillales	Niveispirillum

NA: NCBI classification was not available for Phylum, Class, Order or Genus.

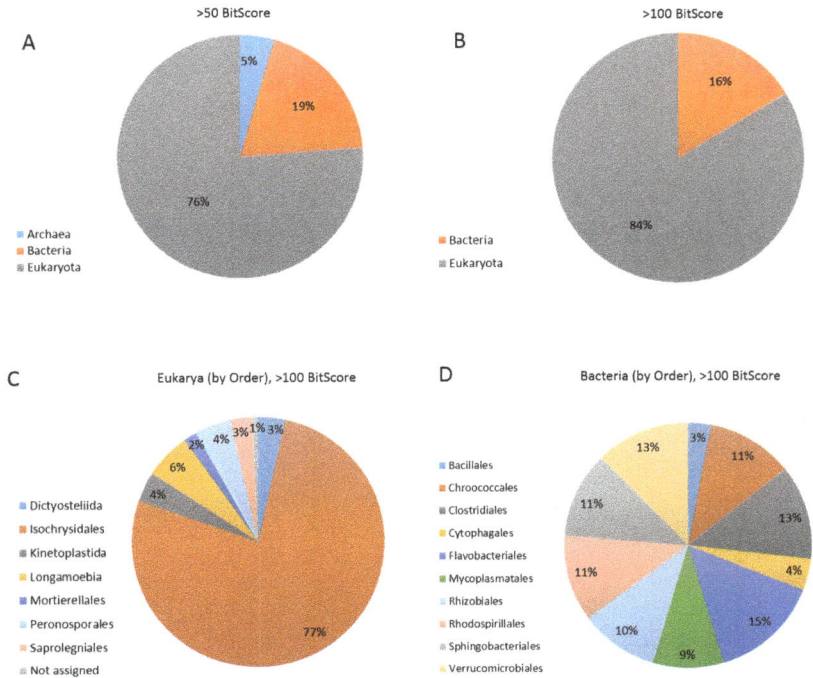

Figure 3. "Best BLAST hit" analysis of coccolithovirus CDSs in relation to the three domains of life: Eukarya, Bacteria and Archaea. Predicted genes within EhV genomes were BLASTp analyzed against possible hits in the three domains of life using a gene BitScore of >50 (**A**); and >100 (**B**). Further EhV gene hits analysis to the taxonomic level of "order" in Eukarya (**C**); and Bacteria (**D**) was performed using a BitScore of >100.

3.6. Coccolithovirus Gene Similarities to Bacteria

There are close ecological relationships between *E. huxleyi* and some bacterial groups. A previous study reports that *E. huxleyi* blooms in the North Atlantic are dominated by *Roseobacter*, SAR86, SAR11, and other *Alpha* and *Gammaproteobacteria* [76]. In their study, bacteria belonging to the *Roseobacter* genus and SAR86 and SAR11 clades account, together, for >50% of the bacterial rDNA in surface waters, whereas a cyanobacterium and members of the *Alphaproteobacteria* are associated with chlorophyll a-rich waters in the euphotic zone (0–50 m in depth), the typical niche at which coccolithophores are found [76]. A link between coccolithophores and bacteria is also found in cultures of *E. huxleyi* and *Coccolithus pelagicus*, which are enriched in hydrocarbon-degrading bacteria belonging to *Marinobacter* and *Marivita* [77]. These bacteria can colonize exopolymeric substances exuded by microalgae, but the ecological role of this phenomenon is unknown [78]. This tight evolutionary coexistence involving *E. huxleyi*, its viral community, and associated bacteria may have resulted in HGT across the domains.

In our more conservative "Best BLAST hit" analysis (using a bit-score of >100) we saw that 16% of the EhV genes also shared strong similarities to bacterial genes (Figure 3), involved in nucleotide, lipid, and carbohydrate transport and metabolism; and replication, recombination and repair (Table 6). Interestingly, almost half of the bacterial-like genes in the Chlorella virus NY2A, and a fifth of those in Mimivirus were proposed to be involved in DNA replication and repair [55,56]. In our analysis, top hits to bacteria include genes that code for sialidase, deoxycytidylate deaminase and fatty acid desaturase. These hits were against members of the genus *Clostridium*,

Sphingobacterium and *Niveispirillum* respectively (Table 6, and Tables S8–S10). Bacteria from these genera are consistently found in all coccolithophore cultures studied by Green et al. [77]. The hits against the *Sphingobacteriales* are particularly intriguing as members of this order are characterized by the presence of cellular lipid components that are comprised of high concentrations of sphingophospholipids [79]. The rate limiting step in sphingolipid biosynthesis in *S. multivorum* is serine palmitoyltransferase (SPT) [80], as in the *E. huxleyi*—virus system [81]. Although the EhV SPT is more similar to the host gene (as indicated above), the closest crystal structure in the PDB database to the EhV SPT is *Sphingobacterium multivorum* [67]. This structure was used to model the EhV derived SPT enzyme and to infer its catalytic capabilities during infection [36,67].

Moreover, in *S. multivorum*, sialidase specifically hydrolyzes deaminated neuraminic acid linkages and is localized in the periplasm [82,83]. Viruses such as influenza, adenoviruses, and polyomaviruses [84,85], can use virus-derived sialidase to target host sialic acid glycosphingolipids located on the host membrane. A similar process could be fundamental for EhV host recognition and entry, notably because susceptible *E. huxleyi* strains are enriched in glycosphingolipids (GSLs) with a sialic acid-modified glycosyl head-group [12,14]. These GSLs are proposed as a target for hydrolysis by EhV-encoded sialidases or as a ligand for the attachment of EhV lectin proteins [23]. Indeed, virus-derived sialidase transcripts are detected during EhV-86 infection 6–24 h post infection [21].

Surprisingly, we observed that the putative sialidase gene is severely truncated (up to 95% of its extension) in some EhV strains. Its full form is present in 1999 isolates (as well as EhV-164 and EhV-145), whereas those isolated in 2001 (as well as in EhV-18 and EhV-156) have a truncated form. In the genomes with the truncated form, the reduction is from a full gene of 1122 bp to 180 bp. Further reduction is seen in EhV-18 and EhV-156 where only a 60 bp fragment of the original gene remains. To date, EhV infectivity data do not show an association between the lack of a full length sialidase gene and a reduced ability to infect *E. huxleyi*. Hence, the selection against this gene in many EhVs and the fact that infection still occurs (presumably alongside other viruses able to utilize the active gene), suggests that its presence is not essential. Previous studies of Influenza A viruses show that a lack of functional sialidase is not essential to all viruses and that some viruses can adapt and grow in its absence in tissue cultures, mice, and embryonic eggs [86].

The role of two other bacterial-like genes, fatty acid desaturase and deoxycytidylate deaminase (Table 6) identified in EhVs requires further investigation. Fatty acid desaturases catalyze the desaturation of fatty acids. During EhV infection, a significant remodeling of the fatty acid profile of *E. huxleyi* occurs and this gene most likely plays an important role [30,87]. Deoxycytidylate deaminase is an enzyme involved in the biosynthesis of deoxyribonucleoside triphosphates (dNTPs). Its role, however, during infection remains elusive, as its expression has not been confirmed [1].

3.7. Possible Mode for HGT with Bacteria

Bacterial homologues in the genomes of nucleocytoplasmic large DNA viruses (NCLDVs) have been previously reported [55,56]. To our knowledge, our study is the first time that this is specifically identified in coccolithoviruses. While it is possible that some of the aforementioned genes were incorporated in the EhV genomes before the evolutionary divergence of eukaryotes and prokaryotes (a detailed "Best BLAST hit" analysis against ancestors of *E. huxleyi* such as *Tisochrysis lutea* and *Prymnesium parvum*, as well as against other primitive lineages of bacteria should reveal that), other explanations are also plausible. HGT could have occurred in situations where the genetic material of EhVs and bacteria co-occurred during virus replication. It is possible that EhVs and bacteria share, at moments, the same intracellular environment within *E. huxleyi*, particularly in light of recent data highlighting the importance of mixotrophy in phytoplankton [88]. For instance, in another NCLDV (i.e., *Cafeteria roenbergensis* virus), a 38-kb genomic region was suggested to be bacterial in origin and possibly acquired during an infection of its microflagellated grazer host that often contains a phagocytosed bacteria in its cytoplasm [89]. While there is no direct evidence for the engulfment of bacteria by *E. huxleyi*, a recent study of the different life stages of this alga shows that both diploid calcifying and motile haploid cells can engulf microbeads as big as 0.5 μm in diameter [90]. Also, a high abundance of transcripts linked to the digestive apparatus and those related to endocytosis are detected in diploid *E. huxleyi* cells, suggestive of an inherent ability of coccolithophores to participate in mixotrophy [90]. Endocytotic vesicles may be an additional mechanism for EhV infection [22], and *E. huxleyi* feeding on cohabiting bacteria might sometimes be infected by lytic viruses. Such mixotrophic-viral infection events may facilitate HGT across the domains.

Whether these events have occurred multiple times during which individual bacterial genes from different co-habiting bacteria were acquired by EhVs throughout their evolutionary history, or there have been only a few events during which large "chunks" of bacterial DNA (comprising of several adjacent genes, possibly sharing a similar function) were incorporated into EhVs, is unknown. However, the EhV genes with the highest bit-scores to Bacteria in our study were not genomically-located near each other in the analyzed genomes and did not appear to hit the same bacterial taxa. This is therefore consistent with the occurrence of multiple independent gene transfer events.

4. Conclusions

In light of the increasing body of literature that focuses on high-throughput metagenomics data and direct sequencing of virus genomes from environmental samples, it is important to remember the utility of model host-virus systems and single genome analysis. Here, we provided a review of the genomic features of 13 coccolithovirus strains and highlighted the differences and similarities in their respective genomes. We showed that the classification of coccolithoviruses using whole genome analysis mimics that of conserved gene based phylogeny. However, this approach does not allow the elucidation of functionally important differences among closely related viruses, and a detailed gene-by-gene examination may sometimes be more suitable. In addition, although many coccolithovirus genes were most similar to their known host and other eukaryotes, we highlighted the close and often ignored relationships that microalgae and their viruses have with bacteria in the marine environment. We revealed that for EhVs, as for other NCLDVs, HGT does not necessarily follow the traditionally expected pattern of host association. This analysis, performed across the three domains of life, contributes to some central questions in current viral ecology and evolution: (1) what defines a viral gene? (2) Where do viral genes come from? (3) are viruses able to serve as gene transfer agents across the domains of life? The existence of clear links to both Eukarya and Bacteria in this giant virus genomic assembly enhances the idea that viruses play a central role in the transfer of genetic information among the different life domains. Unwinding the relationships within the viral "domain", understanding how it came to be, and how it has and continues to shape the three modern domains that we can easily recognize today remains a challenge.

Supplementary Materials: The following are available online at http://www.mdpi.com/1999-4915/9/3/52/s1, Table S1: BLASTp analysis of the EhV DNA polymerase gene against the non-redundant protein database on NCBI, Table S2: BLASTp analysis of the EhV DNA serine palmitoyltranferase gene against the non-redundant protein database on NCBI, Table S3: BLASTp analysis of the EhV phosphate permease gene against the non-redundant protein database on NCBI, Table S4: BLASTp analysis of the EhV ribonucleoside-diphosphate reductase gene against the non-redundant protein database on NCBI, Table S5: BLASTp analysis of the EhV DNA polyubiquitin gene against the non-redundant protein database on NCBI, Table S6: BLASTp analysis of the EhV DNA-directed RNA polymerase subunit B gene against the non-redundant protein database on NCBI, Table S7: BLASTp analysis of the EhV DNA ligase gene against the non-redundant protein database on NCBI, Table S8: BLASTp analysis of the EhV deoxycytidylate deaminase gene against the non-redundant protein database on NCBI, Table S9: BLASTp analysis of the EhV sialidase gene against the non-redundant protein database on NCBI, Table S10: BLASTp analysis of the EhV fatty acid desaturase gene against the non-redundant protein database on NCBI.

Acknowledgments: This work was funded by the NERC Oceans 2025 program, Plymouth Marine Laboratory's Research Program, and a NERC PhD grant awarded to J.I.N. supervised by M.J.A. and S.A.K. at Plymouth Marine Laboratory. We would also like to thank the reviewers in pointing us towards some crucial literature and suggesting significant improvements to this manuscript, as well as Maeve Eason Hubbard for proof-reading the final version of this manuscript.

Author Contributions: J.I.N. isolated originally some of the EhVs in this paper, prepared them for sequencing, analyzed the data, did the genome comparisons, and wrote the initial version of the manuscript; A.P. provided insight into the genomic characteristics of EhV-99B1 and advised on some of the evolutionary aspects of these viruses; F.M., S.C, and D.D.D. conceived the idea and conducted the "Best BLAST hit" analysis, as well as helped in writing the manuscript, S.A.K. provided insight on the ecological aspects of these viruses, and M.J.A. obtained the funding for the work as well as provided insight into the molecular biology and evolution of these viruses. All authors provided critical input into the design, writing, and editorial revisions of the final manuscript.

Conflicts of Interest: The authors declare no conflict of interest.

References

1. Wilson, W.H.; Schroeder, D.C.; Allen, M.J.; Holden, M.T.G.; Parkhill, J.; Barrell, B.G.; Churcher, C.; Hamlin, N.; Mungall, K.; Norbertczak, H.; et al. Complete genome sequence and lytic phase transcription profile of a Coccolithovirus. *Science* **2005**, *309*, 1090–1092. [CrossRef] [PubMed]
2. Schroeder, D.C.; Oke, J.; Malin, G.; Wilson, W.H. Coccolithovirus (Phycodnaviridae): Characterisation of a new large dsDNA algal virus that infects *Emiliana huxleyi*. *Arch. Virol.* **2002**, *147*, 1685–1698. [CrossRef] [PubMed]
3. Wilson, W.H.; Tarran, G.A.; Schroeder, D.; Cox, M.J.; Oke, J.; Malin, G. Isolation of viruses responsible for the demise of an *Emiliania huxleyi* bloom in the English Channel. *J. Mar. Biol. Assoc.* **2002**, *82*, 369–377. [CrossRef]
4. Allen, M.J.; Schroeder, D.C.; Holden, M.T.G.; Wilson, W.H. Evolutionary history of the Coccolithoviridae. *Mol. Biol. Evol.* **2006**, *23*, 86–92. [CrossRef] [PubMed]
5. Dunigan, D.D.; Fitzgerald, L.A.; Van Etten, J.L. Phycodnaviruses: A peek at genetic diversity. *Virus Res.* **2006**, *117*, 119–132. [CrossRef] [PubMed]
6. Wilson, W.H.; Van Etten, J.L.; Allen, M.J. The Phycodnaviridae: The story of how tiny giants rule the world. *Curr. Top. Microbiol. Immunol.* **2009**, *328*, 1–42. [PubMed]
7. Martínez, J.M.; Schroeder, D.C.; Larsen, A.; Bratbak, G.; Wilson, W.H. Molecular dynamics of *Emiliania huxleyi* and cooccurring viruses during two separate mesocosm studies. *Appl. Environ. Microbiol.* **2007**, *73*, 554–562. [CrossRef] [PubMed]
8. Martínez, J.M.; Schroeder, D.C.; Wilson, W.H. Dynamics and genotypic composition of *Emiliania huxleyi* and their co-occurring viruses during a coccolithophore bloom in the North Sea. *FEMS Microbiol. Ecol.* **2012**, *81*, 315–323. [CrossRef] [PubMed]
9. Pagarete, A.; Allen, M.J.; Wilson, W.H.; Kimmance, S.A.; de Vargas, C. Host-virus shift of the sphingolipid pathway along an *Emiliania huxleyi* bloom: Survival of the fattest. *Environ. Microbiol.* **2009**, *11*, 2840–2848. [CrossRef] [PubMed]
10. Vardi, A.; Haramaty, L.; Van Mooy, B.A.S.; Fredricks, H.F.; Kimmance, S.A.; Larsen, A.; Bidle, K.D. Host-virus dynamics and subcellular controls of cell fate in a natural coccolithophore population. *Proc. Natl. Acad. Sci. USA* **2012**, *109*, 19327–19332. [CrossRef] [PubMed]
11. Kimmance, S.A.; Allen, M.J.; Pagarete, A.; Martínez Martínez, J.; Wilson, W.H. Reduction in photosystem II efficiency during a virus-controlled *Emiliania huxleyi* bloom. *Mar. Ecol. Prog. Ser.* **2014**, *495*, 65–76. [CrossRef]
12. Fulton, J.M.; Fredricks, H.F.; Bidle, K.D.; Vardi, A.; Kendrick, B.J.; DiTullio, G.R.; Van Mooy, B.A.S. Novel molecular determinants of viral susceptibility and resistance in the lipidome of *Emiliania huxleyi*. *Environ. Microbiol.* **2014**, *16*, 1137–1149. [CrossRef] [PubMed]
13. Nissimov, J.I.; Napier, J.A.; Allen, M.J.; Kimmance, S.A. Intragenus competition between coccolithoviruses: An insight on how a select few can come to dominate many. *Environ. Microbiol.* **2015**, *18*, 133–145. [CrossRef] [PubMed]
14. Hunter, J.E.; Frada, M.J.; Fredricks, H.F.; Vardi, A.; Van Mooy, B.A.S. Targeted and untargeted lipidomics of *Emiliania huxleyi* viral infection and life cycle phases highlights molecular biomarkers of infection, susceptibility, and ploidy. *Front. Mar. Sci.* **2015**, *2*. [CrossRef]
15. Schroeder, D.C.; Oke, J.; Hall, M.; Malin, G.; Wilson, W.H. Virus succession observed during an *Emiliania huxleyi* bloom. *Appl. Environ. Microbiol.* **2003**, *69*, 2484–2490. [CrossRef] [PubMed]
16. Lehahn, Y.; Koren, I.; Schatz, D.; Frada, M.; Sheyn, U.; Boss, E.; Efrati, S.; Rudich, Y.; Trainic, M.; Sharoni, S.; et al. Decoupling Physical from Biological Processes to Assess the Impact of Viruses on a Mesoscale Algal Bloom. *Curr. Biol.* **2014**, *24*, 2041–2046. [CrossRef] [PubMed]

17. Bidle, K.D. The Molecular Ecophysiology of Programmed Cell Death in Marine Phytoplankton. *Ann. Rev. Mar. Sci.* **2015**, *7*, 341–375. [CrossRef] [PubMed]

18. Allen, M.J.; Lanzén, A.; Bratbak, G. Characterisation of the coccolithovirus intein. *Mar. Genomics* **2011**, *4*, 1–7. [CrossRef] [PubMed]

19. Allen, M.J.; Wilson, W.H. The coccolithovirus microarray: An array of uses. *Brief. Funct. Genomic. Proteom.* **2006**, *5*, 273–279. [CrossRef] [PubMed]

20. Allen, M.J.; Martinez-Martinez, J.; Schroeder, D.C.; Somerfield, P.J.; Wilson, W.H. Use of microarrays to assess viral diversity: from genotype to phenotype. *Environ. Microbiol.* **2007**, *9*, 971–982. [CrossRef] [PubMed]

21. Kegel, J.U.; Blaxter, M.; Allen, M.J.; Metfies, K.; Wilson, W.H.; Valentin, K. Transcriptional host–virus interaction of *Emiliania huxleyi* (Haptophyceae) and EhV-86 deduced from combined analysis of expressed sequence tags and microarrays. *Eur. J. Phycol.* **2010**, *45*, 1–12. [CrossRef]

22. Mackinder, L.C.M.; Worthy, C.A.; Biggi, G.; Hall, M.; Ryan, K.P.; Varsani, A.; Harper, G.M.; Wilson, W.H.; Brownlee, C.; Schroeder, D.C. A unicellular algal virus, *Emiliania huxleyi* virus 86, exploits an animal-like infection strategy. *J. Gen. Virol.* **2009**, *90*, 2306–2316. [CrossRef] [PubMed]

23. Rose, S.L.; Fulton, J.M.; Brown, C.M.; Natale, F.; Van Mooy, B.A.S.; Bidle, K.D. Isolation and characterization of lipid rafts in *Emiliania huxleyi*: A role for membrane microdomains in host-virus interactions. *Environ. Microbiol.* **2014**, *16*, 1150–1166. [CrossRef] [PubMed]

24. Evans, C.; Malin, G.; Mills, G.P.; Wilson, W.H. Viral Infection of *Emiliania huxleyi* (Prymnesiophyceae) Leads to Elevated Production of Reactive Oxygen Species. *J. Phycol.* **2006**, *42*, 1040–1047. [CrossRef]

25. Bidle, K.D.; Haramaty, L.; Barcelos, E.; Ramos, J.; Falkowski, P. Viral activation and recruitment of metacaspases in the unicellular coccolithophore, *Emiliania huxleyi*. *Proc. Natl. Acad. Sci. USA* **2007**, *104*, 6049–6054. [CrossRef] [PubMed]

26. Bidle, K.D.; Kwityn, C.J. Assessing the Role of Caspase Activity and Metacaspase Expression on Viral Susceptibility of the Coccolithophore, *Emiliania Huxleyi* (Haptophyta). *J. Phycol.* **2012**, *48*, 1079–1089. [CrossRef] [PubMed]

27. Schatz, D.; Shemi, A.; Rosenwasser, S.; Sabanay, H.; Wolf, S.G.; Ben-Dor, S.; Vardi, A. Hijacking of an autophagy-like process is critical for the life cycle of a DNA virus infecting oceanic algal blooms. *New Phytol.* **2014**, *204*, 854–863. [CrossRef] [PubMed]

28. Vardi, A.; Van Mooy, B.A.S.; Fredricks, H.F.; Popendorf, K.J.; Ossolinski, J.E.; Haramaty, L.; Bidle, K.D. Viral glycosphingolipids induce lytic infection and cell death in marine phytoplankton. *Science* **2009**, *326*, 861–865. [CrossRef] [PubMed]

29. Bidle, K.D.; Vardi, A. A chemical arms race at sea mediates algal host-virus interactions. *Curr. Opin. Microbiol.* **2011**, *14*, 449–457. [CrossRef] [PubMed]

30. Rosenwasser, S.; Mausz, M.A.; Schatz, D.; Sheyn, U.; Malitsky, S.; Aharoni, A.; Weinstock, E.; Tzfadia, O.; Ben-Dor, S.; Feldmesser, E.; et al. Rewiring Host Lipid Metabolism by Large Viruses Determines the Fate of *Emiliania huxleyi*, a Bloom-Forming Alga in the Ocean. *Plant Cell* **2014**, *26*, 2689–2707. [CrossRef] [PubMed]

31. Ziv, C.; Malitsky, S.; Othman, A.; Ben-Dor, S.; Wei, Y.; Zheng, S.; Aharoni, A.; Hornemann, T.; Vardi, A. Viral serine palmitoyltransferase induces metabolic switch in sphingolipid biosynthesis and is required for infection of a marine alga. *Proc. Natl. Acad. Sci. USA* **2016**, *113*, E1907–E1916. [CrossRef] [PubMed]

32. Frada, M.J.; Schatz, D.; Farstey, V.; Ossolinski, J.E.; Sabanay, H.; Ben-Dor, S.; Koren, I.; Vardi, A. Zooplankton May Serve as Transmission Vectors for Viruses Infecting Algal Blooms in the Ocean. *Curr. Biol.* **2014**, *24*, 2592–2597. [CrossRef] [PubMed]

33. Sharoni, S.; Trainic, M.; Schatz, D.; Lehahn, Y.; Flores, M.J.; Bidle, K.D.; Ben-Dor, S.; Rudich, Y.; Koren, I.; Vardi, A. Infection of phytoplankton by aerosolized marine viruses. *Proc. Natl. Acad. Sci. USA* **2015**, *112*, 6643–6647. [CrossRef] [PubMed]

34. Coolen, M.J.L. 7000 years of *Emiliania huxleyi* viruses in the Black Sea. *Science* **2011**, *333*, 451–452. [CrossRef] [PubMed]

35. Rowe, J.M.; Fabre, M.-F.; Gobena, D.; Wilson, W.H.; Wilhelm, S.W. Application of the major capsid protein as a marker of the phylogenetic diversity of *Emiliania huxleyi* viruses. *FEMS Microbiol. Ecol.* **2011**, *76*, 373–380. [CrossRef] [PubMed]

36. Nissimov, J.I.; Jones, M.; Napier, J.A.; Munn, C.B.; Kimmance, S.A.; Allen, M.J. Functional inferences of environmental coccolithovirus biodiversity. *Virol. Sin.* **2013**, *28*, 291–302. [CrossRef] [PubMed]

37.   Pagarete, A.; Kusonmano, K.; Petersen, K.; Kimmance, S.A.; Martínez Martínez, J.; Wilson, W.H.; Hehemann, J.H.; Allen, M.J.; Sandaa, R.A. Dip in the gene pool: Metagenomic survey of natural coccolithovirus communities. *Virology* **2014**, *466–467*, 129–137. [CrossRef] [PubMed]

38.   Highfield, A.; Evans, C.; Walne, A.; Miller, P.I.; Schroeder, D.C. How many Coccolithovirus genotypes does it take to terminate an *Emiliania huxleyi* bloom? *Virology* **2014**, *466-467*, 138–145. [CrossRef] [PubMed]

39.   Monier, A.; Pagarete, A.; de Vargas, C.; Allen, M.J.; Read, B.; Claverie, J.; Ogata, H. Horizontal gene transfer of an entire metabolic pathway between a eukaryotic alga and its DNA virus. *Genome Res.* **2009**, *19*, 1441–1449. [CrossRef] [PubMed]

40.   Pagarete, A.; Le Corguillé, G.; Tiwari, B.; Ogata, H.; de Vargas, C.; Wilson, W.H.; Allen, M.J. Unveiling the transcriptional features associated with coccolithovirus infection of natural *Emiliania huxleyi* blooms. *FEMS Microbiol. Ecol.* **2011**, *78*, 555–564. [CrossRef] [PubMed]

41.   Lindell, D.; Sullivan, M.B.; Johnson, Z.I.; Tolonen, A.C.; Rohwer, F.; Chisholm, S.W. Transfer of photosynthesis genes to and from *Prochlorococcus* viruses. *Proc. Natl. Acad. Sci. USA* **2004**, *101*, 11013–11018. [CrossRef] [PubMed]

42.   Monier, A.; Welsh, R.M.; Gentemann, C.; Weinstock, G.; Sodergren, E.; Armbrust, E.V.; Eisen, J.A.; Worden, A.Z. Phosphate transporters in marine phytoplankton and their viruses: Cross-domain commonalities in viral-host gene exchanges. *Environ. Microbiol.* **2011**, *14*, 162–176. [CrossRef] [PubMed]

43.   Allen, M.J.; Schroeder, D.C.; Donkin, A.; Crawfurd, K.J.; Wilson, W.H. Genome comparison of two Coccolithoviruses. *Virol. J.* **2006**, *3*, 15. [CrossRef] [PubMed]

44.   Pagarete, A.; Lanzén, A.; Puntervoll, P.; Sandaa, R.A.; Larsen, A.; Larsen, J.B.; Allen, M.J.; Bratbak, G. Genomic sequence and analysis of EhV-99B1, a new coccolithovirus from the Norwegian fjords. *Intervirology* **2013**, *56*, 60–66. [CrossRef] [PubMed]

45.   Nissimov, J.I.; Worthy, C.A.; Rooks, P.; Napier, J.A.; Kimmance, S.A.; Henn, M.R.; Ogata, H.; Allen, M.J. Draft genome sequence of the coccolithovirus EhV-84. *Stand. Genomic Sci.* **2011**, *5*, 1–11. [CrossRef] [PubMed]

46.   Nissimov, J.I.; Worthy, C.A.; Rooks, P.; Napier, J.A.; Kimmance, S.A.; Henn, M.R.; Ogata, H.; Allen, M.J. Draft Genome Sequence of Four Coccolithoviruses: *Emiliania huxleyi* Virus EhV-88, EhV-201, EhV-207, and EhV-208. *J. Virol.* **2012**, *86*, 2896–2897. [CrossRef] [PubMed]

47.   Nissimov, J.I.; Worthy, C.A.; Rooks, P.; Napier, J.A.; Kimmance, S.A.; Henn, M.R.; Ogata, H.; Allen, M.J. Draft Genome Sequence of the Coccolithovirus *Emiliania huxleyi* Virus 202. *J. Virol.* **2012**, *86*, 2380–2381. [CrossRef] [PubMed]

48.   Nissimov, J.I.; Worthy, C.A.; Rooks, P.; Napier, J.A.; Kimmance, S.A.; Henn, M.R.; Ogata, H.; Allen, M.J. Draft genome sequence of the Coccolithovirus *Emiliania huxleyi* virus 203. *J. Virol.* **2011**, *85*, 13468–13469. [CrossRef] [PubMed]

49.   Nissimov, J.I.; Napier, J.A.; Kimmance, S.A.; Allen, M.J. Permanent draft genomes of four new coccolithoviruses: EhV-18, EhV-145, EhV-156 and EhV-164. *Mar. Genomics* **2014**, *15*, 7–8. [CrossRef] [PubMed]

50.   Azam, F.; Fenchel, T.; Field, J. The ecological role of water-column microbes in the sea. *Mar. Ecol. Prog. Ser.* **1983**, *10*, 257–263. [CrossRef]

51.   Azam, F.; Fandino, L.; Grossart, H.; Long, R. Microbial loop: Its significance in oceanic productivity and global change. *Rapp. Comrn. inl. Mer Médit.* **1998**, *35*, 1–3.

52.   Mayers, T.J.; Bramucci, A.R.; Yakimovich, K.M.; Case, R.J. A Bacterial Pathogen Displaying Temperature-Enhanced Virulence of the Microalga *Emiliania huxleyi*. *Front. Microbiol.* **2016**, *7*, 892. [CrossRef] [PubMed]

53.   Egan, S.; Fernandes, N.D.; Kumar, V.; Gardiner, M.; Thomas, T. Bacterial pathogens, virulence mechanism and host defence in marine macroalgae. *Environ. Microbiol.* **2014**, *16*, 925–938. [CrossRef] [PubMed]

54.   Lachnit, T.; Thomas, T.; Steinberg, P. Expanding our understanding of the seaweed holobiont: RNA Viruses of the Red Alga *Delisea pulchra*. *Front. Microbiol.* **2016**, *6*, 1–12. [CrossRef] [PubMed]

55.   Filée, J.; Siguier, P.; Chandler, M. I am what I eat and I eat what I am: acquisition of bacterial genes by giant viruses. *Trends Genet.* **2007**, *23*, 10–15. [CrossRef] [PubMed]

56.   Filée, J.; Pouget, N.; Chandler, M. Phylogenetic evidence for extensive lateral acquisition of cellular genes by Nucleocytoplasmic large DNA viruses. *BMC Evol. Biol.* **2008**, *8*, 320. [CrossRef] [PubMed]

57.   Tamura, K.; Stecher, G.; Peterson, D.; Filipski, A.; Kumar, S. MEGA6: Molecular Evolutionary Genetics Analysis version 6.0. *Mol. Biol. Evol.* **2013**, *30*, 2725–2729. [CrossRef] [PubMed]

58.   Saitou, N.; Nei, M. The Neighbor-joining Method: A New Method for Reconstructing Phylogenetic Trees. *Mol. Biol. Evol.* **1987**, *4*, 406–425. *Mol. Biol. Evol.* [PubMed]

59. Carver, T.J.; Rutherford, K.M.; Berriman, M.; Rajandream, M.-A.; Barrell, B.G.; Parkhill, J. ACT: The Artemis Comparison Tool. *Bioinformatics* **2005**, *21*, 3422–3423. [CrossRef] [PubMed]

60. Darling, A.E.; Mau, B.; Perna, N.T. progressiveMauve: Multiple genome alignment with gene gain, loss and rearrangement. *PLoS ONE* **2010**, *5*, e11147. [CrossRef] [PubMed]

61. Markowitz, V.M.; Mavromatis, K.; Ivanova, N.N.; Chen, I.-M.A.; Chu, K.; Kyrpides, N.C. IMG ER: A system for microbial genome annotation expert review and curation. *Bioinformatics* **2009**, *25*, 2271–2278. [CrossRef] [PubMed]

62. Konstantinidis, K.T.; Tiedje, J.M. Genomic insights that advance the species definition for prokaryotes. *Proc. Natl. Acad. Sci. USA* **2005**, *102*, 2567–2572. [CrossRef] [PubMed]

63. Klappenbach, J.A.; Goris, J.; Vandamme, P.; Coenye, T.; Konstantinidis, K.T.; Tiedje, J.M. DNA–DNA hybridization values and their relationship to whole-genome sequence similarities. *Int. J. Syst. Evol. Microbiol.* **2007**, *57*, 81–91.

64. Altschul, S.F.; Gish, W.; Miller, W.; Myers, E.W.; Lipman, D.J. Basic local alignment search tool. *J. Mol. Biol.* **1990**, *215*, 403–410. [CrossRef]

65. Louie, B.; Higdon, R.; Kolker, E. A statistical model of protein sequence similarity and function similarity reveals overly-specific function predictions. *PLoS ONE* **2009**, *4*, e7546. [CrossRef] [PubMed]

66. Read, B.A.; Kegel, J.; Klute, M.J.; Kuo, A.; Lefebvre, S.C.; Maumus, F.; Mayer, C.; Miller, J.; Monier, A.; Salamov, A.; et al. Pan genome of the phytoplankton *Emiliania* underpins its global distribution. *Nature* **2013**, *499*, 209–213. [CrossRef] [PubMed]

67. Nissimov, J.I. Competitive ecology of coccolithoviruses as revealed through biochemical diversity of serine palmitoyltransferase. **2017**, unpublished, in review.

68. Allen, M.J.; Schroeder, D.C.; Wilson, W.H. Preliminary characterisation of repeat families in the genome of EhV-86, a giant algal virus that infects the marine microalga *Emiliania huxleyi*. *Arch. Virol.* **2006**, *151*, 525–535. [CrossRef] [PubMed]

69. Villarreal, L.P.; DeFilippis, V.R. A hypothesis for DNA viruses as the origin of eukaryotic replication proteins. *J. Virol.* **2000**, *74*, 7079–7084. [CrossRef] [PubMed]

70. Maat, D.S.; Blok, R.; De Brussaard, C.P.D. Combined Phosphorus Limitation and Light Stress Prevent Viral Proliferation in the Phytoplankton Species *Phaeocystis globosa*, but Not in *Micromonas pusilla*. *Front. Micr.* **2016**, *3*. [CrossRef]

71. Wilson, W.H.; Carr, N.G.; Mann, N.H. The effect of phosphate status on the kinetics of cyanophage infection in the oceanic cyanobacterium *Synechococcus* sp. wh7803. *J. Phycol.* **1996**, *32*, 506–516. [CrossRef]

72. Bratbak, G.; Jacobsen, A.; Heldal, M.; Nagasaki, K.; Thingstad, F. Virus production in *Phaeocystis pouchetii* and its relation to host cell growth and nutrition. *Aquat. Microb. Ecol.* **1998**, *16*, 1–9. [CrossRef]

73. Coleman, M.L.; Chisholm, S.W. Ecosystem-specific selection pressures revealed through comparative population genomics. *Proc. Natl. Acad. Sci. USA* **2010**, *107*, 18634–18639. [CrossRef] [PubMed]

74. Balch, W.M. Re-evaluation of the physiological ecology of coccolithophores. In *Coccolithophores*; Springer: Berlin, Germany, 2004; pp. 165–190.

75. Johannessen, T.V.; Bratbak, G.; Larsen, A.; Ogata, H.; Egge, E.S.; Edvardsen, B.; Eikrem, W.; Sandaa, R.-A. Characterisation of three novel giant viruses reveals huge diversity among viruses infecting Prymnesiales (Haptophyta). *Virology* **2015**, *476*, 180–188. [CrossRef] [PubMed]

76. Gonza´lez, J.M.; Simo, R.; Massana, R.; Covert, J.S.; Casamayor, E.O.; Pedro, C.; Moran, M.A. Bacterial Community Structure Associated with a North Atlantic Algal Bloom. *Appl. Environ. Microb.* **2000**, *66*, 4237–4246. [CrossRef]

77. Green, D.H.; Echavarri-Bravo, V.; Brennan, D.; Hart, M.C.; Green, D.H.; Echavarri-Bravo, V.; Brennan, D.; Hart, M.C. Bacterial Diversity Associated with the Coccolithophorid Algae *Emiliania huxleyi* and *Coccolithus pelagicus f. braarudii. Biomed Res. Int.* **2015**, 1–15.

78. Carrias, J.-F.; Serre, J.-P.; Sime-Ngando, T.; Amblard, C. Distribution, size, and bacterial colonization of pico- and nano-detrital organic particles (DOP) in two lakes of different trophic status. *Limnol. Oceanogr.* **2002**, *47*, 1202–1209. [CrossRef]

79. Yabuuchi, E.; Kaneko, T.; Yano, I.; Moss, C.W.; Miyoshi, N. Sphingobacterium gen. nov., Sphingobacterium spiritivorum comb. nov., Sphingobacterium multivorum comb. nov., Sphingobacterium mizutae sp. nov., and Flavobacterium indologenes sp. nov.: Glucose-Nonfermenting Gram-Negative Rods in CDC Groups IIK-2 and IIb. *Int. J. Syst. Bacteriol.* **1983**, *33*, 580–598. [CrossRef]

80. Ikushiro, H.; Islam, M.M.; Tojo, H.; Hayashi, H. Molecular characterization of membrane-associated soluble serine palmitoyltransferases from *Sphingobacterium multivorum* and *Bdellovibrio stolpii*. *J. Bacteriol.* **2007**, *189*, 5749–5761. [CrossRef] [PubMed]

81. Han, G.; Gable, K.; Yan, L.; Allen, M.J.; Wilson, W.H.; Moitra, P.; Harmon, J.M.; Dunn, T.M. Expression of a novel marine viral single-chain serine palmitoyltransferase and construction of yeast and mammalian single-chain chimera. *J. Biol. Chem.* **2006**, *281*, 39935–39942. [CrossRef] [PubMed]

82. Kitajima, K.; Kuroyanagi, H.; Inoue, S.; Ye, J.; Troy, F.A.; Inoue, Y. Discovery of a new type of sialidase, "KDNase," which specifically hydrolyzes deaminoneuraminyl (3-deoxy-D-glycero-D-galacto-2-nonulosonic acid) but not N-acylneuraminyl linkages. *J. Biol. Chem.* **1994**, *269*, 21415–21419. [PubMed]

83. Nishino, S.; Kuroyanagi, H.; Terada, T.; Inoue, S.; Inoue, Y.; Troy, F.A.; Kitajima, K. Induction, localization, and purification of a novel sialidase, deaminoneuraminidase (KDNase), from *Sphingobacterium multivorum*. *J. Biol. Chem.* **1996**, *271*, 2909–2913. [CrossRef] [PubMed]

84. Stray, S.J.; Cummings, R.D.; Air, G.M. Influenza virus infection of desialylated cells. *Glycobiology* **2000**, *10*, 649–658. [CrossRef] [PubMed]

85. Neu, U.; Bauer, J.; Stehle, T. Viruses and sialic acids: rules of engagement. *Curr. Opin. Struct. Biol.* **2011**, *21*, 610–618. [CrossRef] [PubMed]

86. Hughes, M.T.; Matrosovich, M.; Rodgers, M.E.; McGregor, M.; Kawaoka, Y. Influenza A viruses lacking sialidase activity can undergo multiple cycles of replication in cell culture, eggs, or mice. *J. Virol.* **2000**, *74*, 5206–5212. [CrossRef] [PubMed]

87. Evans, C.; Pond, D.; Wilson, W. Changes in *Emiliania huxleyi* fatty acid profiles during infection with *E. huxleyi* virus 86: physiological and ecological implications. *Aquat. Microb. Ecol.* **2009**, *55*, 219–228. [CrossRef]

88. Stoecker, D.K.; Hansen, P.J.; Caron, D.A.; Mitra, A. Mixotrophy in the Marine Plankton. *Ann. Rev. Mar. Sci.* **2016**, *9*, 311–335. [CrossRef] [PubMed]

89. Fischer, M.G.; Allen, M.J.; Wilson, W.H.; Suttle, C.A. Giant virus with a remarkable complement of genes infects marine zooplankton. *Proc. Natl. Acad. Sci. USA* **2010**, *107*, 19508–19513. [CrossRef] [PubMed]

90. Rokitta, S.D.; de Nooijer, L.J.; Trimborn, S.; de Vargas, C.; Rost, B.; John, U. Transcriptome analyses reveal differential gene expression patterns between the life-cycle stages of *Emiliania huxleyi* (haptophyta) and reflect specialization to different ecological niches. *J. Phycol.* **2011**, *47*, 829–838. [CrossRef] [PubMed]

![viruses logo] *viruses*

MDPI

Article

Two *Synechococcus* genes, Two Different Effects on Cyanophage Infection

Ayalla Fedida and Debbie Lindell *

Faculty of Biology, Technion—Israel Institute of Technology, Haifa 32000, Israel; aya.shlos@gmail.com
* Correspondence: dlindell@tx.technion.ac.il; Tel.: +972-4829-5831

Academic Editors: Mathias Middelboe and Corina Brussaard
Received: 7 April 2017; Accepted: 23 May 2017; Published: 2 June 2017

Abstract: *Synechococcus* is an abundant marine cyanobacterium that significantly contributes to primary production. Lytic phages are thought to have a major impact on cyanobacterial population dynamics and evolution. Previously, an investigation of the transcriptional response of three *Synechococcus* strains to infection by the T4-like cyanomyovirus, Syn9, revealed that while the transcript levels of the vast majority of host genes declined soon after infection, those for some genes increased or remained stable. In order to assess the role of two such host-response genes during infection, we inactivated them in *Synechococcus* sp. strain WH8102. One gene, SYNW1659, encodes a domain of unknown function (DUF3387) that is associated with restriction enzymes. The second gene, SYNW1946, encodes a PIN-PhoH protein, of which the PIN domain is common in bacterial toxin-antitoxin systems. Neither of the inactivation mutations impacted host growth or the length of the Syn9 lytic cycle. However, the DUF3387 mutant supported significantly lower phage DNA replication and yield of phage progeny than the wild-type, suggesting that the product of this host gene aids phage production. The PIN-PhoH mutant, on the other hand, allowed for significantly higher Syn9 genomic DNA replication and progeny production, suggesting that this host gene plays a role in restraining the infection process. Our findings indicate that host-response genes play a functional role during infection and suggest that some function in an attempt at defense against the phage, while others are exploited by the phage for improved infection.

Keywords: cyanophage; marine *Synechococcus*; host-virus interactions; host defenses; stress-response genes; gene inactivation; burst-size; PIN-PhoH

1. Introduction

Marine unicellular cyanobacteria belonging to the genus *Synechococcus* are highly abundant in the oceans, where they play a major role in primary production and carbon fixation [1,2]. They are constantly exposed to infection by phages which impact their population dynamics by killing a fraction of the population on a daily basis (estimated to be between 0.005% and 30% daily) [3–5]. Cyanophages are also thought to greatly impact the diversity and evolution of their cyanobacterial hosts [6–12].

One abundant cyanophage group in the oceans comprises the T4-like cyanophages, tailed double-stranded DNA phages that resemble the T4 coliphage archetype, both in virion morphology and core gene content [13–15]. Syn9 is a representative of this group and has a relatively broad host range [4]. It infects multiple *Synechococcus* strains that belong to different phylogenetic clades, occupy different ecological niches, and differ in the gene content of their flexible genome [1,16].

Recently, we found that Syn9 underwent a near identical infection and transcriptional program in multiple *Synechococcus* hosts (*Synechococcus* sp. strains WH8102, ,WH8109, and WH7803), despite the above-mentioned differences [16]. In response to Syn9 infection, the transcript levels of the vast majority of host genes (>90%) in each of the three hosts declined significantly [16]. However, transcript levels of a small group of host genes increased or remained unchanged during the phage latent period, and are considered host-response genes [16]. While these genes belong to the same general function groups in the different hosts (cell envelope, DNA repair, carbon fixation, respiration, and nutrient utilization), the actual genes are highly host-specific, making up part of the flexible genome, with many located in hypervariable genomic islands in their respective hosts [16]. This phenomenon is not unique to infection by Syn9. Indeed, a similar response was found during the infection of *Prochlorococcus* MED4 by the T7-like cyanophage, P-SSP7 [8]. Furthermore, other bacteria also display the upregulation of a limited number of host-response genes after phage infection [17–20].

Little is known about the functional role of these host-response genes during the interaction with the infecting phage. Some of them may serve as host stress-response genes, while others may constitute a host attempt at defense against phage infection. Alternatively, they may be induced by the phage for its own needs. Here, we began testing these hypotheses by investigating the impact of the independent inactivation of two host-response genes in *Synechococcus* sp. strain WH8102 (referred to from here as *Synechococcus* WH8102) on the Syn9 infection process. We chose two genes that may be involved in mounting a host defense, seen by the presence of potential host defense-related domains according to homology-based annotation. Both of the genes are the first in two-gene operons and thus, the two genes in each operon may have related activities that function together.

The first two-gene operon is SYNW1659 and SYNW1658. The SYNW1659 gene consists of a domain of unknown function, DUF3387, that is often associated with restriction enzymes [16], a well-known mechanism of defense against phages [21], as well as with helicases, which is itself a common domain in restriction enzymes. This gene will be referred to as a DUF3387 gene from here on. The SYNW1658 gene consists of a different domain of unknown function (DUF1651) that is also found in other host-response genes in *Synechococcus* sp. strains WH8102 and WH8109 [16], in addition to *Prochlorococcus* sp. strain MED4 [8]. The transcript levels of these genes increased in response to infection by Syn9 in *Synechococcus* WH8102 [16].

The second two-gene operon may form a toxin-antitoxin module [22,23], which is also a known anti-phage defense mechanism [21,24,25]. The first gene in the operon, SYNW1946, contains a single-stranded RNA nuclease PIN domain [16], which is commonly found in toxins from bacterial toxin-antitoxin operons [22]. This gene also encodes a PhoH ATPase domain. This gene will be referred to as PIN-PhoH from here on. The second gene, SYNW1947, has a DNA binding domain which is a common feature of antitoxins [26]. The transcript levels of these genes remain unchanged for 1.5–3 h after Syn9 infection. All four of these genes are located in genomic islands that appear to be important in mediating the cyanobacterial response to phage infection [8,16].

We hypothesized that if these genes are defense related, their inactivation in the host would lead to a shortening of the infection cycle and/or an increase in phage progeny production. Here, we report that, indeed, the PIN-PhoH mutant produced more Syn9 progeny than the wild-type host. However, contrary to our expectations, the DUF3387 mutant produced a lower yield of phage progeny, suggesting that this two-gene operon is beneficial to the phage in the wild-type host.

2. Materials and Methods

2.1. Growth of Cultures

The *Synechococcus* sp. WH8102 wild-type and mutant strains were grown in artificial seawater medium (ASW) [27], with modifications as described in Lindell et al. [28]. The cultures were grown at 22 °C under cool white light with a 14:10 h light:dark cycle at an intensity of 30 μmol photon·m^{-2}·s^{-1} during the light period. These are the same culturing conditions as previously used [16], except that the culture volumes were 30 mL in this study (rather than 800 mL). Growth in liquid medium was monitored by measuring chlorophyll *a* autofluorescence as a proxy for biomass using a Turner Designs 10-AU flourometer (excitation/emission: 340–500/>665 nm) (Turner, San Jose, CA, USA) or the BioTek Synergy 2 microplate reader (excitation/emission: 440 ± 20/680 ± 20 nm) (BioTek, Winooski, VT, USA).

Growth on semi-solid medium to produce colonies was done using a pour plating method [29–31]. Cells were mixed with medium containing Invitrogen Ultra Pure low melting point (LMP) agarose (ThermoFisher Scientific, Waltham, MA, USA) at a final concentration of 0.28%, poured into plastic petri dishes, and grown under the conditions described above. An antibiotic resistant heterotrophic helper strain, *Alteromonas* sp. strain EZ80, was added to the cells for plating colonies after conjugation (see below), to ensure a high plating efficiency [32].

The Syn9 phage lysate was prepared by infecting large volumes of *Synechococcus* WH8102. After complete lysis of the culture, cell debris was removed by centrifugation (13,131× *g* at 21 °C for 15 min) and filtration over a 0.2 μm filter (Nalge Nunc international, Rochester, NY, USA). The filtered lysate was concentrated 100-fold using Centricon Plus 70 centrifugal filters (100 kDa NMWL, Millipore, Billerica, MA, USA), to enable the infection of cultures with small (negligible) volume additions of the phage lysate.

2.2. Insertional Inactivation of Synechococcus WH8102 Genes

The predicted function of the conserved domains of the genes for inactivation was determined from conserved domain searches using the NCBI Blast conserved domains database (CDD) search and Pfam [16].

Insertional inactivation of the first gene in each of the two operons was done following Brahamsha [29]. An internal 192 bp fragment of the SYNW1659 gene was amplified by polymerase chain reaction (PCR) from *Synechococcus* WH8102 with primers that contain a *Bam*HI restriction site (shown in italics): SYNW1659ia2_FW (5'-ATATAT*GGATCCC*CTGCTGATCTGGCGGGTATTTG-3') and SYNW1659ia2_Rv (5'-ATATAT*GGATCC*GCCTTGGCAGACAACCCGTC-3'), and was cloned into the *Bam*HI site on the pMUT100 cargo plasmid. Due to the small size of this gene, the primers were designed to introduce stop codons on both sides of the SYNW1659 gene. For inactivation of the SYNW1946 gene, an internal 350 bp fragment was PCR amplified from *Synechococcus* WH8102 using the following primers that also contain a *Bam*HI restriction site: SYNW1946ia_FW (5'-ATATAT*GGATCCC*CAGGCCCATGCTCTTGACGC-3') and SYNW1946ia_Rv (5'-ATATAT*GGATCC*AGCACCACGCCTTCATTTGC-3'), and was cloned into the pDS3 plasmid. The pMUT100 and pDS3 plasmids are derivatives of pBR322 that carry a kanamycin-resistance gene and can be mobilized into the *Synechococcus* WH8102, but cannot replicate in this host. pDS3 differs from pMUT100 in that the tetracycline gene was replaced with a chloramphenicol gene optimized for expression in *Prochlorococcus* [33]. The resulting plasmids were mobilized into *Synechococcus* WH8102 by conjugation using *Escherichia coli* MC1061, carrying the RP4 derivative conjugative plasmid pRK24 and the helper plasmid pRL528 as a donor [29]. Gene interruption occurs when the plasmid is integrated into the host chromosome by homologous recombination through a single crossover event. Exconjugants were selected for kanamycin resistance (25 μg·mL^{-1}) on semi-solid medium. Verification of the complete segregation of chromosomes in the mutant (i.e., the absence of an intact gene in all of the chromosome copies) was done by PCR using primers which flank the target gene: SYNW1659ia_Fw (5'-ATATATGGATCCTCGCCCAAGGTCTCTGCCTG-3') and SYNW1659ia_Rv (5'-ATATATGGATCCAGAGGAACTGGAGCGTGGCG-3') for SYNW1659 and IAver_02_1946_Fw (5'-GATGCCTTGCCGATGGTGTTC-3'), and IAver_02_1946_Rv (5'-GTTTCCTTGACGCCGGGCAAG-3') for SYNW1946. Verification that the plasmid was inserted at the desired location in the *Synechococcus* chromosome was done using one primer within the vector: pMUT_tet218F (5'-GCCCAGTCCTGCTCGCTTCG-3'), and one of the above verification primers within the chromosome external to the target gene [34].

2.3. Characterization of Infection Dynamics

One-step-growth curves of the Syn9 phage were carried out on exponentially growing cultures (30 mL) of each inactivation mutant, as well as the wild-type strain at the same cell concentration (~2 × 10^7 cells·mL^{-1}) without antibiotic selection. Syn9 was added to the cultures at a multiplicity of infection (MOI) of three infective phages per cell. For determination of the length of the latent period and lytic cycle, phage DNA in the extracellular medium was measured from samples collected every two hours from 0 to 12 h, as well as at 5 h after phage addition. For characterization of the replication of phage DNA inside the cell, intracellular phage DNA was measured from samples collected at 0, 0.5, 1, 2, 3, 4, 5, 6, and 8 h after phage addition.

2.4. Quantification of Intracellular and Extracellular Phage Genomic DNA

Intracellular and extracellular phage genomic DNA (gDNA) was quantified using quantitative real-time PCR (qPCR), as described previously [8]. Extracellular phage gDNA was determined from filtrates containing phage particles after filtration over a 0.2 μm Acrodisc Syringe Filter (Pall Corporation) and dilution 100-fold in 10 mM Tris pH 8. Aliquots of 10 μL were frozen at −80 °C in triplicate and used directly for qPCR assays (see below). Intracellular phage DNA was determined from cells collected on 0.2 μm pore-sized polycarbonate filters (GE Healthcare Life Sciences, Boston, MA, USA) by filtration at a vacuum pressure of 7–10 inch Hg. Filters were washed three times with sterile seawater, once with 3 mL preservation solution (10 mM Tris, 100 mM EDTA, 0.5 M NaCl; pH 8), and were frozen at −80 °C. The DNA was extracted from the cells using a heat lysis method [35]. The polycarbonate filter with the cells was immersed in 10 mM Tris pH 8, and agitated in a mini-bead beater for 2 min at 5000 rpm, without beads. The sample was removed from the shards of filter, heated at 95 °C for 15 min, and 10 μL was used in triplicate qPCR reactions.

2.5. Quantitative PCR Protocol

Assays for qPCR were carried out for the Syn9 portal protein gene ($g20$), as described previously [16]. Each qPCR reaction contained 1× Roche universal probe library (UPL) master mix (LightCycler® 480 Probes Master, Roche, Penzberg, Germany), 100 nM UPL84 hydrolysis probe (Roche), 200 nM of HPLC-purified primers (Syn9_gp20_UPL84_F: 5′-TCGTTTAGAAACAGAAACCACATTT-3′ and Syn9_gp20_UPL84_R: 5′-AACTTTTGGAATTTAACTTCGTCAC-3′), and a 10 μL template in a total volume of 25 μL. Reactions were carried out on a LightCycler 480 Real-Time PCR System (Roche). The cycling program consisted of an initial denaturation step of 95 °C for 15 min followed by 45 cycles of amplification, each including 10 s of denaturation at 95 °C, a 30 s combined annealing and elongation step at 60 °C, and a fluorescence plate read (Ex/Em 465/510 nm). The crossing point was used to determine the amount of initial target using the absolute quantification/2nd-derivative maximum analysis with the LightCycler 480 software (release 1.5.0) (Roche). For intracellular gDNA determination, standard curves were produced using a serial dilution of purified phage DNA of a known quantity. For extracellular gDNA determination, standard curves of phage particles in 10 mM Tris pH 8, that had been enumerated by epifluorescence microscopy after SYBR staining [36], were used.

2.6. Burst Size and Virulence Determination

Burst size and virulence assays were carried out as described by Kirzner et al. [37]. Exponentially growing cultures were diluted to the same concentration ($\approx 4 \times 10^7$ cells per mL) and infected with the Syn9 phage at MOI = 3 in the morning hours. At 4 h after infection, when maximal adsorption had occurred ($\approx 90\%$), but before the end of the latent period and the onset of cell lysis, the cultures were diluted 1000-fold and single cells were dispensed into individual wells of 96 well-plates using the FACSAria-IIIu cell sorter (Becton Dickenson, Franklin Lakes, NJ, USA). For the virulence assay, the cells were dispensed into wells containing the host culture (100 μL). The plates were incubated in growth conditions and chlorophyll *a* fluorescence was measured daily using the Synergy 2 microplate reader (BioTek). Lysis was determined by a significant decrease in fluorescence relative to the control plate containing only the host culture. The relative number of cleared wells in the infected versus the control plate is the percentage of cell lysis caused by the phage. For the burst size assay, cells were dispensed into individual wells containing only growth medium and incubated in growth conditions for 16–18 h after sorting. This allowed sufficient time for the completion of the infection cycle and the exit of all phage progeny. The contents of each well were then plated on a lawn of host cells and the number of plaques produced over a 10 day incubation period was monitored and indicated the number of progeny phages produced by that particular cell. Burst size was determined from plaque-containing plates (with more than one plaque) from four independent experiments for each strain. Each biological replicate consisted of phages arising from 60 to 96 cells each.

2.7. Statistical Analysis

In order to test the significance of the differences between the results obtained for the wild-type and mutant strains (for growth rate, virulence, average burst sizes, and phage gDNA assays), a two tailed *t*-test for independent samples was carried out. This was done after ensuring that the data were normally distributed ($p > 0.05$) using the Kolmogorov-Smirnov or Shapiro-Wilk tests. A repeated measure ANOVA was used to assess whether significant differences existed in the timing of different stages of the infection cycle during Syn9 infection of the mutants relative to the wild-type strains. Since there were significant differences in the level of genomic DNA replication, that data were normalized to maximal levels in each strain before testing for differences in timing. The PASW statistics 17 package was used for these analyses (Rel. 17.0.3. September 2009. Chicago, IL, USA: SPSS Inc.).

3. Results

In order to investigate the effect of host response genes on the infection cycle, we generated two independent *Synechococcus* WH8102 mutants by the insertion of an antibiotic-carrying plasmid into the gene of interest by a single crossover [29] (see Methods). This physically interrupts the gene, rendering it inactive. For two-gene operons, such as in both of our cases, this insertion is expected to also prevent transcription of the downstream gene as it becomes separated from the promoter by the plasmid. Thus, the results presented in this study for each mutant likely relate to the effective inactivation of both genes in the two-gene operons. For simplicity, however, we refer to the mutants by the name of the insertionally inactivated gene: DUF3387 for SYNW1659-SYNW1658 and PIN-PhoH for SYNW1946-SYNW1947.

Before investigating the effect of the insertional inactivation of the DUF3387 and PIN-PhoH genes on phage infection, we tested whether they affected the growth rate of the mutants under normal growth conditions. This was important since the efficiency of phage replication can be intimately linked to the growth rate of its host [8,38,39]. No significant differences were found between the growth rate of the mutants relative to the wild-type strain, nor were there differences in growth between the two mutants (Figure 1). Therefore, any differences observed in the Syn9 infection process in the two mutant strains relative to the wild-type strain cannot be attributed to intrinsic differences in host growth.

We began our investigation of the impact of the host mutations on the phage infection process by assessing phage virulence, as determined from the ability of the phage to infect and lyse the different host strains [37]. This was determined from the percentage of cells lysed by the Syn9 phage when infecting each of the inactivation mutants compared to infection of the wild-type *Synechococcus* strain. Virulence was not significantly different in either of the mutants relative to the wild-type strain ($n = 3$) and was approximately 70% for all three strains (Figure 2). This suggests that mutations in the DUF3387 and PIN-PhoH genes do not impact the ability of the phage to infect the *Synechococcus* host.

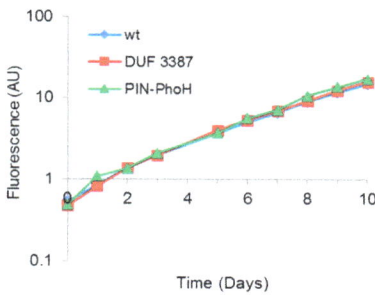

Figure 1. Growth of wild-type and mutant strains of *Synechococcus* WH8102. Representative growth curves are shown on the left and a table on the right presents the mean and standard deviation (S.D.) of the specific growth rate of four biological replicates. No significant differences were found in growth rates between the mutants (DUF3387 and PIN-PhoH) and the wild-type (wt) strains, nor between the two mutants.

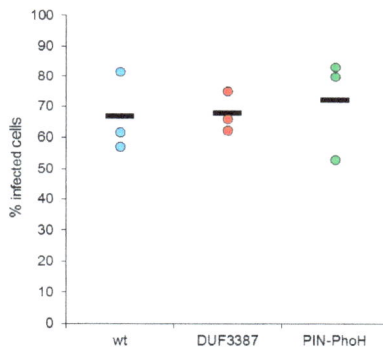

Figure 2. Virulence of the Syn9 cyanophage on wild-type and mutant strains of *Synechococcus* WH8102. The percentage of infected cells that were lysed in cultures of the two mutant strains (DUF3387 and PIN-PhoH) were compared to the wild-type (wt) strain. No significant differences were found. The bar denotes the mean of three biological replicates.

Next, we determined the effect of the host mutations on the length of the Syn9 lytic cycle. One-step-growth curves were carried out by determining the timing of phage release using a qPCR assay for the Syn9 portal protein gene (*g20*) in the extracellular medium. The length of the phage latent period was 5 h and the length of the lytic cycle was 8–10 h during infection of the two mutants, as well as during infection of the wild-type strain (Figure 3). These results are typical of previous findings for Syn9 on *Synechococcus* WH8102 [16]. Thus, the inactivation of the DUF3387 and PIN-PhoH genes did not affect the length of the phage infection cycle.

Following this, we asked whether the extent and timing of phage genome replication was altered during infection of the mutant strains. We analyzed phage genome replication, using the same qPCR assay as above, but on intracellular DNA extracted from infected cells. Here, the timing of phage DNA replication in the mutants was similar to that found in the wild-type host, beginning 1–2 h after phage addition (Figure 4a). However, clear differences in the number of phage genome copies were apparent for both mutants. Significantly more Syn9 phage genome copies were replicated in the PIN-PhoH mutant than in the wild-type strain ($p < 0.05$, $n = 6$) (Figure 4b). In contrast, the phage gDNA levels were significantly lower in the DUF3387 mutant relative to the wild-type strain ($p < 0.01$, $n = 6$) (Figure 4b).

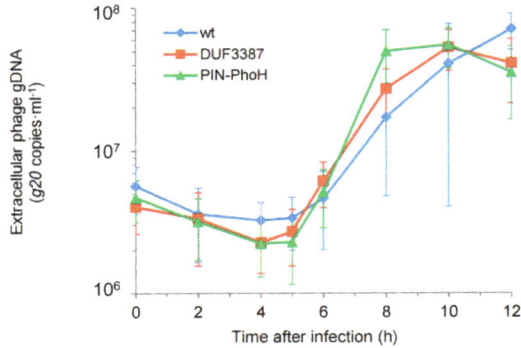

Figure 3. Infection dynamics of the Syn9 phage on wild-type and mutant strains of *Synechococcus* WH8102. One-step growth curves of the Syn9 phage were carried out to determine the length of the latent period and the lytic cycle during infection on the two mutant strains (DUF3387 and PIN-PhoH) and the wild-type (wt) strains. No differences were found in the timing of the infection cycle when comparing the Syn9 infection of the two mutant strains relative to its infection of the wild-type strain ($p = 0.289$ for DUF3387 and $p = 0.071$ for PIN-PhoH, each compared to the wt). Extracellular phage concentrations were determined from qPCR of the *g20* phage gene. Average and standard deviation of six biological replicates. gDNA: genomic DNA.

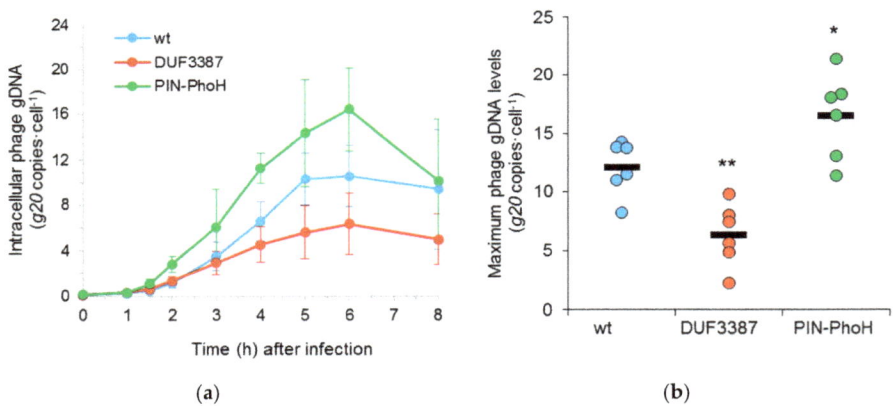

(a)   (b)

Figure 4. Intracellular phage gDNA replication during infection of wild-type and mutant strains of *Synechococcus* WH8102. (**a**) The timing and level of intracellular Syn9 genomic replication (determined by qPCR for the *g20* portal protein gene and normalized per cell) during infection of the two mutant (DUF3387 and PIN-PhoH) and wild-type (wt) strains. Mean and standard deviation of six biological replicates. No differences in the timing of DNA replication were found during the first 6 h of infection of the mutant strains relative to the wild-type strain ($p = 0.61$ for DUF3387 and $p = 0.125$ for PIN-PhoH, each compared to the wt). (**b**) Syn9 gDNA yield per host cell at the maximum amount of phage gDNA produced in that strain. The yield of Syn9 gDNA produced in the mutant strains was compared to the that for the wild-type strain. (* $p < 0.05$, ** $p < 0.01$). The bar denotes the mean of six biological replicates.

In order to assess whether these differences in phage genome replication translated into changes in phage fitness, we investigated the number of infective phages produced per cell using a single-cell burst size assay [37]. Similar to phage genome replication, the median burst size of the Syn9 phage on the PIN-PhoH mutant (79 phages·cell^{-1}) was significantly higher than on the wild-type host (52 phages·cell^{-1}) (p = 0.001, n = 189 cells for the PIN-PhoH mutant and 174 cells for the wild-type host) (Figure 5). In contrast, the median burst size of Syn9 on the DUF3387 mutant (35 phages·cell^{-1}) was significantly lower than that found for the wild type strain (Figure 5) (p < 0.001, n = 164 cells of the DUF3387 mutant). These findings suggest that the product of the PIN-PhoH gene serves the wild-type host during infection by restraining phage genome replication and phage progeny production. However, the lower phage gDNA levels and smaller burst size in the DUF3387 mutant suggests that, in this case, the host gene assists phage genome replication and progeny production when infecting the wild-type host.

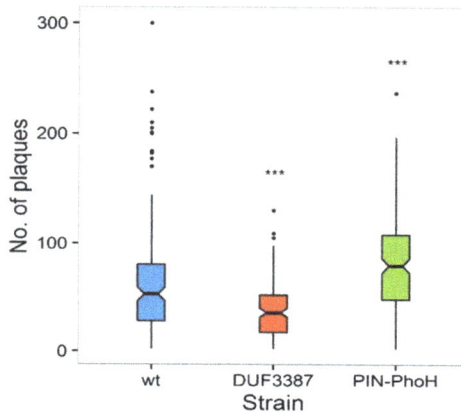

Figure 5. Distribution of the number of infective Syn9 phages produced per cell when infecting the wild-type and mutant strains of *Synechococcus* WH8102. Box plot of single cell burst sizes. Burst sizes were significantly lower on the DUF3387 mutant than on the wild-type (wt) strain (p < 0.001, n = 164 cells for DUF3387 mutant and 174 cells for the wild-type strain), but were significantly higher on the PIN-PhoH mutant than on the wild-type strain (p = 0.001, n = 189 cells for the PIN-PhoH mutant and 174 cells for the wild-type strain). The middle line of the box plot denotes the median burst size and the boxes surrounding the median correspond to the 25th (lower) and 75th (upper) percentiles. Outliers are plotted as individual points. *** $p \leq 0.001$.

Our findings indicate that the insertional inactivation of the two genes that responded transcriptionally to Syn9 infection in the wild-type strain [16] impacts phage fitness. This was manifested at the level of phage genome replication and the number of infective phage progeny produced per cell. However, these mutations had no effect on the ability of the phage to infect the host, nor did they impact the timing of the infection process.

4. Discussion

Over the past decade, a number of whole-genome transcriptional studies have shown that phage infection causes a discernable transcriptional response in different bacterial hosts [17–20,40,41], including in marine cyanobacteria [8,16]. Previously, though, it was thought that phage infection led to a complete and immediate shut-down of host transcription [42,43]. It has been suggested, from homology-based annotations, that some of these genes function as host defenses against infection, while others may be utilized by the phage to enhance reproduction [8,16,20,40,41]. However, the function of such host response genes during infection and their impact on the infection process has rarely been tested (but see [44,45]). Here, we show that at least two cyanobacterial response genes influence the phage infection process at the stage of phage genome replication and impact phage fitness.

The induction of such host-response genes could be the cyanobacterial cell's attempt at defense against phage infection. We initially hypothesized that this is the case for the two sets of genes investigated in this study since they have, or are associated with, domains found in known bacterial defense systems against phage infection. The potential toxin-antitoxin system in *Synechococcus* WH8102, as exemplified by the PIN-PhoH operon (SYNW1946 and SYNW1947), is an example in hand. Our findings for the PIN-PhoH mutant support this hypothesis as it produced significantly more phages than the wild-type strain and thus, this operon likely limits phage replication in the wild-type *Synechococcus* WH8102. Similar findings were reported for the P1 phage in a deletion mutant of the *mazEF* toxin-antitoxin system in *E. coli* [25] and the expression of the *hok/sok* system on a plasmid in *E. coli* led to a reduced burst size of T4 [24].

The vast majority of PIN-domain containing proteins are the toxic components of toxin-antitoxin operons in bacteria [22]. These PIN domains function as sequence specific single stranded RNases [46]. Recently, a PIN-PhoH protein was found to be the RNase toxin in a toxin-antitoxin system from *Myocbacterium tuberculosis* [23]. Thus, the PIN-PhoH protein in *Synechococcus* WH8102 may limit phage progeny production by acting as an RNase toxin which uses the energy provided from the hydrolysis of ATP by the PhoH ATPase domain to cleave phage RNA during infection. Since the mutant was found to impact phage genome replication, the RNA target of the protein may be mRNA needed to produce replication proteins or perhaps the RNA primers required for replication itself. It should be noted that these PIN-PhoH genes are distinct from the *phoH* genes found in many bacterial and phage genomes, including those of cyanobacteria and cyanophages.

If indeed the PIN-PhoH protein is a defense system in *Synechococcus* WH8102, it is ultimately unsuccessful against Syn9 as this phage kills the wild-type host, and no increase in the number of cells killed (its virulence) was observed in the mutant. It is possible that the Syn9 phage encodes genes that interfere with the activity of this toxin-antitoxin system, as is known for T4 [47,48], although no evidence currently exists to support this possibility in Syn9. Furthermore, it may be more successful in defense against other phages. A homology search found that homologues of the same two genes arranged in the same order are also found in *Synechococcus* WH8109 and CC9605, but these genes were not part of the host response gene repertoire during *Synechococcus* WH8109 infection by the Syn9 phage [16].

Unlike the PIN-PhoH operon, our results argue against the DUF3387 operon (SYNW1659 and SYNW1658 operon) being a host defense mechanism. The lower yield of phage progeny produced on the DUF3387 mutant than on the wild-type suggests that these genes facilitate and increase the phage yield in the wild-type *Synechococcus*. While we do not know the mechanism by which this gene enhances phage reproduction, the association of the DUF3387 domain with helicase genes (also beyond those found in restriction enzymes), coupled with the decrease in phage genome levels in the mutant, provides the intriguing possibility that this gene is directly involved in the process of DNA replication. Thus, this gene may be induced as part of the cyanobacterium's stress response and is exploited by the phage for its replication or is directly upregulated by the phage. Our findings do not allow us to discern between these two possibilities. However, it is well known that cellular stress response proteins are utilized by various phages for their replication in *E. coli* [49–52] .

These findings were initially surprising as the DUF3387 domain is associated with type I and type III restriction enzymes, well established as potent mechanisms of defense against phage infection that degrade unmodified phage DNA upon entry into the bacterium (see review in Labrie et al. [21]). However, other domains carry out the endonucleotyic activity of restriction enzymes [53]. Thus, this domain, whose function remains unknown, appears to aid the phage when it is disconnected from the endonucleolytic domain.

Alternatively, the phenotype of this mutant may not be related to the restriction enzyme associated domain, but to the adjacent gene, SYNW1658, which encodes a different domain of unknown function (DUF1651). Intriguingly, this protein domain is limited to, but widespread, in marine cyanobacteria among the bacteria, but is also found in a single known coliphage (the ECBP5 podovirus, NC_027330 [54]). Furthermore, five other DUF1651 domain containing genes in three different marine cyanobacteria are part of the host response to phage infection. These include two other *Synechococcus* WH8102 genes (SYNW2106 and SYNW1944) and a *Synechococcus* WH8109 gene (Syncc8109_0491) in response to Syn9 infection [16], as well as two genes in *Prochlorococcus* MED4 (PMM0684 and PMM0819), whose transcript levels increase in response to infection by the T7-like podovirus, P-SSP7 [8]. It thus appears likely that DUF1651 domain containing proteins are important for cyanophage during infection, and may be responsible for increasing the yield of infective cyanophage in multiple distinct marine cyanobacterial hosts. Why such a gene would be retained by cyanobacteria is unclear. Perhaps it is a stress response gene that provides an advantage to the cell when exposed to other stressors that the phage has evolved to utilize.

5. Conclusions

The findings presented here indicate that host-response genes play a functional role in the phage infection process. They further show that this functionality is not unidirectional: In certain cases, they are in service of the host as an attempt at defense against infection. In other cases, however, they can be exploited by the phage, even though their role in the host may well be a bona-fide response to the abiotic or biotic stressors that they are exposed to in the oceans. Since these genes are located in genomic islands and are often part of the flexible genome, these results continue to highlight the importance of such genomic regions and their gene content for host-phage interactions and the coevolutionary process between cyanobacteria and the phages that infect them.

Acknowledgments: We thank Irena Pekarsky, Gazalah Sabehi, and Sophia Zborowsky for their help with experimentation; Daniel Schwartz for the pDS3 plasmid and help with graphics; Sarit Avrani for help with statistics; Erik Zinser for providing the antibiotic resistant *Alteromonas* helper strain EZ80; and Lindell lab members for discussions. The research was funded by the European Research Council (Starting Grant 203406) to D.L.

Author Contributions: A.F. and D.L. conceived the project and designed the experiments; A.F. performed the experiments and analyzed the data; A.F. and D.L. wrote the paper.

Conflicts of Interest: The authors declare no conflict of interest.

References

1. Scanlan, D.J.; Ostrowski, M.; Mazard, S.; Dufresne, A.; Garczarek, L.; Hess, W.R.; Post, A.F.; Hagemann, M.; Paulsen, I.; Partensky, F. Ecological genomics of marine picocyanobacteria. *Microbiol. Mol. Biol. Rev.* **2009**, *73*, 249–299. [CrossRef] [PubMed]
2. Flombaum, P.; Gallegos, J.L.; Gordillo, R.A.; Rincón, J.; Zabala, L.L.; Jiao, N.; Karl, D.M.; Li, W.K.W.; Lomas, M.W.; Veneziano, D.; et al. Present and future global distributions of the marine cyanobacteria *Prochlorococcus* and *Synechococcus*. *Proc. Natl. Acad. Sci. USA* **2013**, *110*, 9824–9829. [CrossRef] [PubMed]
3. Proctor, L.M.; Fuhrman, J.A. Viral mortality of marine bacteria and cyanobacteria. *Nature* **1990**, *343*, 60–62. [CrossRef]
4. Waterbury, J.B.; Valois, F.W. Resistance to co-occurring phages enables marine *Synechococcus* communities to coexist with cyanophage abundant in seawater. *Appl. Environ. Microbiol.* **1993**, *59*, 3393–3399. [PubMed]
5. Suttle, C.A.; Chan, A.M. Dynamics and distribution of cyanophages and their effects on marine *Synechococcus* spp. *Appl. Environ. Microbiol.* **1994**, *60*, 3167–3174. [PubMed]
6. Lindell, D.; Sullivan, M.B.; Johnson, Z.I.; Tolonen, A.C.; Rohwer, F.; Chisholm, S.W. Transfer of photosynthesis genes to and from *Prochlorococcus* viruses. *Proc. Natl. Acad. Sci. USA* **2004**, *101*, 11013–11018. [CrossRef] [PubMed]
7. Zeidner, G.; Bielawski, J.P.; Shmoish, M.; Scanlan, D.J.; Sabehi, G.; Beja, O. Potential photosynthesis gene recombination between *Prochlorococcus* and *Synechococcus* via viral intermediates. *Environ. Microbiol.* **2005**, *7*, 1505–1513. [CrossRef] [PubMed]

8. Lindell, D.; Jaffe, J.D.; Coleman, M.L.; Futschik, M.E.; Axmann, I.M.; Rector, T.; Kettler, G.; Sullivan, M.B.; Steen, R.; Hess, W.R.; et al. Genome-wide expression dynamics of a marine virus and host reveal features of co-evolution. *Nature* **2007**, *449*, 83–86. [CrossRef] [PubMed]

9. Avrani, S.; Wurtzel, O.; Sharon, I.; Sorek, R.; Lindell, D. Genomic island variability facilitates *Prochlorococcus*-virus coexistence. *Nature* **2011**, *474*, 604–608. [CrossRef] [PubMed]

10. Marston, M.F.; Pierciey, F.J.; Shepard, A.; Gearin, G.; Qi, J.; Yandava, C.; Schuster, S.C.; Henn, M.R.; Martiny, J.B.H. Rapid diversification of coevolving marine *Synehcococcus* and a virus. *Proc. Natl. Acad. Sci. USA* **2012**, *109*, 4544–4549. [CrossRef] [PubMed]

11. Martiny, J.B.H.; Riemann, L.; Marston, M.F.; Middelboe, M. Antagonistic coevolution of marine planktonic viruses and their hosts. *Ann. Rev. Mar. Sci.* **2014**, *6*, 393–414. [CrossRef] [PubMed]

12. Avrani, S.; Lindell, D. Convergent evolution toward an improved growth rate and a reduced resistance range in *Prochlorococcus* strains resistant to phage. *Proc. Natl. Acad. Sci. USA* **2015**, *112*, E2191–E2200. [CrossRef] [PubMed]

13. Mann, N.H.; Clokie, M.R.; Millard, A.; Cook, A.; Wilson, W.H.; Wheatley, P.J.; Letarov, A.; Krisch, H.M. The genome of S-PM2, a "photosynthetic" T4-type bacteriophage that infects marine *Synechococcus*. *J. Bacteriol.* **2005**, *187*, 3188–3200. [CrossRef] [PubMed]

14. Sullivan, M.B.; Coleman, M.; Weigele, P.; Rohwer, F.; Chisholm, S.W. Three *Prochlorococcus* cyanophage genomes: Signature features and ecological interpretations. *PLoS Biol.* **2005**, *3*, 790–806. [CrossRef] [PubMed]

15. Weigele, P.R.; Pope, W.H.; Pedulla, M.L.; Houtz, J.M.; Smith, A.L.; Conway, J.F.; King, J.; Hatfull, G.F.; Lawrence, J.G.; Hendrix, R.W. Genomic and structural analysis of Syn9, a cyanophage infecting marine *Prochlorococcus* and *Synechococcus*. *Environ. Microbiol.* **2007**, *9*, 1675–1695. [CrossRef] [PubMed]

16. Doron, S.; Fedida, A.; Hernandez-Prieto, M.A.; Sabehi, G.; Karunker, I.; Stazic, D.; Feingersch, R.; Steglich, C.; Futschik, M.; Lindell, D.; et al. Transcriptome dynamics of a broad host-range cyanophage and its hosts. *ISME J.* **2016**, *10*, 1437–1455. [CrossRef] [PubMed]

17. Poranen, M.M.; Ravantti, J.J.; Grahn, A.M.; Gupta, R.; Auvinen, P.; Bamford, D.H. Global changes in cellular gene expression during bacteriophage PRD1 infection. *J. Virol.* **2006**, *80*, 8081–8088. [CrossRef] [PubMed]

18. Ravantti, J.J.; Ruokoranta, T.M.; Alapuranen, A.M.; Bamford, D.H. Global transcriptional responses of *Pseudomonas aeruginosa* to phage PRR1 infection. *J. Virol.* **2008**, *82*, 2324–2329. [CrossRef] [PubMed]

19. Lavigne, R.; Lecoutere, E.; Wagemans, J.; Cenens, W.; Aertsen, A.; Schoofs, L.; Landuyt, B.; Paeshuyse, J.; Scheer, M.; Schobert, M.; et al. A multifaceted study of *Pseudomonas aeruginosa* shutdown by virulent podovirus LUZ19. *MBio* **2013**, *4*, e00061-13. [CrossRef] [PubMed]

20. Mojardin, L.; Salas, M. Global transcriptional analysis of virus-host interactions between phage 29 and *Bacillus subtilis*. *J. Virol.* **2016**, *90*, 9293–9304. [CrossRef] [PubMed]

21. Labrie, S.J.; Samson, J.E.; Moineau, S. Bacteriophage resistance mechanisms. *Nat. Rev. Microbiol.* **2010**, *8*, 317–327. [CrossRef] [PubMed]

22. Anantharaman, V.; Aravind, L. New connections in the prokaryotic toxin-antitoxin network: Relationship with the eukaryotic nonsense-mediated RNA decay system. *Genome Biol.* **2003**, *4*, R81. [CrossRef] [PubMed]

23. Andrews, E.S.V.; Arcus, V.L. The mycobacterial PhoH2 proteins are type II toxin antitoxins coupled to RNA helicase domains. *Tuberculosis* **2015**, *95*, 385–394. [CrossRef] [PubMed]

24. Pecota, D.C.; Wood, T.K. Exclusion of T4 phage by the *hok/sok* killer locus from plasmid R1. *J. Bacteriol.* **1996**, *178*, 2044–2050. [CrossRef] [PubMed]

25. Hazan, R.; Engekberg-Kulka, H. *Escherichia coli mazEF*-mediated cell death as a defense mechanism that inhibits the spread of phage P1. *Mol. Genet. Genom.* **2004**, *272*, 227–234. [CrossRef] [PubMed]

26. Chan, W.T.; Espinosa, M.; Yeo, C.C. Keeping the wolves at bay: Antitoxins of prokaryotic type II toxin-antitoxin systems. *Front. Microbiol.* **2016**, *3*, 9. [CrossRef] [PubMed]

27. Wyman, M.; Gregory, R.P.F.; Carr, N.G. Novel role for phycoerythrin in a marine cyanobacterium, *Synechococcus* strain DC2. *Science* **1985**, *230*, 818–820. [CrossRef] [PubMed]

28. Lindell, D.; Padan, E.; Post, A.F. Regulation of *ntcA* expression and nitrite uptake in the marine *Synechococcus* sp. strain WH 7803. *J. Bacteriol.* **1998**, *180*, 1878–1886. [PubMed]

29. Brahamsha, B. A genetic manipulation system for oceanic cyanobacteria of the genus, *Synechococcus*. *Appl. Environ. Microbiol.* **1996**, *62*, 1747–1751. [PubMed]

30. Moore, L.R.; Coe, A.; Zinser, E.R.; Saito, M.A.; Sullivan, M.B.; Lindell, D.; Frois-Moniz, K.; Waterbury, J.B.; Chisholm, S.W. Culturing the marine cyanobacterium *Prochlorococcus*. *Limnol. Oceanogr. Methods* **2007**, *5*, 353–362. [CrossRef]

31. Lindell, D. The genus *Prochlorococcus*, phylum Cyanobacteria. In *The Prokaryotes*; Rosenberg, E., DeLong, E.S.E., Lory, S., Thompson, F.L., Eds.; Springer: Berlin, Germany, 2014.

32. Morris, J.J.; Kirkegaard, R.; Szul, M.J.; Johnson, J.I.; Zinser, E.R. Facilitation of robust growth of *Prochlorococcus* colonies and dilute liquid cultures by "helper" heterotrophic bacteria. *Appl. Environ. Microbiol.* **2008**, *74*, 4530–4534. [CrossRef] [PubMed]

33. Schwartz, D.A. *Generalized Transduction in Marine Cyanobacteria*; Technion—Israel Institute of Technology: Haifa, Israel, 2010.

34. Tetu, S.G.; Brahamsha, B.; Johnson, D.A.; Tai, V.; Phillippy, K.; Palenik, B.; Paulsen, I.T. Microarray analysis of phosphate regulation in the marine cyanobacterium *Synechococcus* sp. WH8102. *ISME J.* **2009**, *3*, 835–849. [CrossRef] [PubMed]

35. Zinser, E.R.; Coe, A.; Johnson, Z.I.; Martiny, A.C.; Fuller, N.J.; Scanlan, D.J.; Chisholm, S.W. *Prochlorococcus* ecotype abundances in the North Atlantic Ocean as revealed by an improved quantitative PCR method. *Appl. Environ. Microbiol.* **2006**, *72*, 723–732. [CrossRef] [PubMed]

36. Patel, A.; Noble, R.T.; Steele, J.A.; Schwalbach, M.S.; Hewson, I.; Fuhrman, J.A. Virus and prokaryote enumeration from planktonic aquatic environments by epifluorescnce microscopy with SYBR Green I. *Nat. Protoc.* **2007**, *2*, 269–276. [CrossRef] [PubMed]

37. Kirzner, S.; Barak, E.; Lindell, D. Variability in progeny production and virulence of cyanophages determined at the single-cell level. *Environ. Microbiol. Rep.* **2016**, *8*, 605–613. [CrossRef] [PubMed]

38. Wilson, W.H.; Carr, N.G.; Mann, N.H. The effect of phosphate status on the kinetics of cyanophage infection in the oceanic cyanobacterium *Synechococcus* sp. WH7803. *J. Phycol.* **1996**, *32*, 506–516. [CrossRef]

39. Hadas, H.; Einav, M.; Fishov, I.; Zaritsky, A. Bacteriophage T4 development depends on the physiology of its host *Escherichia coli*. *Microbiology* **1997**, *143*, 179–185. [CrossRef] [PubMed]

40. Howard-Varona, C.; Roux, S.; Dore, H.; Solonenko, N.E.; Holmfeldt, K.; Markillie, L.M.; Orr, G.; Sullivan, M.B. Regulation of infection efficiency in a globally abundant marine *Bacteriodetes* virus. *ISME J.* **2017**, *11*, 284–295. [CrossRef] [PubMed]

41. Lavysh, D.; Sokolova, M.; Slashcheva, M.; Forstner, K.U.; Severinov, K. Transcriptional profiling of *Bacillus subtilis* cells infected with AR9, a giant phage encoding two multisubunit RNA polymerases. *MBio* **2017**, *8*, e02041-16. [CrossRef] [PubMed]

42. Koerner, J.F.; Snustad, D.P. Shutoff of host macromolecular synthesis after T-even bacteriophage infection. *Microbiol. Rev.* **1979**, *43*, 199–223. [PubMed]

43. Roucourt, B.; Lavigne, R. The role of interactions between phage and bacterial proteins within the infected cell: A diverse and puzzling interactome. *Environ. Microbiol.* **2009**, *11*, 2789–2805. [CrossRef] [PubMed]

44. Wei, D.H.; Zhang, Z.B. Proteomic analysis of interactions between a deep-sea thermophilic bacteriophage and its host at high temperature. *J. Virol.* **2010**, *84*, 2365–2373. [CrossRef] [PubMed]

45. Chamakura, K.R.; Tran, J.S.; Young, R. MS2 lysis of *Escherichia coli* depends on host chaperone DnaJ. *J. Bacteriol.* **2017**. [CrossRef] [PubMed]

46. Clissold, P.M.; Ponting, C.P. PIN domains in nonsense-mediated mRNA decay and RNAi. *Curr. Biol.* **2000**, *10*, R888–R890. [CrossRef]

47. Otsuka, Y.; Koga, M.; Iwamato, A.; Yonesaki, T. A role of Rn1A in the RNase LS activity from *Escherichia coli*. *Genes Genet. Syst.* **2007**, *82*, 291–299. [CrossRef] [PubMed]

48. Alawneh, A.M.; Qi, D.; Yonesaki, T.; Otsuka, Y. An ADP-ribosyltransferase Alt of bacteriophage T4 negatively regulates the *Escherichia coli* MazF toxin of a toxin-antitoxin module. *Mol. Microbiol.* **2016**, *99*, 188–198. [CrossRef] [PubMed]

49. Georgopoulos, C.P.; Hendrix, R.W.; Kaiser, A.D.; Wood, W.B. Role of the host cell in bacteriophage morphogenesis: Effects of a bacterial mutation on T4 head assembly. *Nat. New Biol.* **1972**, *239*, 38–41. [CrossRef] [PubMed]

50. Ang, D.; Chandrasekhar, G.N.; Georgopoulos, C.P. *Escherichia coli grpE* gene codes for heat shock protein B25.3, essential for both λ DNA replication at all temperatures and host growth at high temperature. *J. Bacteriol.* **1986**, *167*, 25–29. [CrossRef] [PubMed]

51. Tabor, S.; Huber, H.E.; Richardson, C.C. *Escherichia coli* thioredoxin confers processivity on the DNA polymerase activity of the gene 5 protein of bacteriophage T7. *J. Biol. Chem.* **1987**, *262*, 16212–16223. [PubMed]

52. Hanninen, A.L.; Bamford, D.H.; Bamford, J.K. Assembly of membrane-containing bacteriophage PRD1 is dependent on GroEL and GroES. *Virology* **1997**, *227*, 207–210. [CrossRef] [PubMed]

53. Murray, N.E. Type I restriction systems: Sophisticated molecular machines (a legacy of Bertani and Weigle). *Microbiol. Mol. Biol. Rev.* **2000**, *64*, 412–434. [CrossRef] [PubMed]

54. Lee, J.S.; Jang, H.B.; Kim, K.S.; Kim, T.H.; Im, S.P.; Kim, S.W.; Lazarte, J.M.; Kim, J.S.; Jung, T.S. Complete genomic and lysis-cassette characterization of the novel phage, KBNP1315, which infects avian pathogenic *Escherichia coli* (APEC). *PLoS ONE* **2015**, *10*, e0142504. [CrossRef] [PubMed]

Article

Schrödinger's Cheshire Cat: Are Haploid *Emiliania huxleyi* Cells Resistant to Viral Infection or Not?

Gideon J. Mordecai, Frederic Verret, Andrea Highfield and Declan C. Schroeder *

Marine Biological Association of the UK, Citadel Hill, Plymouth PL1 2PB, UK; gidmord@gmail.com (G.J.M.); fverret@imbb.forth.gr (F.V.); ancba@mba.ac.uk (A.H.)
* Correspondence: dsch@mba.ac.uk; Tel.: +44-1752-426-484

Academic Editors: Corina Brussaard and Mathias Middelboe
Received: 31 January 2017; Accepted: 14 March 2017; Published: 18 March 2017

Abstract: *Emiliania huxleyi* is the main calcite producer on Earth and is routinely infected by a virus (EhV); a double stranded DNA (dsDNA) virus belonging to the family *Phycodnaviridae*. *E. huxleyi* exhibits a haplodiploid life cycle; the calcified diploid stage is non-motile and forms extensive blooms. The haploid phase is a non-calcified biflagellated cell bearing organic scales. Haploid cells are thought to resist infection, through a process deemed the "Cheshire Cat" escape strategy; however, a recent study detected the presence of viral lipids in the same haploid strain. Here we report on the application of an *E. huxleyi* CCMP1516 EhV-86 combined tiling array (TA) that further confirms an EhV infection in the RCC1217 haploid strain, which grew without any signs of cell lysis. Reverse transcription polymerase chain reaction (RT-PCR) and PCR verified the presence of viral RNA in the haploid cells, yet indicated an absence of viral DNA, respectively. These infected cells are an alternative stage of the virus life cycle deemed the haplococcolithovirocell. In this instance, the host is both resistant to and infected by EhV, i.e., the viral transcriptome is present in haploid cells whilst there is no evidence of viral lysis. This superimposed state is reminiscent of Schrödinger's cat; of being simultaneously both dead and alive.

Keywords: *Emiliania huxleyi*; virus; *Phycodnaviridae*; EhV; transcriptome; tiling array

---

1. Introduction

Coccolithophores are unicellular marine algae that produce a coccosphere made up of calcified platelets commonly referred to as coccoliths [1]. *Emiliania huxleyi* (Haptophyta) is the most ubiquitous and abundant coccolithophorid in modern oceans [2], and forms extensive coastal and mid-oceanic mesoscale blooms at temperate latitudes [3]. Consequently, *E. huxleyi* is a vital contributor to global marine calcium carbonate precipitation and due to the extensive blooms that it produces, plays an important role in the flux of CO_2 between the atmosphere and the oceans [4,5].

As with most prymnesiophytes, *E. huxleyi* possesses a haplodiplontic life cycle, with distinct heteromorphic differentiation between both ploidy levels, both of which are capable of independent asexual growth [6,7]. The diploid (2N) phase consists of a calcified non-motile cell, whereas the haploid (1N) phase is a non-calcified biflagellated cell bearing organic scales. Blooms of diploid *E. huxleyi* are typically terminated abruptly; releasing coccoliths to the sea surface and organic biomass to the ocean floor, a process which is one of the largest long-term sinks of carbon on earth [5]. Viral infection and subsequent lysis is one of the primary mechanisms for bloom termination and is attributed to the giant, double stranded DNA (dsDNA) *Coccolithovirus, Emiliania huxleyi virus* (EhV) [8–10].

Previous researchers have shown that haploid *E. huxleyi* cells do not undergo cellular lysis in the presence of EhV, naming this process the Cheshire cat (CC) escape strategy [11], compared to that of the Red Queen (RQ) evolutionary strategy which predicts an "evolutionary arms race" between virus and host [12]. CC dynamics propose that the host can evade infection, and therefore focus resources on interacting with direct ecological competitors [11].

The genome of EhV-86 (407,339 bp), the type species of the genus *Coccolithovirus*, encodes 472 protein coding sequences (CDSs) that were verified through an EhV-86 microarray [10]. The majority of the EhV-86 CDSs have an unknown function; only 66 of 472 (13.9%) were originally annotated with functional protein predictions on the basis of sequence similarity and protein domain matches. More recently the role of some of these protein domains, such as the virally encoded sphingolipid pathway, has been intensely studied [13]. The EhV-86 genome contains a sphingolipid biosynthetic pathway, which previously had never been seen in a viral genome. Glycosphingolipids (GSLs) are part of the building blocks of membrane lipids and lipid rafts and it has been suggested that these genes have undergone horizontal gene transfer between the host [14] and virus [15]. Sphingolipid biosynthesis is implicated in programmed cell death (PCD) [16,17] as it leads to ceramide production, which is an inducer of PCD [18]. Moreover, lytic EhV infection of *E. huxleyi* not only activates the PCD biochemical machinery but also actively recruits and requires GSLs for successful viral replication [19]. In addition, viral GSLs in particular regulate host–virus interactions by inducing host PCD in non-infected cells and facilitating viral production [17].

Virus like particles have been previously described in a haploid *E. huxleyi* dominated bloom [20]. Additionally, a recent study detected novel glycosphingolipids in the haploid *E. huxleyi* RCC1217 strain, which were similar to the viral GSLs observed in the EhV-infected diploid RCC1216 strain; but the authors did not explain the implications of this observation [21]. Although these GSLs are not yet fully characterized, we suggest that haploid cells are not resistant to infection from viruses. To support this hypothesis, we present EhV sequence data obtained via Reverse Transcription-PCR (RT-PCR) as well as the detection of EhV transcripts in a haploid *E. huxleyi* RCC1217 culture by the use of an *E. huxleyi* CCMP1516 EhV-86 combined tiling array (TA). As with the case of Schrödinger's principle, we suggest that the Cheshire Cat dynamic proposed by Frada et al. [11] is in a superimposed state, in which cells are simultaneously infected and able to avoid lysis.

2. Materials and Methods

2.1. Strains and Culture Conditions

E. huxleyi strain RCC1216 (Table 1) was originally isolated by micropipette isolation of a single diploid cell. The haploid stage of this strain, RCC1217 (Table 1) was isolated after a diploid to haploid shift in a sub-culture of the original strain. These strains were grown in batch cultures in F/2 minus Si medium at 15 °C with a 16:8 light:dark cycle. *E. huxleyi* CCMP1516 (Table S1) was used in EhV-86 infection experiments and to propagate EhV-86 [9], and EhV-86 DNA was used as a positive control for the PCR reaction. The infection dynamic was monitored by flow cytometry as described in Jacquet et al. [22].

Table 1. Host and virus strains used in the current study.

Strain	Origin	Date of Isolation
Emiliania huxleyi RCC1216 (TQ26 2N)	Tasman Sea	October 1998
E. huxleyi RCC1217 (TQ26 1N)	from strain *E. huxleyi* RCC1216	July 1999
E. huxleyi CCMP1516	South Pacific	July 1991
E. huxleyi Virus 86	English Channel [9]	July 1999

1N: haploid; 2N: diploid.

2.2. Nucleic Acid Extraction

DNA and RNA were extracted from *Emiliania huxleyi* strains (Table 1) when the cultures reached a concentration of at least 10^5 cells·mL^{-1}, during their exponential growth stage. Total genomic DNA was extracted using the DNeasy blood and tissue kit (Qiagen, Valencia, CA, USA) according to the manufacturer's recommendations. *E. huxleyi* CCMP1516 (50 mL) was inoculated with EhV-86 (50 μL) and DNA was extracted as detailed above at approximately 48 h post infection (p.i.) to use as a control in PCR. DNA was quantified using a NanoDrop 1000 Spectrophotometer (Thermo Scientific, Willington, DE, USA).

Total RNA was extracted using the RNeasy mini-kit (Qiagen), according to the manufacturer's recommendations, apart from the elution step, in which the RNA was eluted in 50 μL of RNase-free water. The quality of the RNA samples was determined with a Bioanalyzer 2100 (Agilent Technologies, Cheshire, UK). RNA samples were stored at −80 °C.

2.3. cDNA Synthesis

To remove contaminating DNA prior to RT-PCR or complementary DNA (cDNA) synthesis, total RNA was treated with RNase-free DNase (Promega, Madison, WI, USA) according to the manufacturer's instructions. Initially, first-strand cDNA synthesis was performed using the SuperScript III first-strand synthesis system for RT-PCR (Thermo Fisher Scientific, Waltham, MA, USA). Reverse transcription was carried out according to the manufacturer's instructions using random hexamer primers. The major capsid protein (MCP) gene product was amplified from the Superscript III first strand cDNA.

Double-stranded cDNA transcripts were synthesized using the SMARTer PCR cDNA Synthesis Kit (Clontech, Mountain View, CA, USA), which generates full-length double-stranded cDNA. The kit was used according to the manufacturer's instructions, apart from an extended incubation time during first strand synthesis (90 min) to create full-length cDNA. The helicase putative viral sequences from the haploid culture were amplified from the cDNA created by the SMARTer PCR cDNA Synthesis Kit.

2.4. PCR and RT-PCR Amplification

PCR amplification of the *E. huxleyi* genomic marker coding for a calcium binding protein (CBP) with high Glutamic acid, Proline, and Alanine amino acid content, termed GPA [23], was carried out using the qCBP primers [24] (Table 2). PCR amplification of virus-related genes was undertaken using primers designed to amplify four viral genes: DNA polymerase (DNA pol, ehV030), Helicase (hel, ehv440), proliferating cell nuclear antigen protein (PCNA, ehv440), and the MCP (ehv085) (Table 2).

One-step RT-PCR was used to amplify the haploid specific inner and outer arm dynein heavy chain (DHC) genes using the DHC1b and DHCb primers (Table 2). One-step RT-PCR reactions were carried out in a Rotorgene 6000 QPCR machine (Qiagen). The size of all PCR products was verified by gel electrophoresis on a 1% (*w/v*) agarose gel stained with ethidium bromide in 1× TAE buffer viewed on a UV transilluminator (Syngene, Cambridge, UK).

2.5. Sequencing and Alignment

PCR products were purified using the QIAquick PCR purification kit (Qiagen) according to the manufacturer's recommendations. The purified products were either sequenced directly or cloned using an Invitrogen cloning kit (Thermo Fisher Scientific). Sequencing was performed by Geneservice, Cambridge, UK. The ClustalW function within BioEdit (Ibis Biosciences, Carlsbad, CA. USA) was used for all sequence alignments [25].

Table 2. Primers used in this study.

Primer	Sequence (5′ to 3′)	Target (CDS)	Ta (°C)	Reference
qCBP_F	AGTCTCTCGACGCTGCCTC	GPA	60	[24]
qCBP_R	TGGCCTAGCACCAGTCTTTGG			
MCP_F2	TTCGCGCTCGAGTCGATC	MCP (ehv085)	60	[8]
MCP_R2	GACCTTTAGGCCAGGGAG			
EhVhel_F	GCCAACTGGTACAGGGAAAA	Helicase (ehv104)	54	[26]
EhVhel_R	CATCCATGCATGTGTCACAA			
EhVPCNA_F	GGGCATTTCATTTGCCATAC	PCNA (ehv440)	54	[26]
EhVPCNA_R	ATTCTCCGTCGACAAACGC			
EhVpol_F	TATAATGCACGCCAACTTGC	DNA pol (ehv030)	54	[26]
EhVpol_R	GCAATTGCACCAAGTGGATA			
DHC1b_F	GCTTTCTCACTGCGCTCAT	Flagellar inner DHC	55	[27]
DHC1b_R	GTAGAGCGGGCACGAGTACA			
DHCb_F	TGAACCTCGTCCTCAACACA	Flagellar outer DHC	55	[27]
DHCb_R	GAATCATCGGCATCACTGG			

CDS: coding sequences; DHC: dynein heavy chain protein; GPA: glutamic acid, proline and alanine rich protein; MCP: major capsid protein; PCNA: proliferating cell nuclear antigen protein.

2.6. Tiling Array

Fifty millilitre subsamples taken from a batch culture of *E. huxleyi* CCMP1516 at a cell density of 5×10^5 cells per mL (C4) were used as negative viral infection controls in our TA experiment. Simultaneously, identical subsample volumes from replicate batch cultures of CCMP1516 (C1) were also taken after an EhV-86 inoculum (final concentration of 1×10^6 viruses per mL) 30 min (S5) and 3.5 h (S6) p.i.; these served as positive viral infection controls. In addition, single extraction time points at cell densities of 5×10^5 cells per mL were taken from *E. huxleyi* RCC1217 (TQ26 1N) and RCC1216 (TQ26 2N) batch cultures. Both of these cultures were not exposed to EhV-86 in this experiment.

RNA extractions were carried out with the RNeasy extraction kit (Qiagen). The first strand complementary RNA (cRNA) Ambion amplification kit (Thermo Fisher Scientific) was used to generate cRNA and the same kit was used in a second step to create double-stranded cDNA. The quality of the RNA samples was verified with a Bioanalyzer 2100 (Agilent Technologies). All these protocols were carried out according to the manufacturer's recommendations.

The customised TA includes 2,076,726 oligo-nucleotide 50 mer probes spaced by 20 nt on average, designed on one strand of the *E. huxleyi* strain CCMP1516 genome [14]. In addition to *E. huxleyi* probes, the TA includes 18,654 oligo-nucleotide 50 mer probes designed on one strand of the *E. huxleyi* infecting virus EhV-86 genome [10]. A total of 4 two-color fluorescent labelling and hybridisations were carried out by NimbleGen (Table 3). The tiling array hybridization was visualised using SignalMap (NimbleGen, Roche, Madison, WI, USA). The EhV CDSs expressed by RCC1217 and EhV-86 (3.5 h p.i.) (Table S1) were recorded by noting the start and stop of each peak and labelled based on the EhV-86 complete genome annotation available on Genbank [10]. Fluorescence values greater than 3000 (empirically determined when compared to negative controls, RCC1216 and uninfected CCMP1516, expression profiles) was used as a threshold value to determine gene expression. If individual probes were also expressed in CDSs in the negative controls, they were removed and not included. Once the CDSs were recorded, they were compared with previous transcriptional studies [10,28,29]

Table 3. Tiling Array hybridization design (081216_Ehux_DS_CGH_HX1) carried out by NimbleGen (design ID 8739, ORD_ID 29312).

CHIP_ID	Image Name	Dye	Sample Description *	Virus Inoculation
57501502	57501502_532.tif	Cy3	1516 S6C1	3.5 h p.i.
	57501502_635.tif	Cy5	1516 S5C1	0.5 h p.i.
57501802	57501802_532.tif	Cy3	1516 S5C1	0.5 h p.i.
	57501802_635.tif	Cy5	1516 S5C4	No
58219802	58219802_532.tif	Cy3	1516 S6C4	No
	58219802_635.tif	Cy5	1516 S6C1	3.5 h p.i.
57488502	57488502_532.tif	Cy3	TQ26 1N	No
	57488502_635.tif	Cy5	TQ26 2N	No

*: S indicates the sample time point, while C indicates the culture replicate; p.i.: post infection.

3. Results

3.1. Tiling Array

The EhV-86 TA was used to assess the presence of the EhV transcriptome in RCC1216 (2N), RCC1217 (1N), and CCMP1516 (2N) pre- and post-infection with EhV-86. The CCMP1516 EhV-86 infection dynamic followed the typical culture crash three days post inoculation [9]. The CCMP1516 cell count and EhV-86 virus count were 5×10^5 cells per mL and 0.5×10^6 viruses per mL, respectively, as determined by flow cytometry at 30 min and 3.5 h post infection (data not shown). Tiling arrays allow the identification of novel transcribed sequences, as the arrays are designed to cover the whole genome independent of annotation data. Nonetheless, the expressed regions of the tiling array were labelled according to their location on the EhV-86 genome, using the EhV-86 coding sequence annotation [10]. This enabled the expression profile of the TA data to be compared with previous transcriptional studies (Figure 1, Table S1). Of the total 472 EhV-86 CDSs [10], 389 (82.2%) are expressed during a CCMP1516 EhV-86 lytic infection cycle. No virus expression was detected 30 min p.i. (Figure 2); producing an expression profile identical to an uninfected culture of either CCMP1516 or RCC1216 (data not shown). The CCMP1516 EhV-86 TA data revealed 37 expressed CDSs that were not previously detected, although the expression of two of these (ehv361 and ehv402) have previously been detected using expressed sequence tags ESTs (Table S1).

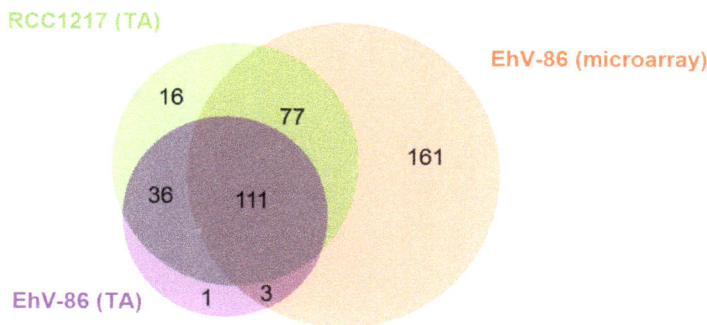

Figure 1. Venn diagram illustrating the comparative Emiliania *huxleyi virus* (EhV) coding sequence (CDS) expression in different strains and between treatments. In green, RCC1217 (1N) complementary cDNA (cDNA) hybridised onto a CCMP1516 + EhV-86 tiling array (TA) (this study). In purple CCMP1516 + EhV-86 3.5 h post infection cDNA hybridised onto the TA (this study). In orange, CCMP1516 EhV-86 cDNA hybridised onto conventional microarrays [10,29].

Figure 2. *Cont.*

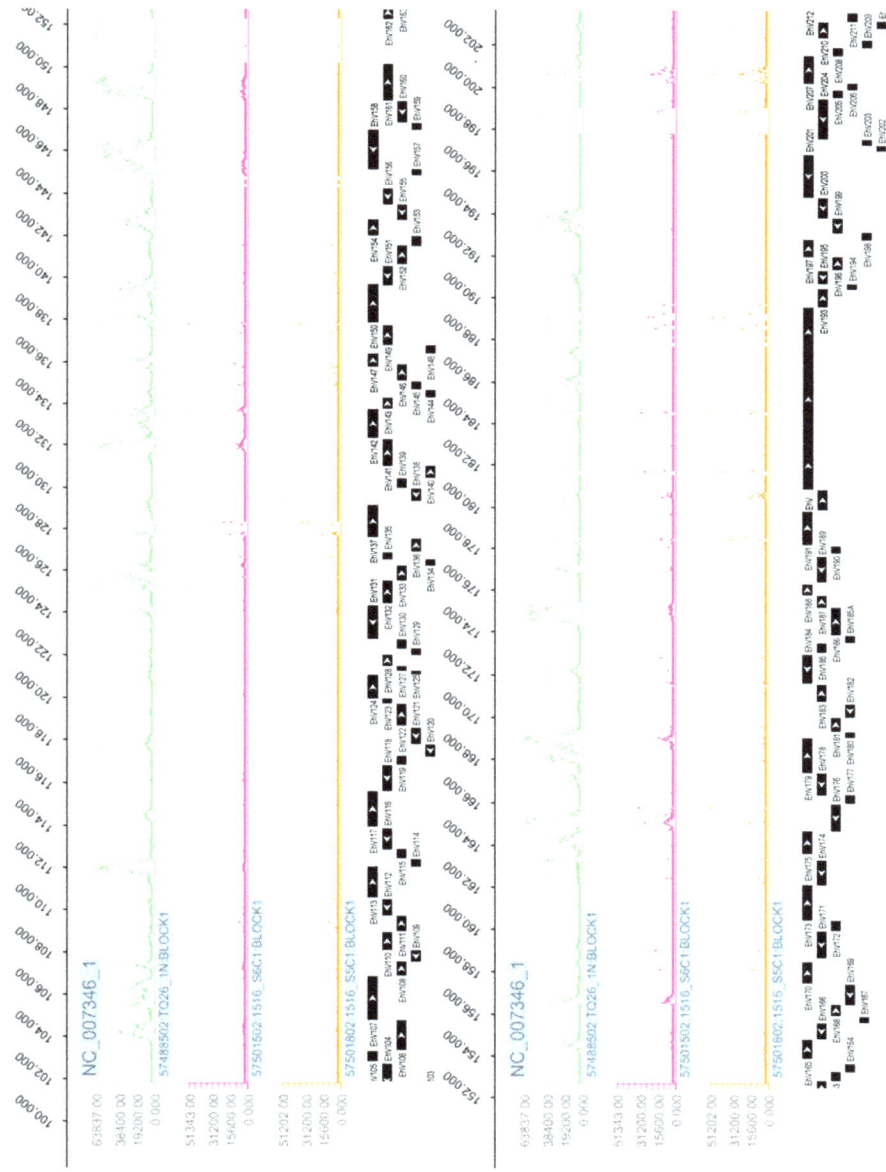

Figure 2. *Cont.*

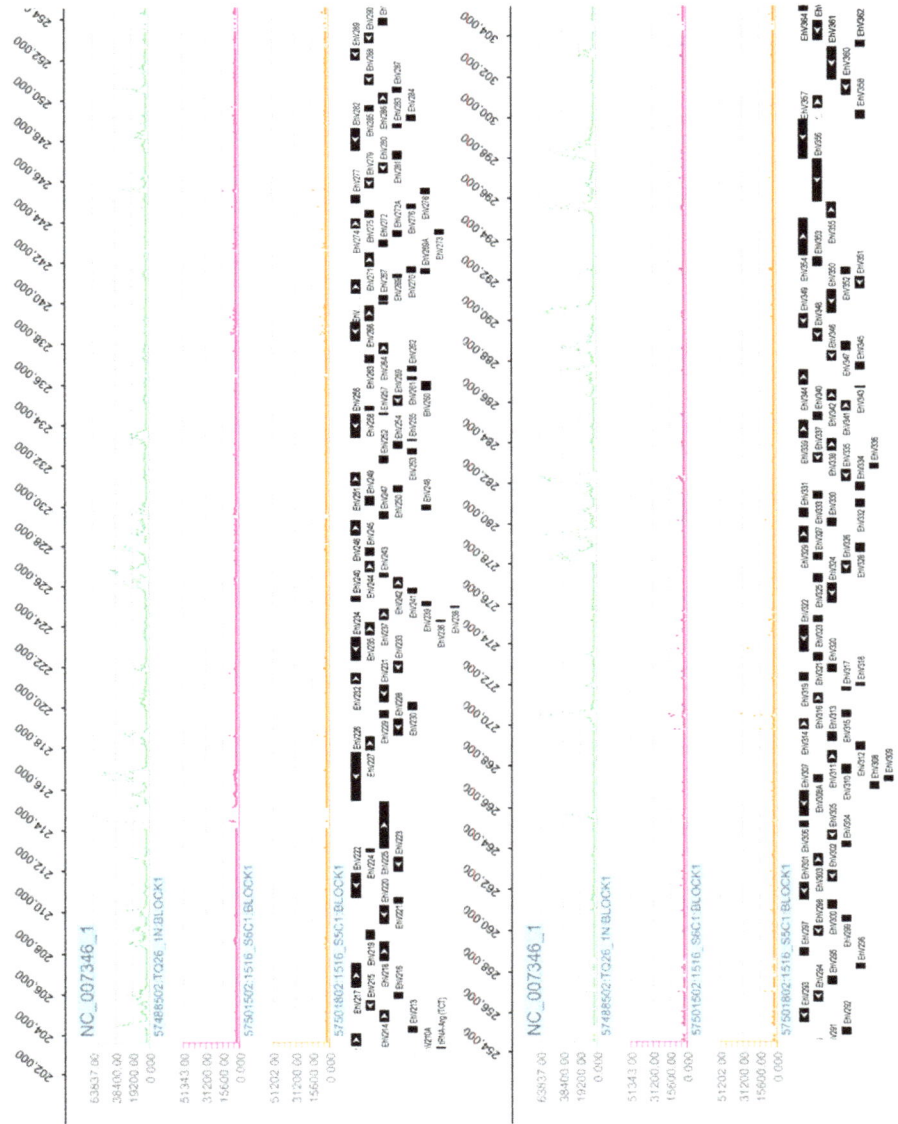

Figure 2. *Cont.*

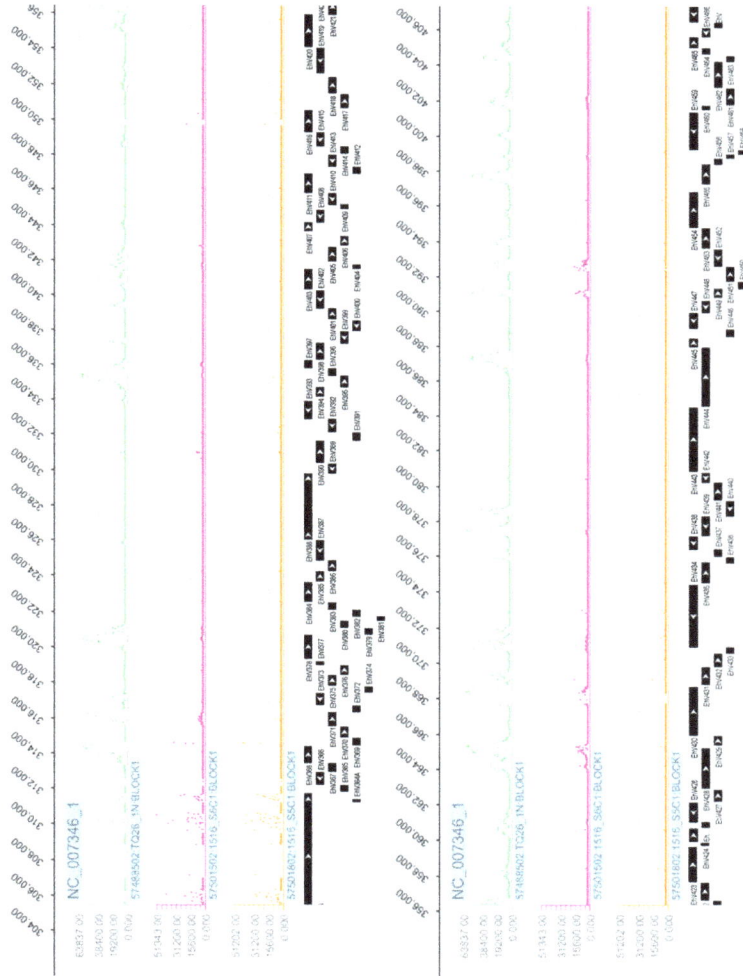

Figure 2. Nimblegen SignalMap displays of fluorescence values from TA for the whole EhV-86 genome (NC_007346_1) in base pairs with corresponding CDS map created by The National Center for Biotechnology Information (NCBI) [30] shown below each set of genome tracks. The first, second, and third tracks in green, purple, and orange display the expression profiles obtained from RCC1217 (1N) from chip 57288502, CCMP1516 infected with EhV-86 3.5 h p.i. from chip 57501502, and CCMP1516 infected with EhV-86 30 min p.i. from chip 57501802 (Table 3), respectively. Individual CDSs are illustrated as solid black rectangles with the internal white arrow head indicating the 5′ to 3′ coding direction.

The EhV-86 TA revealed the strong expression of EhV transcripts in an uninoculated culture of the haploid RCC1217 (Figures 1 and 2, Table S1). The detection of EhV transcripts expressed by the RCC1217 (1N) cells brings the total number expressed up to 85.9% (405 of a total 472). Uninfected cultures of diploid CCMP1516 and RCC1216 confirmed the specificity of the hybridization conditions (Figure 2), only showing small areas of non-specific hybridization, likely due to genes shared between the host and virus which are a result of horizontal gene transfer [31].

The hybridisation profile of the haploid RCC1217 is distinct from that of CCMP1516 infected with EhV-86, indicating that the expression is not a result of EhV-86 cross contamination (Figure 2). In addition, oligo deoxythymine (dT) primed expression assays are susceptible to 3′ bias due to the direction of cDNA synthesis (3′ to 5′) by the enzyme Reverse Transcriptase. The effect of this is seen in the TA hybridisation (Figure 2), in which low probe hybridisation occurred at the 5′ end of the genes. The total number of viral transcripts in the haploid RCC1217 culture (240 CDSs) exceeds that of CCMP1516 + EhV-86, 3.5 h p.i., which expressed a total of 151 CDSs (Figure 1, Table S1). However, the CCMP1516 EhV-86 microarray reported the expression of a total of 352 CDSs at four time points up to 33 h p.i. (Figure 1, Table S1).

Four genes in the EhV-86 genome were identified to be involved in sphingolipid biosynthesis: ehv031, ehv050, ehv077, and ehv079 [10]. All of these were detected in the haploid RCC1217 transcriptome (Table S1). Probe hybridisation of the TA reveals 16 CDSs transcribed by the haploid strain which were not expressed by the CCMP1516 EhV-86 tiling and conventional single probe microarray transcriptomes (Figure 1, Table S1). These 16 CDSs are all unique to the haploid transcriptome apart from ehv093 and ehv415, the expression of which was detected at 24 h p.i. using an EST approach [29]. Only 2 of 16 CDS have a known function assigned to them. Ehv415 is predicted to code for a fatty acid desaturase, a protein involved in membrane lysis, while ehv093 codes for a HNH endonuclease protein, which is involved in DNA homing, restriction, repair, or chromosome degradation [32].

The haploid RCC1217 culture expressed 22 of a possible 25 of the nucleo-cytoplasmic large double-stranded DNA virus (NCLDV) core genes that are present in the EhV-86 genome [33] (Table S2). The haploid transcriptome presents the first recorded expression of the CDS ehv128 by a virus infecting *E. huxleyi* which was not detected in the lytic-phase transcriptional profile of CCMP1516 EhV-86 infection. The CDS ehv128 encodes for a hypothetical ERV1/ALR protein belonging to a large family of proteins that includes the *Saccharomyces cerevisiae* ERV1 (Essential for Respiration and Vegatative growth) protein, which is required for mitochondrial biogenesis, and its homologs in other organisms, the mammalian hematopoetin (alternatively named ALR for its role as an Augmenter of Liver Regeneration), and animal and plant quiescins, so called because of their up-regulation in quiescent cells [34]. The precise functions of these proteins, however, remain unknown. Searches on GenBank on the remainder of the haploid RCC1217 unique transcripts revealed that three of the transcripts contain conserved sequence domains (Table S3), suggesting there are functional viral protein units within the hypothetical protein CDSs expressed by the haploid strain.

The expression of the host sphingolipid biosynthesis pathway was examined using the *E. huxleyi* TA, but no difference was apparent between samples (e.g., Serine palmitoyltransferase (SPT), and Ceramide glycosyltranferase (UGCG), Figure S1). The GSLs detected in *E. huxleyi* are a result of the sphingolipid biosynthesis pathway, which is encoded by both the host and virus as a result of horizontal gene transfer [10,17,35]. Rosenwasser et al. previously demonstrated that these genes are expressed by the host but down regulated upon lytic infection [13]. However, in their experiments down-regulation was observed 24 h p.i., which could explain why our TA expression of this pathway was unresponsive.

3.2. E. huxleyi Strain Confirmation

A fragment of the GPA gene, a common molecular marker for genotyping *E. huxleyi* [23], was amplified and sequenced to confirm the identity of the strains. Multiple sequence alignment of the amplified GPA fragment confirmed the identity of the RCC1217 haploid strain in culture (Figure S2).

Additionally, the pronounced transcriptional differentiation between haploid and diploid *E. huxleyi* cells [27] was employed to verify that the haploid culture was indeed in the gametal stage of the haplodiploid life cycle. RT-PCR was used to assess the haploid culture for the presence of 1N specific transcripts. The primer sets DHCb and DHC1b (Table 2) target the cytoplasmic DHC outer arm (OA) and inner arm (IA) genes, respectively, which are linked with a flagella function. The IA and OA flagella genes are solely expressed by haploid *E. huxleyi* cells [27,36], and accordingly were used to validate that RCC1217 was in the haploid form. The IA and OA amplicon sequences amplified from RCC1217 were identical to the previously obtained sequences from the same strain (Figure S3).

In addition to the molecular evidence, light microscopy confirmed the haploid cells to be highly motile (data not shown) and cultures had cells that remained in suspension rather than settling to the bottom as seen with diploid cells. Under normal culturing conditions the RCC1217 culture showed no symptoms of a lytic infection; reaching and maintaining a cell density (typical of a healthy culture) of at least 1×10^5 cells mL^{-1}, approximately 7 days post-transfer to fresh media.

3.3. Detection of EhV DNA

During the lytic infection of EhV, viral DNA is readily amplified by PCR [9]. PCR of four viral genes (PCNA, MCP, helicase, and DNA polymerase) was carried out on DNA extracted from RCC1216, RCC1217, and CCMP1516 infected with EhV-86. Viral DNA was amplified from infected CCMP1516 but not from RCC1216 and RCC1217 (Figure 3). To validate this negative PCR result, a conservative estimate of the sensitivity of the PCR reaction was calculated by amplifying the MCP gene using serial dilutions of the positive control DNA (CCMP1516 + EhV-86 3.5 h p.i.). At 3.5 h p.i. the infected CCMP1516 culture are yet to complete crash; however, replication and budding occurs at this stage of the infection [37]. The number of viral particles produced at 3.5 h p.i. was (over-) estimated to be 400, to give a conservative estimate of the equivalent number of viruses that the DNA was extracted from. Positive amplification of the MCP gene was achieved from a 1 in 1,000,000 dilution of the DNA, a volume of DNA equivalent to that extracted from three virions (data not shown).

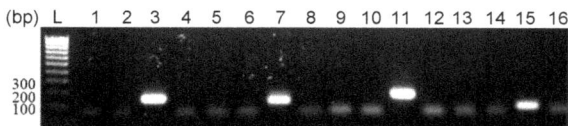

Figure 3. PCR amplification of EhV PCNA (lanes 1–4), MCP (lanes 5–8), helicase (9–12), and DNA polymerase (lanes 13–16) from DNA extracted from RCC1217 (1N) (lanes 1, 5, 9, 13), RCC1216 (lanes 2, 6, 10, 14), and CCMP1516 + EhV-86 (lanes 3, 7, 11, 15). Lanes 4, 8, 12, 16 are negative DNA (no template) controls.

3.4. Detection of EhV RNA

RT-PCR was undertaken on regions of the major capsid protein (ehv085) and helicase (ehv430) genes from RNA extracted from RCC1217 and CCMP1516 infected with EhV-86. MCP-amplified gene fragments from RCC1217 were sequenced, showing homology to the previously characterized genus *Coccolithovirus*, family *Phycodnaviridae* (Figure 4). The MCP sequences obtained on two separate occasions over the period of several years (first amplified in November 2010, then again in April 2012) were different, with 10 nucleotide polymorphisms across the region. However, the substitutions are "silent mutations", with no change to the resulting amino acid sequence. The helicase amplicon from RCC1217 was identical to the helicase fragment from the EhV-86 genome (Figure 5). Helicase is involved in DNA replication and is consequently a highly conserved gene within the NCLDVs (Table S2).

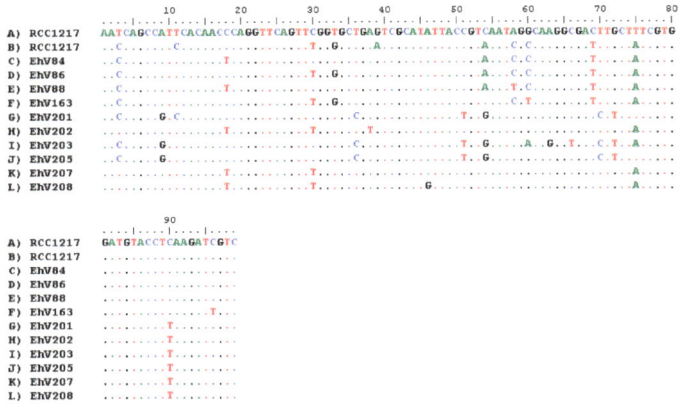

Figure 4. Multiple nucleotide sequence alignment of a 100 bp region of the viral MCP gene (ehv085). The MCP gene was amplified from RCC1217 using reverse transcription polymerase chain reaction (RT-PCR) on two separate occasions (A,B) and in both cases the sequence differed from all previously sequenced MCP genes fragments (C–L) obtained from GenBank [30]. Dots represent positions where the same nucleotides are present as in the top sequence, and letters represent nucleotide substitutions.

The presence of viral RNA and hence the state of viral infection is not permanent in the RCC1217 strain. Five years on since the RT amplification in 2012, the RCC1217 strain is virus free (data not shown).

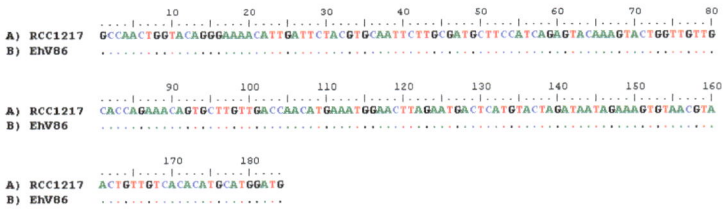

Figure 5. Pairwise nucleotide sequence alignment of a 180 bp fragment of the viral helicase gene (ehv430) amplified using RT-PCR from RCC1217 (1N) aligned with the helicase gene from EhV-86. Dots represent positions where the same nucleotides are present as in the top sequence, and letters represent nucleotide substitutions.

4. Discussion

The combination of tiling array data, RT-PCR of amplified NCLDV core genes, sequence data, and the viral-like glycosphingolipids present in the haploid cells as discovered by Hunter et al. [21] strongly indicates that an active EhV infection is present in the haploid *E. huxleyi* RCC1217 cells. To establish whether EhV, a DNA virus, is able to exist solely in an RNA form, we searched the EhV genome for the RNA replicase gene. Viruses with solely RNA-based genomes encode RNA replicase (RNA-dependant RNA polymerase, RdRP), which catalyzes the replication of RNA from an RNA template (the host does not produce an enzyme with this capability). A nucleotide BLAST search of the EhV-86 genome for an RNA replicase nucleotide sequence yields a negative result, implying RNA replicase is absent from the EhV-86 genome. However, EhV-86 codes for six different RNA polymerases (ehv064, ehv105, ehv108, ehv167, ehv399 and ehv434), and it is plausible that one of these or one of the 406 hypothetical proteins could function as an RNA replicase [10].

The majority of the haploid specific transcripts are annotated as hypothetical proteins with an unknown function. Three of these contain conserved protein domains (Table S3). Interestingly, ehv131 contains a calcium binding motif (beta/gamma crystalline), the role of which is unknown, but similarly there are proline rich proteins with calcium binding proteins encoded in the EhV-86 genome in the family B repeat region [38]. Although the function of these calcium binding motifs is not known, they are of interest as calcification in *E. huxleyi* is closely coordinated with cellular metabolism and photosynthesis [39].

The ERV1/ALR protein (ehv128) was the only coding sequence to be solely expressed in the haploid transcriptome (Table S2). The ERV1/ALR protein has been attributed to a cytoplasmic pathway of disulfide bond formation and is thought to represent a class of cellular thiol oxidoreductases [34]. Thiols are a functional group of the amino acid cysteine, and are produced by phytoplankton in response to increased copper concentrations [40]. These organic ligands bind to Cu, reducing free metal concentrations. Copper was shown to disrupt the lytic infection cycle of EhV-86 [26]. Therefore, it would be advantageous for the virus to express genes involved in the thiol biosynthesis pathway, playing a role in a switch from viral persistence to a more acute infection.

The putative haploid RCC1217 virus also expressed a fatty acid desaturase (ehv415) gene and a gene encoding a sterol desaturase conserved protein domain (ehv088) (Table S1), both of which have roles in lipid metabolism. The role of lipids during EhV-86 infection is not well understood, but it is thought that the control of lipid production could be associated with host membrane interactions, which facilitate the intracellular transport and subsequent budding of virus particles [41,42]. The HNH endonuclease family protein (ehv093) (Table S1), a type of restriction endonuclease, was present in the haploid transcriptome. It has been suggested that a virally encoded endonuclease by EhV-86 could lead to the degradation of host DNA [28]. Interestingly, the putative haploid virus expresses a methyltransferase encoding domain (ehv432) (Table S1). In eukaryotes, methylation of DNA bases is involved in epigenetic gene silencing and also plays a defensive role, protecting host DNA against the activity of restriction endonucleases, which cleave foreign DNA as a defense mechanism [43]. A potential function of the virally encoded methyltransferase (ehv452) is to protect viral genetic material from virally encoded endonucleases which degrade host nucleic acids [44].

Wilson proposed that the *Coccolithovirus-E. huxleyi* pairing (the coccolithovirocell, CLVC) exemplifies this concept due to the unique physiological interactions of EhV and its host [45]. We propose that the viral lipids [21] and EhV transcriptome detected in the haploid infer that haploid cells are a unique part of the *Coccolithovirus* lifecycle, which we propose naming as the Haplococcolithovirocell (HCLVCs) (Figure 6).

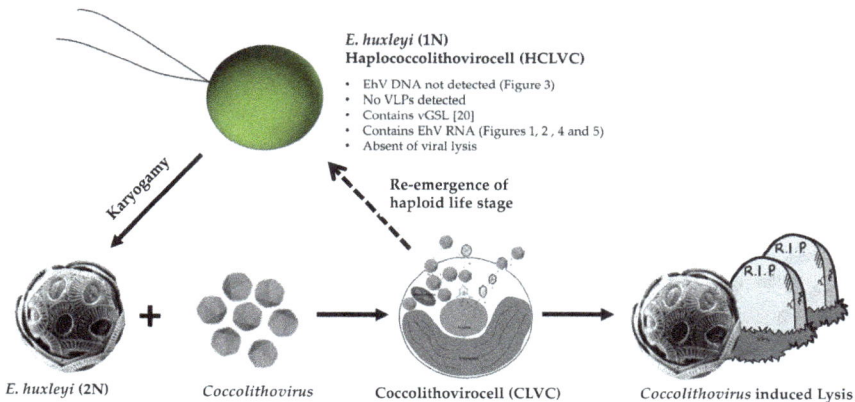

Figure 6. Proposed new *Coccolithovirus-E. huxleyi* life cycle incorporating the Haplococcolithovirocell. Infected diploid cells either undergo viral induced lysis or re-emerge as haploid cells containing viral RNA and lipids. Images: CLCV adapted from Mackinder et al. [37], *Coccolithoviruses* adapted from ViralZone [46]. vGSL: viral glycosphingolipids; VLP: virus like particles.

Although the viral GSL (vGSL) detected in haploid cells is yet to be fully characterised, judging by the number of EhV transcripts detected in our haploid culture, we propose that these GSLs may contribute to part of the *Coccolithovirus* lifecycle. Viral GSLs are vital for EhV infection as they induce the production of reactive species (oxygen and nitrogen) which initiate programmed cell death. We propose that the HCLVCs help maintain vGSL levels between EhV infection cycles, maintaining an advantage for the virus in the co-evolutionary arms race between virus and host.

Karyogamy (fusion of two haploid cells) has never been observed in haploid *E. huxleyi* cells and we propose that in the case of this culture, RCC1217, this is due to the viral signature in the cells causing a dead end to this stage of the life cycle. However, the RNA viral signature is not permanent; at least not in vitro. Nonetheless, the release of the 1N cell from this putative latent infection state is possible, which in turn would then allow karyogamy to take place. The identity of the vGSL detected in haploid cells is distinct from those found in CLVCs [21]. These differences are consistent with the tiling array data presented here, as the transcriptomic profile of the HCLVCs and CLVCs are distinct.

The majority of giant algal viruses exhibit an acute lytic *r*-selected life strategy characterised by high reproduction rates which eventually lead to cellular lysis. The exception to this rule are the phaeoviruses, which have a K-selected life strategy. Phaeoviruses exhibit viral persistence, and integrate their genomes into the gametal and spore life stages of their hosts, Ectocarpales brown algae. The infected gametes contain a latent provirus, which is then spread throughout the algae during adult development [47]. Overt disease symptoms are seen in the adult reproductive organs, which are deformed and produce virus particles. Although the details of the haploid infecting virus are not identical to the phaeoviruses (specifically, the absence of viral DNA), there are some interesting analogies, namely the switch to viral latency in the gametal life stages. We propose that, similarly to phaeoviruses, coccolithoviruses also have a secondary life cycle strategy in gametal *E. huxleyi* cells. However, there is no evidence of genome integration as EhV DNA was not detected in the haploid cells. We suggest that the EhV transcriptome is present in haploid cells in a novel mechanism of viral persistence.

Viruses infecting *E. huxleyi* exist in a world of two halves; during a bloom, host abundance is extremely high, and there are plenty of opportunities for re-infection and horizontal transmission. Conversely, between blooms host abundance is lower and the viral decay rate is high [48], and as a result there is less opportunity for horizontal transmission. We propose that within coccolithoviruses, two distinct infection strategies exist, which occupy different ecological niches and each is suited to the boom and bust ecology of the host. The infection of the haploid cells by an RNA viral life stage may help to explain the stability of the coexistence of virus and host in marine algae.

Flow cytometric analysis has shown the emergence of a sub-population of haploid cells after virus-mediated bloom demise [22]. Additionally, gene expression studies found that infection by EhV increased expression of pathways related to spermatogenesis [13], suggesting that the virus initiates gametogenesis and is an integral part of the *E. huxleyi* lifecycle (Figure 6). Previous work used PCR to show that EhV does not replicate in haploid *E. huxleyi* cells [11]. In HCLVCs we have detected the absence of EhV DNA, but the presence of RNA. The mechanisms behind this remain unexplored and there is no other example of a DNA virus with a separate RNA life stage where DNA is not detectable. The data presented here in combination with the viral GSLs detected in haploid cells by Hunter et al. [21] depict a different scenario to the Cheshire Cat dynamic. Future metabolomic and transcriptional studies will hopefully ascertain the breadth and mechanism of this infection.

5. Conclusions

Coccolithoviruses drive *E. huxleyi* bloom dynamics and therefore play an integral role in the global ocean carbon pump and climate [45]. The realisation that the host-virus dynamic is not always lytic and that covert infection can arise, highlights our limited understanding into the full extent in which they control their host. This study clearly demonstrates the potential of NCDLVs to interact with and manipulate the ecology of their hosts; the life cycle of which is not fully understood for both the host and their viruses. Although persistence has been revealed in other phycodnaviruses [47], the potential mechanism seen in haploid *E. huxleyi* cells, in which solely the viral transcriptome is detected in gametes, is entirely novel. Despite our limited knowledge, coccolithoviruses are clearly extraordinary viruses which deserve further research.

Supplementary Materials: The following are available online at www.mdpi.com/1999-4915/9/3/51/s1, Figure S1: CCMP1516 tiling array, Figure S2: Multiple sequence alignment of PCR-amplified GPA sequence from RCC1217 (1N) compared to previously sequenced GPA amplicons, Figure S3: Pairwise nucleotide sequence alignment of fragments of the amplified RT-PCR products of the dynein heavy chain (DHC) (A,B) inner arm (DHC1b) and (C,D) outer arm (DHCb) genes, Table S1: Unique EhV-86 CDSs expression data, Table S2: Presence of NCLDV Core genes in the EhV-86 genome and in the transcriptome of the virus infecting RCC1217, Table S3: Haploid unique transcripts in the EhV-86 genome and in the transcriptome of the virus infecting RCC1217.

Acknowledgments: Declan Schroeder was funded by an Marine Biological Association (MBA) research fellowship. A special thanks to Cecilia Balestreri for the electron micrograph images used in Figure 6.

Author Contributions: Declan Schroeder conceived and designed the experiments; Gideon Mordecai, Andrea Highfield, and Fred Verret performed the experiments; Gideon Mordecai. and Declan Schroeder analyzed the data; Declan Schroeder contributed reagents/materials/analysis tools; Gideon Mordecai and Declan Schroeder wrote the paper.

Conflicts of Interest: The authors declare no conflict of interest.

References

1. Paasche, E. A review of the coccolithophorid *Emiliania huxleyi* (prymnesiophyceae), with particular reference to growth, coccolith formation, and calcification-photosynthesis interactions. *Phycologia* **2001**, *40*, 503–529. [CrossRef]

2. Westbroek, P.; Brown, C.W.; Vanbleijswijk, J.; Brownlee, C.; Brummer, G.J.; Conte, M.; Egge, J.; Fernandez, E.; Jordan, R.; Knappertsbusch, M. A model system approach to biological climate forcing—The example of *Emiliania huxleyi*. *Glob. Planet. Chang.* **1993**, *8*, 27–46. [CrossRef]

3. Holligan, P.M.; Viollier, M.; Harbour, D.S.; Camus, P.; Champagnephilippe, M. Satellite and ship studies of coccolithophore production along a continental-shelf edge. *Nature* **1983**, *304*, 339–342. [CrossRef]

4. Elderfield, H. Climate change: Carbonate mysteries. *Science* **2002**, *296*, 1618–1621. [CrossRef] [PubMed]

5. Riebesell, U.; Schulz, K.G.; Bellerby, R.G.J.; Botros, M.; Fritsche, P.; Meyerhofer, M.; Neill, C.; Nondal, G.; Oschlies, A.; Wohlers, J.; et al. Enhanced biological carbon consumption in a high CO_2 ocean. *Nature* **2007**, *450*, 545–548. [CrossRef] [PubMed]

6. Green, J.C.; Course, P.A.; Tarran, G.A. The life-cycle of *Emiliania huxleyi*: A brief review and a study of relative ploidy levels analysed by flow cytometry. *J. Mar. Syst.* **1996**, *9*, 33–44. [CrossRef]

7. Klaveness, D. *Coccolithus huxleyi* (lohm.) kamptn ii. The flagellate cell, aberrant cell types, vegetative propagation and life cycles. *Eur. J. Phycol.* **1972**, *7*, 309–318. [CrossRef]

8. Schroeder, D.C.; Oke, J.; Hall, M.; Malin, G.; Wilson, W. Virus succession observed during an *Emiliania huxleyi* bloom. *Appl. Environ. Microbiol.* **2003**, *69*, 2484–2490. [CrossRef] [PubMed]

9. Schroeder, D.C.; Oke, J.; Malin, G.; Wilson, W.H. Coccolithovirus (*phycodnaviridae*): Characterisation of a new large dsdna algal virus that infects *Emiliania huxleyi*. *Arch. Virol.* **2002**, *147*, 1685–1698. [CrossRef] [PubMed]

10. Wilson, W.H.; Schroeder, D.C.; Allen, M.J.; Holden, M.T.; Parkhill, J.; Barrell, B.G.; Churcher, C.; Hamlin, N.; Mungall, K.; Norbertczak, H. Complete genome sequence and lytic phase transcription profile of a coccolithovirus. *Science* **2005**, *309*. [CrossRef] [PubMed]

11. Frada, M.; Probert, I.; Allen, M.J.; Wilson, W.H.; de Vargas, C. The "cheshire cat" escape strategy of the coccolithophore *Emiliania huxleyi* in response to viral infection. *Proc. Natl. Acad. Sci. USA* **2008**, *105*, 15944–15949. [CrossRef] [PubMed]

12. VanValen, L. A new evolutionary law. *Evolut. Theory* **1973**, *1*.

13. Rosenwasser, S.; Mausz, M.A.; Schatz, D.; Sheyn, U.; Malitsky, S.; Aharoni, A.; Weinstock, E.; Tzfadia, O.; Ben-Dor, S.; Feldmesser, E.; et al. Rewiring host lipid metabolism by large viruses determines the fate of *Emiliania huxleyi*, a bloom-forming alga in the ocean. *Plant Cell* **2014**, *26*, 2689–2707. [CrossRef] [PubMed]

14. Read, B.A.; Kegel, J.; Klute, M.J.; Kuo, A.; Lefebvre, S.C.; Maumus, F.; Mayer, C.; Miller, J.; Monier, A.; Salamov, A.; et al. Pan genome of the phytoplankton *Emiliania* underpins its global distribution. *Nature* **2013**, *499*, 209–213. [CrossRef] [PubMed]

15. Monier, A.; Pagarete, A.; de Vargas, C.; Allen, M.J.; Read, B.; Claverie, J.-M.; Ogata, H. Horizontal gene transfer of an entire metabolic pathway between a eukaryotic alga and its DNA virus. *Genome Res.* **2009**, *19*, 1441–1449. [CrossRef] [PubMed]

16. Bidle, K.D. The molecular ecophysiology of programmed cell death in marine phytoplankton. *Annu. Rev. Mar. Sci.* **2015**, *7*, 1–596. [CrossRef] [PubMed]

17. Vardi, A.; Van Mooy, B.A.S.; Fredricks, H.F.; Popendorf, K.J.; Ossolinski, J.E.; Haramaty, L.; Bidle, K.D. Viral glycosphingolipids induce lytic infection and cell death in marine phytoplankton. *Science* **2009**, *326*, 861–865. [CrossRef] [PubMed]

18. Hannun, Y.A.; Obeid, L.M. Principles of bioactive lipid signalling: Lessons from sphingolipids. *Nat. Rev. Mol. Cell Biol.* **2008**, *9*, 139–150. [CrossRef] [PubMed]

19. Bidle, K.D.; Haramaty, L.; Barcelos e Ramos, J.; Falkowski, P. Viral activation and recruitment of metacaspases in the unicellular coccolithophore, *Emiliania huxleyi*. *Proc. Natl. Acad. Sci. USA* **2007**, *104*, 6049–6054. [CrossRef] [PubMed]

20. Brussaard, C.P.D.; Kempers, R.S.; Kop, A.J.; Riegman, R.; Heldal, M. Virus-like particles in a summer bloom of *Emiliania huxleyi* in the north sea. *Aquat. Microb. Ecol.* **1996**, *10*, 105–113. [CrossRef]

21. Hunter, J.E.; Frada, M.J.; Fredricks, H.F.; Vardi, A.; Van Mooy, B.A.S. Targeted and untargeted lipidomics of *Emiliania huxleyi* viral infection and life cycle phases highlights molecular biomarkers of infection, susceptibility and ploidy. *Front. Mar. Sci.* **2015**, *2*, 81. [CrossRef]

22. Jacquet, S.; Heldal, M.; Iglesias-Rodriguez, D.; Larsen, A.; Wilson, W.H.; Bratbak, G. Flow cytometric analysis of an *Emiliana huxleyi* bloom terminated by viral infection. *Aquat. Microb. Ecol.* **2002**, *27*, 111–124. [CrossRef]

23. Schroeder, D.C.; Biggi, G.F.; Hall, M.; Davy, J.; Martinez-Martinez, J.; Richardson, A.J.; Malin, G.; Wilson, W.H. A genetic marker to separate *Emiliania huxleyi* (Prymnesiophyceae) morphotypes. *J. Phycol.* **2005**, *41*, 874–879. [CrossRef]

24. Krueger-Hadfield, S.A.; Balestreri, C.; Schroeder, J.; Highfield, A.; Helaouët, P.; Allum, J.; Moate, R.; Lohbeck, K.T.; Miller, P.I.; Riebesell, U.; et al. Genotyping an *Emiliania huxleyi* (Prymnesiophyceae) bloom event in the north sea reveals evidence of asexual reproduction. *Biogeosciences* **2014**, *11*, 5215–5234. [CrossRef]

25. Bioedit. Available online: http://www.mbio.ncsu.edu/bioedit/bioedit.html (accessed 14 March 2017).

26. Gledhill, M.; Devez, A.; Highfield, A.; Singleton, C.; Achterberg, E.P.; Schroeder, D. Effect of metals on the lytic cycle of the coccolithovirus, EhV86. *Front. Microbiol.* **2012**, *3*, 155. [CrossRef] [PubMed]

27. von Dassow, P.; Ogata, H.; Probert, I.; Wincker, P.; Da Silva, C.; Audic, S.; Claverie, J.-M.; de Vargas, C. Transcriptome analysis of functional differentiation between haploid and diploid cells of *Emiliania huxleyi*, a globally significant photosynthetic calcifying cell. *Genome Biol.* **2009**, *10*, R114. [CrossRef] [PubMed]

28. Kegel, J.; Blaxter, M.; Allen, M.J.; Metfies, K.; Wilson, W.; Valentin, K. Transcriptional host-virus interaction of *Emiliania huxleyi* (Haptophyceae) and EhV-86 deduced from combined analysis of expressed sequence tags and microarrays. *Eur. J. Phycol.* **2010**, *45*, 1–12. [CrossRef]

29. Allen, M.J.; Forster, T.; Schroeder, D.C.; Hall, M.; Roy, D.; Ghazal, P.; Wilson, W.H. Locus-specific gene expression pattern suggests a unique propagation strategy for a giant algal virus. *J. Virol.* **2006**, *80*, 7699–7705. [CrossRef] [PubMed]

30. The National Center for Biotechnology Information. Available online: https://www.ncbi.nlm.nih.gov/ (accessed 14 March 2017).

31. Pagarete, A.; Allen, M.J.; Wilson, W.H.; Kimmance, S.A.; De Vargas, C. Host–virus shift of the sphingolipid pathway along an *Emiliania huxleyi* bloom: Survival of the fattest. *Environ. Microbiol.* **2009**, *11*, 2840–2848. [CrossRef] [PubMed]

32. Hsia, K.-C.; Chak, K.-F.; Liang, P.-H.; Cheng, Y.-S.; Ku, W.-Y.; Yuan, H.S. DNA binding and degradation by the hnh protein cole7. *Structure* **2004**, *12*, 205–214. [CrossRef] [PubMed]

33. Schroeder, D.C.; Park, Y.; Yoon, H.-M.; Lee, Y.S.; Kang, S.W.; Meints, R.H.; Ivey, R.G.; Choi, T.-J. Genomic analysis of the smallest giant virus—*Feldmannia* sp. Virus 158. *Virology* **2009**, *384*, 223–232. [CrossRef] [PubMed]

34. Senkevich, T.G.; White, C.L.; Koonin, E.V.; Moss, B. A viral member of the ERV1/ALR protein family participates in a cytoplasmic pathway of disulfide bond formation. *Proc. Natl. Acad. Sci. USA* **2000**, *97*, 12068–12073. [CrossRef] [PubMed]

35. Rosenwasser, S.; Graff van Creveld, S.; Schatz, D.; Malitsky, S.; Tzfadia, O.; Aharoni, A.; Levin, Y.; Gabashvili, A.; Feldmesser, E.; Vardi, A. Mapping the diatom redox-sensitive proteome provides insight into response to nitrogen stress in the marine environment. *Proc. Natl. Acad. Sci. USA* **2014**, *111*, 2740–2745. [CrossRef] [PubMed]

36. Frada, M.J.; Bidle, K.D.; Probert, I.; de Vargas, C. In situ survey of life cycle phases of the coccolithophore *Emiliania huxleyi* (Haptophyta). *Environ. Microbiol.* **2012**, *14*, 1558–1569. [CrossRef] [PubMed]

37. Mackinder, L.C.M.; Worthy, C.A.; Biggi, G.; Hall, M.; Ryan, K.P.; Varsani, A.; Harper, G.M.; Wilson, W.H.; Brownlee, C.; Schroeder, D.C. A unicellular algal virus, *Emiliania huxleyi virus* 86, exploits an animal-like infection strategy. *J. Gen. Virol.* **2009**, *90*, 2306–2316. [CrossRef] [PubMed]

38. Allen, M.J.; Schroeder, D.C.; Wilson, W.H. Preliminary characterisation of repeat families in the genome of EhV-86, a giant algal virus that infects the marine microalga *Emiliania huxleyi. Arch. Virol.* **2006**, *151*, 525–535. [CrossRef] [PubMed]

39. Mackinder, L.; Wheeler, G.; Schroeder, D.; Riebesell, U.; Brownlee, C. Molecular mechanisms underlying calcification in coccolithophores. *Geomicrobiol. J.* **2010**, *27*, 585–595. [CrossRef]

40. Kawakami, S.K.; Gledhill, M.; Achterberg, E.P. Production of phytochelatins and glutathione by marine phytoplankton in response to metal stress1. *J. Phycol.* **2006**, *42*, 975–989. [CrossRef]

41. Schatz, D.; Shemi, A.; Rosenwasser, S.; Sabanay, H.; Wolf, S.G.; Ben-Dor, S.; Vardi, A. Hijacking of an autophagy-like process is critical for the life cycle of a DNA virus infecting oceanic algal blooms. *New Phytol.* **2014**, *204*, 854–863. [CrossRef] [PubMed]

42. Malitsky, S.; Ziv, C.; Rosenwasser, S.; Zheng, S.; Schatz, D.; Porat, Z.; Ben-Dor, S.; Aharoni, A.; Vardi, A. Viral infection of the marine alga *Emiliania huxleyi* triggers lipidome remodeling and induces the production of highly saturated triacylglycerol. *New Phytol.* **2016**, *210*, 88–96. [CrossRef]

43. Iyer, L.M.; Tahiliani, M.; Rao, A.; Aravind, L. Prediction of novel families of enzymes involved in oxidative and other complex modifications of bases in nucleic acids. *Cell Cycle* **2009**, *8*, 1698–1710. [CrossRef] [PubMed]

44. Xia, Y.N.; Burbank, D.E.; Uher, L.; Rabussay, D.; Van Etten, J.L. Restriction endonuclease activity induced by PBCV-1 virus infection of a chlorella-like green alga. *Mol. Cell. Biol.* **1986**, *6*, 1430–1439. [CrossRef] [PubMed]

45. Wilson, W.H. Coccolithovirus-emiliania huxleyi dynamics: An introduction to the coccolithovirocell. *Perspect. Phycol.* **2015**, *2*, 91–103.

46. ViralZone. Available online: http://viralzone.expasy.org/ (accessed 14 March 2017).

47. Stevens, K.; Weynberg, K.; Bellas, C.; Brown, S.; Brownlee, C.; Brown, M.T.; Schroeder, D.C. A novel evolutionary strategy revealed in the phaeoviruses. *PLoS ONE* **2014**, *9*, e86040. [CrossRef] [PubMed]

48. Suttle, C.A.; Feng, C. Mechanisms and rates of decay of marine viruses in seawater. *Appl. Environ. Microbiol.* **1992**, *58*, 3721–3729. [PubMed]

Article

Virus Resistance Is Not Costly in a Marine Alga Evolving under Multiple Environmental Stressors

Sarah E. Heath [1,*], Kirsten Knox [2], Pedro F. Vale [1] and Sinead Collins [1]

[1] Institute of Evolutionary Biology, School of Biological Sciences, University of Edinburgh,
    Ashworth Laboratories, The King's Buildings, Charlotte Auerbach Road, Edinburgh EH9 3FL, UK;
    Pedro.Vale@ed.ac.uk (P.F.V.); s.collins@ed.ac.uk (S.C.)
[2] Institute of Molecular Plant Sciences, School of Biological Sciences, University of Edinburgh,
    Rutherford Building, Max Born Crescent, Edinburgh EH9 3BF, UK; kirsten.knox@ed.ac.uk
*   Correspondence: s.heath-2@sms.ed.ac.uk; Tel.: +44-131-651-7112

Academic Editors: Mathias Middelboe and Corina Brussaard
Received: 25 January 2017; Accepted: 28 February 2017; Published: 8 March 2017

Abstract: Viruses are important evolutionary drivers of host ecology and evolution. The marine picoplankton *Ostreococcus tauri* has three known resistance types that arise in response to infection with the Phycodnavirus OtV5: susceptible cells (S) that lyse following viral entry and replication; resistant cells (R) that are refractory to viral entry; and resistant producers (RP) that do not all lyse but maintain some viruses within the population. To test for evolutionary costs of maintaining antiviral resistance, we examined whether *O. tauri* populations composed of each resistance type differed in their evolutionary responses to several environmental drivers (lower light, lower salt, lower phosphate and a changing environment) in the absence of viruses for approximately 200 generations. We did not detect a cost of resistance as measured by life-history traits (population growth rate, cell size and cell chlorophyll content) and competitive ability. Specifically, all R and RP populations remained resistant to OtV5 lysis for the entire 200-generation experiment, whereas lysis occurred in all S populations, suggesting that resistance is not costly to maintain even when direct selection for resistance was removed, or that there could be a genetic constraint preventing return to a susceptible resistance type. Following evolution, all S population densities dropped when inoculated with OtV5, but not to zero, indicating that lysis was incomplete, and that some cells may have gained a resistance mutation over the evolution experiment. These findings suggest that maintaining resistance in the absence of viruses was not costly.

Keywords: evolution; trade-off; cost of resistance; Phycodnavirus; Prasinovirus; environmental change; virus-host interactions; marine viral ecology; *Ostreococcus tauri*

1. Introduction

Viruses are the most abundant biological entities in the oceans, with an estimated 10^{30} particles globally [1]. Viruses play a key role in marine food webs, partially because viral infection of unicellular organisms often results in cell lysis, where the infected cell bursts to release the new viruses; products of lysis feed back into the microbial loop and provide organic matter to organisms at the base of the food web daily [2]. In addition to being a large cause of mortality to their hosts, viruses can exert strong selection on host immune defense, leading to the evolution of host resistance mechanisms. Strong immune defenses, in turn, impose strong selection on viruses to evade these resistance responses leading to an ongoing co-evolutionary process between hosts and viruses [3]. Experimental evidence of host-virus coevolution has come mainly from bacteria-phage systems [3,4]. Viruses evolve rapidly due to their small size and high mutation rates [5] which can strongly influence the evolution of their hosts. However, in addition to infection, hosts are also subject to other selection pressures, such as severe or stressful environmental changes. In the case of marine hosts, they will be subject to natural selection both from their viruses, and from, for example, the changes in nutrients, temperature and light associated with global change in the oceans [6], which opens up the possibility that the genetic and physiological changes associated with resistance may affect host evolution in response to challenges other than the virus itself. This in turn has the potential to affect how primary productivity at the base of marine food webs evolves in response to global change. Studies have examined environmental effects on interactions between microalgae and their viruses under a range of conditions including changes in temperature [7,8], nutrients [9–13], UV radiation [14], light intensity [11,15,16], and CO_2 levels [13,17,18]. Environmental change can have direct effects on marine viruses, for example by damaging and/or deactivating the particles through UV exposure or extreme temperatures [8,14]. However, viral abundance is thought to be mainly dependent on host availability and, therefore, the effects of environmental change on viruses are expected to be mainly indirect (e.g., [19]). Here we focus on host evolution rather than viral selection.

Hosts are capable of evolving resistance to their viruses, though resistance often entails a fitness cost, which can vary in form and magnitude [20]. Costs of resistance that have been reported in microorganisms include reduced competitive ability [20,21], reduced growth rate [22,23], reduced original function of a receptor protein [24,25], and increased susceptibility to other viruses [26–28]. If the cost of resistance is substantial and related to growth or competitive ability, resistance might be lost when the selection pressure for it is removed (i.e., when viruses are absent) [29]. For example, under conditions where viruses are present and able to interact with their host cells, resistant hosts should have a selective advantage over susceptible hosts by avoiding lysis. However, in the absence of viruses, the selection pressure for resistance is removed and costs of resistance, if present and substantial, should reduce host fitness, so that there is an advantage to losing resistance. Most studies have focused on costs of resistance in bacteria (e.g., [22,28,30,31]), however data for eukaryotic microalgae are lacking, which limits our ability to translate the literature on host-virus interactions to primary producers in the oceans. Because marine algae are the dominant primary producers in oceans [32], changes in the abundance, distribution and composition of microalgal assemblages in response to climate change are likely to have important implications for marine communities.

The marine picoeukaryote *Ostreococcus tauri* and its viruses, *Ostreococcus tauri* viruses (OtVs), are abundant in Mediterranean lagoons [33]. OtVs are lytic viruses belonging to the family Phycodnaviridae that cause susceptible (S) host *O. tauri* cells to burst following infection [34]. However, two resistant host types have been observed [35,36]. In the first type, viruses can attach to the resistant (R) host cells but are unable to replicate and cause lysis. In the second type, resistant producer (RP) populations consist mainly of resistant cells with a minority of susceptible cells (<0.5%) that maintains a population of viruses. These two resistance mechanisms have been observed repeatedly and remain resistant to lysis over many generation of sub-culturing [35,36]. Previous work found that there was no difference in growth rates between the three resistance types when they were maintained separately under standard laboratory culturing conditions, although long term competitions indicated a cost of resistance with susceptible cells outcompeting resistant cells and resistant cells outcompeting resistant producers after 100 and 200 days, respectively [35].

This study examined whether a cost of resistance could be detected in *O. tauri* in terms of the ability to adapt to different environmental conditions, and whether the evolutionary responses to environmental change were affected by resistance type. Populations of S, R and RP *O. tauri* were evolved under different environmental conditions in the absence of viruses for 200 generations to answer whether resistance type was maintained and how resistance type affected evolutionary responses, even in the absence of coevolutionary dynamics imposed by the presence of viruses. We found that all R and RP populations remained resistant to OtV5 inoculation across all environments, whereas S populations had a lower proportion of cell lysis at the end than at the start of the evolution experiment. Additionally, resistance type affected cell division rates, size and chlorophyll content, whereas selection environment affected cell division rates and competitive ability.

2. Materials and Methods

2.1. Susceptible and Resistant Lines

O. *tauri* lines were obtained from N. Grimsley, Observatoire Océanologique, Banyuls-sur-Mer, France. Three susceptible lines (NG'2, NG'3 and NG'4), three resistant lines (NG5, NG'13 and NG26) and three resistant producer lines (NG'10, NG'16 and NG27) were used. All lines were derived from a single clone of O. *tauri* (RCC4221) and therefore had the same starting genotype.

2.2. Culturing Conditions

For each of the nine lines described above, three biological replicates were evolved per environment (27 independent populations in total per environment). We refer to each independent replicate as a population. Populations were grown in batch culture. Culture medium was prepared using 0.22 μm filtered Instant Ocean artificial seawater (salinity 30 ppt) supplemented with Keller and f/2 vitamins [37]. Control cultures were maintained in a 14:10 light:dark cycle at 85 μmol photon m^{-2} s^{-1} at a constant temperature of 18 °C (Table 1). Each population was grown in 20 mL media and each week, 200 μL was transferred to fresh media to ensure populations were always growing exponentially. Cultures were resuspended by gentle shaking every 2–3 days to prevent cells sticking to the bottom of the flask. For the evolution experiment, O. *tauri* populations were grown either in the control environment as described above, in low light, low phosphate, low salinity or high temperature (Table 1), or a changing environment (random) in which one of the environments from those listed was chosen at random at each transfer. We refer to the environments where the populations evolved as "selection environments". Populations were grown in the absence of viruses for 32 weeks, corresponding to approximately 200 generations.

Table 1. A comparison of the control environment and the treatments used for each selection environment used in this study.

Selection Environment	Control	Treatment
Light (μmol m^{-2} s^{-1})	85	60
Phosphate (μM)	10	5
Salinity (ppt)	30	25
Temperature (°C)	18	20

For the low light environment, culture flasks were wrapped in 0.15 neutral density foil to reduce light intensity. For the low phosphate environment, phosphate was reduced by preparing Keller medium with half the amount of β-glycerophosphate present in the control media. For low salt, Instant Ocean was added to reach a salinity of 25 ppt. Cultures in the high temperature environment were maintained on a heat mat (Exo Terra Heat Wave substrate heat mat, Yorkshire, UK) set at 20 °C. These selection environments were chosen so that the populations responded to them by changing their growth rates relative to the control environment—in batch culture rapid growth is favored by natural selection, so any environment that decreases growth rates should then result in natural selection for traits that will allow cell division rates to recover in that environment. However, the selection environments were not extreme, so that populations were still able to grow at a measurable rate and survive the dilution rate of the experiment. This is in part so that a similar number of generations elapse in all environments over the course of the experiment.

2.3. Testing RP Lines for Viral Production

All resistant producer (RP) lines were tested for viral production prior to the start of the experiment. To check whether the three producing lines (NG'10, NG'16 and NG27) were releasing infectious viruses, we used the supernatant to infect susceptible O. *tauri* strain RCC4221. Two milliliters of each population were transferred to an Eppendorf tube and centrifuged at $8000 \times g$ for 15 min. Four hundred milliliters of the supernatant were removed carefully without drawing up any of the cells from the pellet at the bottom of the tube, and used to inoculate 1 mL of susceptible *O. tauri*. OtV5 was used as a positive control and Keller media was used as a negative control. Eight replicates were performed before the experiment was started. The test was performed every four weeks with three replicates per population. Samples were checked for lysis either by observing by eye whether they were green or clear, or by measuring cell densities using a BD FACSCanto II (BD Biosciences, Oxford, UK) flow cytometer.

In addition to liquid lysis tests, frozen stocks of RP supernatant were made by adding dimethyl sulfoxide (DMSO) (final concentration 10%) and storing at $-80\,°C$. We tested these samples for viruses using the plaque assay technique [34]. A 1.5% agarose suspension was made and 5 mL aliquots were prepared in Falcon tubes and held at $70\,°C$ in a water bath. In a 50 mL Falcon tube, 30 mL exponentially growing *O. tauri* culture, 15 mL Keller media and 5 mL agarose were mixed rapidly but gently by inverting the tube (final agarose concentration 0.15%). The agarose was poured into a 12 cm square petri dish and left to set. Tenfold serial dilutions of the RP supernatant were made in 96-well plates using one row per sample. A Boekel Replicator was used to transfer all of the serial dilutions from one 96-well plate to one square petri dish. The replicator was sterilized between each use using ethanol and a flame. Petri dishes were checked daily for lysis plaques for a maximum of 10 days.

2.4. Testing Resistance Type Using OtV5 Inoculation

OtV5 inoculum was prepared prior to the start of the experiment and stored at $-80\,°C$ in 10% DMSO (final concentration) and inoculations were performed from the frozen stocks. The experiment did not include a co-evolving virus which allowed us to measure host evolution relative to the ancestral virus. After 32 weeks of evolution, each population was inoculated with a suspension of OtV5 particles to test whether it was susceptible or resistant to viral lysis. Samples were tested by inoculating 1 mL cell culture at a density of 10^5 with 10 µL OtV5 in 48-well plates with three replicates for each sample. Negative controls that were not inoculated with OtV5 were used as a comparison of cell growth. Cell density was measured using a FACSCanto flow cytometer 3 days after inoculation. Samples were run on 96-well plates by counting the total number of cells in 10 µL with a flow rate of 2.0 µL per second.

Data were analyzed with linear mixed effects models using the statistical packages lme4 [38] and lmerTest [39] in R (version 3.2.0, R Core Team, Vienna, Austria) to identify differences in cell densities after OtV5 inoculation compared to controls that were not inoculated. Selection environment, resistance type and treatment (inoculated or not inoculated) were set as fixed effects with population as a random effect. Post hoc Tukey tests were performed using lsmeans to confirm where significant differences occurred within the different effects.

2.5. Population Growth Rates, Cell Size and Cell Chlorophyll Content after Evolution

At the end of the evolution experiment, we quantified evolutionary responses by measuring average cell division rates and by measuring cell size and chlorophyll content for each population. All evolved populations were assayed in their selection environment and in the control environment, and all control populations were assayed in all selection environments except high temperature, since all populations in the high temperature environment went extinct and therefore there were no high temperature evolved strains. The populations that had evolved in a random environment for each transfer were only assayed in the control environment, which was not one of the environments they had been exposed to during the experiment, meaning only a correlated response (rather than a direct response) to selection could be obtained. Each population was assayed in triplicate. Due to the size of the experiment, assays were divided randomly into seven time blocks. This was factored into the statistical analysis.

Average cell division rates, which we refer to as "growth rates" are the average number of cell divisions per day over seven days, which corresponds to one transfer cycle. All populations were first maintained in their assay environment for an acclimation period of one week, which was one full transfer cycle, prior to measuring growth rates. After acclimation, cells were counted using a FACSCanto flow cytometer before the transfer into the assay environment (to calculate the number of cells transferred into fresh media) and again after seven days of growth. Each sample was counted in triplicate. The cell counts were converted to cells per milliliter and the number of divisions per day was calculated using Equation (1).

$$\mu\left(d^{-1}\right) = \frac{\log_2\left(\frac{N_t}{N_0}\right)}{t - t_0} \tag{1}$$

where μ is population growth rate, and N_t and N_0 are the cell densities (cells mL^{-1}) at times t and t_0 (days), respectively. This measures the average number of cell divisions per ancestor over a single growth cycle and allows a comparison of offspring production between environments even if there are differences in the shape of the population growth curve, or in cases where r cannot be accurately estimated. To avoid biases of cell divisions being dependent on the time of the cell cycle, cells were always measured at the same time of day (at the beginning of the light period when cells are in G1 phase).

Cell size was inferred from FSC (forward scatter), which was calibrated using beads of known sizes (1 µm, 3 µm and 6.6 µm). Chlorophyll fluorescence was inferred by measuring PerCP-Cy5.5 emission with excitation at 488 nm. Relative chlorophyll was analyzed by taking the average chlorophyll fluorescence for all susceptible strains in the control environment and setting this to a value of 1, with chlorophyll measurements of all other strains relative to this value.

Data were analyzed with linear mixed effects models. To analyze differences in growth rate, cell size and chlorophyll under different environments, selection environment, assay environment and resistance type were fixed effects and population and block ware random effects that were treated as un-nested. An additional model was fitted to examine whether there was a difference in growth rate when populations were assayed in their selection environment or when they were assayed in a different environment, with assay as the only fixed effect and population and block set as random effects.

2.6. Competition Assay

To measure competitive fitness, all evolved populations were competed against a green fluorescent protein (GFP) line of *O. tauri*. A Gateway enabled entry clone containing roGFP2 was obtained by linearizing pH2GW7-roGFP2 [40] with EcoRV. The linearized vector was recombined with pDONR207, creating a pDONR207-roGFP2 clone. A pOtOX binary vector [41] was adapted to become a Gateway® destination vector and pDON207-roGFP2 was recombined into the vector, downstream of the high-affinity phosphate transporter (HAPT) promoter [41]. The pOtOx-roGFP2 vector was subsequently transformed into *O. tauri* using the procedure previously described [42].

All evolved populations competed in the selection environment that they evolved in, and all control populations competed in the control environment as well as in each selection environment to measure plastic response. All of the random populations competed in the control environment. All populations, including the roGFP line, were acclimated for one week in the corresponding assay environment prior to the assay. Equal starting densities of 5×10^5 of each evolved population and the roGFP line were grown in 20 mL media for one week, after which cells were counted using a FACSCanto flow cytometer. GFP and non-GFP populations were distinguished by measuring fluorescein isothiocyanate A (FITC-A) emission at 519 nm with excitation at 495 nm. Competitiveness of the evolved populations was measured relative to the roGFP line as fold change in cell density. Data were analyzed with a linear mixed effects model, with selection environment, assay environment and resistance type as fixed effects and population and assay replicate as random effects.

3. Results

3.1. Susceptibility to OtV5 after Evolution

3.1.1. Host Resistance Type Was Maintained during Evolution

After 200 generations of evolution in the selection environments, all surviving R and RP populations remained resistant to OtV5 lysis and all S populations remained susceptible to viral lysis in those environments (Figure 1). A significant interaction between selection environment, resistance type and treatment (OtV5 inoculation) affected susceptibility of *O. tauri* to OtV5 (ANOVA environment × resistance type × treatment, $F_{8,238} = 15.22$, $p < 0.0001$). A post hoc Tukey test showed that this was due to cell lysis of S populations ($t_{8,238} = 10.66$, $p < 0.001$), whereas cell density of R and RP lines did not decrease compared to controls that were not inoculated. The highest cell densities were observed in the low salt (post hoc Tukey test, $t_{8,238} = -29.90$, $p < 0.0001$) and random (post hoc Tukey test, $t_{8,238} = -7.54$, $p < 0.0001$) environments. The OtV5-inoculated S populations in low phosphate were the only populations where cell density fell below the starting cell density across all populations, indicating almost complete cell lysis and no cell growth for this combination of resistance type and selection environment. R and RP lines did not show decreases in cell density after inoculation with OtV5 compared to controls that were not inoculated, whereas S lines did.

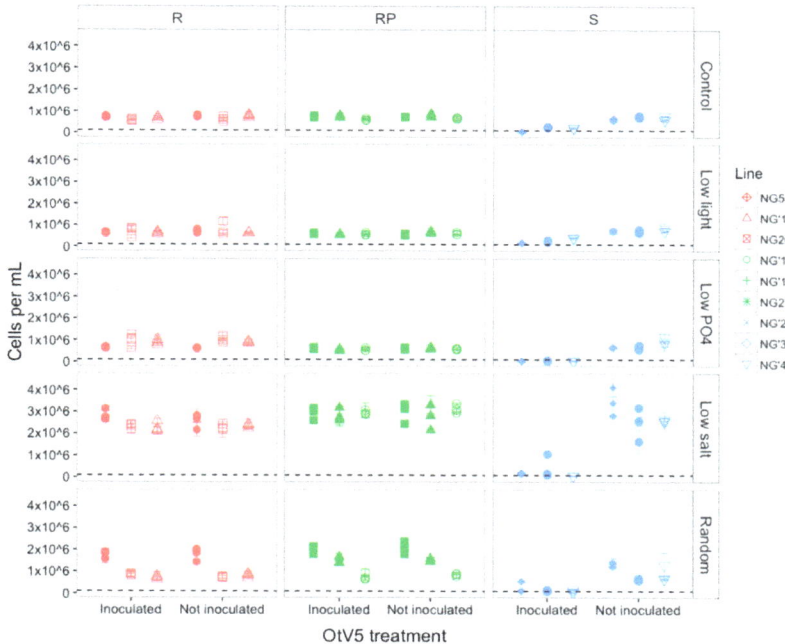

Figure 1. Mean (± SE) cell density mL^{-1} of resistant (R), resistant producer (RP) and susceptible (S) *O. tauri* lines three days after OtV5 inoculation in five environments. Points represent the average of the three assay replicates for each evolved population. Inoculated = populations inoculated with OtV5, Not inoculated = negative control populations that were grown for the same period without OtV5 inoculation. There were three evolved populations of each line. The dashed line represents the starting cell density at 100,000 cell mL^{-1}.

R and RP populations did not show a significant difference in cell density between populations that had been inoculated with OtV5 and populations that had not (Figure S1). In contrast, all S populations inoculated with OtV5 showed a change in cell density relative to non-inoculated S populations in the same environments (ANOVA effect of resistance type on difference $F_{2,125} = 66.51$, $p < 0.0001$). The largest differences in cell densities between inoculated and non-inoculated populations were observed in S populations evolved in the low salt environment, showing that whilst all populations in this environment were able to reach high densities in the absence of viruses, they were unable to grow in the presence of OtV5 (Figure 1). The large difference in S populations in low salt was due to the high growth rate of populations that had not been inoculated, since inoculated populations did not fall to lower densities than inoculated S populations in any other environments.

3.1.2. OtV5-Mediated Lysis Decreased in Susceptible Populations

Although S populations remained sensitive to viral lysis at the end of the evolution experiment, complete lysis was not observed in all populations, with a small proportion of populations able to reach numbers above the starting density of 100,000 cells mL^{-1} (Figure 1). This was in contrast to the beginning of the evolution experiment, when all susceptible populations fell below 100,000 cells mL^{-1} after inoculation with OtV5, indicating near-complete lysis (ANOVA effect of time point on cell density, $F_{1,65} = 21.87$, $p < 0.0001$) (Figure 2). The highest proportion of S cells that did not lyse was found in low salt evolved populations, suggesting that resistance mutations had been maintained in this environment, despite no selection by OtV5. To eliminate the possibility that the infection dynamics had changed and that the population decline was still in process, we measured the population density seven days after inoculation and did not observe any further decrease in population density (Figure S2).

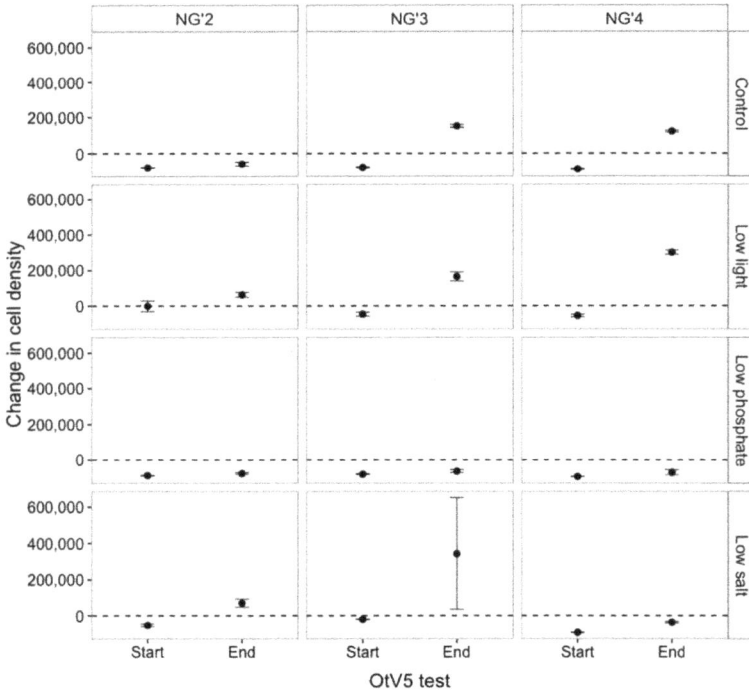

Figure 2. Change in cell density of the susceptible lines NG′2, NG′3 and NG′4 after OtV5 inoculation one week into the selection experiment (Start) and after 32 transfer cycles of evolution (End). The dashed line represents no change.

3.1.3. RPs Stopped Producing Viruses Early in the Evolution Experiment

During the evolution experiment, RP populations (NG27, NG'10 and NG'16) were tested to check that they were still producing viruses. Seven transfer cycles into the evolution experiment, all NG27 populations in all environments were still producing infectious viruses, as observed by cell lysis when their supernatant was used to inoculate the susceptible *O. tauri* strain RCC4221. In contrast, RCC4221 cultures that were inoculated with the supernatant of all populations of NG'10 and NG'16 continued growing, showing that no observable lysis had occurred. After 17 transfers in the selection environments, all RP populations in all environments had stopped producing infectious viruses (Figure S3), as observed by flow cytometric cell counts of RCC4221 populations inoculated with the supernatant of RP populations. When it was clear that all RP populations had stopped producing infectious viruses, frozen supernatant samples collected at transfers 9, 12, 14 and 15 were tested using the plaque assay method. No plaques were observed in any samples tested, thus we concluded that all RP populations in all environments had stopped producing viruses within nine weeks of the selection experiment.

3.2. Changes in Trait Values after Evolution

3.2.1. Changes in Cell Division Rate and Population Persistence during the Selection Experiment

Here, we focus on how growth rates vary with resistance type, selection environment and the number of transfer cycles in the selection environment. Growth rates of all populations were measured as the number of cell divisions per day, at four time points during the experiment (including at the beginning and end) (Figure 3). When comparing these time points, growth was significantly affected by environment, resistance type and time point ($p < 0.0001$ for all effects). In the first transfer cycle, which measured the population growth rates at the very start of the experiment following one week of acclimation, two out of the three RP lines (NG'10 and NG'16) had increased growth rates across all environments except for low phosphate (ANOVA effect of growth rate on cell divisions, $F_{3,5} = 17.19$, $p = 0.046$). These results are reported in [43].

After 14 transfer cycles, growth rates of all populations were approximately one division per day in the high salt, low phosphate, low light and random environments (Figure 3). In the control environment, growth rate varied across all S lines, even between populations of the same starting line, ranging from 0.18 to 0.87 divisions per day. The increased growth of all lines evolving in low phosphate to one division per day, which is the normal growth rate reported for *O. tauri* in phosphate-replete media, is consistent with adaptation to low phosphate in less than 100 generations. Additionally, RP lines that had been dividing more rapidly at transfer 1 were dividing at the same rate as other lines within each environment (Figure 3). This may be because the RP populations had stopped producing viruses and shifted to the R resistance type (see Section 3.1.3), thereby losing the growth advantage associated with the RP resistance type early on in this experiment. By transfer 24, all populations in the high temperature environment had gone extinct. RP populations went extinct more quickly than S and R populations, with 66% of RP lines extinct by T14 compared to 33% and 22% of S and R, respectively (Figure 3). At transfer 20, only three high temperature populations remained: one S (NG'4) and two R (NG'13 and NG26).

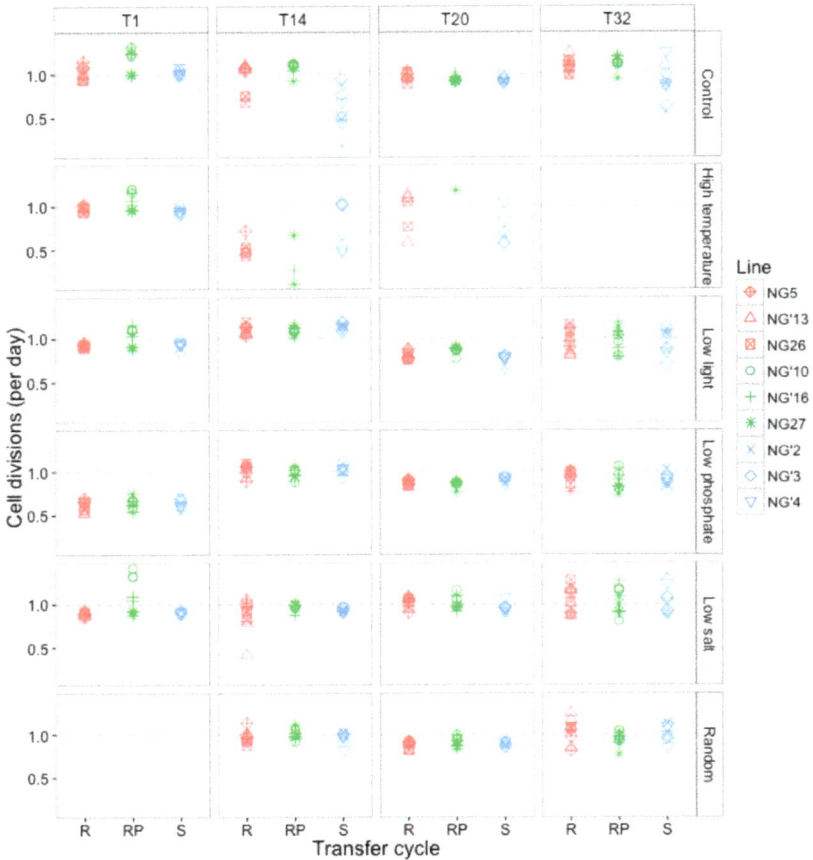

Figure 3. Growth rates as measured by mean cell divisions per day for each evolving population over four time points (1, 14, 20 and 32 transfer cycles). The dashed line represents one cell division per day. T1 is the growth rate following acclimation at the beginning of the experiment. There are no growth measurements for the randomized environment at T1 because lines had only been growing for one transfer cycle.

3.2.2. Growth Rates Varied with Selection Environment and Assay Environment after Evolution

After approximately 200 generations of evolution in each environment, a transplant assay was performed to quantify environmental effects on population growth rate, cell size and cell chlorophyll content for each evolved population. Here we define the selection environment as the environment that the population evolved in, and the assay environment as the environment in which measurements were taken. The direct response to selection compares the growth rate of a population evolved in a given selection environment with the growth rate of a population evolved in the control environment when both are grown (separately) in that given selection environment. The effect of selection environment on the direct response to evolution was large, and driven by the direct response to selection in the low phosphate environment (ANOVA effect selection environment on direct response, $F_{2,228} = 9.26$, $p = 0.0001$), whereas the effect of resistance type was smaller (ANOVA effect of resistance type on direct response, $F_{2,228}$ 2.87, $p = 0.06$).

Selection environment alone and assay environment alone both had a significant effect on population growth rate (ANOVA effect of selection environment on growth, $F_{4,200} = 19.92$, $p < 0.0001$; ANOVA effect of assay environment on growth, $F_{3,758} = 32.43$, $p < 0.0001$), which shows that environment affected growth rates. Resistance type also had an effect on growth rate (ANOVA effect of resistance type on growth, $F_{2,195} = 4.21$, $p = 0.02$), with R populations having the fastest cell division rates and S populations having the slowest cell division rates. Additionally, an interaction between selection environment and assay environment affected growth rate, indicating that the way in which selection environment affected growth differed between assay environments (ANOVA selection environment × assay environment, $F_{3,757} = 2.89$, $p = 0.03$). The fastest growth rates were seen in the evolved control populations that were assayed in low salt (Figure 4). Better performance was not due to being assayed in the same selection environment that the populations had evolved in (ANOVA effect of being assayed in selection environment on growth, $F_{1,831} = 1.70$, $p = 0.19$).

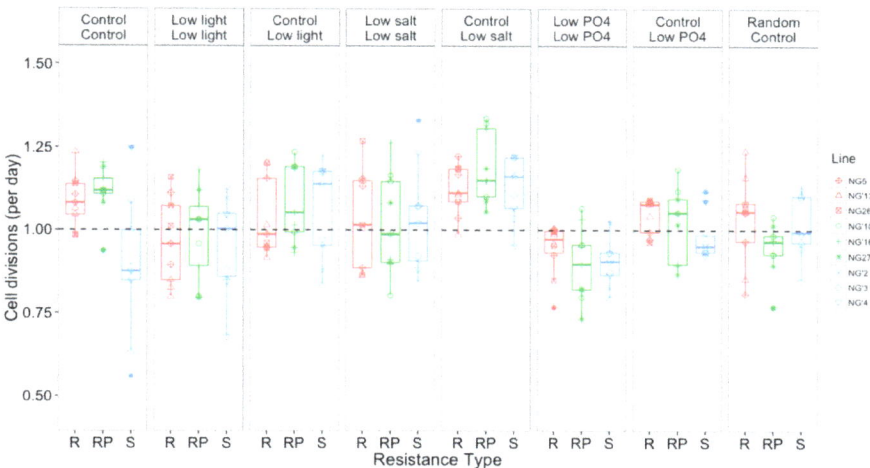

Figure 4. Mean cell divisions per day (±SEM). R = resistant, RP = resistant producer, S = susceptible. Each panel represents a growth assay, with cells evolved in the selection environment (top label) and growth rates measured in the assay environment (bottom label). The dashed line indicates, for reference, one cell division per day.

3.2.3. Resistance Type Affected Cell Size and Chlorophyll Content

Cells from different resistance types had different cell sizes (ANOVA effect of resistance type on size, $F_{2,140} = 9.49$, $p = 0.0001$) (Figure S4) and this was not affected during evolution in any of the environments (ANOVA effect of selection environment on size, $F_{4,155} = 0.66$, $p = 0.62$; ANOVA effect of assay environment on size, $F_{3,735} = 1.60$, $p = 0.19$). The greatest variation in cell size between populations was observed when control-evolved cells were assayed in low salt (0.86–0.97 µm) across all resistance types. Less variation was found in the control-evolved cells assayed in low phosphate (0.82–0.97 µm).

The environment in which populations were assayed had a significant effect on the relative chlorophyll content per cell volume (ANOVA effect of assay environment on chlorophyll, $F_{3,744} = 17.83$, $p < 0.0001$). However, selection environment did not (ANOVA effect of selection environment on chlorophyll, $F_{4,168} = 0.90$, $p = 0.47$). Resistance type affected chlorophyll content (ANOVA effect of resistance type on chlorophyll, $F_{2,153} = 8.54$, $p < 0.0001$). Susceptible populations that had been evolving in the control environment contained high amounts of chlorophyll relative to their cell size when assayed under all three selection environments (low light, low salt and low phosphate) (Figure S5).

3.3. Selection and Assay Environments Affect Competitive Ability of O. tauri

In addition to measuring growth rate, size and chlorophyll content, we also tested if costs of resistance could be observed during pairwise competition between each population of S, R, and RP. We measured relative competitive ability, by competing each population against a common competitor harboring a GFP reporter, which allowed us to distinguish between the evolved population and the GFP line. Both selection environment and assay environment affected competitive ability against a roGFP-labeled strain (ANOVA effect of selection environment on competitiveness, $F_{4,622} = 16.41$, $p = < 0.0001$; ANOVA effect of assay environment on competitiveness, $F_{3,622} = 10.96$, $p < 0.0001$). Most populations were poor competitors relative to the roGFP line (Figure 5). Lines evolved in low light and low salt were the best competitors. Lines that were assayed in the same environment that they had evolved in were better competitors than control lines that were assayed in the selection environments. This shows that these lines adapted to their selection environment and that growth rate is not necessarily the most appropriate measure of adaptation in this study, which is consistent with other studies in *Ostreococcus* spp. [44]. Interestingly, populations in the control environment were the worst competitors, regardless of resistance type, with a 0.56 mean fold change, showing that all populations were out-competed by the roGFP line. This indicates that the control environment did in fact exert less selection on the populations than did the other environments.

Resistance type alone did not affect competitive ability (ANOVA effect of resistance type of competitiveness, $F_{2,622} = 1.22$, $p = 0.30$). Although competitive ability differed between resistance types, the response was not consistent across assay environments, with no one resistance type consistently being a better or poorer competitor.

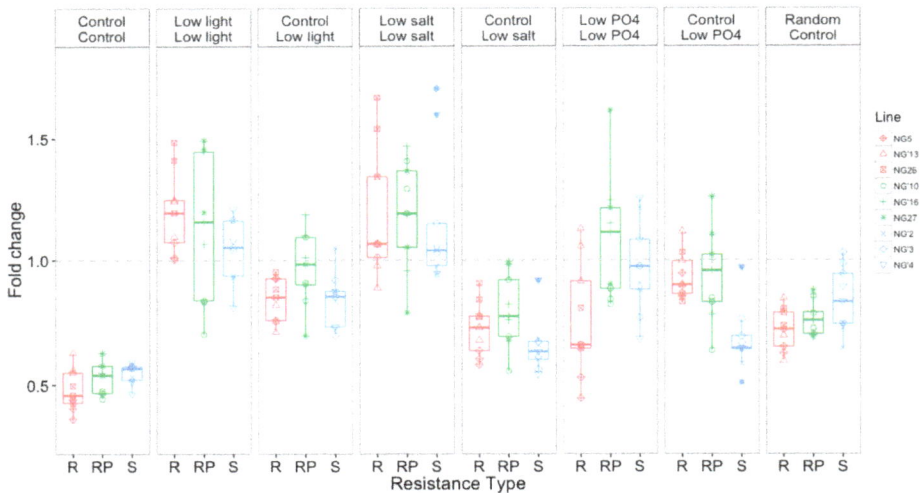

Figure 5. Competitive ability, as measured by fold difference in growth relative to a roGFP-modified *O. tauri* line, of evolved populations and control populations assayed in the selection environments. R = resistant, RP= resistant producer, S = susceptible. Each panel represents one assay, with populations evolved in the selection environment (top label) and competitiveness measured in the assay environment (bottom label). The dashed line represents no change (i.e., equal proportions of roGFP and competitor populations).

4. Discussion

We examined whether cost of resistance varied with the abiotic environment in which *O. tauri* populations evolved. A cost of resistance can manifest in different ways depending on the interaction between host and virus and on the way in which resistance is acquired (e.g., entry of the virus into the cell, and ability of the virus to replicate within the cell and cause lysis). This means that it is often difficult to detect a cost of resistance, so we measured three host responses: ability to maintain resistance, population growth rate and competitive ability.

4.1. Susceptibility to OtV5 Did Not Change after Evolution

After evolution in a new environment, OtV5 was still able to lyse susceptible (S) *O. tauri* populations under all environmental conditions tested, whereas R and RP populations remained resistant under all environments, despite the absence of selection pressure for viral resistance (Figure 1). Resistance to pathogens often comes at a fitness cost, such that a proportion of susceptible individuals remain in the population, thereby allowing viruses to persist [21]. If resistance does carry a fitness cost, populations should revert to susceptibility over time, in the prolonged absence of viruses, even if that cost is low, because susceptible cells have a fitness advantage in the absence of viruses [29]. Our study indicated that if there is a cost to simply maintaining resistance in *O. tauri*, it is small. Over the time scale of our experiment, the fitness advantage of susceptible types in the absence of viruses would have to be about 0.005 for a mutation conferring susceptibility in a resistant background to be fixed in the population following a spontaneous reversion of a resistant cell (where we calculate s from $\frac{1}{2}s/(1-e^{-2sN})$, and assume a starting frequency of $1/N$ [45]).

It is possible that there is a genetic constraint preventing the loss of resistance, making the transition from resistant to susceptible phenotypes rare even if resistance is costly. This is consistent with recent studies showing that the resistance mechanism in *O. tauri* is an intracellular response [35] and probably also involves rearrangements of chromosome 19 [36]. The presence of a genetic constraint on losing resistance would favor compensatory mutations that lead to alleles being selected that reduce the cost of resistance [46,47]. Studies evolving *E. coli* in the absence of bacteriophage observed that the cost of resistance to the T4 bacteriophage decreased after 400 generations due to compensatory adaptations [46]. A second possibility is that the cost of resistance to one strain of OtV means increased susceptibility to other virus strains. For example, cyanobacteria can rapidly evolve viral resistance when coevolving with viruses, however increased resistance to one virus can lead to a narrower resistance range thereby making cells more susceptible to other virus strains [27,28]. *O. tauri*-virus interactions can be complex with some OtVs being very specific to host *O. tauri* strains while others are generalists that can infect many strains [26,48]. Our experiment focused only on OtV5 and did not examine evolution of host resistance range.

At the end of the evolution experiment, OtV5 lysed susceptible (S) populations in all environments, but the extent of lysis differed between environments (Figure 1). This could be because one or more resistance mutations had appeared and risen to a detectable frequency in some populations. It is unclear whether incomplete lysis was due to some resistant cells evolving in the susceptible populations, or whether susceptible populations had evolved to make virus entry harder but still possible. Inoculations were performed from frozen stocks, thus OtV5 was not coevolving with the host, enabling us to measure evolution in the *O. tauri* populations relative to the ancestral virus population. We cannot rule out the possibility that there was a slow loss of infective virus titer in the cryopreserved stock, leading to fewer infectious viruses in the inoculum and therefore a lower multiplicity of infection. Physiological changes in susceptible populations arising as an adaptive response to abiotic environmental change did not prevent viral lysis, indicating that viral adsorption was not completely inhibited. This was even evident in the control populations, suggesting that although these populations did not experience a change in environment, they may have evolved changes in cell surface proteins, since were still evolving for the full length of the experiment. However, the biotic environment plays a larger role in resistance acquisition, since resistance to viruses is selected for by the virus [49]. Chemostat experiments to monitor population dynamics in *Chlorella* and *Paramecium bursaria Chlorella* Virus 1 (PBCV-1) showed that control populations maintained in the absence of viruses did not evolve resistance to the ancestor virus, suggesting that resistance arises from host-virus interactions [23]. In contrast, sensitive *E. coli* cells evolved complete resistance to λ infection and resistant cells increased susceptibility to T6* infection after 45,000 generations in the absence of phage [29]. In our experiment, low phosphate was the only environment in which the cell numbers of all lines fell below the starting cell density (Figure 1), suggesting that this environment either affected the infectivity of OtV5 directly or the cells' response to infection. Other studies report the opposite, with reduced virus infection of algae under low phosphate, possibly due to the requirement of phosphate for viral replication [9,10,13]. Though phosphate levels were low in our experiment, they were sufficient for population growth to be positive, and were higher than found in the Mediterranean Sea [50]. Conflicting results highlight the complexity of host-virus interactions in different study systems as well as different growth conditions.

There was a selection pressure against viral production on RP lines, but not on host resistance across all RP lines in all environments. Similarly, Yau et al. reported that over a two year period RP populations maintained under standard laboratory conditions stopped producing viruses [36]. If RP populations are indeed made up of a majority of resistant cells with a small proportion of susceptible cells arising that lyse upon OtV5 infection, thus maintaining the production of viruses in the media, then we would expect resistance to be selected for in the presence of viruses. Resistance in *O. tauri* is expected to be caused by over-expression of glycosyltransferase genes on chromosome 19 [36]. In this study, the selection environment did not affect the time it took for a selective sweep of resistance to occur in the RP lines, supporting the conclusion that there was little or no selection against resistance, that there is a genetic constraint on losing resistance, or that compensatory mutations enabled resistance to be maintained.

4.2. Resistance Type and Environment Affect Evolutionary Response of O. tauri *to Environmental Change*

We did not observe a growth cost of *O. tauri* being resistant to viral lysis, since R populations had the fastest growth overall whereas S populations had the slowest growth. Data on the growth effects of resistance in marine algae are rare. A 20% reduction in growth was reported in the ubiquitous cyanobacterium *Synechococcus* [22], however it is unknown whether viral resistance generally carries a growth cost in eukaryotic algae. Even with no or minimal costs of resistance, the chromosomal rearrangement associated with resistance in *O. tauri* means that the different resistance types have different genetic backgrounds. Therefore, evolution could take different trajectories in hosts with different resistance types due to epistatic interactions between resistance and adaptive changes. For example, trade-off shape varied in response to environmental change and physiological changes of bacteriophage resistant *E. coli*, leading to variation in sensitivity to environmental change across different strains [51]. In our study, when considering the direct response to evolution (which compares the growth rate of the evolved population in its selection environment with the plastic response of the control line in that selection environment), resistance type did not drive direct response. This indicates that the growth response of the three resistance types was similar within environments. If there is an effect of genetic background being introduced by resistance, it is not evident at the level of growth rate under these conditions.

Selection environment affected population growth, with populations evolved in the control environment having the highest growth rates in all assay environments (Figure 4). The decrease in growth in response to our selection environments is consistent with them being of lower quality than the control environment, by design, so that selection was stronger in the non-control environments. Variation in the direct response to evolution was explained by selection environment. Populations evolved in low phosphate had the lowest growth, which is expected when cells are nutrient limited. Interestingly, populations that had evolved in the control environment grew more rapidly in low phosphate than populations that had evolved in low phosphate. This may be because populations that had been evolving in the control environment had enough phosphate reserves within the cell to grow normally for a short period, since growth was only assayed for seven days. Overall, growth rates of populations evolved in the control environment were greater when assayed in the selection environments than the populations that had evolved in those environments, showing that increased growth could be initiated as a stress response, and that cells in the control environment (which was nutrient-replete, and at the optimal temperature and usual salinity for these lines of *O. tauri*) were in better condition overall. The extent of a cost of resistance can be highly dependent on environment. For example, cost of resistance differs when fitness of *E. coli* is measured under different nutrient resources and concentrations [20,52]. We show here that growth rate measurements may not be sensitive enough to detect very small differences between populations conferring a cost of resistance in *O. tauri*, as has also been observed in short term experiments using a single [35] and multiple environments [43]. Studies in bacteria also found that resistant strains grew at the same rate as susceptible strains [21,46]. Our results indicate that, regardless of resistance type, *O. tauri* is able to adapt to environmental change including low light, low salt and low phosphate. However, all populations in the high temperature environment went extinct, despite the modest (2 °C) increase, suggesting that although *O. tauri* can tolerate and grow at higher temperatures over the short-term, sustained temperature increases may exert stronger selection than predicted from short-term studies. It is not possible to infer as of yet whether resistance affects growth rate in natural habitats or whether a cost of resistance is instead associated with tradeoffs that are not related to the abiotic environment, such as resistance to other viral strains.

In contrast to cell division rates, resistance type affected cell size and chlorophyll content, but selection environment did not. Cells in RP populations were sometimes larger in size and S populations were slightly smaller. Often, small size is associated with a response to nutrient limitation, increased temperature and light limitation in phytoplankton [53–57], however all lines in this study showed slightly increased cell size in low phosphate. An increased cell volume has been observed in coccolithophores in response to phosphate limitation suggesting the adaptive strategy is to reduce phosphorous requirements rather than increasing surface area to volume ratio [58].

RP populations had less chlorophyll in most environments, however overall there was substantial variation in chlorophyll content, especially in S populations. When assayed in the control environment, populations that had evolved in low light, low salt, low phosphate and the random environment had lower chlorophyll than did control populations assayed in these same environments. The response of populations evolved in the control environment increasing their relative chlorophyll content when assayed in low light is consistent with responses to light limitation in other green algae [59–61]. Here, we show that response of chlorophyll content to environmental change is variable, both with environment and with resistance type. Previous studies in marine microalgae have reported lower reduced chlorophyll content under nutrient limitation [62] and higher chlorophyll content under some optimal salinities [63,64].

4.3. Resistance Type Did Not Affect Competitive Ability Regardless of Environment

Reduced competitive ability is often one of the main restrictions for resistance spreading through a population, however resistance type did not affect the competitive ability of evolved populations in our experiment. We found that environment did affect competitive ability, and similarly in bacteria, the environment that populations evolve in, such as the limiting sugar source or spatial heterogeneity, can affect competitive ability, both with and without coevolving phage [51,52,65]. Other studies have reported a trade-off between competitive ability and resistance, whereas here we found no evidence for reduced resistance with increased competitive ability. The nature of a cost of resistance will depend on the genetic or physiological changes to the cell. For example, *E. coli* mutants showed high variability in competitiveness which was associated with resistance strategy, with cross-resistance to phage T7 significantly decreasing competitive fitness by approximately 3-fold [21]. In contrast, competitions with cyanobacteria showed that total resistance (the total number of viruses to which a host strain was resistant) did not affect competitive ability [22]. These examples reveal that the magnitude of the reduced competitiveness trade-off can depend on the specific resistance strategy.

Evolved populations in the non-control environments were better competitors than control populations that had been exposed to the selection environments for the first time (plastic response), indicating that all lines had adapted to their selection environment. Thus, growth rate is not the most appropriate measure of adaptation in *O. tauri*, since the plastic response was to increase population growth rates, and the evolutionary response was to reverse this plastic increase in growth rates, and this strategy was associated with an increase in competitive fitness. Similar results have been reported previously in *Ostreococcus* spp. where populations with high growth rates in monoculture were poorer competitors than those with lower growth rates in monoculture [44].

5. Conclusions

Here, we show that there was no detectable cost of resistance to OtV5 as measured by growth rate or competitive ability for *O. tauri* evolved in several different environments, and that resistance to viruses did not affect adaptation to environmental change. Additionally, we found no reversion of R or RP populations to S as tested by exposure to OtV5, whereas lysis occurred in all S populations. Additionally, all RP lines stopped producing viruses within nine weeks of the experiment. This suggests that a shift from susceptibility to resistance is more common than a shift from resistance to susceptibility, regardless of selection environment, at least for the range of environments used here. Our experiment shows that the conditions under which a cost of resistance may occur or affect adaptation in *O. tauri* are not clear in the laboratory. More work is needed to understand the factors that affect host–virus interactions in the marine environment to better understand evolutionary and ecological responses of marine eukaryotic microalgae to environment change.

Supplementary Materials: Supplementary Information is available online at www.mdpi.com/1999-4915/9/3/39/s1. The pHAPT-roGFP line (CCAP 157/4) and its untransformed parent strain (CCAP 157/2) will be publicly distributed by CCAP https://www.ccap.ac.uk/.

Data: All data is available from DataDryad doi:10.5061/dryad.vr3hk.

Acknowledgments: This work was supported by a BBSRC EASTBIO Doctoral Training Programme grant [BB/J01446X/1] awarded to SEH and BBSRC awards [BB/D019621 and BB/J009423] to Andrew Millar and others. SC is supported by a Royal Society University Research Fellowship and an ERC starting grant (260266). PFV is supported by a strategic award from the Wellcome Trust for the Centre for Immunity, Infection and Evolution (http://ciie.bio.ed.ac.uk; grant reference no. 095831), a Chancellor's Fellowship from the School of Biological Sciences—University of Edinburgh, and by a Society in Science—Branco Weiss fellowship ([http://www.society-in-science.org)] http://www.society-in-science.org), administered by ETH Zürich. We thank the lab group at Observatoire Océanologique, Banyuls-sur-Mer, for help with virus culturing techniques and for providing the *O. tauri* lines. We thank A. Millar for facilities and funding to create the roGFP line and K. Kis for technical assistance and maintaining the roGFP line. We are grateful to R. Lindberg and H. Kuehne for help with laboratory work and useful discussions.

Author Contributions: S.E.H., P.F.V. and S.C. conceived and designed the experiments; S.E.H. and K.K. performed the experiments; and S.E.H., P.F.V. and S.C. analyzed the data. All authors contributed to writing the paper.

Conflicts of Interest: The authors declare no conflict of interest.

References

1. Suttle, C.A. Marine viruses—Major players in the global ecosystem. *Nat. Rev. Microbiol.* **2007**, *5*, 801–812. [CrossRef] [PubMed]
2. Wilhelm, S.W.; Suttle, C.A. Viruses and nutrient cycles in the sea. *Bioscience* **1999**, *49*, 781–788. [CrossRef]
3. Koskella, B.; Brockhurst, M.A. Bacteria-phage coevolution as a driver of ecological and evolutionary processes in microbial communities. *FEMS Microbiol. Rev.* **2014**, *38*, 916–931. [CrossRef] [PubMed]
4. Dennehy, J.J. What Can Phages Tell Us about Host-Pathogen Coevolution? *Int. J. Evol. Biol.* **2012**, *2012*, 396165. [CrossRef] [PubMed]
5. Flint, S.J.; Enquist, L.W.; Krug, L.; Racaniella, V.; Skalka, A. *Principles of Virology: Molecular Biology, Pathogenesis, Virus Ecology*, 4th ed.; ASM Press: Washington, DC, USA, 2000.
6. Doney, S.C.; Ruckelshaus, M.; Emmett Duffy, J.; Barry, J.P.; Chan, F.; English, C.A.; Galindo, H.M.; Grebmeier, J.M.; Hollowed, A.B.; Knowlton, N.; et al. Climate Change Impacts on Marine Ecosystems. *Ann. Rev. Mar. Sci.* **2012**, *4*, 11–37. [CrossRef] [PubMed]

7. Nagasaki, K.; Yamaguchi, M. Effect of temperature on the algicidal activity and the stability of HaV (Heterosigma akashiwo virus). *Aquat. Microb. Ecol.* **1998**, *15*, 211–216. [CrossRef]

8. Wells, L.E.; Deming, J.W. Effects of temperature, salinity and clay particles on inactivation and decay of cold-active marine Bacteriophage 9A. *Aquat. Microb. Ecol.* **2006**, *45*, 31–39. [CrossRef]

9. Bellec, L.; Grimsley, N.; Derelle, E.; Moreau, H.; Desdevises, Y. Abundance, spatial distribution and genetic diversity of *Ostreococcus tauri* viruses in two different environments. *Environ. Microbiol. Rep.* **2010**, *2*, 313–321. [CrossRef] [PubMed]

10. Bratbak, G.; Egge, J.K.; Heldal, M. Viral mortality of the marine alga *Emiliania huxleyi* (Haptophyceae) and termination of algal blooms. *Mar. Ecol. Prog. Ser.* **1993**, *93*, 39–48. [CrossRef]

11. Bratbak, G.; Jacobsen, A.; Heldal, M.; Nagasaki, K.; Thingstad, F. Virus production in *Phaeocystis pouchetii* and its relation to host cell growth and nutrition. *Aquat. Microb. Ecol.* **1998**, *16*, 1–9. [CrossRef]

12. Wilson, W.H.; Carr, N.G.; Mann, N.H. The effect of phosphate status on the kinetics of cyanophage infection in the oceanic cyanobacterium *Synechococcus* sp. WH7803. *J. Phycol.* **1996**, *32*, 506–516. [CrossRef]

13. Maat, D.; Crawfurd, K.; Timmermans, K.; Brussard, C. Elevated CO_2 and phosphate limitation favor *Micromonas pusilla* through stimulated growth and reduced viral impact. *Appl. Environ. Microbiol.* **2014**, *80*, 3119–3127. [CrossRef] [PubMed]

14. Jacquet, S.; Bratbak, G. Effects of ultraviolet radiation on marine virus-phytoplankton interactions. *FEMS Microbiol. Ecol.* **2003**, *44*, 279–289. [CrossRef]

15. Jacquet, S.; Heldal, M.; Iglesias-Rodriguez, D.; Larsen, A.; Wilson, W.; Bratbak, G. Flow cytometric analysis of an *Emiliana huxleyi* bloom terminated by viral infection. *Aquat. Microb. Ecol.* **2002**, *27*, 111–124. [CrossRef]

16. Thyrhaug, R.; Larsen, A.; Brussaard, C.P.D.; Mcfadden, P. Cell Cycle Dependent Virus Production in Marine Phytoplankton 1. *Cell Cycle* **2002**, *343*, 338–343.

17. Larsen, J.B.; Larsen, A.; Thyrhaug, R.; Bratbak, G.; Sandaa, R.-A. Response of marine viral populations to a nutrient induced phytoplankton bloom at different pCO2 levels. *Biogeosci. Discuss.* **2007**, *4*, 3961–3985. [CrossRef]

18. Chen, S.; Gao, K.; Beardall, J. Viral attack exacerbates the susceptibility of a bloom-forming alga to ocean acidification. *Glob. Chang. Biol.* **2014**, *21*, 629–636. [CrossRef] [PubMed]

19. Danovaro, R.; Corinaldesi, C.; Dell'Anno, A.; Fuhrman, J.A.; Middelburg, J.J.; Noble, R.T.; Suttle, C.A. Marine viruses and global climate change. *FEMS Microbiol. Rev.* **2011**, *35*, 993–1034. [CrossRef] [PubMed]

20. Bohannan, B.J.M.; Kerr, B.; Jessup, C.M.; Hughes, J.B.; Sandvik, G. Trade-offs and coexistence in microbial microcosms. *Antonie van Leeuwenhoek* **2002**, *81*, 107–115. [CrossRef] [PubMed]

21. Lenski, R.E. Experimental Studies of Pleiotropy and Epistasis in *Escherichia coli*. I. Variation in Competitive Fitness Among Mutants Resistant to Virus T4. *Evolution* **1988**, *42*, 425–432. [CrossRef]

22. Lennon, J.T.; Khatana, S.A.M.; Marston, M.F.; Martiny, J.B.H. Is there a cost of virus resistance in marine cyanobacteria? *ISME J.* **2007**, *1*, 300–312. [CrossRef] [PubMed]

23. Frickel, J.; Sieber, M.; Becks, L. Eco-evolutionary dynamics in a coevolving host-virus system. *Ecol. Lett.* **2016**, *19*, 450–459. [CrossRef] [PubMed]

24. Seed, K.D.; Faruque, S.M.; Mekalanos, J.J.; Calderwood, S.B.; Qadri, F.; Camilli, A. Phase Variable O Antigen Biosynthetic Genes Control Expression of the Major Protective Antigen and Bacteriophage Receptor in Vibrio cholerae O1. *PLoS Pathog.* **2012**, *8*, e1002917. [CrossRef] [PubMed]

25. León, M.; Bastías, R. Virulence reduction in bacteriophage resistant bacteria. *Front. Microbiol.* **2015**, *6*, 1–7. [CrossRef] [PubMed]

26. Clerissi, C.; Desdevises, Y.; Grimsley, N. Prasinoviruses of the marine green alga *Ostreococcus tauri* are mainly species specific. *J. Virol.* **2012**, *86*, 4611–4619. [CrossRef] [PubMed]

27. Marston, M.F.; Pierciey, F.J.; Shepard, A.; Gearin, G.; Qi, J.; Yandava, C.; Schuster, S.C.; Henn, M.R.; Martiny, J.B.H. Rapid diversification of coevolving marine Synechococcus and a virus. *Proc. Natl. Acad. Sci. USA* **2012**, *109*, 4544–4549. [CrossRef] [PubMed]

28. Avrani, S.; Wurtzel, O.; Sharon, I.; Sorek, R.; Lindell, D. Genomic island variability facilitates *Prochlorococcus*-virus coexistence. *Nature* **2011**, *474*, 604–608. [CrossRef] [PubMed]

29. Meyer, J.R.; Agrawal, A.A.; Quick, R.T.; Dobias, D.T.; Schneider, D.; Lenski, R.E. Parallel changes in host resistance to viral infection during 45,000 generations of relaxed selection. *Evolution* **2010**, *64*, 3024–3034. [CrossRef] [PubMed]

30. Avrani, S.; Lindell, D. Convergent evolution toward an improved growth rate and a reduced resistance range in *Prochlorococcus* strains resistant to phage. *Proc. Natl. Acad. Sci. USA* **2015**, *112*, E2191–E2200. [CrossRef] [PubMed]

31. Avrani, S.; Schwartz, D.A.; Lindell, D. Virus-host swinging party in the oceans: Incorporating biological complexity into paradigms of antagonistic coexistence. *Mob. Genet. Elem.* **2012**, *2*, 88–95. [CrossRef] [PubMed]

32. Field, C.B. Primary Production of the Biosphere: Integrating Terrestrial and Oceanic Components. *Science* **1998**, *281*, 237–240. [CrossRef] [PubMed]

33. Bellec, L.; Grimsley, N.; Moreau, H.; Desdevises, Y. Phylogenetic analysis of new Prasinoviruses (*Phycodnaviridae*) that infect the green unicellular algae *Ostreococcus*, *Bathycoccus* and *Micromonas*. *Environ. Microbiol. Rep.* **2009**, *1*, 114–123. [CrossRef] [PubMed]

34. Derelle, E.; Ferraz, C.; Escande, M.-L.; Eychenié, S.; Cooke, R.; Piganeau, G.; Desdevises, Y.; Bellec, L.; Moreau, H.; Grimsley, N. Life-cycle and genome of OtV5, a large DNA virus of the pelagic marine unicellular green alga *Ostreococcus tauri*. *PLoS ONE* **2008**, *3*, e2250. [CrossRef] [PubMed]

35. Thomas, R.; Grimsley, N.; Escande, M.-L.; Subirana, L.; Derelle, E.; Moreau, H. Acquisition and maintenance of resistance to viruses in eukaryotic phytoplankton populations. *Environ. Microbiol.* **2011**, *13*, 1412–1420. [CrossRef] [PubMed]

36. Yau, S.; Hemon, C.; Derelle, E.; Moreau, H.; Piganeau, G.; Grimsley, N. A Viral Immunity Chromosome in the Marine Picoeukaryote, *Ostreococcus tauri*. *PLoS Pathog.* **2016**, *12*, e1005965. [CrossRef] [PubMed]

37. Keller, M.D.; Selvin, R.C.; Claus, W.; Guillard, R.R.L. Media for the culture of oceanic ultraphytoplankton. *J. Phycol.* **1987**, *23*, 633–638.

38. Bates, D.; Mächler, M.; Bolker, B.; Walker, S. Fitting Linear Mixed-Effects Models using lme4. *J. Stat. Softw.* **2014**, *67*, 51.

39. Kuznetsova, A.; Brockhoff, P.B.; Christensen, R.H.B. lmerTest: Tests in Linear Mixed Effects Models. Cran, R Package. 2015. Available online: http://CRAN.R-project.org/package=lmerTest (accessed on 15 January 2017).

40. Schwarzländer, M.; Fricker, M.; Müller, C.; Marty, L.; Brach, T.; Novak, J.; Sweetlove, L.; Hell, R.; Meyer, A. Confocal imaging of glutathione redox potential in living plant cells. *J. Microsc.* **2008**, *231*, 299–316. [CrossRef] [PubMed]

41. Corellou, F.; Schwartz, C.; Motta, J.-P.; Djouani-Tahri, E.B.; Sanchez, F.; Bouget, F.-Y. Clocks in the green lineage: Comparative functional analysis of the circadian architecture of the picoeukaryote *Ostreococcus*. *Plant Cell* **2009**, *21*, 3436–3449. [CrossRef] [PubMed]

42. Van Ooijen, G.; Knox, K.; Kis, K.; Bouget, F.-Y.; Millar, A.J. Genomic Transformation of the Picoeukaryote *Ostreococcus tauri*. *J. Vis. Exp.* **2012**, *65*, e4074. [CrossRef] [PubMed]

43. Heath, S.E.; Collins, S. Mode of resistance to viral lysis affects host growth across multiple environments in the marine picoeukaryote *Ostreococcus tauri*. *Environ. Microbiol.* **2016**, *18*, 4628–4639. [CrossRef] [PubMed]

44. Schaum, E.; Collins, S. Plasticity predicts evolution in a marine alga. *Proc. R. Soc. B* **2014**, *281*. [CrossRef] [PubMed]

45. Bell, G. *Selection: The Mechanism of Evolution*, 2nd ed.; Oxford University Press: Oxford, UK, 2008.

46. Lenski, R.E. Experimental Studies of Pleiotropy and Epistasis in *Escherichia coli*. II. Compensation for Maldaptive Effects Associated with Resistance to Virus T4. *Evolution* **1988**, *42*, 433–440. [CrossRef]

47. Björkman, J.; Nagaev, I.; Berg, O.G.; Hughes, D.; Andersson, D.I. Effects of environment on compensatory mutations to ameliorate costs of antibiotic resistance. *Science* **2000**, *287*, 1479–1482. [PubMed]

48. Bellec, L.; Clerissi, C.; Edern, R.; Foulon, E.; Simon, N.; Grimsley, N.; Desdevises, Y. Cophylogenetic interactions between marine viruses and eukaryotic picophytoplankton. *BMC Evol. Biol.* **2014**, *14*, 59. [CrossRef] [PubMed]

49. Luria, S.; Delbrück, M. Mutations of Bacteria from Virus Sensitivity to Virus Resistance. *Genetics* **1943**, *28*, 491–511. [PubMed]

50. Karafistan, A.; Martin, J.M.; Rixen, M.; Beckers, J.M. Space and time distributions of phosphate in the Mediterranean Sea. *Deep. Sea Res. I* **2002**, *49*, 67–82. [CrossRef]

51. Jessup, C.M.; Bohannan, B.J.M. The shape of an ecological trade-off varies with environment. *Ecol. Lett.* **2008**, *11*, 947–959. [CrossRef] [PubMed]

52. Bohannan, B.J.M.; Travisano, M.; Lenski, R.E. Epistatic Interactions Can Lower the Cost of Resistance to Multiple Consumers. *Evolution* **1999**, *53*, 292–295. [CrossRef]

53. Finkel, Z.V.; Beardall, J.; Flynn, K.J.; Quigg, A.; Rees, T.A.V.; Raven, J.A. Phytoplankton in a changing world: Cell size and elemental stoichiometry. *J. Plankton Res.* **2010**, *32*, 119–137. [CrossRef]

54. Peter, K.H.; Sommer, U. Interactive effect of warming, nitrogen and phosphorus limitation on phytoplankton cell size. *Ecol. Evol.* **2015**, *5*, 1011–1024. [CrossRef] [PubMed]
55. Atkinson, D.; Ciotti, B.J.; Montagnes, D.J.S. Protists decrease in size linearly with temperature: Ca. 2.5% °C^{-1}. *Proc. Biol. Sci.* **2003**, *270*, 2605–2611. [CrossRef] [PubMed]
56. Morán, X.A.G.; López-Urrutia, Á.; Calvo-Díaz, A.; Li, W.K.W. Increasing importance of small phytoplankton in a warmer ocean. *Glob. Chang. Biol.* **2010**, *16*, 1137–1144. [CrossRef]
57. Geider, R.; Platt, T.; Raven, J. Size dependence of growth and photosynthesis in diatoms: A synthesis. *Mar. Ecol. Ser.* **1986**, *30*, 93–104. [CrossRef]
58. Šupraha, L.; Gerecht, A.C.; Probert, I.; Henderiks, J. Eco-physiological adaptation shapes the response of calcifying algae to nutrient limitation. *Sci. Rep.* **2015**, *5*, 16499. [CrossRef] [PubMed]
59. Ryther, J.; Menzel, D. Light adaptation by marine phytoplankton. *Limnol. Oceanogr.* **1959**, *4*, 492–497. [CrossRef]
60. Wozniak, B.; Hapter, R.; Dera, J. Light curves of marine plankton photosynthesis in the Baltic. *Oceanologia* **1989**, *29*, 61–78.
61. Renk, H.; Ochocki, S. Photosynthetic rate and light curves of phytoplankton in the southern Baltic. *Oceanologia* **1998**, *40*, 331–344.
62. Riemann, B.; Simonsen, P.; Stensgaard, L. The carbon and chlorophyll content of phytoplankton from various nutrient regimes. *J. Plankton Res.* **1989**, *11*, 1037–1045. [CrossRef]
63. McLachlan, J. The effect of salinity on growth and chlorophyll content in representative classes of unicellular marine algae. *Can. J. Microbiol.* **1961**, *7*, 399–406. [CrossRef]
64. Sigaud, T.C.S.; Aidar, E. Salinity and temperature effects on the growth and chlorophyll-a content of some planktonic aigae. *Bol. Inst. Oceanogr.* **1993**, *41*, 95–103. [CrossRef]
65. Brockhurst, M.A.; Rainey, P.B.; Buckling, A. The effect of spatial heterogeneity and parasites on the evolution of host diversity. *Proc. Biol. Sci.* **2004**, *271*, 107–111. [CrossRef] [PubMed]

Article

Emerging Interaction Patterns in the *Emiliania huxleyi*-EhV System

Eliana Ruiz [1,*], **Monique Oosterhof** [1,2], **Ruth-Anne Sandaa** [1], **Aud Larsen** [1,3] and **António Pagarete** [1]

[1] Department of Biology, University of Bergen, Bergen 5006, Norway; Ruth.Sandaa@uib.no (R.-A.S.); Antonio.Pagarete@uib.no (A.P.)
[2] NRL for fish, Shellfish and Crustacean Diseases, Central Veterinary Institute of Wageningen UR, Lelystad 8221 RA, The Nederlands; Monique.oosterhof@wur.nl
[3] Uni Research Environment, Nygårdsgaten 112, Bergen 5008, Norway; Aud.Larsen@uni.no
* Correspondence: Eliana.Martinez@uib.no; Tel.: +47-5558-8194

Academic Editors: Mathias Middelboe and Corina Brussaard
Received: 30 January 2017; Accepted: 16 March 2017; Published: 22 March 2017

Abstract: Viruses are thought to be fundamental in driving microbial diversity in the oceanic planktonic realm. That role and associated emerging infection patterns remain particularly elusive for eukaryotic phytoplankton and their viruses. Here we used a vast number of strains from the model system *Emiliania huxleyi*/Emiliania huxleyi Virus to quantify parameters such as growth rate (μ), resistance (R), and viral production (Vp) capacities. Algal and viral abundances were monitored by flow cytometry during 72-h incubation experiments. The results pointed out higher viral production capacity in generalist EhV strains, and the virus-host infection network showed a strong co-evolution pattern between *E. huxleyi* and EhV populations. The existence of a trade-off between resistance and growth capacities was not confirmed.

Keywords: *Phycodnaviridae*; Coccolithovirus; Coccolithophore; *Haptophyta*; Killing-the-winner; cost of resistance; infectivity trade-offs; algae virus; marine viral ecology; viral-host interactions

1. Introduction

Since the discovery of high viral concentrations in the marine environment, normally ranging between 10^7 and 10^{11} virions/L [1], hypotheses regarding the potential impact those viruses could have on their microbial host populations, have been put forward. Viral-induced microbial lysis in Earth's oceans could amount to an impressive 10^{23} new infections per second, releasing up to 10^9 tons of cellular carbon every day [2,3]. Consequently, viral lysis contributes greatly to marine biogeochemical cycling of nutrients as well as reducing the transport of organic matter to upper trophic levels in a process known as viral shunt [4–6]. Through horizontal gene transfer and the lysis of their hosts, marine viruses contribute to structuring the diversity and composition of microbial communities [7–11].

Viral activity has been suggested as a plausible mechanism contributing to explain Hutchinson's paradox, which questions the existence of highly diverse planktonic communities in nutrient limited environments [4,12,13]. Viral strain or species-specific lysis may potentially explain the coexistence of cells with different growth and resistance capacities [14,15]. This scenario is contemplated in the Killing-the-Winner (KtW) hypothesis, notably with the concept that resistance has an inherent cost. This trade-off, also known as cost of resistance (COR), ultimately regulate the co-existence of competition specialists (with higher growth rates) and defence specialists (with higher immune capacity against viral infection), respectively [16].

COR can be detected by analysing the virus-host infection network patterns (VHINs) that emerge after cross-infectivity experiments [17–19]. The most frequently tested VHIN patterns are nestedness and modularity [17,20]. Nested patterns are characterized by specialist viruses tending to infect the most susceptible hosts, while the viruses with broader host-range infect hosts that are more resistant [21]. On the other hand, in modular patterns the interactions tend to occur within different groups of viruses and hosts, but not between groups [17,22].

The role of viruses as an important driver of microbial diversity has become clear in prokaryotic-virus systems [23–27] such as the *Pseudoalteromonas* [28] and the *Pseudomonas aeruginosa* host-virus systems, in which resistant cells emerging after infection were less competitive than the sensitive ones [24]. In other prokaryote-virus systems that role remains elusive [29–34]. The very few examples of trade-off between resistance and growth rate in eukaryotic hosts include studies on the prasinophyte *Ostreococcus tauri* [35] and the trebouxiophyte *Chlorella variabilis* [36].

Here we aim at getting insight on the main emerging patterns that result from eukaryotic host-virus interactions in the planktonic realm by focusing on *Emiliania huxleyi*, the most abundant and widely distributed calcifying haptophyte in our oceans [37], and its lytic viruses. Mostly known for its impressive blooms [38,39] this microalga is an important player in global geochemical cycles [40,41]. This photosynthetic unicellular eukaryote is infected by *Emiliania huxleyi* viruses (EhV), lytic giant viruses belonging to the genus *Coccolithovirus*, within the *Phycodnaviridae* family. These viruses are ubiquitous in the marine environment [42] and abundant, reaching 10^7/mL in natural seawater during bloom conditions and from 10^8 to 10^9/mL in laboratory cultures [43]. Genomic and metagenomic EhV characterizations show both a global consistency of this viral genome on a planetary scale as well as the maintenance of specific localized genetic traits. For example, despite the high levels of sequence similarity (>95%) between EhV isolates from a Norwegian fjord and the English Channel, these viral populations also contain distinctive genetic traits [44–50]. It is surprising that these genetic traits have been maintained through decades although no geographical isolation and speciation have occurred to date [45], allowing these viral communities to infect hosts from distant geographic places [44,51].

Taking advantage of the large number of *E. huxleyi* cell and EhV lines available for this host-virus system, from diverse geographical origins that include the major oceanic regions, an extensive array of cross-infectivity experiments was conducted in order to investigate parameters such as growth rate (μ), resistance (R), and viral production (Vp). We then confronted possible existence of correlations between those parameters with the theoretical hypotheses (Table 1) that delimit our conception of virus-microbe interactions in the oceans and the way we model those interactions.

Table 1. Hypotheses tested in the current study based on outcome of previous virus-host interaction studies. μ: growth rate; R: resistance; Vp: viral production.

Number	Hypothesis	Reference
1	Resistance is associated with reduced growth rates (COR).	Prokaryotes: [23–25,27–29,52–55] Eukaryotes: [35,36,56,57]
2	Host strains with higher μ produce more viruses.	[58–66]
3	Host strains with higher μ are infected by more viral strains.	[36]
4	Host strains with higher R produce fewer viruses.	[56,67]
5	Specialist viruses have higher Vp than generalists.	[14,68]

2. Materials and Methods

2.1. Emiliania Huxleyi and EhV Strains

Algal strains were obtained from the Roscoff Culture Collection, France; and from the University of Bergen, Norway. A total of 49 *E. huxleyi* strains (Table S1) were maintained in 30 mL polystyrene flasks with IMR $\frac{1}{2}$ medium [69] at 16 °C and a 14:10 h light:dark illumination cycle at 155 μmol photon m^{-2}/s irradiance.

A total number of 13 viral strains were obtained from the Plymouth Marine Laboratory, UK; and from the University of Bergen, Norway (Table S2). For all viral isolates, viral stocks were produced by infection of exponentially growing *E. huxleyi* RCC1257 strain. Viral lysates were centrifuged at

12,000 × g for 20 min and the supernatant was filtered through a 0.45 μm syringe filter (Whatman plc, GE Healthcare Life Sciences, Kent, UK) to remove cellular debris. Viral stocks were kept at 4 °C in the dark and were renewed so often as to never be more than 2 weeks old before inoculation in order to preserve the agent's viability. Plaque assays were not conducted as haptophytes in general do not grow on agar plates and have only been achieved for a few *E. huxleyi* strains [70,71].

2.2. Cross-Infectivity Experiments

Cross-infectivity experiments were performed between all the *E. huxleyi* and EhV strains (Table S3). Prior to each experiment, *E. huxleyi* cultures were maintained in exponential growth phase with cell concentrations ranging from 10^5 to 10^6 cells/mL. The experiments were performed in 24 well culture plates under the same temperature and light conditions as the general culturing conditions described above. Triplicates of 2 mL of each algal culture (1×10^5 cells/mL) were inoculated with each of the 13 viral strains at a concentration of 1×10^6 viral particles/mL, resulting in a virus to host ratio (VHR) of 10. Three replicates of uninfected culture were also used as a control for each *E. huxleyi* strain. An incubation time of 72 h was chosen because this is consistent with the time scales reported for *E. huxleyi*/EhV selection dynamics observed in the natural environment [72–74]. Moreover, preliminary growth tests [75] performed on several *E. huxleyi* strains, did not indicate that prolonged incubation period would contribute essential knowledge on the growth capacity of each strain.

2.3. Enumeration of Algae and Viruses

At times 0 h and 72 h, 500 μL was subsampled from each well to determine algae and virus concentrations using a FACSCalibur BC flow cytometer (Becton–Dickinson, Biosciences, Franklin Lakes, NJ, USA) [76–78] provided with an air-cooled laser procuring 15 mW at 488 nm. Viral samples were fixed with 20 μL of glutaraldehyde (25%) for 30 min at 4 °C, and frozen at −80 °C until further use. For flow cytometry analysis, samples were thawed, diluted 500-fold in TE buffer (10:1 mM Tris:EDTA, pH 8, filtered through 0.2 μm), and stained with SYBR Green I 100× diluted (Invitrogen, 1600 Faraday Avenue, PO Box 6482, Carlsbad CA, 92008 United States) for 10 min at 80 °C before analysis. Algal enumeration was conducted on fresh samples, and cell populations were discriminated using chlorophyll auto-fluorescence (670 LP) and SSC signals. Virus populations were determined and enumerated on basis of their green fluorescence (530/30) and SSC signals.

2.4. Growth Rate, Resistance, Viral Production

Growth rates (μ) were calculated for each *E. huxleyi* strain using the control non-inoculated incubations according to the following formula [79]:

$$\mu = \text{Ln} (N2/N1)/t \tag{1}$$

where N1 and N2 were the cell concentrations at the beginning and end of the experiment, respectively, and t was the incubation time in days.

The level of resistance of each *E. huxleyi* strain to viral infection was measured in two manners. The first manner (R_1) was based on the difference of cells that were not lysed after incubation with viruses, compared to the non-inoculated controls. For each *E. huxleyi* strain a resistance value was hence calculated against each of the 13 EhV strains and the 13 resistance values were then averaged to obtain an overall resistance capacity for each alga strain (R_1). Resistance was also estimated as the number of EhV strains that successfully produced progeny on that host (R_2).

A value of viral production (Vp), corresponding to the capacity of each viral strain to produce new progeny on a certain host, was calculated for each virus — host pair as the difference between final and initial viral concentrations. These values were averaged to obtain a global infectivity capacity for each viral strain, per algal strain. The maximum amount of viruses that each EhV strain, per algal strain, produced was registered as "Maximum viral production".

Potential correlations between the different parameters (growth rate, resistance, and viral production) were investigated with regression slopes and statistical probability analyses, using either Anova (*F*) or Pearson analysis.

A potential impact of domestication on these parameters was also investigated. An analysis was performed on two groups of *E. huxleyi* strains, which were isolated in different periods of time. The periods before and after 2009, respectively, were chosen for an apparent increase in Vp was preliminary observed in strains old or younger than 2009 (Figure S1).

2.5. Host-Virus Network Analysis

In order to test the structure of the infection network, we used the BiMat package for Matlab [21]. This network-based analysis was applied on a binary matrix where 0 referred to no lysis and 1 to lysis. The NODF algorithm was used to measure nestedness and is based on overlap and decreasing fill [80]. It returns a score between 0 and 1, where 1 corresponds to a perfectly nested structure. Modularity (Qb) was calculated using the Leading-Eigenvector algorithm [81]. The value Qb, introduced by Barber [82], is calculated using the standard bipartite modularity function. To quantify the statistical significance of the nestedness (NODF) and modularity (Qb), 100 null random matrices (for each) were created with the null model Equiprobable (a random matrix in which all the interactions are uniformly permuted).

3. Results

Forty-nine *E. huxleyi* strains were characterized according to their ability to grow under a standard set of nutrients, light and temperature conditions. Growth rate (μ) varied significantly among *E. huxleyi* strains, ranging from 0.12 (SD $\pm$ 0.01) to 1.11 (SD $\pm$ 0.02)/d (registered in strains RCC4533 and RCC1744, respectively) (Figure S2). The difference in growth rate among the algal strains was not related to the ocean they were isolated from (one-way ANOVA F (2, 37) = 0.275, $p = 0.76$).

We confronted the observed differences in resistance capacity with the parameters growth rate and viral production, respectively. The level of resistance of *E. huxleyi* to EhV infection was accessed in two manners: (R_1) percentage of cells that were not lysed after incubation with viruses (Figure S3) and (R_2) the number of EhV strains that successfully produced progeny on that host, meaning that lower R_2 levels indicate higher resistance capacity. A trade-off between resistance and growth rate capacities (hypotheses 1 and 3 in Table 1) was not confirmed with our results. Neither types of resistance, R_1 and R_2, were significantly correlated to growth rate (Pearson's r = -0.131, $p = 0.370$, and Pearson's r = -0.0959, $p = 0.512$; respectively) (Figures 1 and 2). R_1 was indirectly correlated with viral production (Figure 3) (Pearson's r = -0.499, $p > 0.01$), in accordance with hypothesis 4. R_2 was significantly and positively correlated with maximum viral production (Pearson's r = 0.614, $p < 0.01$), which means that the *E. huxleyi* strains that were susceptible to more EhV types were also the ones that presented higher maximum viral production (Figure 4). Viral production and growth rate did not correlate significantly (Pearson's r = 0.1, $p = 0.494$) (Figure S4) and hence did not confirm hypothesis 2 (Table 1).

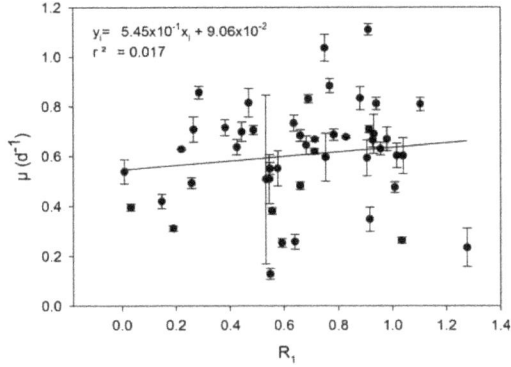

Figure 1. Resistance capacity R_1 (calculated as the ratio between the number of cells that did not lyse after incubation with viruses and the number of cells in the non-inoculated controls) plotted against growth rate (μ). Error bars show standard deviation ($n = 3$).

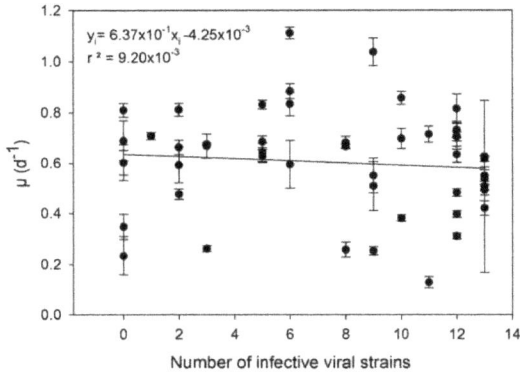

Figure 2. Resistance capacity R_2 (number of viral strains infecting each algal strain) plotted against growth rate (μ). Error bars show standard deviation ($n = 3$).

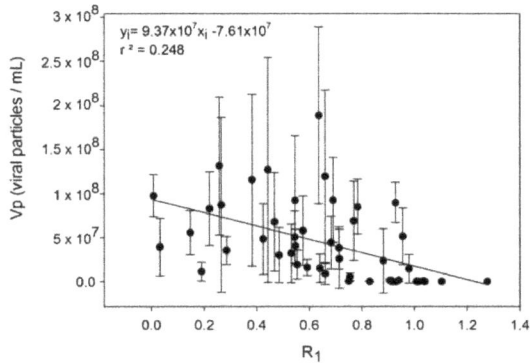

Figure 3. Viral production (Vp) plotted against resistance capacity R_1. Error bars show standard deviation ($n = 13$).

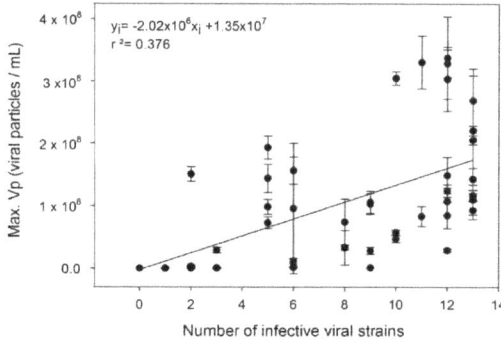

Figure 4. Number of viral strains infecting each algal strain and maximum viral production correlation. Error bars show standard deviation ($n = 3$).

Seven out of the 49 *E. huxleyi* strains (RCC1259, RCC1269, RCC3856, 371, P847, PERU15-40 and SO52) were susceptible to infection by all the EhV strains tested, while 6 *E. huxleyi* strains (RCC1211, RCC1218, RCC1235, RCC1256, RCC1276 and RCC3548) were resistant to infection by all the EhV strains tested. When analysing these two groups of *E. huxleyi* strains, no significant differences in growth rate were found (one-way ANOVA F $(1, 11) = 0.01592$, $p = 0.90188$), while their R_1 values were significantly different (one-way ANOVA F $(1, 11) = 36.8593$, $p = 8.1 \times 10^{-5}$).

A significant higher viral production was found in the most recently isolated algal strains (one-way ANOVA F $(1, 47) = 30.36$, $p = 1.5 \times 10^{-6}$). For the other parameters (growth rate, R_1 and R_2) there were no significant differences between younger and older strains (one-way ANOVA F $(1, 47) = 1.094$, $p = 0.30$; one-way ANOVA F $(1, 47) = 0.106$, $p = 0.745$; one-way ANOVA F $(1, 47) = 0.909$, $p = 0.345$; respectively).

We observed significant variation in "Maximum viral production" capacity among the different EhV strains (Figure 5). Those differences did not translate into significant differences in "Average Viral Production" (Figure S5), as the capacity of each EhV to produce progeny depended very much on which host strain it was infecting. Host-ranges among EhV strains also proved very variable, from generalists that infected up to 36 host strains (e.g., EhV-207) to specialists capable of infecting only 1 strain (e.g., EhV-99b1). Surprisingly, and against the prediction in hypothesis 5, generalist viral strains (EhV-164, EhV-202, EhV-208, EhV-201 and EhV-207) produced significantly more virus progeny viral production (one-way ANOVA *F* $(1, 8) = 8.123$, $p = 0.021$) than specialist strains (EhV-99b1, EhV-203, EhV-156, EhV-86, and EhV-145).

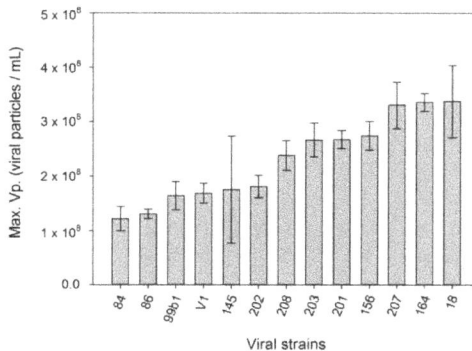

Figure 5. Differences between maximum viral production among EhV strains. Error bars show standard deviation ($n = 49$).

The bipartite network analysis applied to the whole host-range matrix displayed a nested structure (Figure 6) with a NODF value of 0.60. In that nested pattern there was a tendency for hosts with higher resistance to only be infected by more generalist viruses, while specialist viruses tend to infect the most sensitive hosts.

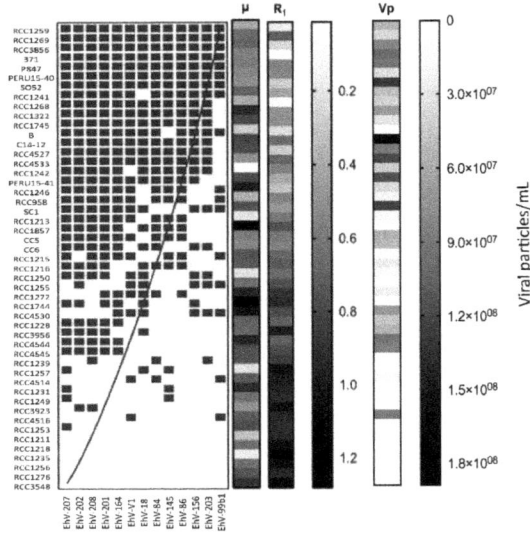

Figure 6. Viral-host infectivity network with a clear nested pattern (NODF value of 0.60) where specialist viruses tend to infect the most susceptible hosts, while viruses with broader host-range infect hosts with higher resistance. ■: infection; □: no infection. Sidebars represent μ, R_1 and Vp parameters, respectively.

4. Discussion

Since Hutchinson first stated the Paradox of the phytoplankton in the early sixties, many hypotheses explaining the high diversity in the oceans have been postulated [13]. Among these, viral activity has proven to be a potential disrupter on equilibrium in planktonic communities [4,12]. Due to the lack of quantitative data for viral-host interactions, especially in marine micro-eukaryotic organisms, we therefore decided to perform a vast survey on strains of the ubiquitous and environmentally relevant coccolithophorid *Emiliania huxleyi* sp. (*E. huxleyi*) (Lohman) and its virus, *Emiliania huxleyi* virus (EhV), and investigate for emerging patterns resulting from this arms race.

Among the different hypotheses tested (Table 1) was the existence, or not, of a clear trade-off between resistance and growth rate (COR). COR has been previously confirmed in some bacteria-virus systems [26–30], and is fundamental in the formulation of the Killing the Winner model [19]. In our study we did not observe a clear COR trade-off in the *E. huxleyi*-EhV system. Instead, we found that highly resistant algal strains were capable of growing at high rates. This indicates that, at least in this system, viruses may not be the main selective force acting upon their hosts or, that if they are, their impact is camouflaged by antagonic impacts from other selective factors (e.g., different adaptation to the standard culture conditions used). However, it could be that viral-imposed selection was so strong that it would result in an emerging global cost of resistance observable on *E. huxleyi* strains independently of their inherent local adaptations. An approximation to such global "cost of resistance" is precisely the parameter value used when trying to model the interactions between viruses and their hosts [83]. Its prominence in current models justified the present attempt to evaluate its real extension.

When Avrani and colleagues [29] observed a similar response in viral resistant *Prochlorococcus* strains, they also found that the reduced growth rates increased after 7 months and that these strains reduced their resistance against the viruses [84]. The changes in growth rate and resistance occurred as independent events, indicating that the selection pressure on these phenotypes was decoupled. Decoupled selective pressure for growth rate and resistance may be the reason for the lack of correlation between these parameters in our study as well.

COR not being observed for the *E. huxleyi*-EhV system using our approach is not necessarily proving it does not exist or that it is irrelevant. As also tried in the current study, COR is often measured as reduction of growth rates in the resistant host [23–27], but other CORs, like altered susceptibility to other viruses and possibly also to some bacteria [85], have also been argued [29,84,86–88]. Trade-off might also emerge when strains with different resistance capacities are put under competition for a limited level of nutrients [30,33,89,90], and this is the logical follow up to our study. Another aspect to take into account is the potential impact that domestication has on the isolated strains [91]. In vitro growing conditions (nutrients, light, temperature) are inevitably different from what the cells would be experiencing in the natural environment. Particularly, in vitro cells are released from viral pressure, a situation that, with time, could potentially erase the selective traits that viruses might impose on cells in the natural environment. A sign of domestication-related effects in our case was the lower viral production capacity observed for "older" strains (isolated before 2009).

Patterns other than COR that shed light on the global interaction between *E. huxleyi* and EhV did, however, emerge in this study. Contrary to our expectations [14,68], we observed a tendency for generalist viruses (e.g., EhV-207) to produce more progeny than the specialists (e.g., EhV-86). It was recently reported that a generalist EhV strain could outcompete a specialist 8 h post infection [92]. One explanation for this apparent difference in infective success between generalist and specialist viruses may thus be a trade-off where high host-range/replication rates are associated with hindered progeny (new virions) fitness [64,93–96]. An alternative possibility could be the presence of an "un-costly" strong adaptive potential to new hosts, as shown for the Tobacco etch potyvirus (TEV) [97]. It also has to be taken into consideration that viral infective performance; such as viral adsorption coefficient and burst size also depends strongly on host traits. In the current study, a set of *E. huxleyi* strains were the ones that presented the higher viral production, independently of the EhV strain that was infecting them. Such added levels of complexity create niches for different strains of viruses and hosts with different infection and resistance capacities, respectively, to coexist. The patterns emerging from the interaction between *E. huxleyi* and EhV indicate that there's a plethora of niches that create the possibility for co-existence of viruses and hosts with unexpected trait capacities. Notably, viral strains with narrower host-ranges and smaller virion production competing with generalist strains. Future studies should try to evaluate the possibility of take-over in the case of two specialist or generalist strains.

The emerging virus-host interaction network (VHIN) pattern showed a significant nestedness match between viral strains and their hosts. A nested structure like this is considered to result from sequential gene-for-gene (GFG) adaptations [98,99]. In the GFG model one genotype replaces another leading to continued fitness improvements of both, host and virus populations, resulting in an everlasting arms race dynamics. Different mesocosm studies on natural *E. huxleyi*/EhV communities [73,74] have shown that host and viral strain diversity can co-change in very short periods of just a few days during *E. huxleyi* blooms. This supports the Arms Race dynamics indicated by our VHIN. Future studies should evaluate the potential for strains with similar host-range capacity to take-over one another. The currently observed cross-infection network did not however have a perfect nested structure. An alternative co-evolution mechanism, termed diffuse co-evolution, appears to be more adequate for multi-species and/or multi-strain communities where selection pressures due to one species, can change in the presence of other species [17,100]. In order to predict diffuse co-evolution, however, experiments in which the different species/strains could interact, allowing real fitness costs associate to both, viruses and hosts, to arise [100] are necessary.

As also previously shown, the same *E. huxleyi* viruses (isolated in the English Channel and the Norwegian fjords) proved able to infect *E. huxleyi* hosts isolated in a large spatio-temporal scale [44,51], indicating a strong genomic consistency between geographically distant EhV strains. Nonetheless, and despite high abundance of conserved genomic sequences among these strains, significant genomic variety is also documented [44,73,101]. As EhVs are enveloped viruses [102], their entry mechanism should be endocytosis or fusion of the envelope with the host's membrane and the progeny release through a budding mechanism [103]. Such an infection mechanism potentially generates a highly lipid-specific contact between host and virus. The host, *E. huxleyi*, has high phenotypic plasticity [104–108] and adaptation capacity [104,109–112] that could result in ecotypes that respond differently to viral infection [37,108,112]. Even if genes associated with virus susceptibility have been found within non-core regions of the *E. huxleyi* genome [108], our results did not show significant differences in growth rate, resistance, or viral production in hosts from very distant geographical locations. Hence, despite the recognized genetic variability in both host and virus, our results suggest a globally, non-segregated evolution process between *E. huxleyi* and EhV [113].

In conclusion, and despite a lack of supporting evidence of a trade-off between resistance and growth capacities, our results did indeed, through the nested host-virus interaction pattern, demonstrate a strong co-evolution pattern between *E. huxleyi* and EhV populations. The absence of trade-off between growth rate and resistance, invites us to think that EhVs may not be the main force driving the *E. huxleyi* selection, and that other fitness costs, which passed unnoticeably in the present study, exist. Further work should aim at unravelling these.

Supplementary Materials: The following are available online at www.mdpi.com/1999-4915/9/3/61/s1, Table S1: *E. huxleyi* strain information, in blank = No information; Table S2: EhV strain information; Table S3: Measured parameters from the cross-infectivity experiments between each *E. huxleyi*-EhV pair. fC = final concentration of *E. huxleyi* cells in cells/mL, μ = *E. huxleyi* growth rate, R_1 = percentage of cells that were not lysed after incubation with viruses, compared to the non-inoculated controls, Vp = viral production in viral particles/mL, R2 = number of EhV strains that successfully produced progeny on that host, R1 AV = averaged R1 for each algal strain, Vp AV = averaged Vp for each algal strain in viral particles/mL, Max. Vp AV = averaged maximum Vp for each algal strain in viral particles/mL, SD = standard deviation; Figure S1: Correlation between viral production per host cell (Vp) and isolation year of the algal strains. Error bars show standard deviation (n = 13); Figure S2: Growth rates (μ/d) measured for control samples of all of the *E. huxleyi* strains (see Table S1 for strain information) used in the infection experiment measured over a period of x days and, calculated according to Levasseur et al. (1993). Values correspond to the control samples. Error bars show standard deviation (n = 3); Figure S3: Resistance strategy (R) for each *E. huxleyi* strain. Error bars show standard deviation (n = 13); Figure S4: Correlation between growth rate (μ) and viral production per host cell (Vp), in viral particles/mL. Error bars show standard deviation (n = 13); Figure S5: Average viral production per EhV strain, for all the algal strains. Error bars show standard deviation (n = 49).

Acknowledgments: The authors are grateful to William Wilson and Mike Allen for providing the EhV strains. We are also thankful to Tron Frede Thingstad for the helpful discussions and advice, and to the reviewers for their comments and suggestions that contributed to improve the manuscript. This work was supported by the Research Council of Norway, through project VIMPACT, no. 234197.

Author Contributions: A.P., E.R., and R.-A.S. conceived and designed the experiments; E.R. and M.O. performed the experiments; E.R. and A.P. analyzed the data; E.R., A.P., A.L., and R.-A.S. wrote the paper.

References

1. Wilhelm, S.W.; Suttle, C.A. Viruses and nutrient cycles in the sea—Viruses play critical roles in the structure and function of aquatic food webs. *Bioscience* **1999**, *49*, 781–788. [CrossRef]
2. Suttle, C.A. Marine viruses—Major players in the global ecosystem. *Nat. Rev. Microbiol.* **2007**, *5*, 801–812. [CrossRef] [PubMed]
3. Breitbart, M. Marine viruses: Truth or dare. *Annu. Rev. Mar. Sci.* **2012**, *4*, 425–448. [CrossRef] [PubMed]
4. Fuhrman, J.A. Marine viruses and their biogeochemical and ecological effects. *Nature* **1999**, *399*, 541–548. [CrossRef] [PubMed]

5. Wommack, K.E.; Colwell, R.R. Virioplankton: Viruses in aquatic ecosystems. *Microbiol. Mol. Biol. Rev.* **2000**, *64*, 69–114. [CrossRef] [PubMed]

6. Weinbauer, M.G.; Rassoulzadegan, F. Are viruses driving microbial diversification and diversity? *Environ. Microbiol.* **2004**, *6*, 1–11. [CrossRef] [PubMed]

7. Brussaard, C.P.D. Viral control of phytoplankton populations—A review. *J. Eukaryot. Microbiol.* **2004**, *51*, 125–138. [CrossRef] [PubMed]

8. Suttle, C.A.; Chan, A.M. Dynamics and distribution of cyanophages and their effect on marine *Synechococcus* spp. *Appl. Environ. Microbiol.* **1994**, *60*, 3167–3174. [PubMed]

9. Waterbury, J.B.; Valois, F.W. Resistance to cooccurring phages enables marine synechococcus communities to coexist with cyanophages abundant in seawater. *Appl. Environ. Microbiol.* **1993**, *59*, 3393–3399. [PubMed]

10. Weinbauer, M.G. Ecology of prokaryotic viruses. *FEMS Microbiol. Rev.* **2004**, *28*, 127–181. [CrossRef] [PubMed]

11. Jiang, S.C.; Paul, J.H. Gene transfer by transduction in the marine environment. *Appl. Environ. Microbiol.* **1998**, *64*, 2780–2787. [PubMed]

12. Fuhrman, J.A.; Suttle, C.A. Viruses in marine planktonic systems. *Oceanography* **1993**, *6*, 51–63. [CrossRef]

13. Hutchinson, G.E. The paradox of the plankton. *Am. Nat.* **1961**, *95*, 137–145. [CrossRef]

14. Winter, C.; Bouvier, T.; Weinbauer, M.G.; Thingstad, T.F. Trade-offs between competition and defense specialists among unicellular planktonic organisms: The "killing the winner" hypothesis revisited. *Microbiol. Mol. Biol. Rev.* **2010**, *74*, 42–57. [CrossRef] [PubMed]

15. Thingstad, T.F.; Vage, S.; Storesund, J.E.; Sandaa, R.A.; Giske, J. A theoretical analysis of how strain-specific viruses can control microbial species diversity. *Proc. Natl. Acad. Sci. USA* **2014**, *111*, 7813–7818. [CrossRef] [PubMed]

16. Sheldon, B.C.; Verhulst, S. Ecological immunology: Costly parasite defences and trade-offs in evolutionary ecology. *Trends Ecol. Evol.* **1996**, *11*, 317–321. [CrossRef]

17. Weitz, J.S.; Poisot, T.; Meyer, J.R.; Flores, C.O.; Valverde, S.; Sullivan, M.B.; Hochberg, M.E. Phage-bacteria infection networks. *Trends Microbiol.* **2013**, *21*, 82–91. [CrossRef] [PubMed]

18. Poullain, V.; Gandon, S.; Brockhurst, M.A.; Buckling, A.; Hochberg, M.E. The evolution of specificity in evolving and coevolving antagonistic interactions between a bacteria and its phage. *Evolution* **2008**, *62*, 1–11. [CrossRef] [PubMed]

19. Meyer, J.R.; Dobias, D.T.; Weitz, J.S.; Barrick, J.E.; Quick, R.T.; Lenski, R.E. Repeatability and contingency in the evolution of a key innovation in phage lambda. *Science* **2012**, *335*, 428–432. [CrossRef] [PubMed]

20. Fortuna, M.A.; Stouffer, D.B.; Olesen, J.M.; Jordano, P.; Mouillot, D.; Krasnov, B.R.; Poulin, R.; Bascompte, J. Nestedness versus modularity in ecological networks: Two sides of the same coin? *J. Anim. Ecol.* **2010**, *79*, 811–817. [CrossRef] [PubMed]

21. Flores, C.O.; Poisot, T.; Valverde, S.; Weitz, J.S. Bimat: A matlab package to facilitate the analysis of bipartite networks. *Methods Ecol. Evol.* **2016**, *7*, 127–132. [CrossRef]

22. Gomez, P.; Buckling, A. Bacteria-phage antagonistic coevolution in soil. *Science* **2011**, *332*, 106–109. [CrossRef] [PubMed]

23. Benmayor, R.; Buckling, A.; Bonsall, M.B.; Brockhurst, M.A.; Hodgson, D.J. The interactive effects of parasitesf disturbance, and productivity on experimental adaptive radiations. *Evolution* **2008**, *62*, 467–477. [CrossRef] [PubMed]

24. Brockhurst, M.A.; Buckling, A.; Rainey, P.B. The effect of a bacteriophage on diversification of the opportunistic bacterial pathogen, pseudomonas aeruginosa. *Proc. R. Soc. B-Biol. Sci.* **2005**, *272*, 1385–1391. [CrossRef] [PubMed]

25. Brockhurst, M.A.; Rainey, P.B.; Buckling, A. The effect of spatial heterogeneity and parasites on the evolution of host diversity. *Proc. R. Soc. B-Biol. Sci.* **2004**, *271*, 107–111. [CrossRef] [PubMed]

26. Middelboe, M.; Holmfeldt, K.; Riemann, L.; Nybroe, O.; Haaber, J. Bacteriophages drive strain diversification in a marine flavobacterium: Implications for phage resistance and physiological properties. *Environ. Microbiol.* **2009**, *11*, 1971–1982. [CrossRef] [PubMed]

27. Riemann, L.; Grossart, H.-P. Elevated lytic phage production as a consequence of particle colonization by a marine flavobacterium (*Cellulophaga* sp.). *Microb. Ecol.* **2008**, *56*, 505–512. [CrossRef] [PubMed]

28. Middelboe, M. Bacterial growth rate and marine virus-host dynamics. *Microb. Ecol.* **2000**, *40*, 114–124. [PubMed]

29. Avrani, S.; Wurtzel, O.; Sharon, I.; Sorek, R.; Lindell, D. Genomic island variability facilitates prochlorococcus-virus coexistence. *Nature* **2011**, *474*, 604–608. [CrossRef] [PubMed]

30. Meaden, S.; Paszkiewicz, K.; Koskella, B. The cost of phage resistance in a plant pathogenic bacterium is context-dependent. *Evolution* **2015**, *69*, 1321–1328. [CrossRef] [PubMed]

31. Lennon, J.T.; Khatana, S.A.M.; Marston, M.F.; Martiny, J.B.H. Is there a cost of virus resistance in marine cyanobacteria? *ISME J.* **2007**, *1*, 300–312. [CrossRef] [PubMed]

32. Lythgoe, K.A.; Chao, L. Mechanisms of coexistence of a bacteria and a bacteriophage in a spatially homogeneous environment. *Ecol. Lett.* **2003**, *6*, 326–334. [CrossRef]

33. Bohannan, B.J.M.; Lenski, R.E. The relative importance of competition and predation varies with productivity in a model community. *Am. Nat.* **2000**, *156*, 329–340. [CrossRef]

34. Mizoguchi, K.; Morita, M.; Fischer, C.R.; Yoichi, M.; Tanji, Y.; Unno, H. Coevolution of bacteriophage pp01 and escherichia coli o157: H7 in continuous culture. *Appl. Environ. Microbiol.* **2003**, *69*, 170–176. [CrossRef] [PubMed]

35. Thomas, R.; Grimsley, N.; Escande, M.L.; Subirana, L.; Derelle, E.; Moreau, H. Acquisition and maintenance of resistance to viruses in eukaryotic phytoplankton populations. *Environ. Microbiol.* **2011**, *13*, 1412–1420. [CrossRef] [PubMed]

36. Frickel, J.; Sieber, M.; Becks, L. Eco-evolutionary dynamics in a coevolving host-virus system. *Ecol. Lett.* **2016**, *19*, 450–459. [CrossRef] [PubMed]

37. Brown, C.W.; Yoder, J.A. Coccolithophorid blooms in the global ocean. *J. Geophys. Res.-Oceans* **1994**, *99*, 7467–7482. [CrossRef]

38. Tyrrell, T.; Taylor, A.H. A modelling study of *Emiliania huxleyi* in the ne atlantic. *J. Mar. Syst.* **1996**, *9*, 83–112. [CrossRef]

39. Egge, J.K.; Heimdal, B.R. Blooms of phytoplankton including *Emiliania-huxleyi* (haptophyta)—Effects of nutrient supply in different n-p ratios. *Sarsia* **1994**, *79*, 333–348. [CrossRef]

40. Broerse, A.T.C.; Ziveri, P.; van Hinte, J.E.; Honjo, S. Coccolithophore export production, species composition, and coccolith-caco3 fluxes in the ne atlantic (34 degrees n 21 degrees w and 48 degrees n 21 degrees w). *Deep-Sea Res. Part II Top. Stud. Oceanogr.* **2000**, *47*, 1877–1905. [CrossRef]

41. Burkill, P.H.; Archer, S.D.; Robinson, C.; Nightingale, P.D.; Groom, S.B.; Tarran, G.A.; Zubkov, M.V. Dimethyl sulphide biogeochemistry within a coccolithophore bloom (disco): An overview. *Deep-Sea Res. Part II Top. Stud. Oceanogr.* **2002**, *49*, 2863–2885. [CrossRef]

42. Cottrell, M.T.; Suttle, C.A. Wide-spread occurrence and clonal variation in viruses which cause lysis of a cosmopolitan, eukaryotic marine phytoplankter, micromonas-pusilla. *Mar. Ecol. Prog. Ser.* **1991**, *78*, 1–9. [CrossRef]

43. Nissimov, J.I.; Napier, J.A.; Kimmance, S.A.; Allen, M.J. Permanent draft genomes of four new coccolithoviruses: Ehv-18, ehv-145, ehv-156 and ehv-164. *Mar. Genom.* **2014**, *15*, 7–8. [CrossRef] [PubMed]

44. Allen, M.J.; Martinez-Martinez, J.; Schroeder, D.C.; Somerfield, P.J.; Wilson, W.H. Use of microarrays to assess viral diversity: From genotype to phenotype. *Environ. Microbiol.* **2007**, *9*, 971–982. [CrossRef] [PubMed]

45. Pagarete, A.; Kusonmano, K.; Petersen, K.; Kimmance, S.A.; Martinez, J.M.; Wilson, W.H.; Hehemann, J.H.; Allen, M.J.; Sandaa, R.A. Dip in the gene pool: Metagenomic survey of natural coccolithovirus communities. *Virology* **2014**, *466*, 129–137. [CrossRef] [PubMed]

46. Pagarete, A.; Lanzen, A.; Puntervoll, P.; Sandaa, R.A.; Larsen, A.; Larsen, J.B.; Allen, M.J.; Bratbak, G. Genomic sequence and analysis of ehv-99b1, a new coccolithovirus from the norwegian fjords. *Intervirology* **2013**, *56*, 60–66. [CrossRef] [PubMed]

47. Nissimov, J.I.; Worthy, C.A.; Rooks, P.; Napier, J.A.; Kimmance, S.A.; Henn, M.R.; Ogata, H.; Allen, M.J. Draft genome sequence of the coccolithovirus ehv-84. *Stand. Genom. Sci.* **2011**, *5*, 1–11. [CrossRef] [PubMed]

48. Nissimov, J.I.; Worthy, C.A.; Rooks, P.; Napier, J.A.; Kimmance, S.A.; Henn, M.R.; Ogata, H.; Allen, M.J. Draft genome sequence of four coccolithoviruses: *Emiliania huxleyi* virus ehv-88, ehv-201, ehv-207, and ehv-208. *J. Virol.* **2012**, *86*, 2896–2897. [CrossRef] [PubMed]

49. Nissimov, J.I.; Worthy, C.A.; Rooks, P.; Napier, J.A.; Kimmance, S.A.; Henn, M.R.; Ogata, H.; Allen, M.J. Draft genome sequence of the coccolithovirus *Emiliania huxleyi* virus 202. *J. Virol.* **2012**, *86*, 2380–2381. [CrossRef] [PubMed]

50. Nissimov, J.I.; Worthy, C.A.; Rooks, P.; Napier, J.A.; Kimmance, S.A.; Henn, M.R.; Ogata, H.; Allen, M.J. Draft genome sequence of the coccolithovirus *Emiliania huxleyi* virus 203. *J. Virol.* **2011**, *85*, 13468–13469. [CrossRef] [PubMed]

51. Pagarete, A.J. Functional Genomics of Coccolithophore Viruses. Ph.D. Thesis, Pierre and Marie Curie University, Paris, France, 2010; p. 222.

52. Middelboe, M.; Jacquet, S.; Weinbauer, M. Viruses in freshwater ecosystems: An introduction to the exploration of viruses in new aquatic habitats. *Freshw. Biol.* **2008**, *53*, 1069–1075. [CrossRef]

53. Bouvier, T.; del Giorgio, P.A. Key role of selective viral-induced mortality in determining marine bacterial community composition. *Environ. Microbiol.* **2007**, *9*, 287–297. [CrossRef] [PubMed]

54. Chao, L.; Levin, B.R.; Stewart, F.M. A complex community in a simple habitat: An experimental study with bacteria and phage. *Ecology* **1977**, *58*, 369–378. [CrossRef]

55. Lenski, R.E.; Levin, B.R. Constraints on the coevolution of bacteria and virulent phage: A model, some experiments, and predictions for natural communities. *Am. Nat.* **1985**, *125*, 585–602. [CrossRef]

56. Thyrhaug, R.; Larsen, A.; Thingstad, T.F.; Bratbak, G. Stable coexistence in marine algal host-virus systems. *Mar. Ecol. Prog. Ser.* **2003**, *254*, 27–35. [CrossRef]

57. Haaber, J.; Middelboe, M. Viral lysis of phaeocystis pouchetii: Implications for algal population dynamics and heterotrophic c, n and p cycling. *ISME J.* **2009**, *3*, 430–441. [CrossRef] [PubMed]

58. Moebus, K. Marine bacteriophage reproduction under nutrient-limited growth of host bacteria.2. Investigations with phage-host system h3:H3/1. *Mar. Ecol. Prog. Ser.* **1996**, *144*, 13–22. [CrossRef]

59. Moebus, K. Marine bacteriophage reproduction under nutrient-limited growth of host bacteria.1. Investigations with six phage-host systems. *Mar. Ecol. Prog. Ser.* **1996**, *144*, 1–12. [CrossRef]

60. Parada, V.; Herndl, G.J.; Weinbauer, M.G. Viral burst size of heterotrophic prokaryotes in aquatic systems. *J. Mar. Biol. Assoc. UK* **2006**, *86*, 613–621. [CrossRef]

61. Motegi, C.; Nagata, T. Enhancement of viral production by addition of nitrogen or nitrogen plus carbon in subtropical surface waters of the South Pacific. *Aquat. Microb. Ecol.* **2007**, *48*, 27–34. [CrossRef]

62. Bratbak, G.; Jacobsen, A.; Heldal, M.; Nagasaki, K.; Thingstad, F. Virus production in phaeocystis pouchetii and its relation to host cell growth and nutrition. *Aquat. Microb. Ecol.* **1998**, *16*, 1–9. [CrossRef]

63. Baudoux, A.-C.; Brussaard, C.P.D. Influence of irradiance on virus-algal host interactions. *J. Phycol.* **2008**, *44*, 902–908. [CrossRef] [PubMed]

64. Demory, D.; Arsenieff, L.; Simon, N.; Six, C.; Rigaut-Jalabert, F.; Marie, D.; Ge, P.; Bigeard, E.; Jacquet, S.; Sciandra, A.; et al. Temperature is a key factor in micromonas–virus interactions. *ISME J.* **2017**, *13*, 601–612. [CrossRef] [PubMed]

65. Maat, D.S.; Crawfurd, K.J.; Timmermans, K.R.; Brussaard, C.P.D. Elevated CO_2 and phosphate limitation favor micromonas pusilla through stimulated growth and reduced viral impact. *Appl. Environ. Microbiol.* **2014**, *80*, 3119–3127. [CrossRef] [PubMed]

66. Maat, D.S.; de Blok, R.; Brussaard, C.P.D. Combined phosphorus limitation and light stress prevent viral proliferation in the phytoplankton species phaeocystis globosa, but not in micromonas pusilla. *Front. Mar. Sci.* **2016**, *3*. [CrossRef]

67. Kendrick, B.J.; DiTullio, G.R.; Cyronak, T.J.; Fulton, J.M.; van Mooy, B.A.S.; Bidle, K.D. Temperature-induced viral resistance in *Emiliania huxleyi* (prymnesiophyceae). *PLoS ONE* **2014**, *9*, e112134. [CrossRef] [PubMed]

68. MacArthur, R.H.; Wilson, E.O. *The Theory of Island Biogeography*; Princeton University Press: Princeton, NJ, USA, 1967; p. 203.

69. Klochkova, T.A.; Kang, S.H.; Cho, G.Y.; Pueschel, C.M.; West, J.A.; Kim, G.H. Biology of a terrestrial green alga, *Chlorococcum* sp. (chlorococcales, chlorophyta), collected from the miruksazi stupa in Korea. *Phycologia* **2006**, *45*, 349–358. [CrossRef]

70. Wilson, W.H.; Tarran, G.A.; Schroeder, D.; Cox, M.; Oke, J.; Malin, G. Isolation of viruses responsible for the demise of an *Emiliania huxleyi* bloom in the english channel. *J. Mar. Biol. Assoc. UK* **2002**, *82*, 369–377. [CrossRef]

71. Vaughn, J.M.; Balch, W.M.; Novotny, J.F.; Vining, C.L.; Palmer, C.D.; Drapeau, D.T.; Booth, E.; Vaughn, J.M.; Kneifel, D.M.; Bell, A.L. Isolation of *Emiliania huxleyi* viruses from the gulf of maine. *Aquat. Microb. Ecol.* **2010**, *58*, 109–116. [CrossRef]

72. Highfield, A.; Evans, C.; Walne, A.; Miller, P.I.; Schroeder, D.C. How many coccolithovirus genotypes does it take to terminate an *Emiliania huxleyi* bloom? *Virology* **2014**, *466*, 138–145. [CrossRef] [PubMed]

73. Martinez, J.M.; Schroeder, D.C.; Larsen, A.; Bratbak, G.; Wilson, W.H. Molecular dynamics of *Emiliania huxleyi* and cooccurring viruses during two separate mesocosm studies. *Appl. Environ. Microbiol.* **2007**, *73*, 554–562. [CrossRef] [PubMed]

74. Sorensen, G.; Baker, A.C.; Hall, M.J.; Munn, C.B.; Schroeder, D.C. Novel virus dynamics in an *Emiliania huxleyi* bloom. *J. Plankton Res.* **2009**, *31*, 787–791. [CrossRef] [PubMed]

75. Ruiz, E. *Growth Rate Experiments in Emiliania Huxleyi*; Universitetet I Bergen: Bergen, Norway, 2014.

76. Brussaard, C.P.D.; Thyrhaug, R.; Marie, D.; Bratbak, G. Flow cytometric analyses of viral infection in two marine phytoplankton species, micromonas pusilla (prasinophyceae) and phaeocystis pouchetii (prymnesiophyceae). *J. Phycol.* **1999**, *35*, 941–948. [CrossRef]

77. Brussaard, C.P.D.; Marie, D.; Bratbak, G. Flow cytometric detection of viruses. *J. Virol. Methods* **2000**, *85*, 175–182. [CrossRef]

78. Brussaard, C.P.D.; Marie, D.; Thyrhaug, R.; Bratbak, G. Flow cytometric analysis of phytoplankton viability following viral infection. *Aquat. Microb. Ecol.* **2001**, *26*, 157–166. [CrossRef]

79. Levasseur, M.; Thompson, P.A.; Harrison, P.J. Physiological acclimation of marine-phytoplankton to different nitrogen-sources. *J. Phycol.* **1993**, *29*, 587–595. [CrossRef]

80. Almeida-Neto, M.; Guimaraes, P.; Guimaraes, P.R.; Loyola, R.D.; Ulrich, W. A consistent metric for nestedness analysis in ecological systems: Reconciling concept and measurement. *Oikos* **2008**, *117*, 1227–1239. [CrossRef]

81. Newman, M.E.J. Modularity and community structure in networks. *Proc. Natl. Acad. Sci. USA* **2006**, *103*, 8577–8582. [CrossRef] [PubMed]

82. Barber, M.J. Modularity and community detection in bipartite networks. *Phys. Rev. E* **2007**, *76*. [CrossRef] [PubMed]

83. Våge, S.; Storesund, J.E.; Thingstad, T.F. Adding a cost of resistance description extends the ability of virus-host model to explain observed patterns in structure and function of pelagic microbial communities. *Environ. Microbiol.* **2013**, *15*, 1842–1852. [CrossRef] [PubMed]

84. Avrani, S.; Lindell, D. Convergent evolution toward an improved growth rate and a reduced resistance range in prochlorococcus strains resistant to phage. *Proc. Natl. Acad. Sci. USA* **2015**, *112*, E2191–E2200. [CrossRef] [PubMed]

85. Segev, E.; Wyche, T.P.; Kim, K.H.; Petersen, J.; Ellebrandt, C.; Vlamakis, H.; Barteneva, N.; Paulson, J.N.; Chai, L.; Clardy, J.; et al. Dynamic metabolic exchange governs a marine algal-bacterial interaction. *Elife* **2016**, *5*, e17473. [CrossRef] [PubMed]

86. Marston, M.F.; Pierciey, F.J.; Shepard, A.; Gearin, G.; Qi, J.; Yandava, C.; Schuster, S.C.; Henn, M.R.; Martiny, J.B.H. Rapid diversification of coevolving marine synechococcus and a virus. *Proc. Natl. Acad. Sci. USA* **2012**, *109*, 4544–4549. [CrossRef] [PubMed]

87. Avrani, S.; Schwartz, D.A.; Lindell, D. Virus-host swinging party in the oceans: Incorporating biological complexity into paradigms of antagonistic coexistence. *Mob. Genet. Elements* **2012**, *2*, 88–95. [CrossRef] [PubMed]

88. Castillo, D.; Christiansen, R.H.; Espejo, R.; Middelboe, M. Diversity and geographical distribution of flavobacterium psychrophilum isolates and their phages: Patterns of susceptibility to phage infection and phage host range. *Microb. Ecol.* **2014**, *67*, 748–757. [CrossRef] [PubMed]

89. Holt, R.D.; Grover, J.; Tilman, D. Simple rules for interspecific dominance in systems with exploitative and apparent competition. *Am. Nat.* **1994**, *144*, 741–771. [CrossRef]

90. Leibold, M.A. A graphical model of keystone predators in food webs: Trophic regulation of abundance, incidence, and diversity patterns in communities. *Am. Nat.* **1996**, *147*, 784–812. [CrossRef]

91. Lakeman, M.B.; von Dassow, P.; Cattolico, R.A. The strain concept in phytoplankton ecology. *Harmful Algae* **2009**, *8*, 746–758. [CrossRef]

92. Nissimov, J.I.; Napier, J.A.; Allen, M.J.; Kimmance, S.A. Intragenus competition between coccolithoviruses: An insight on how a select few can come to dominate many. *Environ. Microbiol.* **2016**, *18*, 133–145. [CrossRef] [PubMed]

93. Duffy, S.; Turner, P.E.; Burch, C.L. Pleiotropic costs of niche expansion in the RNA bacteriophage phi 6. *Genetics* **2006**, *172*, 751–757. [CrossRef] [PubMed]

94. Elena, S.F.; Agudelo-Romero, P.; Lalic, J. The evolution of viruses in multi-host fitness landscapes. *Open Virol. J.* **2009**, *3*, 1–6. [CrossRef] [PubMed]

95. Nikolin, V.M.; Osterrieder, K.; von Messling, V.; Hofer, H.; Anderson, D.; Dubovi, E.; Brunner, E.; East, M.L. Antagonistic pleiotropy and fitness trade-offs reveal specialist and generalist traits in strains of canine distemper virus. *PLoS ONE* **2012**, *7*, e50955. [CrossRef]

96. Keen, E.C. Tradeoffs in bacteriophage life histories. *Bacteriophage* **2014**, *4*, e28365. [CrossRef] [PubMed]

97. Bedhomme, S.; Lafforgue, G.; Elena, S.F. Multihost experimental evolution of a plant rna virus reveals local adaptation and host-specific mutations. *Mol. Biol. Evol.* **2012**, *29*, 1481–1492. [CrossRef] [PubMed]

98. Agrawal, A.F.; Lively, C.M. Modelling infection as a two-step process combining gene-for-gene and matching-allele genetics. *Proc. R. Soc. B-Biol. Sci.* **2003**, *270*, 323–334. [CrossRef] [PubMed]

99. Flor, H.H. Host-parasite interaction in flax rust—Its genetics and other implications. *Phytopathology* **1955**, *45*, 680–685.

100. Inouye, B.; Stinchcombe, J.R. Relationships between ecological interaction modifications and diffuse coevolution: Similarities, differences, and causal links. *Oikos* **2001**, *95*, 353–360. [CrossRef]

101. Rowe, J.M.; Fabre, M.F.; Gobena, D.; Wilson, W.H.; Wilhelm, S.W. Application of the major capsid protein as a marker of the phylogenetic diversity of *Emiliania huxleyi* viruses. *FEMS Microbiol. Ecol.* **2011**, *76*, 373–380. [CrossRef] [PubMed]

102. Mackinder, L.C.M.; Worthy, C.A.; Biggi, G.; Hall, M.; Ryan, K.P.; Varsani, A.; Harper, G.M.; Wilson, W.H.; Brownlee, C.; Schroeder, D.C. A unicellular algal virus, *Emiliania huxleyi* virus 86, exploits an animal-like infection strategy. *J. Gen. Virol.* **2009**, *90*, 2306–2316. [CrossRef] [PubMed]

103. Allen, M.J.; Howard, J.A.; Lilley, K.S.; Wilson, W.H. Proteomic analysis of the ehv-86 virion. *Proteome Sci.* **2008**, *6*. [CrossRef] [PubMed]

104. Medlin, L.K.; Barker, G.L.A.; Campbell, L.; Green, J.C.; Hayes, P.K.; Marie, D.; Wrieden, S.; Vaulot, D. Genetic characterisation of *Emiliania huxleyi* (haptophyta). *J. Mar. Syst.* **1996**, *9*, 13–31. [CrossRef]

105. Iglesias-Rodriguez, M.D.; Schofield, O.M.; Batley, J.; Medlin, L.K.; Hayes, P.K. Intraspecific genetic diversity in the marine coccolithophore *Emiliania huxleyi* (prymnesiophyceae): The use of microsatellite analysis in marine phytoplankton population studies. *J. Phycol.* **2006**, *42*, 526–536. [CrossRef]

106. Blanco-Ameijeiras, S.; Lebrato, M.; Stoll, H.M.; Iglesias-Rodriguez, D.; Muller, M.N.; Mendez-Vicente, A.; Oschlies, A. Phenotypic variability in the coccolithophore *Emiliania huxleyi*. *PLoS ONE* **2016**, *11*, e0157697. [CrossRef] [PubMed]

107. Paasche, E. A review of the coccolithophorid *Emiliania huxleyi* (prymnesiophyceae), with particular reference to growth, coccolith formation, and calcification-photosynthesis interactions. *Phycologia* **2001**, *40*, 503–529. [CrossRef]

108. Kegel, J.U.; John, U.; Valentin, K.; Frickenhaus, S. Genome variations associated with viral susceptibility and calcification in *Emiliania huxleyi*. *PLoS ONE* **2013**, *8*, e80684. [CrossRef] [PubMed]

109. Young, J.R.; Westbroek, P. Genotypic variation in the coccolithophorid species *Emiliania-huxleyi*. *Mar. Micropaleontol.* **1991**, *18*, 5–23. [CrossRef]

110. Young, J.R. *A Guide to Extant Coccolithophore Taxonomy*; International Nannoplankton Association: Bremerhaven, Germany, 2003; p. 121.

111. Cook, S.S.; Whitlock, L.; Wright, S.W.; Hallegraeff, G.M. Photosynthetic pigment and genetic differences between two southern ocean morphotypes of *Emiliania huxleyi* (haptophyta). *J. Phycol.* **2011**, *47*, 615–626. [CrossRef] [PubMed]

112. Hagino, K.; Bendif, E.; Young, J.R.; Kogame, K.; Probert, I.; Takano, Y.; Horiguchi, T.; de Vargas, C.; Okada, H. New evidence for morphological and genetic variation in the cosmopolitan coccolithophore *Emiliania huxleyi* (prymnesiophyceae) from the cox1b-atp4 genes. *J. Phycol.* **2011**, *47*, 1164–1176. [CrossRef] [PubMed]

113. Coolen, M.J.L. 7000 years of *Emiliania huxleyi* viruses in the black sea. *Science* **2011**, *333*, 451–452. [CrossRef] [PubMed]

viruses

MDPI

Review

Marine Prasinoviruses and Their Tiny Plankton Hosts: A Review

Karen D. Weynberg [1,*], Michael J. Allen [2] and William H. Wilson [3]

[1]   Australian Institute of Marine Science, PMB 3, Townsville, Queensland 4810, Australia
[2]   Plymouth Marine Laboratory, Prospect Place, Plymouth PL1 3DH, UK; mija@pml.ac.uk
[3]   Sir Alister Hardy Foundation for Ocean Science, The Laboratory, Citadel Hill, Plymouth PL1 2PB, UK;
      wilwil@sahfos.ac.uk
*   Correspondence: kweynberg@gmail.com; Tel.: +61-07-4754-4444

Academic Editors: Corina P.D. Brussaard and Mathias Middelboe
Received: 31 January 2017; Accepted: 8 March 2017; Published: 15 March 2017

Abstract: Viruses play a crucial role in the marine environment, promoting nutrient recycling and biogeochemical cycling and driving evolutionary processes. Tiny marine phytoplankton called prasinophytes are ubiquitous and significant contributors to global primary production and biomass. A number of viruses (known as prasinoviruses) that infect these important primary producers have been isolated and characterised over the past decade. Here we review the current body of knowledge about prasinoviruses and their interactions with their algal hosts. Several genes, including those encoding for glycosyltransferases, methyltransferases and amino acid synthesis enzymes, which have never been identified in viruses of eukaryotes previously, have been detected in prasinovirus genomes. The host organisms are also intriguing; most recently, an immunity chromosome used by a prasinophyte in response to viral infection was discovered. In light of such recent, novel discoveries, we discuss why the cellular simplicity of prasinophytes makes for appealing model host organism–virus systems to facilitate focused and detailed investigations into the dynamics of marine viruses and their intimate associations with host species. We encourage the adoption of the prasinophyte *Ostreococcus* and its associated viruses as a model host–virus system for examination of cellular and molecular processes in the marine environment.

Keywords: virus–host interactions; marine virus ecology; virus-driven evolution

1. Introduction

Viruses are the most diverse and abundant biological entities in the world's oceans, with estimates often reaching in excess of 10^8 viruses per millilitre (mL) of seawater [1,2]. The advent of techniques such as epifluorescence microscopy and flow cytometry initially helped to reveal the sheer abundance of viruses in the world's ocean, and allowed the 'viral shunt' to be incorporated into the classic food web scenario [3–6]. Building on these observations, isolation, molecular and physiological studies have shown that marine viruses exert a great influence on nutrient and energy cycling, biogeochemistry, population dynamics, genetic exchange and evolution in the marine environment [4,5].

Marine microbes comprise more than 90% of the living biomass in the global oceans [4,5,7] and the phytoplankton component of this biomass are responsible for approximately half of the world's photosynthetic activity [8,9]. Phytoplankton have very high turnover rates, with the global phytoplankton population estimated to be replaced on average once every week [9]. Thus, phytoplankton have the potential to adapt to rapidly changing environmental conditions, which in light of current climate change predictions [10,11] will likely be of great significance. The simple cell structure and biology of these tiny ocean-dwelling autotrophs, coupled with rapidly increasing knowledge data on their associated viruses, has led to the establishment of new host–virus model systems [12]. Through infection, disruption and manipulation of cellular function and resulting cellular mortality, viruses influence the dynamics of key processes including primary production in the oceans. Host–virus model systems are proving increasingly important since classic oceanic food webs and nutrient cycling models have, until recently, tended to overlook the role of viruses. However, there are some notable exceptions, with recent marine microbial models now beginning to incorporate the role of viruses [13].

Over the last decade, the field of marine virology has widened to include studies on viruses that infect the smallest of the eukaryotic phytoplankton (typically less than 3 microns). Although the cyanobacteria *Synechococcus* and *Prochlorococcus* are two of the most productive genera in the open ocean [14–16], microalgae of the class Prasinophyceae, particularly of the family Mamiellaceae, and notably the genera *Ostreococcus*, *Micromonas* and *Bathycoccus*, are significant contributors in coastal and estuarine waters [8,17]. The class Prasinophyceae is viewed as the most basal in the green plant lineage and is believed to be the root from which all other green algae and land plants emerged [18]. Phylogenetic analyses have revealed Prasinophyceae is a paraphyletic early branching lineage of the Chlorophyta [19]. The non-monophyletic classification of this class of algae can complicate the study of their coevolution with viruses.

These microalgae are generically referred to as picophytoplankton based on their size range (<3 microns). The majority of viruses isolated and described that infect eukaryotic phytoplankton have been large dsDNA viruses assigned to the family *Phycodnaviridae*, comprised of five genera: *Prasinovirus*, *Chlorovirus*, *Phaeovirus*, *Raphidovirus*, *Coccolithovirus*. The *Phycodnaviridae* are believed to share a common evolutionary ancestor with other viral families, namely the *Asfarviridae*, *Poxviridae*, *Iridoviridae*, *Ascoviridae*, *Mimiviridae*, *Marseilleviridae*, *Megaviridae*, *Pithoviridae*, and *Pandoraviridae* [20–22]. This collective of virus families infect a diverse range of eukaryotic hosts and are grouped as the Nucleo-Cytoplasmic Large DNA Viruses (NCLDVs) [20,21,23,24]. The NCLDVs are characterised as possessing large dsDNA genomes of greater than 100 kilobases (kb) with replication predominately occurring solely in the cytoplasm or, in certain interactions, initiating transcription in the nucleus before completion of assembly in the cytoplasm. Previously, it was proposed that the NCLDVs form a new order of giant viruses, the 'Megavirales' [25], although, to date, this nomenclature has not yet been widely adopted.

Genome sequences are now available for the host prasinophyte picophytoplankton species *Ostreococcus tauri* [26,27], *Ostreococcus lucimarinus* [27], *Micromonas pusilla* [28] and *Bathycoccus prasinos* [29], as well as 12 whole prasinovirus genome sequences [30–34] (Table 1). Analysis of the NCLDVs present in microbial metagenomes collected during the TARA Oceans survey indicated that prasinoviruses were the most abundant NCLDVs, outnumbering the next most abundant members of the Megavirales, Mimivirus [35]. High abundances of prasinoviruses have also been described in other waters [36]. Clearly, the role of prasinoviruses in the natural environment is significant, and the study of their host interactions will provide insight on evolutionary and population dynamics, as well as intracellular physiological function when viruses rewire biochemical pathways as part of their manipulation during infection. For example, certain prasinoviruses have been identified as encoding novel genes never before reported in viruses e.g., a prasinoviral genome, OtV-2, encodes the only cytosolic, non-membrane-bound haem-binding protein cytochrome b_5 [37]. Despite the precise metabolic role still remaining elusive, the very identification of this gene in the genome of a virus, which infects a host with such a small physical structure, is intriguing. Indeed, it suggests that within smaller and simpler physiological cellular systems, such as *Ostreococcus*, organizational constraints associated with membrane and cellular function and compartmentalisation may not necessarily be as stringent as in more complex cells.

Table 1. Key features of characterised prasinophytes and their prasinoviruses.

Host Species	Host Genome Size	Host GC Content (%)	Number of Chromosomes	Number of Host Genes	Virus Genome Size	Virus Genome GC %	Number of Viral ORFs	Average Viral ORF Length (bp)	Viral tRNAs
O. tauri clade C	12.6 Mb	59	20	8116	OtV1 = 191,761 bp OtV5 = 186,713 bp	OtV1 = 45 OtV5 = 45	232	750	OtV1 = 4 OtV5 = 5
Ostreococcus sp. RCC809 clade B	13.3 Mb	60	20	7492	OtV2 = 184,409 bp	OtV2 = 42.15	237	725	OtV2 = 4
O. lucimarinus clade A	13.2 Mb	60	21	7805	OlV1 = 194,022 bp	OlV1 = 41	268	732	OlV1 = 5
O. mediterraneus clade D	-	-	-	-	OmV1 = 193,301	OmV1 = 44.6	251	730	OmV1 = 5
Micromonas sp. RCC299	21.96 Mb	64	17	10,286	MpV1 = 184,095 bp	MpV1 = 39	203	793	MpV1 = 6
B. prasinos RCC 1105	151 Mb	48	19	7847	BpV1 = 198,519 bp BpV2 = 187,069 bp	BpV1 = 37 BpV2 = 38	210 244	746 715	BpV1 = 4 BpV2 = 4

Bathycoccus species virus had large predicted protein removed for the purpose of this table. The gene comprises 14% of the total genome. ORFs, open reading frames; tRNA, transfer RNA; bp, base pair; OlV, *Ostreococcus lucimarinus* virus; MpV, *Micromonas pusilla* virus; BpV, *Bathycoccus prasinos* virus.

Over recent decades, our knowledge about the role of viruses in the marine environment has increased, with a growing number of virus–host systems isolated from the environment now established under laboratory conditions. Genome data arising from studies on such systems help inform investigations of key interactions, including horizontal gene transfer (HGT), between algal hosts and their viruses. A high degree of conservation has been reported between the genomes of prasinoviruses that infect all three genera [31], hereon referred to as OVs (*Ostreococcus* viruses—OtV refers to *Ostreococcus tauri* virus; OlV refers to *Ostreococcus lucimarinus* virus; OmV refers to *Ostreococcus mediterraneus* virus); MpVs (*Micromonas pusilla* viruses); BpVs (*Bathycoccus prasinos* viruses).

In light of their significant contribution and abundance in the marine environment, this review assesses what has been revealed in recent studies about prasinoviruses and their host interactions. We also argue for the global adoption of a Prasinovirus model host–virus system to help unveil more about the dynamics of these intimate and important associations.

2. Prasinophyte Host Genomes and Ecology

To date, six Prasinophyceae genomes, all within the order Mamiellales, have been sequenced. The sequenced genomes of the Mamilellophyceae show a gene-dense genome structure with minimal redundancy, indicating a simple organisation and highly reduced gene copy numbers, often down to just a single gene copy. This reflects positively in terms of these eukaryotic algae being suitable model host organisms for virus studies. Quirks in host metabolism of these species may also aid enquiries into host–virus metabolic mechanisms. A complex history of gene acquisition appears to have occurred in prasinophytes, which possess relatively small eukaryotic genomes with evidently high levels of HGT. Although sexual reproduction has not been reported in Mamelliophyceae strains, there are genes present in both *Ostreococcus* and *Micromonas* for meiosis control and it would appear that genetic recombination does occur between strains within the same clade [38]. Sequenced *Micromonas* genomes are more divergent than the *Ostreococcus* genomes [28] and *Ostreococcus tauri* has fewer than 8000 genes. The genetic distances between different clades within a genus of these prasinophytes revealed two *Ostreococcus* strains belonging to two clades actually only showed a divergence of less than 0.5% but approximately 25% divergence in amino-acid identity in their orthologous protein coding genes [34].

Two of the *Ostreococcus* host genomes that have been sequenced (*O. tauri* strain OTH95 [26] and *O. lucimarinus* strain CCE9901 [27]) have been described as high-light adapted coastal strains, whilst the *Ostreococcus* sp. strain RCC809 is viewed as a low-light adapted open-oceans species (genome available at [39,40]. Similar to high-light and low-light ecotypes seen in the cyanobacterium genus *Prochlorococcus* [41,42], there is evidence for niche adaptation within *Ostreococcus* [40], although further evidence is required to confirm this, as the occurrence of different *Ostreococcus* strains at various depths may be attributed to other factors, e.g., iron bioavailability or be indicative of acclimatisation, not adaptation. Three clades of *Ostreococcus*—A, C and D—are classified as high-light adapted, as defined through laboratory photophysiology assessments, and clade B is described as low-light adapted [40]. However, this low-light adapted clade has been reported in surface waters [8,43] and although different ecotypes do not commonly occur at the same geographical location, the drivers behind clade distribution may be more complex than determined by sea surface irradiance levels alone [43]. The putative low-light ecotype strain has increased photosensitivity and an adapted photoprotection mechanism.

Ostreococcus has recently been proposed as a suitable model organism for investigating iron uptake and metabolism in eukaryotic phytoplankton [44]. The bioavailability of iron in the oceans is known to be low, particularly in oligotrophic open ocean waters. The exact mechanisms involved in how marine microalgae manage to acquire iron are poorly understood. *O. tauri* has been shown to have tightly regulated diurnal cycling linked to iron uptake [44]. *Ostreococcus* sequenced genomes lack a suite of genes involved in iron transport with the exception of a multi-copper oxidase found only in *O. tauri* [27] indicating *Ostreococcus* uses an alternative iron acquisition pathway to other phytoplankton. A recent RNASeq analysis was conducted to examine cellular responses of *O. tauri* to a lack of iron over a diurnal cycle, making this the largest transcription study of *O. tauri* to date. Diurnal cycles were seen to be key to *O. tauri*'s metabolism of iron and via a non-reductive pathway, unlike the iron metabolism of *Chlamydomonas*. Instead of copper, zinc plays a major role in iron uptake in *O. tauri*, as many zinc finger proteins are involved in the upregulation of related genes. The mechanisms underpinning the utilisation of iron in this microalgal species differed greatly from those described for the established algal model *Chlamydomonas reinhardtii*. To date, no virus has been described that infects model species such as *Chlamydomonas* and yet we have an easy-to-study system in the unicellular *Ostreococcus* and its viruses that we can explore so much further. The findings that concluded that *Ostreococcus* has adopted an alternative iron uptake system to established model organisms, such as *Chlamydomonas*, add strength to the argument for establishment of this host system. Despite *C. reinhardtii* being a well-established model organism, there is a paucity of available technical tools, for example tightly regulated promoters and targeted gene inactivation and replacement, to study its cell cycle. *O. tauri* can undergo transformation [45] and its viruses could also be used as vectors and tools for manipulation. The complete chloroplast and mitochondrial genomes for *O. tauri* are also available to provide insights into processes such as photosynthesis [46]. The single chloroplast found in *Ostreococcus* cells is simple as it has a reduced size with just three layers of stacked thylakoid membranes.

3. *Micromonas* Viruses: The Roots of Algal Virology

The first report of phycodnavirus isolations was from the prasinophyte *Micromonas pusilla* [47]. These workers described the relatively easy method of isolating the virus from seawater and subsequent lysis of unialgal cultures. However, the results of this work were largely ignored at the time and no major investigations began until the high abundance of viruses in the marine environment was described by Bergh et al. in 1989 [1]. Until recently, *Micromonas*-specific viruses were the only viruses that infect prasinophytes to be described in any detail. Similar to other characterised prasinoviruses, transmission electron microscopy (TEM) reveals MpVs have icosahedral caspids with diameters ~120 nm consisting of electron-dense cores (Figure 1A,B). Most subsequent studies following the initial discovery of MpVs have largely investigated the ecological roles of these viruses and the genetic diversity they harbour [48–53]. It has been shown that *M. pusilla*-specific dsDNA viruses can lyse up to 25% of their host population on a daily basis [54]. The host species survives such high mortality rates through high growth rates and a high host diversity [8].

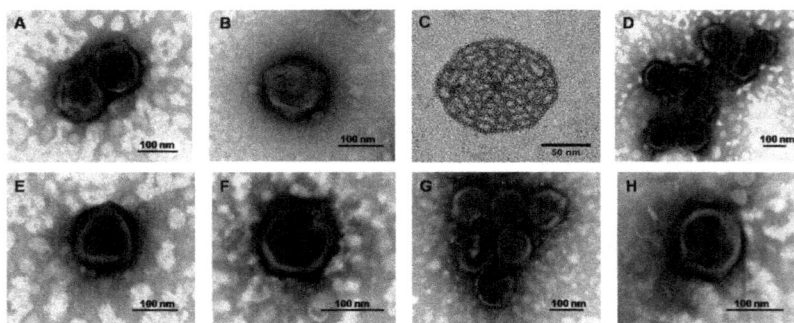

Figure 1. Negatively stained transmission electron microscopy micrographs of (**A**,**B**) *Micromonas pusilla* viruses (MpVs); (**C**) 'Spiderweb'-like plate from exterior of *Bathycoccus prasinos* cell; (**D–H**) *O. tauri* viruses (OtVs).

Transient blooms of *Micromonas* have been monitored and observed and *M. pusilla* has been reported as being the dominant member of the photosynthetic picoeukaryote assemblage all year round in certain locations, such as the Western English Channel, where a recent study revealed the dynamics of bloom-and-bust scenarios and the diversity of viral genotypes involved [55]. Whole genome sequences of two isolates of the *Micromonas* host species have been reported [28], which will help to unveil more about host–virus interactions. Somewhat surprisingly, for a long time only partial gene sequences, mainly of the DNA polymerase gene [56,57], were available in the public databases until Moreau and co-workers sequenced the complete genome of a dsDNA MpV [31] (Table 1).

4. *Bathycoccus* Viruses

The picophytoplankton genus *Bathycoccus* is comprised of a globally distributed green alga, which produces characteristic 'spiderweb-like' plates on the outside of the cells (Figure 1C). The whole genome sequence of the species *B. prasinos* was recently reported [29] and shows approximately 5% HGT, mainly from bacterial and eukaryotic origin with very little of viral origin [29]. Two of the 19 chromosomes found in *B. prasinos*, are described as outlier chromosomes with the small outlier chromosome (SOC) possibly playing a key role in the susceptibility of the alga to viral infection due to its observed hypervariability [29]. Interestingly, as will be discussed in detail, a SOC has been identified in *O. tauri* that has been shown to be key to providing resistance to viral infection [58]. In fact, outlier chromosomes have been identified in all prasinophyte genomes sequenced to date and this is an exciting development in researching immunity in these organisms. Two whole genome sequences have been reported for viruses (BpV-1 and BpV-2) that infect *Bathycoccus prasinos* [31] (Table 1). Unexpectedly, two long protein coding sequences (CDSs) were found within both the BpV-1 (11,202 and 17,067 bp) and BpV-2 (9,378 and 11,028 bp) viral genomes, which have no close taxonomic or functional matches in the public databases. These CDSs constitute between 10% and 15% of the entire genome and their presence raises the question as to what role they perform, particularly as genome reduction is a general feature of viruses.

Bathycoccus virus genomes appear to have acquired a heat shock protein *hsp70* gene from their host [31], the recruitment of which can greatly increase viral survival [59]. During infection, viruses can induce heat shock proteins resulting in enhanced viral infection, which has been demonstrated in detail for Hsp70. In eukaryotic cells, Hsp70 chaperone proteins are known to be involved in functions that support protein folding, translocation and assembly, and preventing apoptosis. In well-characterised viral infections, such as those of adenoviruses, these functions have been shown to aid maturation of viral proteins during infection and delay cell death enabling viral infection to reach completion. If BpVs can express their own Hsp70 proteins, rather than recruit host Hsp70 proteins, then they can regulate the expression of these proteins, and in the late stages of infection can suppress them to enable apoptosis and thus facilitate cell lysis.

Notably, the BpV genomes do not encode for certain amino acid biosynthesis pathway genes, such as a 3-dehydroquinate synthase gene, that are found in the genomes of prasinoviruses that infect *O. tauri* (OtVs), *O. lucimarinus* (OlVs) and *M. pusilla* (MpV-1) [31]. These genes encode for enzymes involved in the synthesis of essential amino acids that are formed in complex pathways when compared to the synthesis of nonessential amino acids. Their absence in these *Bathycoccus*-specific viruses is as yet unexplained. It has been postulated that the *Ostreococcus*-specific and *Micromonas*-specific dsDNA viruses encode certain hydrophilic and aromatic amino acids, as they do not utilise their host's biosynthetic pathways for these important resources during infection. This may reflect differences in the host species protein metabolism or divergences in viral infection and replication strategies. A previous study isolated the OVs and MpV-1 from coastal lagoons but the BpVs were isolated from coastal open ocean seawater [31] and perhaps environmental factors/ecological niches play a role in the replication strategy of these particular viruses.

5. *Ostreococcus*: The Simplest Viral Host of Them All

The prasinophyte *Ostreococcus tauri* was first described in 1995, following isolation from a coastal lagoon in France [60,61]. The host genus *Ostreococcus* is emerging as a suitable model plant organism due to an array of attributions [26,27]. These include its cellular simplicity (haploid with just a single mitochondrion, chloroplast and Golgi body, no cell wall or motility apparatus e.g., flagella). *Ostreococcus*, unlike *Micromonas*, is not flagellated and unlike *Bathycoccus* does not have scales or liths on the outside of the cell); its phylogenetic positioning as an early-diverging green plant lineage [26,27]; ease of keeping in culture; successful genetic transformation [45]; availability of four complete fully annotated genomes [30–33]; complete mitochondrial and chloroplast genomes available for *O. tauri* [46] and proteome analysis [62,63]. *Ostreococcus* has emerged as an ideal model organism in studies of signalling in eukaryotic cells [64] and light and circadian clocks or oscillators [65–67]. The latter is a research field that endeavours to gain understanding about the in-built regulators of physiological responses of organisms to the 24 h daily cycle of changes in the Earth's environment. *Ostreococcus* undergoes binary cell division and this can be manipulated by light/dark cycles in laboratory controlled conditions. *Ostreococcus* is a highly promising candidate for use as a model organism due to its simplicity as a eukaryote that can be used to extrapolate further for more complex eukaryotes. *Ostreococcus* could be used to study how cells respond to light, nutrient changes and to further investigate photosynthesis and cellular metabolism.

6. *Ostreococcus* and Its Viruses as a Host–Virus Model

With the early discovery of viruses that infect bacteria (bacteriophages), research focus was trained on a select number of model systems under laboratory conditions to characterise fully the dynamics of these viruses and their host interactions [68]. For eukaryotic virus systems, *Ostreococcus*-associated viruses form an equally attractive model candidate. The *Chlorella* algal host–virus system has become a classic model in algal virus research [12,69]. However, chlorovirus hosts are endosymbiotic zoochlorellae occurring within a range of hosts, including ciliates and metazoans, and chloroviruses are found in freshwater environments whereas OVs are representative of algal viruses occurring in marine environments.

The very first report of the existence of viruses targeting *Ostreococcus* species was in 2001, during monitoring of a picoplankton coastal bloom [70]. TEM analysis reveals *Ostreococcus*-specific viruses have similar morphology to other prasinoviruses, with capsid diameters ranging between 100–120 nm (Figure 1D–H). Assessments of the life cycle of OtV5 in culture revealed a latent period of 8 h, followed by cell lysis at 12–16 h post-inoculation [30]. Complete viral genomes were observed as early as two hours following inoculation. Host cell chromosomes remain intact throughout the infection cycle and are seen to decrease only as lysis occurs. TEM analysis confirmed viruses are localised to a region of the cytoplasm and do not associate with the nucleus or other organelles [30]. *Ostreococcus* and their viruses are globally distributed in coastal and open-ocean euphotic environments [34,36,43]. Viruses that infect *O. tauri* have been reported to be prevalent in coastal sites [32,33,36], whilst OlVs have been detected in more widespread marine locations, including oligotrophic sites in the Atlantic and Pacific Oceans [34,36,71,72], although this geographical distribution is not related to genetic distance, based on analysis of the *polB* gene [36]. Persistence of OVs and other prasinoviruses is linked to their environments. For example, MpVs are detected year-round, even in the absence of their hosts [53], and although the abundance of OtVs varies over time, they have also been detected throughout an annual sampling period [36]. To date, more than 300 OtVs have been sampled [36,56,71] and three complete OtV genomes [30,32,33], seven OlV genomes and one OmV genome sequenced and described, (with the latter being made publicly available—NCBI accession number NC_028092—but not yet published) [31,34] (Table 1) leading to insights into virus–host interactions, such as HGT events. The OtV-2 genome contains 42 unique genes with predicted functions not found in the genomes of the OtV-1 and OtV-5 viruses that infect high-light adapted hosts [33]. These include a putative cytochrome b_5 gene, the function and structure of which have recently been characterised, that is located in the same region of the OtV-2 genome as putative RNA polymerase sigma factor and high-affinity phosphate transporter genes, all three of which appear to have been acquired via HGT from the eukaryotic host [33].

Apart from members of the giant virus family the *Mimiviridae* [73,74], OtV-2 is the only other virus known to possess a putative cytochrome b_5 gene [37]. Why the largest viruses ever to be described and a virus infecting the smallest known free-living eukaryote all putatively encode a cytosolic cytochrome b_5 protein remains an unsolved mystery, as does the actual function of cytochrome b_5 itself. Cytochrome b_5 is an ubiquitous electron transport protein and can exist in two forms—a soluble enzyme used during photosynthesis in bacteria and a membrane-bound enzyme in animal tissues to reduce hemoglobin [75]. Cytochrome b_5 exists typically as a membrane-bound protein inserted in the outer membrane of mitochondria and the endoplasmic reticulum via an alpha helix at its carboxy-terminus. However, the cytochrome b_5 encoded by OtV-2 lacks this alpha helix and, as the hydrophobic C-terminal anchor is missing, the protein is not membrane-bound. The cytochrome b_5 protein encoded by OtV-2 was cloned, biochemically characterised and crystallography was used to resolve its three-dimensional structure [37] (Figure 2). It was found that the absorption spectra of oxidised and reduced recombinant OtV-2 cytochrome b_5 protein were almost identical to those of purified human cytochrome b_5 (Figure 2A,B). The virally encoded cytochrome b_5 was also substituted for yeast cytochrome b_5 activity to confirm the viral cytochrome b_5 was enzymatically active. Although structurally similar to other known cytochromes b_5, the viral version, by lacking a hydrophobic C-terminal anchor, is the first cytosolic cytochrome b_5 to be characterised (Figure 2C). The function of the viral version has seemingly diverged from that of its host protein to enable a different role during viral infection. The viral cytochrome b_5 protein was seen to have a portion of the haem-binding domain missing when compared to the host version, indicating that the virus and host utilise cytochrome b_5 for different functions. The viral protein lacks a reductase domain that confirms a divergence in function, although this physiological role remains unknown. As the host cell interior is ostensibly at the extremities of physical eukaryotic cell size, perhaps the occurrence of a virally encoded cytochrome b_5 in the cytosol of the host cell enables electron transfer to occur more easily during the virus infection process?

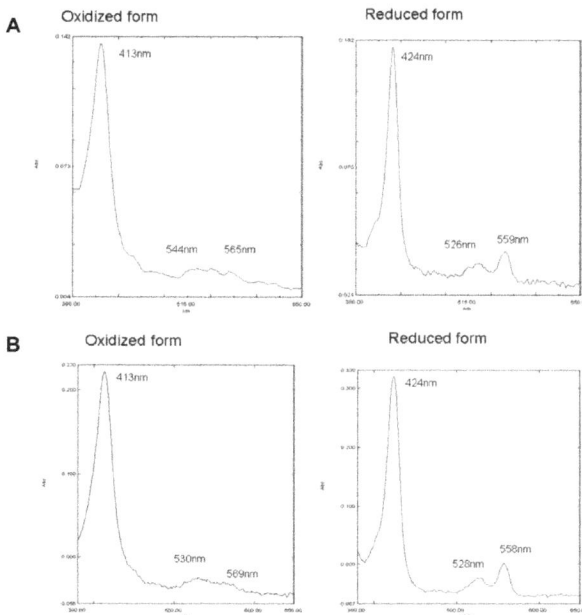

Figure 2. *Cont.*

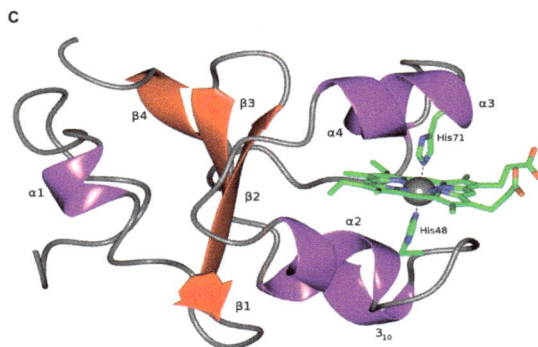

Figure 2. Characterisation of the OtV-2 virally encoded cytochrome b_5 protein. Absorbance spectra for oxidised and reduced forms of (**A**) human cytochrome b_5 protein and (**B**) OtV-2 viral cytochrome b_5 protein and (**C**) structural display of the OtV-2 protein as a ribbon diagram. Adapted from [37].

7. General Features of Prasinovirus Genomes

There is a high degree of evidence of HGT in the prasinovirus genomes, including gene acquisitions from their prasinophyte hosts, as well as other eukaryotes and bacteria. There are some unique genes not found in other viruses and genes that share close homology to host genes, indicating lateral gene transfer events have occurred between host and virus in these systems. Novel genes encoded by prasinovirus genomes include a 3-dehydroquinate synthase, glycosyltransferases, *N*-myristoyltransferase, methyltransferases, 6-phosphofructokinase and prolyl 4-hydroxylase [31–33]. Further, the high number of eight major capsid proteins encoded by these viruses is unique [31–33] and could have implications for host interactions, for example viral adsorption to the cell membrane, although this area has yet to be explored.

Prasinovirus genomes are smaller than most other phycodnavirus genomes, ranging from 184 to 198 kbp (Table 1), compared, for example, to the coccolithoviruses, which can be as large as 415 kbp [12,76,77]. The smaller sizes of prasinovirus genomes can be attributed not only to fewer coding DNA sequences (CDSs) but also to a general trend towards smaller CDSs and intergenic regions. For example OtV-2 [33] contains a similar CDS number to the phycodnavirus *Heterosigma akashiwo* virus, HaV-53 [78], (237 versus 246 CDSs), but has a 110 kbp smaller genome (Table 1). The GC content of prasinovirus genomes is lower than that of their host genomes—37%–45% GC [31,33] compared to 50%–64% GC [27–29], respectively (Table 1).

To date, all sequenced dsDNA prasinovirus genomes have 125 predicted genes in common. A high degree of collinearity is observable between prasinovirus genomes, except for the very ends of the genomes, approximately 10,000 bp at either end, which are the sites of terminal inverted repeats [31]. There is a 32 kbp central inverted region in the genomes of two subgroups (termed type II viruses) of OIV genomes [34], indicating an intriguing evolutionary event in these viruses. Examining the conservation of prasinovirus genomes in comparison to their host genomes, it is worthy to note that the host genomes exhibit greater plasticity, have undergone greater evolutionary divergence in comparison to their associated viruses and there is lower nucleotide variation between virus genomes compared to their host genomes [31]. All members of the Mamiellales have outlier chromosomes with similar predicted gene functions but low sequence homology (11% versus 89% in non-outlier chromosomes between *O. tauri* and *O. lucimarinus*). The Mamiellales need to strike a balance between high genetic variability, short replication time and a rapid response to infection. This may be explained by the selection pressures exerted by viruses and the subsequent response of the host in changes in arrangement of the outlier chromosomes and differential expression rates in response to viral infection. This may also reflect the divergence within the *Ostreococcus* genus as it comprises several clades.

In a study by Moreau and colleagues, the amino acid sequence identity of prasinoviruses and their respective host proteomes were compared [31]. The six prasinoviruses examined shared between 58% (OtV-1 and BpV-2) and 98% (OtV-1 and OtV5) average amino acid identity among their orthologous genes. This identity was consistently lower between the host's orthologous proteins than between their viruses (*O. tauri* versus *O. lucimarinus* was 73.8% similarity compared to 81% between their viruses; 58.4% similarity between *O. tauri* and *Micromonas* sp. compared to 67.6% similarity between their viruses; and 54.8% similarity between *O. tauri* and *Bathycoccus* sp. compared to 58.5% similarity between their virus amino acid identity), indicating the evolutionary distance may be greater between hosts than their viruses. This observation was also reflected in the percentage of common genes between hosts being lower than between their viruses.

The level of divergence between hosts has implications for their cospeciation patterns with their respective viruses. A high degree of cospeciation indicates viruses are more likely to be highly specific to a particular host and less likely to switch between different hosts. A study of cospeciation between prasinoviruses and their Mamiellales hosts isolated from open ocean waters found that although a high degree of host specificity does exist, infection of different hosts species within a genus was also observed [79]. Cophylogenetic analysis revealed a likely complex coevolution of prasinoviruses and their hosts, with indications of host switching and varying susceptibility among host strains, potentially as a result of differing resistance [58,80], which may lead to reduced growth, as well as increasing susceptibility to other viral strains. It was postulated that cospeciation seen in this marine algal virus system was a result of close evolutionary ties, and therefore adaptation, between host and virus, as an open ocean system offers few physical barriers to dispersal with a higher potential for encountering a wide range of hosts, hypothetically affording more opportunities for host switching [79].

Chloroviruses are the prototype model for the *Phycodnaviridae* and 22 core orthologous genes are shared between prasinoviruses and chloroviruses, which range between 66% and 100% in similarity [34]. Of the 22 genes, 19 have a predicted function in DNA replication (nine genes) or protein (four), nucleic acid (four), sugar (one) and lipid (one) metabolism. The average identity of the core genes at nucleotide level was 88% between OlVs and 85% across all *Ostreococcus* viruses but dropped to 62% for prasinoviruses as a whole. Prasinoviruses and chloroviruses shared an average of 30% identity and were identified as the most closely related phylogenetically within the *Phycodnaviridae*. It may therefore be advantageous to build upon the knowledge acquired about chloroviruses by expanding the research conducted on their closest known relatives, the prasinoviruses.

8. Prasinovirus Gene Repertoire Contains Unique Highlights

Viruses can hijack a host-derived pathway and modify it in favour of virus metabolism. A key example of this is the acquisition of a coccolithophore host sphingolipid pathway by the coccolithovirus *Emiliania huxleyi* virus (EhV) [81] to enable a metabolic shift directing sphingolipid synthesis towards virus assembly and infectivity [82]. A number of genes not previously described in viruses are found in prasinovirus genomes. These include a cluster of genes found in the *Ostreococcus*- and *Micromonas*-specific prasinovirus genomes involved in the biosynthesis of amino acids [30–33]. The amino acids valine, leucine and isoleucine are synthesised using the acetolactate synthase gene, which is found in the MpV-1 and OV genomes and appears to have closest homology to bacterial genes but is not present in the BpVs, or in any other known virus [31–33]. Codon usage for leucine and valine are higher in all genomes compared to the BpV genomes but isoleucine is highest in MpV and lowest in OtV-1. The aromatic amino acids tyrosine, phenylalanine and tryptophan are synthesised with involvement of the enzyme 3-dehydroquinate synthase. A 3-dehydroquinate synthase gene has been identified only in OV genomes (with the exception of OtV-2 that infects the low-light ecotype of *Ostreococcus* [33]) and is unique among all viral genomes described to date. Prior to the availability of whole prasinovirus genomes, Mimivirus was the only virus known to encode an asparagine synthase gene [83]. This gene is also present in OVs and MpV-1, but not in BpV genomes [31–33], and was recently described in a virus that infects the prymnesiophyte *Phaeocystis globosa* (PgV) [84]. It has been postulated that most prasinoviral amino acid metabolic genes were acquired from bacteria, with the exception of asparagine synthase that instead appears to have originated from a eukaryotic source within the green plant lineage [31]. Interestingly, it would appear there has been no exchange of the asparagine synthase gene between virus families, or with their respective hosts, since both have seemingly undergone independent evolution [85]. The conversion of aspartate to asparagine occurs in cellular organisms across the tree of life but only in a small representative of two virus families, the *Phycodnaviridae* and the *Megaviridae*. An important question to pose is why such contrastingly different viruses have this particular basic housekeeping gene in common.

MpV-1 is the only prasinovirus, and in fact the only virus reported to date, that encodes for acetaldehyde dehydrogenase and oxovalerate aldolase enzymes [31], which are involved in the conversion of 4-hydroxypentanoate to acetaldehyde and pyruvate, and toxic aldehyde to harmless acetate in acetaldehyde metabolism, respectively. These enzymes perform important functions during oxidative stress and scavenge aldehyde in the cell that is produced during oxidative degradation of lipid membranes. The aldehyde dehydrogenase superfamily of enzymes is well represented in plants and these enzymes are commonly involved in stress response pathways. There are also conserved domains in the putative acetaldehyde dehydrogenase gene in MpV that are found in the TPP enzyme superfamily. Interestingly, these enzymes are involved in the biosynthesis of the amino acids isoleucine, leucine and valine. This adds further evidence pointing to the preferred use of certain amino acids by prasinoviruses indicating these may be important in capsid formation. It would appear prasinoviruses OVs and MpV favour the amino acid ratio towards these particular amino acids. These genes are also found clustered at the 5′ extremities of these viral genomes indicating their possible early transcription in the infection process.

Two distantly related forms of DNA ligase are encoded for by different members of the NCLDV group. All prasinoviruses have ATP dependent DNA ligases, not nicotinamide adenine dinucleotide(NAD)-dependent ligases. The evolution and acquisition of these genes in the NCLDVs is unclear, as phylogenetic analysis has indicated either form of DNA ligase may have appeared in these viruses first [24]. The prasinoviral genomes are linear with terminal repeat regions on their ends. Such repetitive regions are indicated to maintain genome stability [86] and it has been hypothesised that repetitive regions at the extremes of virus genomes were the precursors to telomeres in cellular chromosomes.

Viral replication of large DNA viruses requires a supply of deoxynucleotides. An essential intermediate in the synthesis of the deoxynucleotide dTTP is dUMP, which certain prasinoviruses produce using dCMP deaminase [31]. Deoxycytidine deaminase, thymidylate synthase, thymidine kinase and ribonucleotide reductase are all enzymes involved in dTTP synthesis. The dCMP deaminase enzyme is encoded by all *Chlorella* viruses [87] and is also found in OtV-2 [33], MpV-1, and OlV-1 [31], but not in any of the remaining prasinoviruses. Instead, they utilise ribonucleotide reductase in the pathway to synthesise dUMP for dTTP synthesis. Similar to chloroviruses, prasinoviruses encode for dUTPase and thymidylate synthase but only prasinoviruses also encode thymidine kinase, a key enzyme in the pyrimidine synthesis pathway involved in dTTP synthesis.

Unique to eukaryotes, mRNA capping is a post-transcriptional modification of messenger RNA (mRNA) that viruses can manipulate to ensure their mRNA is efficiently translated, whilst also gaining protection from cellular exonuclease degradation and recognition as foreign RNA by their hosts [88]. All prasinoviruses encode two mRNA capping enzymes, whilst *Chlorella* viruses encode one [31,33]. Methyltransferases can also offer a means for viral DNA to be protected from host cellular defence mechanisms. DNA methyltransferases are not commonly encoded by viruses but are found in bacteriophages and members of the *Phycodnaviridae*, namely chloroviruses [89], phaeoviruses [90] and prasinoviruses [31,33].

All prasinovirus genomes sequenced to date encode for as many as eight putative major capsid proteins (MCP) except the BpVs which lack MCP 1 maybe due to gene loss [30–33]. Most viruses encode for a single MCP and additional minor capsid proteins. The roles of these proteins may be related to structure and assembly but also to adsorption and host cell membrane fusion. Such a surprisingly high number of putative capsid proteins may indicate complexities or subtleties in the structure of the capsids of these viruses, although this has yet to be examined in depth. More complex viruses encode several versions of capsid proteins that are then used to build up the capsid structure surrounding the virus genome. Virus genome size is minimised by the formation of a capsid from multiple copies of a major capsid protein. So why do prasinoviruses encode multiple major capsid genes encoding for multiple capsid subunit proteins? The complexities of the capsid structure may hold functions pertaining to assembly, encapsidation, how entry into the host cell is executed and cues for adsorption and disassembly. TEM has shown OtV-1 capsid fuses with and remains fused to the host cell membrane [32]. Do any of the MCPs play a role in evading a host antiviral response? To begin to resolve the roles of these multiple MCPs in prasinoviruses, techniques such as X-ray crystallography and three-dimensional electron cryo-electron microscopy (cryo-EM) will need to be employed. TEM analysis has confirmed the prasinoviruses have an icosahedral structure (Figure 1). Icosahedrons typically have 12 vertices with 5-fold symmetry, and 20 triangular faces with 3-fold symmetry and 30 edges with 2-fold symmetry [77]. Much structural work has been conducted into viruses (as reviewed in [91]). As capsids are complex structures with hundreds of subunits, not all of the identical structures e.g., pentamers and hexamers, have been characterised in dsDNA virus families such as papillomaviruses and adenoviruses. Prasinoviruses must have very complex and intricate icosahedral architecture in their capsid structures. Considering the prasinoviral capsids are between 100 and 120 nm in diameter and interact within a host cell smaller than 1000 nm (meaning viruses are about one-eighth of the host cell size), their architecture, packaging and maturation must reflect this in the number of capsid proteins required. The burst size for prasinoviruses has been estimated to be less than 100 virions per cell and experimental data has indicated between six and 15 viruses are produced per host cell [30]. Considering that the host nucleus, mitochondrion and chloroplast remain intact throughout virus assembly in the cytoplasm [30], this adds to the global constraints on virus particle structure and the number of different capsid proteins required may reflect this. The internal pressure of packaging a genome inside the capsid will also influence the structure.

9. Viral Sugar Metabolism

Glycovirology is a newly emerging area of interest in virus–host interactions, studying how viruses manipulate the host glycome. The field of glycobiology is important as a potential focal point for antiviral therapy. It has been a point of interest that, like the chloroviruses, the prasinoviruses encode several glycosyltransferases, which are likely involved in glycosylation of viral components usually facilitated by the cellular glycosylation machinery. Sugar manipulation enzymes, such as 6-phosphofructokinase and glycosyltransferases, are encoded by prasinoviruses [31–33]. Glycosyltransferases (GTs) are known to be encoded by certain viruses including bacteriophages, phycodnaviruses, baculoviruses, poxviruses and herpesviruses (reviewed in [92]). GTs are instrumental in the formation of glycans. Viruses use glycans to ensure correct folding and conformation maintenance of viral glycoproteins and also for recognition for attachment to the host cell surface. MpV-1 and BpVs encode for three GTs [31], whilst OtV-2 and the remaining OV genomes contain four and six glycosyltransferase genes, respectively [31,33]. In the chloroviruses, GTs are postulated to be involved in post-translational modification and glycosylation of the capsid [92,93]. It has also been speculated that the chloroviral GT enzymes may have been acquired from a bacterial source and existed prior to formation of the cellular endoplasmic reticulum and Golgi body [93], as these organelles are not involved in chloroviral glycosylation. With the number of capsid proteins and glycosyltransferases encoded for by prasinoviruses, it would appear glycoconjugates are as vital for prasinovirus virion structure as they are in chloroviruses such as the prototype PBCV-1, which has six glycosylated sites on the MCP [93]. Of further note, is the fact that *N*-linked glycans are typically derived from asparagine and the OtVs encode asparagine synthase and, for example, OtV-1 encodes its own asparagine synthase and serine/threonine protein kinase [32].

An intriguing discovery was the presence of a gene, *pfk1*, encoding 6-phosphofructokinase (PFK), in both the MpV-1 and OV genomes [31,33], although not encoded for by the BpVs [31] or, in fact, by any other virus described to date. PFK is the key regulatory enzyme in the glycolysis metabolic pathway that converts glucose to pyruvate and ATP. PFK catalyses the glycolysis step that results in production of fructose-1,6-bisphosphate, which is not simply a metabolic intermediate but plays an important role in cell signalling and has been shown to delay cellular death in animal tissues [94]. The role of PFK in tumourigenic cells is a major focus for medical research at present. It has been shown that with increased PFK activity cell cycle progression is promoted and cell proliferation increases, whilst a decrease in PFK activity is linked to the onset of apoptosis [95]. The role of this crucial enzyme in cell metabolism could now be further revealed by studying the production of this virally encoded enzyme during infection of simple, minimalist algal host cells. Why does only this subset of prasinoviruses encode for such a centrally key metabolic enzyme? It seems plausible the viruses are exploiting the role of this enzyme in driving energy production by harnessing cell proliferation mechanisms to increase virus production, whilst also delaying apoptosis. It would be timely to widen the search for virally encoded PFK enzymes, as viral versions may remain undetected thus far and instead are amongst the countless hypotheticals and ORFans due to the divergence of viral from non-viral versions of the gene. It is also worth highlighting that PFK is the only glycolysis enzyme encoded by the OVs and MpV-1 and these viral genes share homology with bacterial genes, not equivalent host genes.

10. Host Resistance

Researchers have begun to explore viral infection in *Arabidopsis*, the classic terrestrial plant model organism, but not in a marine equivalent that is comparatively simple and known to be open to cellular manipulation e.g., *Ostreococcus*. In this regard, plant–virus coevolution requires established systems for in-depth analysis of how hosts react to infection e.g., development of resistance. It has been reported that *Ostreococcus* experiences high levels of infection by its associated viruses [36] and in response can develop resistance to viral attack [80], similar to that described in species such as *Prochlorococcus* [96]. Resistance to viral infection in *O. tauri* has been observed to be maintained for at least two years [58] and described in all three host genera [80] but can incur a cost, namely reduced growth rates and/or vulnerability to infection by other viruses. It has been proposed that the susceptibility of the *Ostreococcus* host to viral attack is related to the size of two outlier chromosomes [71], one big (BOC) and one small (SOC) similar to the theory purported to explain potential viral susceptibility in *Bathycoccus*.

In a recent transcriptional study of *O. tauri* in response to infection by OtV-5 in culture, the number of differentially expressed genes located on chromosome 19, the identified SOC, was markedly disproportionally higher than on other chromosomes [58]. The SOC has a bipartite structure being divided into an active left side during viral resistance and an active right side during susceptibility to infection. Most of the differentially transcribed genes located on the SOC in resistant hosts were associated with carbohydrate metabolism and transport. Four amino acid biosynthetic pathways were seen to be downregulated whereas post-translational modification and chaperones were upregulated. A transmembrane phosphate transporter was overexpressed and a calcium transporter underexpressed. Differential expression was seen in genes encoding translation, transcription, protein modification and turnover, amino acid transport and modification and other transporters. Viral immunity also appeared to be linked to a downturn in amino acid production. What is the cost of encoding two chromosomes involved in host defence? The cost to the cell may well be slowed host cell growth as ribosomal subunits were underexpressed in resistant cells. Histone modification genes were over expressed indicating likely chromatin restructuring occurred during resistance. By altering transporter activity, the host cell may be altering the available substrates for viral replication to occur.

Higher levels of GTs are located on the SOC and become overexpressed in resistant cells, whilst GTs are downregulated on the BOC. The SOC genes with known function encode methyltransferases, glycosylation-related genes and membrane proteins. Interestingly, both OtV genomes and the SOC in *O. tauri* encode a NAD-dependent epimerase/dehydratase with the host gene bearing closest homology to a gene in higher plants whilst the viral gene is closest to a bacterial gene.

A FkbM methyltransferase gene is present in both viral and host genomes, occurring close to the sugar gene cluster in the SOC, and is predicted to play a role in glycan methylation [32,58]. Glycans are carbohydrate modifications on proteins or lipids that act as ligands that bind carbohydrate-binding proteins called lectins. Glycans are the key at the interface of virus and host. Viruses use their own or their host glycans for replication and infectivity. A virus capsid possessing glycans can act as a ligand for host lectins and host glycoconjugates that can in turn act as receptors for viral lectins and facilitate cell entry. It would appear *O. tauri* encodes its own versions of sugar enzymes and methyltransferases on the SOC that are replicated in response to initial infection events. The resulting events divert virus adsorption and result in glycosylation and methylation to provide protection away from the virus-directed pathways. The host is playing the virus at its own game and, by using different versions of the same gene, the host machinery is regaining control and blocking the virus from the point of adsorption. It has been postulated the OtV5 receptor may only be available at a certain point in the host cell life cycle. However, if the compact intracellular organisation and limited physical space available within the host cell is taken into account, then it would seem more likely the host will try to arrest infection earlier rather than later, so therefore, the point of viral attachment and adsorption would be a more logical stage to create resistance. Do the host flood cell surface receptors with host versions of sugars or does it alter the substrate pool, or perhaps both strategies are adapted simultaneously?

A long inverted repeat region (LIRR) was identified in the SOC and contains all the over-transcribed genes in the resistant cells and is the region that is silenced in the susceptible wild-type state [58]. In the LIRR, clusters of genes involved in carbohydrate metabolism were seen. The *O. tauri* SOC is unique by encoding a rhaman synthesis gene *RgpF* not found in any other eukaryote but only in bacteria. Rhaman synthesis performed by genes located on the SOC seems to be a key weapon in the host arsenal against viral infection. Notably, OtV-1 encodes a dTDP-D-glucose-4,6-dehydratase that shares closest homology with the metazoan *Nematostella*. This protein is also known as rhamnose synthase and is also encoded by the host. OtV5 encodes a GDP-D-mannose 4,6-dehydratase with closest homology to a gene in the chlorovirus PBCV-1 that is involved in rhamnose and fucose synthesis, which are monosaccharides commonly seen in the virion capsid but occur rarely in the host. This must be one of the key points that resistant hosts interfere with. Do prasinoviruses within an already exposed culture attach to host cell surfaces and employ host glycoconjugates to enable entry into the cell but the SOC switches and changes the glycobiology and available conjugates, so as to prevent viral infection? It would appear that the underlying mechanism here is related to glycan-mediated host–virus interactions.

The SOC and BOC are able to swap genetic material between each other and rearrange within themselves, with transposons most likely playing a key role in this activity [58]. Karyotypic changes were observed in the size of the SOC in all resistant lines of host cells via duplications, deletions and possible translocations [58]. Are these changes causative or resultant of resistance? The glycobiology aspects of prasinophyte host–virus interactions represent an exciting area for future research to expand into. The next logical step is to take the analysis of chromosome 19, the SOC, into the natural environment.

11. Nitrogen Metabolism in Prasinophytes and Their Viruses

Ostreococcus and *Micromonas* contain genes for nitrate/nitrite transporters and reductases and *Ostreococcus* also encodes proteins involved in molybdate transport and metabolism [27,28]. *O. tauri*, *O. lucimarinus* and *B. prasinos* encode animal-type nitric oxide (NO) synthase (NOS) enzymes that appear to have been acquired via HGT [97]. This is of interest as NOS is a haem protein similar to cytochrome P450 reductase, and contains a haem domain, as does cytochrome b_5. Interestingly, the NOS unveiled in *O. tauri* shares 45% similarity to human NOS and displays similar protein folding to the human version [98]. This was the first report of a NOS in a plant. The generation of NO in *O. tauri* was seen to be higher under high light irradiance and during exponential growth [98]. The production of NO is seen to increase with high light intensity irradiation, which may reflect clade assignation and ecotype niche partitioning and hence indicates a link between physiology and NOS activity. It would be interesting to explore such metabolic processes further, particularly in regard to active viral infection. It is unique in plants that *O. tauri* encodes NOS with similarities to animal and bacterial NOS [98]. Interestingly, nitric oxide has been implicated to have a role in immune responses to viruses in animals [99]. However, to date no research has been conducted into the role of these enzymes in algal host defence against viruses. Nitrogen, along with elements such as phosphorus, is an important limiting factor in viral infection [100].

12. Inteins in Prasinoviruses

Complete large inteins containing a homing endonuclease have been detected in the DNA polymerase gene of a number of prasinoviruses, including in the OtV-1 genome and other environmental *O. tauri* sequences (isolates OtV06_1, OtV09_561, OtV09_600) and BpV-2 [32,101]. Inteins are genetic selfish elements that insert themselves into conserved regions of conserved genes and are capable of self-splicing following translation [102]. The homing endonuclease facilitates the lateral transfer of their own coding region and flanking sequences between genomes and this process is called 'homing' and is recombination-dependent. It has also been reported that a recombination event occurred between two inteins found in OtV viruses infecting *O. lucimarinus* and *O. tauri* [101]. This finding could have important ecological and evolutionary significance and result in increasing virus diversity, as recombination events occur via this mechanism. Inteins have been adopted as biotechnological tools by exploiting their natural splicing mechanisms and their use in aiding protein purification [102]. Inteins are seen to be widely distributed in nature although their distribution is sporadic indicating lateral transfer events are at play. Recombination events between inteins found in OVs have been reported [101]. Prasinoviruses can seemingly co-infect the same host enabling lateral transfer of inteins between viruses. If prasinoviruses are similar enough genetically then they can coinfect the same host cell, which not only enables lateral transfer of inteins but has implications for the host too, in terms of being assaulted by more than one virus simultaneously. It has been postulated that co-infection is likely widespread and virus–virus interactions are common [102]; although most work has been conducted into bacteriophages, this may well also apply to eukaryotic viruses. If coinfection by two or more viruses can occur in a small host cell such as *Ostreococcus* then this has some major implications for viral evolution, viral exchange and infection dynamics. Clerissi and co-workers speculated that the higher concentration of prasinophyte hosts in lagoon environments may increase the incidence of inteins and their transfer, compared to similar events occurring in coastal and open waters [101].

13. Conclusions and Future Directions

It is now timely to establish the smallest known free-living eukaryotes at the base of the green algal lineage, particularly *Ostreococcus*, and their associated viruses as a suitable model system to examine in-depth the dynamics of marine algal virus infections. There exist large gaps in our knowledge surrounding viral life history and interactions with their hosts. Viruses that infect the prasinophytes can provide novel and incisive insights into how such model systems work and allow us to study processes including primary production, cellular resource allocation, genetic transfer events and evolution. Recent unveiling of the genetic treasures of the prasinoviral genomes and infection strategies, and those of their host responses to these remarkable features, are helping to open a whole new aspect to not only viral–host interactions but also cellular biology, genetics and physiology.

Conflicts of Interest: The authors declare no conflict of interest.

References

1. Bergh, O.; Borsheim, K.Y.; Bratbak, G.; Heldal, M. High abundance of viruses found in aquatic environments. *Nature* **1989**, *340*, 467–468. [CrossRef] [PubMed]
2. Wigington, C.H.; Sonderegger, D.; Brussaard, C.P.D.; Buchan, A.; Finke, J.F.; Fuhrman, J.A.; Lennon, J.T.; Middelboe, M.; Suttle, C.A.; Stock, C.; et al. Re-examination of the relationship between marine virus and microbial cell abundances. *Nat. Microbiol.* **2016**, *1*, 15024. [CrossRef] [PubMed]
3. Danovaro, R.; Dell'Anno, A.; Corinaldesi, C.; Magagnini, M.; Noble, R.; Tamburini, C.; Weinbauer, M. Major viral impact on the functioning of benthic deep-sea ecosystems. *Nature* **2008**, *454*, 1084–1087. [CrossRef] [PubMed]
4. Suttle, C.A. Viruses in the sea. *Nature* **2005**, *437*, 356–361. [CrossRef] [PubMed]
5. Suttle, C.A. Marine viruses—Major players in the global ecosystem. *Nat. Rev. Microbiol.* **2007**, *5*, 801–812. [CrossRef] [PubMed]
6. Wilhelm, S.W.; Suttle, C.A. Viruses and nutrient cycles in the sea—Viruses play critical roles in the structure and function of aquatic food webs. *Bioscience* **1999**, *49*, 781–788. [CrossRef]
7. Whitman, W.B.; Coleman, D.C.; Wiebe, W.J. Prokaryotes: The unseen majority. *Proc. Natl. Acad. Sci. USA* **1998**, *95*, 6578–6583. [CrossRef] [PubMed]
8. Worden, A.Z. Picoeukaryote diversity in coastal waters of the Pacific Ocean. *Aquat. Microb. Ecol.* **2006**, *43*, 165–175. [CrossRef]
9. Field, C.B.; Behrenfeld, M.J.; Randerson, J.T.; Falkowski, P. Primary production of the biosphere: Integrating terrestrial and oceanic components. *Science* **1998**, *281*, 237–240. [CrossRef] [PubMed]
10. Burrows, M.T.; Schoeman, D.S.; Buckley, L.B.; Moore, P.; Poloczanska, E.S.; Brander, K.M.; Brown, C.; Bruno, J.F.; Duarte, C.M.; Halpern, B.S.; et al. The pace of shifting climate in marine and terrestrial ecosystems. *Science* **2011**, *334*, 652–655. [CrossRef] [PubMed]
11. Parmesan, C. Ecological and evolutionary responses to recent climate change. *Annu. Rev. Ecol. Evol. Syst.* **2006**, *37*, 637–669. [CrossRef]
12. Wilson, W.H.; van Etten, J.L.; Allen, M.J. The *Phycodnaviridae*: The story of how tiny giants rule the world. In *Lesser Known Large dsDNA Viruses*; van Etten, J.L., Ed.; Springer: Berlin, Germany, 2009; pp. 1–42.
13. Weitz, J.S.; Stock, C.A.; Wilhelm, S.W.; Bourouiba, L.; Coleman, M.L.; Buchan, A.; Follows, M.J.; Fuhrman, J.A.; Jover, L.F.; Lennon, J.T.; et al. A multitrophic model to quantify the effects of marine viruses on microbial food webs and ecosystem processes. *ISME J.* **2015**, *9*, 1352–1364. [CrossRef] [PubMed]
14. Liu, H.B.; Campbell, L.; Landry, M.R. Growth and mortality rates of *Prochlorococcus* and *Synechococcus* measured with a selective growth inhibitor. *Mar. Ecol. Prog. Ser.* **1995**, *116*, 277–287. [CrossRef]
15. Liu, H.B.; Nolla, H.A.; Campbell, L. *Prochlorococcus* growth rate and contribution to primary production in the equatorial and subtropical North Pacific Ocean. *Aquat. Microb. Ecol.* **1997**, *12*, 39–47. [CrossRef]
16. Vaulot, D.; Marie, D.; Olson, R.J.; Chisholm, S.W. Growth of *Prochlorococcus*, a photosynthetic prokaryote, in the Equatorial Pacific Ocean. *Science* **1995**, *268*, 1480–1482. [CrossRef] [PubMed]
17. Vaulot, D.; Eikrem, W.; Viprey, M.; Moreau, H. The diversity of small eukaryotic phytoplankton (≤ 3 mu m) in marine ecosystems. *Fems Microbiol. Rev.* **2008**, *32*, 795–820. [CrossRef] [PubMed]
18. Sym, S.; Pienaar, R. The class Prasinophyceae. In *Progress in Phycological Research*; Round, F., Chapman, D., Eds.; Biopress Ltd.: Bristol, UK, 1993; pp. 281–376.
19. Ruhfel, B.R.; Gitzendanner, M.A.; Soltis, P.S.; Soltis, D.E.; Burleigh, J.G. From algae to angiosperms–inferring the phylogeny of green plants (Viridiplantae) from 360 plastid genomes. *BMC Evol. Biol.* **2014**, *14*, 23. [CrossRef] [PubMed]
20. Iyer, L.M.; Aravind, L.; Koonin, E.V. Common origin of four diverse families of large eukaryotic DNA viruses. *J. Virol.* **2001**, *75*, 11720–11734. [CrossRef] [PubMed]
21. Yutin, N.; Wolf, Y.I.; Raoult, D.; Koonin, E.V. Eukaryotic large nucleo-cytoplasmic DNA viruses: Clusters of orthologous genes and reconstruction of viral genome evolution. *Virol. J.* **2009**, *6*, 223. [CrossRef] [PubMed]
22. Abergel, C.; Legendre, M.; Claverie, J.-M. The rapidly expanding universe of giant viruses: Mimivirus, Pandoravirus, Pithovirus and Mollivirus. *Fems Microbiol. Rev.* **2015**, *39*, 779–796. [CrossRef] [PubMed]
23. Koonin, E.V.; Yutin, N. Origin and evolution of eukaryotic large nucleo-cytoplasmic DNA viruses. *Intervirology* **2010**, *53*, 284–292. [CrossRef] [PubMed]

24. Yutin, N.; Koonin, E.V. Hidden evolutionary complexity of nucleo-cytoplasmic large DNA viruses of eukaryotes. *Virol. J.* **2012**, *9*, 161. [CrossRef] [PubMed]
25. Colson, P.; de Lamballerie, X.; Yutin, N.; Asgari, S.; Bigot, Y.; Bideshi, D.K.; Cheng, X.-W.; Federici, B.A.; van Etten, J.L.; Koonin, E.V.; et al. "Megavirales", a proposed new order for eukaryotic nucleocytoplasmic large DNA viruses. *Arch. Virol.* **2013**, *158*, 2517–2521. [CrossRef] [PubMed]
26. Derelle, E.; Ferraz, C.; Rombauts, S.; Rouze, P.; Worden, A.Z.; Robbens, S.; Partensky, F.; Degroeve, S.; Echeynie, S.; Cooke, R.; et al. Genome analysis of the smallest free-living eukaryote *Ostreococcus tauri* unveils many unique features. *Proc. Natl. Acad. Sci. USA* **2006**, *103*, 11647–11652. [CrossRef] [PubMed]
27. Palenik, B.; Grimwood, J.; Aerts, A.; Rouze, P.; Salamov, A.; Putnam, N.; Dupont, C.; Jorgensen, R.; Derelle, E.; Rombauts, S.; et al. The tiny eukaryote *Ostreococcus* provides genomic insights into the paradox of plankton speciation. *Proc. Natl. Acad. Sci. USA* **2007**, *104*, 7705–7710. [CrossRef] [PubMed]
28. Worden, A.Z.; Lee, J.H.; Mock, T.; Rouze, P.; Simmons, M.P.; Aerts, A.L.; Allen, A.E.; Cuvelier, M.L.; Derelle, E.; Everett, M.V.; et al. Green evolution and dynamic adaptations revealed by genomes of the marine picoeukaryotes *Micromonas*. *Science* **2009**, *324*, 268–272. [CrossRef] [PubMed]
29. Moreau, H.; Verhelst, B.; Couloux, A.; Derelle, E.; Rombauts, S.; Grimsley, N.; Van Bel, M.; Poulain, J.; Katinka, M.; Hohmann-Marriott, M.F.; et al. Gene functionalities and genome structure in *Bathycoccus prasinos* reflect cellular specializations at the base of the green lineage. *Genome Biol.* **2012**, *13*, R74. [CrossRef] [PubMed]
30. Derelle, E.; Ferraz, C.; Escande, M.L.; Eychenie, S.; Cooke, R.; Piganeau, G.; Desdevises, Y.; Bellec, L.; Moreau, H.; Grimsley, N. Life-cycle and genome of OtV5, a large DNA virus of the pelagic marine unicellular green alga *Ostreococcus tauri*. *PLoS ONE* **2008**, *3*. [CrossRef] [PubMed]
31. Moreau, H.; Piganeau, G.; Desdevises, Y.; Cooke, R.; Derelle, E.; Grimsley, N. Marine prasinovirus genomes show low evolutionary divergence and acquisition of protein metabolism genes by horizontal gene transfer. *J. Virol.* **2010**, *84*, 12555–12563. [CrossRef] [PubMed]
32. Weynberg, K.D.; Allen, M.J.; Ashelford, K.; Scanlan, D.J.; Wilson, W.H. From small hosts come big viruses: The complete genome of a second *Ostreococcus tauri* virus, OtV-1. *Environ. Microbiol.* **2009**, *11*, 2821–2839. [CrossRef] [PubMed]
33. Weynberg, K.D.; Allen, M.J.; Gilg, I.C.; Scanlan, D.J.; Wilson, W.H. Genome sequence of *Ostreococcus tauri* virus OtV-2 throws light on the role of picoeukaryote niche separation in the ocean. *J. Virol.* **2011**, *85*, 4520–4529. [CrossRef] [PubMed]
34. Derelle, E.; Monier, A.; Cooke, R.; Worden, A.Z.; Grimsley, N.H.; Moreau, H. Diversity of viruses infecting the green microalga *Ostreococcus lucimarinus*. *J. Virol.* **2015**, *89*, 5812–5821. [CrossRef] [PubMed]
35. Hingamp, P.; Grimsley, N.; Acinas, S.G.; Clerissi, C.; Subirana, L.; Poulain, J.; Ferrera, I.; Sarment, H.; Villar, E.; Lima-Mendez, G.; et al. Exploring nucleo-cytoplasmic large DNA viruses in Tara Oceans microbial metagenomes. *ISME J.* **2013**, *7*, 1–18. [CrossRef] [PubMed]
36. Bellec, L.; Grimsley, N.; Derelle, E.; Moreau, H.; Desdevises, Y. Abundance, spatial distribution and genetic diversity of *Ostreococcus tauri* viruses in two different environments. *Environ. Microbiol. Rep.* **2010**, *2*, 313–321. [CrossRef] [PubMed]
37. Reid, E.L.; Weynberg, K.D.; Love, J.; Isupov, M.N.; Littlechild, J.A.; Wilson, W.H.; Kelly, S.L.; Lamb, D.C.; Allen, M.J. Functional and structural characterisation of a viral cytochrome b_5. *FEBS Lett.* **2013**, *587*, 3633–3639. [CrossRef] [PubMed]
38. Grimsley, N.; Pequin, B.; Bachy, C.; Moreau, H.; Piganeau, G. Cryptic sex in the smallest eukaryotic marine green alga. *Mol. Biol. Evol.* **2010**, *27*, 47–54. [CrossRef] [PubMed]
39. Rodriguez, F.; Derelle, E.; Guillou, L.; le Gall, F.; Vaulot, D.; Moreau, H. Ecotype diversity in the marine picoeukaryote *Ostreococcus* (Chlorophyta, Prasinophyceae). *Environ. Microbiol.* **2005**, *7*, 853–859. [CrossRef] [PubMed]
40. Moore, L.R.; Rocap, G.; Chisholm, S.W. Physiology and molecular phylogeny of coexisting *Prochlorococcus* ecotypes. *Nature* **1998**, *393*, 464–467. [PubMed]
41. Rocap, G.; Larimer, F.W.; Lamerdin, J.; Malfatti, S.; Chain, P.; Ahlgren, N.A.; Arellano, A.; Coleman, M.; Hauser, L.; Hess, W.R.; et al. Genome divergence in two *Prochlorococcus* ecotypes reflects oceanic niche differentiation. *Nature* **2003**, *424*, 1042–1047. [CrossRef] [PubMed]

42. Demir-Hilton, E.; Sudek, S.; Cuvelier, M.L.; Gentemann, C.L.; Zehr, J.P.; Worden, A.Z. Global distribution patterns of distinct clades of the photosynthetic picoeukaryote *Ostreococcus*. *ISME J.* **2011**, *5*, 1095–1107. [CrossRef] [PubMed]

43. Lelandais, G.; Scheiber, I.; Paz-Yepes, J.; Lozano, J.-C.; Botebol, H.; Pilátová, J.; Žárský, V.; Léger, T.; Blaiseau, P.-L.; Bowler, C.; et al. *Ostreococcus tauri* is a new model green alga for studying iron metabolism in eukaryotic phytoplankton. *BMC Genom.* **2016**, *17*, 1–23. [CrossRef] [PubMed]

44. van Ooijen, G.; Knox, K.; Kis, K.; Bouget, F.-Y.; Millar, A.J. Genomic transformation of the picoeukaryote *Ostreococcus tauri*. *J. Vis. Exp.* **2012**, e4074. [CrossRef] [PubMed]

45. Robbens, S.; Derelle, E.; Ferraz, C.; Wuyts, J.; Moreau, H.; Van de Peer, Y. The complete chloroplast and mitochondrial DNA sequence of *Ostreococcus tauri*: Organelle genomes of the smallest eukaryote are examples of compaction. *Mol. Biol. Evol.* **2007**, *24*, 956–968. [CrossRef] [PubMed]

46. Mayer, J.A.; Taylor, F.J.R. Virus which lyses the marine nanoflagellate *Micromonas pusilla*. *Nature* **1979**, *281*, 299–301. [CrossRef]

47. Cottrell, M.T.; Suttle, C.A. Dynamics of a lytic virus infecting the photosynthetic marine picoflagellate *Micromonas pusilla*. *Limnol. Oceanogr.* **1995**, *40*, 730–739. [CrossRef]

48. Cottrell, M.T.; Suttle, C.A. Genetic diversity of algal viruses which lyse the photosynthetic picoflagellate *Micromonas pusilla* (Prasinophyceae). *Appl. Environ.Microbiol.* **1995**, *61*, 3088–3091. [PubMed]

49. Cottrell, M.T.; Suttle, C.A. Widespread occurrence and clonal variation in viruses which cause lysis of a cosmopolitan, eukaryotic marine phytoplankter, *Micromonas pusilla*. *Mar. Ecol. Prog. Ser.* **1991**, *78*, 1–9. [CrossRef]

50. Sahlsten, E. Seasonal abundance in Skagerrak-Kattegat coastal waters and host specificity of viruses infecting the marine photosynthetic flagellate *Micromonas pusilla*. *Aquat. Microb. Ecol.* **1998**, *16*, 103–108. [CrossRef]

51. Sahlsten, E.; Karlson, B. Vertical distribution of virus-like particles (VLP) and viruses infecting *Micromonas pusilla* during late summer in the Southeastern Skagerrak, North Atlantic. *J. Plankton Res.* **1998**, *20*, 2207–2212. [CrossRef]

52. Zingone, A.; Natale, F.; Biffali, E.; Borra, M.; Forlani, G.; Sarno, D. Diversity in morphology, infectivity, molecular characteristics and induced host resistance between two viruses infecting *Micromonas pusilla*. *Aquat. Microb. Ecol.* **2006**, *45*, 1–14. [CrossRef]

53. Evans, C.; Archer, S.D.; Jacquet, S.; Wilson, W.H. Direct estimates of the contribution of viral lysis and microzooplankton grazing to the decline of a *Micromonas* spp. population. *Aquat. Microb. Ecol.* **2003**, *30*, 207–219. [CrossRef]

54. Baudoux, A.C.; Lebredonchel, H.; Dehmer, H.; Latimier, M.; Edern, R.; Rigaut-Jalabert, F.; Ge, P.; Guillou, L.; Foulon, E.; Bozec, Y.; et al. Interplay between the genetic clades of *Micromonas* and their viruses in the Western English Channel. *Environ. Microbiol. Rep.* **2015**, *7*, 765–773. [CrossRef] [PubMed]

55. Bellec, L.; Grimsley, N.; Moreau, H.; Desdevises, Y. Phylogenetic analysis of new prasinoviruses (*Phycodnaviridae*) that infect the green unicellular algae *Ostreococcus, Bathycoccus* and *Micromonas*. *Environ. Microbiol. Rep.* **2009**, *1*, 114–123. [CrossRef] [PubMed]

56. Chen, F.; Suttle, C.A. Amplification of DNA polymerase gene fragments from viruses infecting microalgae. *Appl. Environ. Microbiol.* **1995**, *61*, 1274–1278. [PubMed]

57. Yau, S.; Hemon, C.; Derelle, E.; Moreau, H.; Piganeau, G.; Grimsley, N. A viral immunity chromosome in the marine picoeukaryote, *Ostreococcus tauri*. *PLoS Pathog* **2016**, *12*, e1005965. [CrossRef] [PubMed]

58. Mayer, M.P. Recruitment of hsp70 chaperones: A crucial part of viral survival strategies. *Rev. Physiolol. Biochem. Pharmacol.* **2005**, *153*, 1–46.

59. Chretiennotdinet, M.J.; Courties, C.; Vaquer, A.; Neveux, J.; Claustre, H.; Lautier, J.; Machado, M.C. A new marine picoeukaryote—*Ostreococcus tauri* gen et sp-nov (Chlorophyta, Prasinophyceae). *Phycologia* **1995**, *34*, 285–292. [CrossRef]

60. Courties, C.; Vaquer, A.; Trousellier, M.; Lautier, J.; Chretiennot-Dinet, M.J.; Neveux, J.; Machado, C.; Claustre, H. Smallest eukaryotic organism. *Nature* **1994**, *370*, 255. [CrossRef]

61. Jancek, S.; Gourbiere, S.; Moreau, H.; Piganeau, G. Clues about the genetic basis of adaptation emerge from comparing the proteomes of two *Ostreococcus* ecotypes (Chlorophyta, Prasinophyceae). *Mol. Biol. Evol.* **2008**, *25*, 2293–2300. [CrossRef] [PubMed]

62. Le Bihan, T.; Martin, S.F.; Chirnside, E.S.; van Ooijen, G.; Barrios-Llerena, M.E.; O'Neill, J.S.; Shliaha, P.V.; Kerr, L.E.; Millar, A.J. Shotgun proteomic analysis of the unicellular alga *Ostreococcus tauri*. *J. Proteom.* **2011**, *74*, 2060–2070. [CrossRef] [PubMed]

63. Hindle, M.M.; Martin, S.F.; Noordally, Z.B.; van Ooijen, G.; Barrios-Llerena, M.E.; Simpson, T.I.; le Bihan, T.; Millar, A.J. The reduced kinome of *Ostreococcus tauri*: Core eukaryotic signalling components in a tractable model species. *BMC Genom.* **2014**, *15*, 1–21. [CrossRef] [PubMed]

64. Eckardt, N.A. Features of the circadian clock in the picoeukaryote *Ostreococcus*. *Plant Cell* **2009**, *21*, 3414. [CrossRef]

65. Pfeuty, B.; Thommen, Q.; Corellou, F.; Djouani-Tahri, E.B.; Bouget, F.-Y.; Lefranc, M. Circadian clocks in changing weather and seasons: Lessons from the picoalga *Ostreococcus tauri*. *Bioessays* **2012**, *34*, 781–790. [CrossRef] [PubMed]

66. Thommen, Q.; Pfeuty, B.; Corellou, F.; Bouget, F.-Y.; Lefranc, M. Robust and flexible response of the *Ostreococcus tauri* circadian clock to light/dark cycles of varying photoperiod. *FEBS J.* **2012**, *279*, 3432–3448. [CrossRef] [PubMed]

67. Kutter, E.; Sulakvelidze, A. (Eds.) *Bacteriophages: Biology and Applications*; CRC Press: Boca Raton, FL, USA, 2005.

68. Van Etten, J.L.; Graves, M.V.; Muller, D.G.; Boland, W.; Delaroque, N. *Phycodnaviridae*—Large DNA algal viruses. *Arch. Virol.* **2002**, *147*, 1479–1516. [CrossRef] [PubMed]

69. O'Kelly, C.J.; Sieracki, M.E.; Thier, E.C.; Hobson, I.C. A transient bloom of *Ostreococcus* (Chlorophyta, Prasinophyceae) in West Neck Bay, Long Island, New York. *J. Phycol.* **2003**, *39*, 850–854. [CrossRef]

70. Clerissi, C.; Desdevises, Y.; Grimsley, N. Prasinoviruses of the marine green alga *Ostreococcus tauri* are mainly species specific. *J. Virol.* **2012**, *86*, 4611–4619. [CrossRef] [PubMed]

71. Bellec, L.; Grimsley, N.; Desdevises, Y. Isolation of prasinoviruses of the green unicellular algae *Ostreococcus* spp. On a worldwide geographical scale. *Appl. Environ. Microbiol.* **2010**, *76*, 96–101. [CrossRef] [PubMed]

72. Claverie, J.M.; Abergel, C.; Ogata, H. Mimivirus. In *Lesser Known Large dsDNA Viruses*; Van Etten, J.L., Ed.; Springer: Berlin, Germany, 2009; Volume 328, pp. 89–121.

73. Suzan-Monti, M.; La Scola, B.; Raoult, D. Genomic and evolutionary aspects of Mimivirus. *Virus Res.* **2006**, *117*, 145–155. [CrossRef] [PubMed]

74. Vergères, G.; Waskell, L. Cytochrome b_5, its functions, structure and membrane topology. *Biochimie* **1995**, *77*, 604–620. [CrossRef]

75. Wilson, W.H.; Schroeder, D.C.; Allen, M.J.; Holden, M.T.G.; Parkhill, J.; Barrell, B.G.; Churcher, C.; Hamlin, N.; Mungall, K.; Norbertczak, H.; et al. Complete genome sequence and lytic phase transcription profile of a Coccolithovirus. *Science* **2005**, *309*, 1090. [CrossRef] [PubMed]

76. King, A.A.; Carstens, E.B.; Lefkowitz, E.J. *Virus Taxonomy: Ninth Report of the International Committee on Taxonomy of Viruses*; Elsevier Academic Press: San Diego, CA, USA, 2011.

77. Ogura, Y.; Hayashi, T.; Ueki, S. Complete genome sequence of a phycodnavirus, *Heterosigma akashiwo* virus strain 53. *Genome Announc.* **2016**, *4*, e01279-16. [CrossRef] [PubMed]

78. Bellec, L.; Clerissi, C.; Edern, R.; Foulon, E.; Simon, N.; Grimsley, N.; Desdevises, Y. Cophylogenetic interactions between marine viruses and eukaryotic picophytoplankton. *BMC Evol. Biol.* **2014**, *14*, 59. [CrossRef] [PubMed]

79. Thomas, R.; Grimsley, N.; Escande, M.L.; Subirana, L.; Derelle, E.; Moreau, H. Acquisition and maintenance of resistance to viruses in eukaryotic phytoplankton populations. *Environ. Microbiol.* **2011**, *13*, 1412–1420. [CrossRef] [PubMed]

80. Monier, A.; Pagarete, A.; de Vargas, C.; Allen, M.J.; Read, B.; Claverie, J.-M.; Ogata, H. Horizontal gene transfer of an entire metabolic pathway between a eukaryotic alga and its DNA virus. *Genome Res.* **2009**, *19*, 1441–1449. [CrossRef] [PubMed]

81. Ziv, C.; Malitsky, S.; Othman, A.; Ben-Dor, S.; Wei, Y.; Zheng, S.; Aharoni, A.; Hornemann, T.; Vardi, A. Viral serine palmitoyltransferase induces metabolic switch in sphingolipid biosynthesis and is required for infection of a marine alga. *Proc. Natl. Acad. Sci. USA* **2016**, *113*, E1907–E1916. [CrossRef] [PubMed]

82. Raoult, D.; Audic, S.; Robert, C.; Abergel, C.; Renesto, P.; Ogata, H.; La Scola, B.; Suzan, M.; Claverie, J.-M. The 1.2-megabase genome sequence of Mimivirus. *Science* **2004**, *306*, 1344–1350. [CrossRef] [PubMed]

83. Santini, S.; Jeudy, S.; Bartoli, J.; Poirot, O.; Lescot, M.; Abergel, C.; Barbe, V.; Wommack, K.E.; Noordeloos, A.A.M.; Brussaard, C.P.D.; et al. Genome of *Phaeocystis globosa* virus PgV-16t highlights the common ancestry of the largest known DNA viruses infecting eukaryotes. *Proc. Natl. Acad. Sci. USA* **2013**, *110*, 10800–10805. [CrossRef] [PubMed]
84. Mozar, M.; Claverie, J.-M. Expanding the Mimiviridae family using asparagine synthase as a sequence bait. *Virology* **2014**, *466–467*, 112–122. [CrossRef] [PubMed]
85. Deng, Z.; Wang, Z.; Lieberman, P.M. Telomeres and viruses: Common themes of genome maintenance. *Front. Oncol.* **2012**, *2*, 201. [CrossRef] [PubMed]
86. Zhang, Y.; Maley, F.; Maley, G.F.; Duncan, G.; Dunigan, D.D.; van Etten, J.L. Chloroviruses encode a bifunctional dCMP-dCTP deaminase that produces two key intermediates in dTTP formation. *J. Virol.* **2007**, *81*, 7662–7671. [CrossRef] [PubMed]
87. Furuichi, Y.; Shatkin, A.J. Viral and cellular mRNA capping: Past and prospects. *Adv. Virus Res.* **2000**, *55*, 135–184.
88. Agarkova, I.V.; Dunigan, D.D.; Van Etten, J.L. Virion-associated restriction endonucleases of chloroviruses. *J. Virol.* **2006**, *80*, 8114–8123. [CrossRef] [PubMed]
89. Müller, D.G.; Knippers, R. Phaeovirus. In *The Springer Index of Viruses*; Tidona, C., Darai, G., Eds.; Springer: New York, NY, USA, 2011; pp. 1259–1263.
90. Prasad, B.V.V.; Schmid, M.F. Principles of virus structural organization. In *Viral Molecular Machines*; Rossmann, M.G., Rao, V.B., Eds.; Springer US: Boston, MA, USA, 2012; pp. 17–47.
91. Markine-Goriaynoff, N.; Gillet, L.; van Etten, J.L.; Korres, H.; Verma, N.; Vanderplasschen, A. Glycosyltransferases encoded by viruses. *J. Gen. Virol.* **2004**, *85*, 2741–2754. [CrossRef] [PubMed]
92. Graves, M.V.; Bernadt, C.T.; Cerny, R.; van Etten, J.L. Molecular and genetic evidence for a virus-encoded glycosyltransferase involved in protein glycosylation. *Virology* **2001**, *285*, 332–345. [CrossRef] [PubMed]
93. Rogido, M.; Husson, I.; Bonnier, C.; Lallemand, M.-C.; Mérienne, C.; Gregory, G.A.; Sola, A.; Gressens, P. Fructose-1,6-biphosphate prevents excitotoxic neuronal cell death in the neonatal mouse brain. *Dev. Brain Res.* **2003**, *140*, 287–297. [CrossRef]
94. Yalcin, A.; Clem, B.F.; Imbert-Fernandez, Y.; Ozcan, S.C.; Peker, S.; O'Neal, J.; Klarer, A.C.; Clem, A.L.; Telang, S.; Chesney, J. 6-phosphofructo-2-kinase (pfkfb3) promotes cell cycle progression and suppresses apoptosis via cdk1-mediated phosphorylation of p27. *Cell Death Dis.* **2014**, *5*, e1337. [CrossRef] [PubMed]
95. Avrani, S.; Wurtzel, O.; Sharon, I.; Sorek, R.; Lindell, D. Genomic island variability facilitates *Prochlorococcus*-virus coexistence. *Nature* **2011**, *474*, 604–608. [CrossRef] [PubMed]
96. Sanz-Luque, E.; Chamizo-Ampudia, A.; Llamas, A.; Galvan, A.; Fernandez, E. Understanding nitrate assimilation and its regulation in microalgae. *Front. Plant Sci.* **2015**, *6*, 899. [CrossRef] [PubMed]
97. Foresi, N.; Mayta, M.L.; Lodeyro, A.F.; Scuffi, D.; Correa-Aragunde, N.; García-Mata, C.; Casalongué, C.; Carrillo, N.; Lamattina, L. Expression of the tetrahydrofolate-dependent nitric oxide synthase from the green alga *Ostreococcus tauri* increases tolerance to abiotic stresses and influences stomatal development in Arabidopsis. *Plant J.* **2015**, *82*, 806–821. [CrossRef] [PubMed]
98. Bogdan, C. Nitric oxide and the immune response. *Nat. Immunol.* **2001**, *2*, 907–916. [CrossRef] [PubMed]
99. Maat, D.S.; Brussaard, C.P.D. Both phosphorus- and nitrogen limitation constrain viral proliferation in marine phytoplankton. *Aquat. Microb. Ecol.* **2016**, *77*, 87–97. [CrossRef]
100. Clerissi, C.; Grimsley, N.; Desdevises, Y. Genetic exchanges of inteins between prasinoviruses (Phycodnaviridae). *Evolution* **2013**, *67*, 18–33. [CrossRef] [PubMed]
101. Gogarten, J.P.; Senejani, A.G.; Zhaxybayeva, O.; Olendzenski, L.; Hilario, E. Inteins: Structure, function, and evolution. *Annu. Rev. Microbiol.* **2002**, *56*, 263–287. [CrossRef] [PubMed]
102. Diaz-Munoz, S.L. Viral coinfection is shaped by host ecology and virus-virus interactions across diverse microbial taxa and environments. *bioRxiv* **2016**. [CrossRef]

viruses

MDPI

Review

A Student's Guide to Giant Viruses Infecting Small Eukaryotes: From *Acanthamoeba* to *Zooxanthellae*

Steven W. Wilhelm *, Jordan T. Bird, Kyle S. Bonifer, Benjamin C. Calfee, Tian Chen, Samantha R. Coy, P. Jackson Gainer, Eric R. Gann, Huston T. Heatherly, Jasper Lee, Xiaolong Liang, Jiang Liu, April C. Armes, Mohammad Moniruzzaman, J. Hunter Rice, Joshua M. A. Stough, Robert N. Tams, Evan P. Williams and Gary R. LeCleir

The Department of Microbiology, The University of Tennessee, Knoxville, TN 37996, USA;
jbird9@tennessee.edu (J.T.B.); kbonifer@tennessee.edu (K.S.B.); bcalfee@tennessee.edu (B.C.C.);
Tchen18@tennessee.edu (T.C.); srose16@utk.edu (S.R.C.); pgainer@utk.edu (P.J.G.);
egann@tennessee.edu (E.R.G.); hheather@tennessee.edu (H.T.H.); jlee175@tennessee.edu (J.L.);
xliang5@tennessee.edu (X.L.); jliu36@tennessee.edu (J.L.); amitch51@tennessee.edu (A.C.A.);
mmoniruz@tennessee.edu (M.M.); jrice18@utk.edu (J.H.R.); jstough@tennessee.edu (J.M.A.S.);
rtams@tennessee.edu (R.N.T.); ewilli99@tennessee.edu (E.P.W.); glecleir@tennessee.edu (G.R.L.)
* Correspondence: wilhelm@utk.edu; Tel.: +1-865-974-0665

Academic Editors: Mathias Middelboe and Corina Brussard
Received: 27 December 2016; Accepted: 9 March 2017; Published: 17 March 2017

Abstract: The discovery of infectious particles that challenge conventional thoughts concerning "what is a virus" has led to the evolution a new field of study in the past decade. Here, we review knowledge and information concerning "giant viruses", with a focus not only on some of the best studied systems, but also provide an effort to illuminate systems yet to be better resolved. We conclude by demonstrating that there is an abundance of new host–virus systems that fall into this "giant" category, demonstrating that this field of inquiry presents great opportunities for future research.

Keywords: giant viruses; nucleocytoplasmic large DNA viruses (NCLDVs); *Mimiviridae*

1. Introduction: Defining Giant Viruses

In their editorial introduction to the "Giant Viruses" special issue of Virology, Fischer and Condit [1] stated "It is commonly agreed upon that these are double-stranded DNA (dsDNA) viruses with genome sizes beyond 200 kb pairs, and particles that do not pass through a 0.2-μm pore-size filter". This definition illustrates the two striking features of giant viruses: their genome and particle size are both larger than has been historically considered for viruses. Beyond their breaking of previous paradigms, how giant viruses are defined remains contentious. Our goal in assembling this synthesis is to provide a "primer" for students of microbiology whom are interested in knowing more about these atypical viruses, and to establish a set of boundaries for their discussion. While not exhaustive, this overview addresses many of the main ideas that, for now, are current within a rapidly expanding field.

Some definitions of giant viruses focus only on genome size with lower limits ranging from undefined [2] to stringent (280 kb or 300 kb) cutoffs [3,4]. Other efforts have focused on the virus particle, suggesting they should be larger than 100 nm [2] or need be easily visible by light microscopy (>300 nm) [5]. One problem with establishing a particular definition for either genome or particle size is that, as additional large viruses are isolated, the rationale may no longer be justified (e.g., Aureococcus anophagefferens virus (AaV), a close phylogenetic relative of *Mimivirus*, is only ~140 nm in diameter) [6]. Indeed, a previous definition proposed a genome minimum of 280 kb due to a notable inflection point in a rank order plot of virus genome size [3]. However, in re-examining the largest 100 complete virus genomes in the National Center for Biotechnology Information's (NCBI) genome database, this gap is no longer present and a change in slope now occurs at ~400 kb (Figure 1A). This undersampling of giant viruses has resulted in a lack of sufficient information to describe their general characteristics [7,8]. While the vagaries of this definition will fade over time, herein we consider viruses 'giant' if their genome is larger than 200 kb. Moreover, this review will focus primarily on giants that infect single-celled eukaryotes.

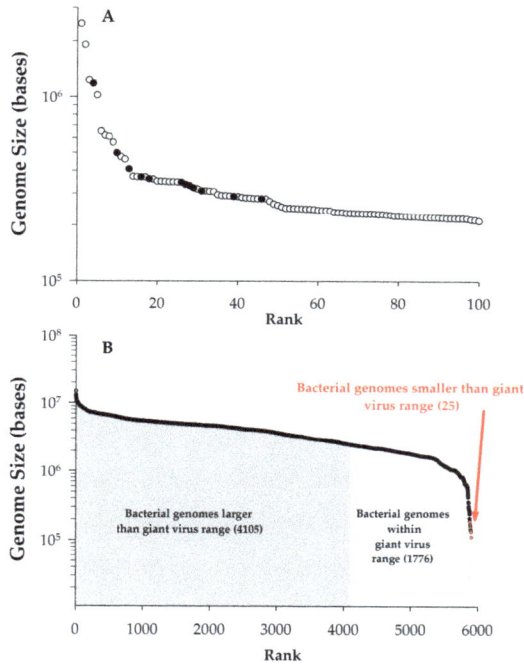

Figure 1. The scale of giant virus genomes. (**A**). Genome size vs. rank plot for the largest 100 complete viral genomes as of January 2016 from National Center for Biotechnology Information (NCBI). Data points noted (●) were previously used in discussion by Claverie et al. [3] to define giants viruses as having genomes > 280 kb, open circles (○) represent additional data; (**B**). Genome size vs. rank order of completed bacterial genomes in NCBI as of January 2016. Sizes are color-coded to match the ranges of giant virus genomes.

Using a cutoff of genomic content >200 kb pairs (kbp), ~2.2% (115/5356) of all of the completed virus genomes in NCBI fall within the realm of giants (Figure 1A). To date, all of these giants have genomes consisting of double-stranded DNA: the largest complete genome for other nucleic acid-type viruses is that of the double-stranded RNA (dsRNA) *Dendrolimus punctatus* cypovirus 22 (32.75 kbp) [9]. Perhaps more surprising is that this genome size range for giant viruses overlaps with more than ~one third of the complete prokaryotic genomes in NCBI (Figure 1B), as well as the genome sizes of several small eukaryotes [10]. This includes the smallest free-living archaeon (*Methanothermus fervidus*, 1.2 Mb) and the smallest free-living bacterium (*Candidatus* Actinomarina minuta, estimated ~700 kbp) [11]. While we will not consider them beyond the occasional passing mention in this article, it should be noted that several bacteriophages have genomes exceeding the 200 kbp genome size (see Table 1), and therefore qualify as giants. These phages infect both Gram-positive and -negative bacteria, including cyanobacteria [12,13].

Table 1. Comparison of host and viral genome size and GC content. All data was collected from the NCBI repository.

Giant Virus	Size Virus (Mb)	Virus GC (%)	ORFs*	Accession	Host	Size Host (Mb)	Host GC (%)	Host-Virus Genome Size	Host-Virus GC	Accession
Pandoravirus salinus	2.5	61.7	2541	NC_022098.1	Acanthamoeba castellanii	46.7	58.3	18.9	-3.4	AHJI00000000.1
Pandoravirus dulcis	1.9	63.7	1487	NC_021858.1	A. castellanii	46.7	58.3	24.5	-5.4	AHJI00000000.1
Acanthamoeba polyphaga mimivirus	1.2	28.0	1018	NC_014649.1	A. polyphaga	120.4	59.3	102.0	31.3	CDFK00000000.1
Acanthamoeba polyphaga moumouvirus	1.0	24.6	915	NC_020104.1	A. polyphaga	120.4	59.3	118.1	34.7	CDFK00000000.1
Madivirus sibericum	0.7	60.1	523	NC_027867.1	A. castellanii	42.0	58.4	64.6	-1.7	AHJI00000000.1
Pithovirus sibericum	0.6	35.8	467	NC_023423.1	A. castellanii	42.0	58.4	68.9	22.6	AHJI00000000.1
Emiliania huxleyi virus 86	0.4	40.2	478	NC_007346.1	Emiliania huxleyi	167.7	65.7	409.0	25.5	AHAL00000000.1
Marseillevirus marseillevirus	0.4	44.7	457	NC_013756.1	A. polyphaga	120.4	59.3	325.5	14.6	CDFK00000000.1
Aureococcus anophagefferens virus	0.4	28.7	384	NC_024697.1	A. anophagefferens	56.7	69.5	153.1	40.8	NZ_ACJI00000000.1
Melbournevirus	0.4	44.7	403	NC_025412.1	A. castellanii	42.0	58.4	113.6	13.7	AHJI00000000.1
Paramecium bursaria Chlorella virus NY2A	0.4	40.7	411	NC_009898.1	Chlorella variabilis NC64A	46.2	67.1	124.8	26.4	ADIC00000000.1
Brazilian marseillevirus	0.4	43.3	491	NC_029692.1	A. castellanii	42.0	58.4	116.7	15.1	AHJI00000000.1
Lausannevirus	0.4	42.9	444	NC_015326.1	A. castellanii	42.0	58.4	120.1	15.5	AHJI00000000.1
Ectocarpus siliculosus virus 1	0.3	51.7	240	NC_002287.1	Ectocarpus siliculosus	195.8	53.5	575.9	1.8	CABU00000000.1
Paramecium bursaria Chlorella virus AR158	0.3	40.8	366	NC_009899.1	C. variabilis NC64A	46.2	67.1	135.8	26.3	ADIC00000000.1
Paramecium bursaria Chlorella virus 1	0.3	40.0	376	NC_000852.5	C. variabilis NC64A	46.2	67.1	139.9	27.1	ADIC00000000.1
Micromonas pusilla virus 12T	0.2	39.8	265	NC_020864.1	Micromonas pusilla	22.0	65.9	104.6	26.1	NZ_ACCP00000000.1
Sample Bacteriophage										
Bacillus phago G	0.5	29.9	694	NC_022719.1	Bacillus megaterium	5.3	38.1	10.7	8.2	NZ_CT009920.1
Prochlorococcus phage P-SSM2	0.3	35.5	335	NC_006883.2	Prochlorococcus marinus	1.8	36.4	7.0	0.9	NC_005042.1
Ralstonia phage RSL1	0.2	58.0	345	NC_010811.2	Ralstonia solanacearum	5.6	66.5	24.3	8.5	NC_003295.1
Sinorhizobium phage phiN3	0.2	49.1	408	NC_028945.1	Sinorhizobium meliloti	3.7	62.7	17.4	13.6	NC_003047.1
Pseudomonas phage EL	0.2	49.3	201	NC_007623.1	Pseudomonas aeruginosa	6.3	66.6	29.8	17.3	NC_002516.2

* ORF = Open reading frame

As with observed ranges in genomic size, there is also a wide range of GC content of these viruses relative to the small eukaryotes they infect (Table 1). On average, mobile elements such as phage and plasmids are more AT-rich than their host, but usually by only ~5% [14]. In contrast, Emiliania huxleyi virus (EhV) and AaV, which infect eukaryotic algae, have GC contents that are 24.3% and 38.7% lower than their hosts nuclear genomes, respectively [15,16], while the chloroviruses (freshwater viruses infecting *Chlorella*) have GC contents that are ~21% lower than their host's nuclear genome. While not a defining feature of all large viruses, this GC difference raises interesting questions concerning the scavenging of nucleotides during the infection cycle. Construction of new viruses is in some cases thought to depend on materials "scavenged" from the host cell, yet in the case of these viruses there would seem to be a discrepancy in terms of what would be available for scavenging. An interest side note to this is that mitochondrial and chloroplast genomes are often observed to have such relative low GC content genomes, similar to these viruses [14,15], implying a potential for scavenged materials from organelles to be important in the construction of new virus particles.

The current size range for giant virus particles varies from our operationally defined ~200 nm to >1500 nm in diameter [5], although as noted, phylogenetic relatives to these giants exist that are only ~140 nm. Indeed, the upper limit of this range is larger than for several bacteria and archaea (Figure 1B), redefining how we think about the relative size of prokaryotes and viruses. These large particle diameters may be needed to house their large genomes (see below), but it has been argued that there are other evolutionary pressures for these virus particles to retain large physical sizes [5]. For example, viruses infecting *Acanthamoeba* are internalized via phagocytosis, and it has been shown that this process works less efficiently on smaller (<600 nm) particles [16]. Additionally, based on standard contact kinetics, a larger particle size may increase the probability of contact between the virus and its host in the environment [17].

In addition to a tremendous variation in genome and particle size, giant viruses also have highly diverse morphologies that can be broadly categorized into two groups: ovoid and icosahedral (Figure 2). These morphological differences correspond to the structural proteins that make the virion capsids; icosahedrons are built by homologous β-barrel jelly-roll Major Capsid Proteins (MCPs) with minor capsid proteins acting as scaffolds connecting trisymmetrons and the outer capsid to the inner membrane surrounding the viral genome [18]. In contrast, ovoid viruses encode phylogenetically distant (*Mollivirus*) to unconvincing (*Pandoravirus* and *Pithovirus*) homologs to MCP [19–21]. It is unclear how the virion shape provides a selective advantage, since both types have been isolated in similar habitats.

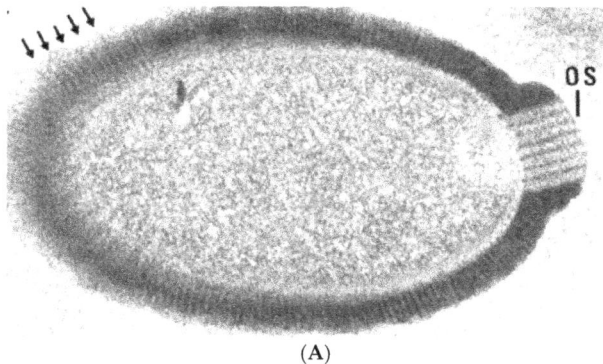

(A)

Figure 2. *Cont.*

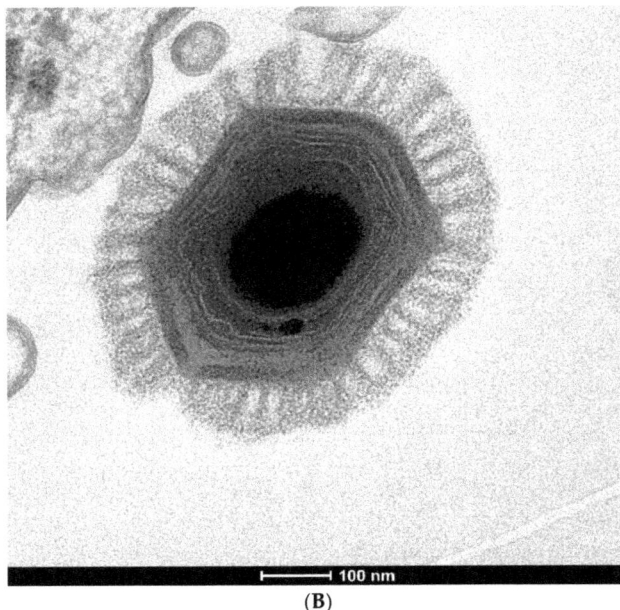

(B)

Figure 2. Transmission electron micrographs of giant virus particles. (**A**) *Pithovirus*, as seen in Michel et al. [22]. Originally identified as a KC5/2 parasite, the image shows the electron dense viral wall consisting of perpendicularly oriented fibers or microtubules (arrows), and a marked ostiole (os) located at the apical end of the cell. Reprinted with permission—original magnification at 85,000×; (**B**) *Megavirus chilensis*. Image courtesy of Professors Chantal Abergel and Jean-Michel Claverie.

Another mysterious aspect of these giant virus particles are the unique biochemical and morphological features. Virus–host interactions are thought to be facilitated in one of two ways: adsorption to the host cell wall, as is typical of algal host–virus systems [23], or phagocytosis by a protist host. These interactions often involve unique structures. For example, *Mimivirus* and its close relatives (*Megavirus*, *Marseillevirus*, *Lausannevirus*, and *Moumouvirus*) are characterized by proteinaceous fibers anchored to the icosahedron capsid [24,25] that are covered in glycolinkages [26–28]. It has been hypothesized that these fibers work in tandem with the large size of the viruses to facilitate phagocytosis, as they appear to have a similar composition to peptidoglycan and thus help mimic a bacterium (indeed, the name *Mimivirus* comes from "Mimicking Microbe" [29]). Additionally, the fibrous glycoproteins enable viral adsorption to diverse organisms ranging from bacteria and fungi to arthropods [30], implying a potential for both environmental dispersion and an incidental infection strategy in amoeba. *Phycodnaviridae* may also use unique structures to gain access to their host, though their mode of entry is typically by adsorption/injection, as opposed to phagocytosis. For example, the Chlorovirus capsid contains one spike located at a unique vertex of the icosahedral capsid that must be oriented towards the host cell surface to initiate infection [31]. Similarly, *Mimivirus* and its relatives utilize a five-pointed vertex called the 'stargate' structure that permits the first step in activating infection. Infection is initiated by fusion of the internal viral lipid membrane to the phagosomal membrane [24,32], which differs from algal viruses that fuse with the host cell membrane. This fusion event is observed in all giant viruses despite differences in structural features or infection strategies [20]. Whether these features are the result of homologous or convergent evolution remains to be determined, though given the breadth of physiological variation in the taxa, conservation of this mechanism is a compelling argument for monophyly.

2. Non-Structural Components of the Virion

An anomaly among the giant viruses are several viruses that include EhV, which have a lipid envelope outside of the capsid. These viruses include a *Phaeocystis globosa* virus (PgV-07T) [33] and several viruses infecting *Micromonas pussila* [34] which allows for a unique mode of infection and provides protection from environmental stressors [35]. This may be vital to the survival and transmission of these viruses, as they are ingested and transported across blooms by copepods [36]. An additional role of this lipid envelope and its associated proteins is an assumed association with recognition of the host and initiation of infection.

The nucleocytoplasmic large DNA viruses (NCLDVs) (described below) package a variety of proteins inside their capsids encoded by either viral or host genomes that are deployed immediately upon infection. For example, the seven proteomes of giant virus particles currently available (below) contain proteins predicted to combat oxidative stress, presumably because viral infections have been shown to generate Reactive Oxygen Species (ROS) that can inhibit viral replication [37]. Interestingly, *Pandoravirus salinus* carries one viral-encoded oxidoreductase, as well as three host-derived proteins predicted to combat ROS [19]. *P. bursaria chlorella* virus-1 (PBCV-1) and the more recently described *Megavirus chilensis* package homologous Cu-Zn superoxide dismutases [38,39]. In *M. chilensis*, this protein is remarkable for having the unique ability to fold and incorporate key metallic cofactors without the aid of chaperone proteins [39]. Additionally, *C. roenbergensis* virus (CroV) and *Mimivirus* both package novel sulfhydryl oxidases that may function in the formation of disulfide bonds [40,41]. These sulfhydryl oxidases as well as other protein disulfide isomerases present in CroV could aid in protein folding or viral entry similar to those found in retroviruses [42–44].

3. Gauging the Host Range of Giant Viruses in Nature

One concern regarding giant virus isolation using *Acanthamoeba* spp. is that while these are permissive, they may not be the natural hosts. Genomic analyses have been used in an attempt to determine natural hosts. In *Mimivirus*, most of the genes horizontally transferred from eukaryotes originated from amoeba, indicating amoebae are most likely the natural host of *Mimivirus*, but alternative hosts are still possible [45]. Indeed, their unique size and independence from host machinery may allow giant viruses to infect a wide range of hosts, which makes the search for the natural host more challenging. In addition to amoeba, NCLDVs have been reported to infect mice [46] and the symbiotic zooxanthelle of corals [47]. Giant viruses have also been isolated from human blood [48] and have been found in the human virome [49], indicating a potential role in human health (or at least a route of exposure). Indeed, the recent finding of *Acanthocystis turfacea* chlorella virus 1 (ATCV-1) from human oropharyngeal samples is intriguing: subsequent analyses have shown consistency between the presence of these viruses and reduced cognitive function in humans and mice [49].

4. Creating (an) Order from the Chaos: The Nucleocytoplasmic Large DNA Viruses

The NCLDV classification was created to define a monophyletic group of families that, when initially conceived, included *Asfarviridae*, *Phycodnaviridae*, *Poxviridae*, and *Iridoviridae* [50]. The rationale for this grouping was based on a conserved core of (1) nine genes hypothesized as representative of a common NCLDV ancestor and (2) a total of 22 more genes found in at least three of the four constituent viral families. The name is a reference to the replicative strategies of the included families as they replicate in both the nucleus and cytoplasm (phycodnaviruses, asfarvavirus and iridovirus) [51–53], or totally within the cytoplasm (poxviruses) [54]. A NCLDV ancestor has been hypothesized to have originated early in evolutionary history, possibly contemporaneously with early eukaryotic evolution, as suggested by the broad host range of NCLDV members [55]. However, the nature of this ancestral NCLDV remains unclear. Due to non-orthologous displacement of core genes [55] and potential reductive evolution [5] it is especially difficult to estimate the approximate genome size of any common ancestor and whether it would qualify as a giant virus when compared with modern giants. Indeed given theories on genome size variability, such as the genomic accordion [56,57], it is likely that predecessors of a variety of genome sizes existed. Moreover, it has been argued that mobile genetic elements encoded by virophages and transpovirions may have contributed significantly to the size of the NCLDV genome [58,59]. Therefore, the ancestral NCLDV may have been much smaller in genome size than modern representatives, and the mechanism by which it expanded its genome may have resulted in the wide range of genome sizes seen in current NCLDV members [60].

The NCLDV classification is not without its shortcomings. As new members are added to the group, the "nucleocytoplasmic" distinction of replicative strategies becomes less useful due to the increasing diversity of virion production. Many NCLDV families utilize a nucleocytoplasmic route for replication, including *Asfarviridae* [52], *Iridoviridae* and *Ascoviridae* [53], *Phycodnaviridae* [51], and *Pandoraviridae*. Other families, like *Poxviridae* [54], *Mimiviridae* [61], *Marseilleviridae* [62] and *Pithoviridae* [5], begin and complete their replication cycles exclusively in the cytoplasm, encoding the replication and transcription machinery necessary to produce virions without nuclear involvement. From a taxonomic perspective, the NCLDV group does not follow the naming conventions of (and is not recognized by) the International Committee on Taxonomy of Viruses (ICTV), as the classification lacks context within a larger hierarchy. To rectify this, Colson et al. proposed to reclassify NCLDVs within the new viral order *Megavirales* [63,64] based on the presence of conserved ancestral genes and a large icosahedral capsid composed of a homologous β-barrel jelly roll protein. This classification scheme, however, excludes the *Poxviridae* and *Ascoviridae*, [65] as well as *Pandoravirus*, *Pithovirus* and *Mollivirus*. In addition, the *Megavirales* classification required the capacity to assemble viral factories within the cytoplasm of host cells [62,66–69], a feature found in RNA viruses [70] but not seen in DNA viruses outside of the NCLDV group [64]. Currently (as of December 2016), the *Megavirales* is not considered a classification by the ICTV.

Most recently, the NCLDV genome size range has expanded to include genomes from 100 kb to 2.77 Mb encoding from 110 to 2556 genes [19,60]. The ten groups of NCLDV (*Phycodnaviridae*, *Poxviridae*, *Asfarviridae*, *Ascoviridae*, *Iridoviridae*, *Mimiviridae*, *Marseilleviridae*, *Pandoraviridae*, *Pithovirus*, and *Mollivirus*) infect a broad spectrum of hosts. In keeping with the NCLDV group's high degree of variability regarding particle size and host range, these viruses also display varying degrees of reliance on host metabolism and machinery, resulting in a limited number of highly conserved or "core" genes (e.g., see [71]). Yet despite these variances in NCLDV traits, common ground does exist. There are genes conserved amongst all available NCLDV genomes that are crucial for viral production or virion structure, such as the D5R packaging ATPase, D13L major capsid protein, and the B family DNA polymerase.

Comparative analyses of the genes conserved amongst different giant virus families has historically supported the monophyletic nature of the NCLDV group, and recent efforts to determine the clusters of orthologous groups (COGs) for giant viruses support their monophyly [72]. The conserved genes further provide potential markers that might be used in the discovery of novel NCLDVs and the determination of phylogenetic relationships between more closely related taxa [60]. For example, Moniruzzaman and colleagues [73] demonstrated an expanded level of diversity of the algal-specific members of the *Mimiviridae* by targeting the conserved MCP gene in this clade. However, this approach has its limits; the three recently discovered representatives of *Pandoravirus* lack the major capsid protein and the D5R helicase, as well as a number of other core NCLDV genes [19]. Indeed only 17 of the 49 inferred ancestral NCLDV genes were found in at least one of the *Pandoravirus* genomes, calling into question their inclusion in the giant virus clade despite their particle and genome size [74].

5. Viruses as a Possible Fourth Domain of Life

Initially viruses were defined by their intrinsic filterability away from cellular life forms [75–77], a definition subsequently refined to include their lack of ribosomes, a host-dependent metabolic strategy, and replication by means other than binary fission [78]. That the unique capabilities of NCLDVs still fit well within the latter definition, after fifty years of discovery and scientific scrutiny, highlights a fundamental difference between cellular organisms and viral particles. However, giant viruses do challenge these distinctions. Independent of their size, which invalidates the informal 0.2-μm filter cutoff, NCLDVs are remarkably cell-like in virion structure and gene content. In addition to their protein coat, membrane, and genome, *Mimivirus* and *Marseillevirus* particles contain messenger RNA molecules, making them the only viruses, to date, that contain both types of nucleic acid [79]. Moreover, several viruses encode genes involved in translational processes, such as varying numbers of aminoacyl-transfer RNA (tRNA) synthetases [69,80]. Indeed, we hypothesize these proteins may be useful in overcoming differences in GC content seen between some viruses and hosts (Table 1), but this has yet to be empirically demonstrated.

The discovery of translational machinery (including that mentioned above) encoded in select virus genomes allows for comparisons to traditionally "cellular" functions normally associated with the three domains of life. Sequence alignments comparing multiple genes involved in DNA replication and repair, transcription, and translation shared between cellular organisms and NCLDVs appear to show deeply branching relationships as ancient as the domain Eukarya. It was subsequently hypothesized that giant viruses evolved from a cellular common ancestor belonging to a currently extinct fourth domain of life, unique from Bacteria, Archaea, and Eukarya [63,81]. Seemingly in support of this hypothesis is the abundance of coding sequences (ORFans) in giant viral genomes with no known homologues in the other domains.

These ideas have proven somewhat controversial, as direct sequence comparison of genes conserved among cellular organisms with virus-encoded homologs is problematic. As selective pressures on similar genes within viruses and their hosts are likely different, accelerated sequence divergence in viruses may exaggerate their perceived distance from the derived gene [82]. Subsequent alignments accounting for compositional heterogeneity and homoplasy place giant virus genes with eukaryotes [83]. While it has been countered that giant virus genes do not evolve more quickly than their cellular counterparts, this has yet to be demonstrated outside of a single example within *Marseilleviridae* [5,84]. Indeed, an overabundance of viral open reading frames (ORFs) without known homologues is not a problem unique to giant viruses [85]. These observations and others have led to the alternative hypothesis: gene content within the different NCLDV families suggests that their genomes have been built up from smaller viruses over time, rather than by loss of unnecessary genes by an ancient cellular ancestor [72]. As some of the current NCLDVs replicate in phagotrophs like *Acanthamoeba* and *Cafeteria roenbergensis*, it was hypothesized that smaller viruses may incorporate genetic material from other organisms phagocytosed by the host.

6. Giant Viruses in the Environment

While surveys are not yet exhaustive, giant viruses appear to be found in all environments. Since the discovery of *Mimivirus* from a water cooling tower [86], giant viruses have been found in locations where amoebae normally thrive, including seawater, soil, aerosols, and man-made aquatic environments such as sewage, fountains and air conditioners [87], in addition to harsh, unexpected ecosystems such as permafrost [20]. Lastly, giant viruses or their DNA sequences have been observed in animals such as dinoflagellate-associated coral [47], arthropods, and humans [49,88].

A powerful tool in the identification of putative new viruses are environmental metagenomic studies (Table 2), though most have not focused specifically on giant viruses until recently [21]. Current research suggests giant viruses only comprise a small percentage of viruses (<1%) in most samples. However, virus densities can fluctuate based on contact with their host: for example, *Chlorella* viruses are much more abundant when their hosts, normally sequestered as endosymbionts of *Paramecium bursaria*, are made available as a consequence of predatory activity on the *Paramecium* [89]. Regardless, it is clear some families tend to be more common than others: in marine metagenomics samples *Phycodnaviridae*-related sequences were found to be highest in abundance, followed by

Mimiviridae [90,91]. It is also clear that these viruses are persistent: the discovery of 30,000-year-old *Mollivirus* particles in permafrost suggests that giant viruses can survive, under the correct conditions, for long periods of time [21]. When combined with other tools such as flow cytometry sorting of either individual particles [92] or infected hosts [93], these new approaches will begin to shed significant light on the natural diversity of these populations.

Table 2. Comparison of giant virus reads to total viral reads in shotgun metagenomic studies from different environments.

Environment	Location	Abundance	Total Reads	Most Common Virus Families Present	Source
Marine	Indian Ocean	0.3%–1.4%	N/A	*Mimiviridae, Phycodnaviridae*	[91]
Antarctic soil	Antarctica	2.82%–7.71%	123/1595-177/6264	*Mimiviridae, Phycodnaviridae*	[94]
Coral	USA	1.2%	744/60485	*Mimiviridae, Phycodnaviridae*	[95]
Human (respiratory system)	Sweden	0.00002%	2/111931	*Mimiviridae*	[96]

To date, much of the focus on giant viruses has been on their genomics rather than their influence on the environments in which they persist. Several large, dsDNA viruses including EhV [97], PgV [98], AaV [6,97] and *Heterosigma akashiwo* virus (HaV) [99] are associated with algal blooms, although only a few have been directly shown to infect and lyse the phytoplankton involved with the bloom in situ [97,100]. Algal blooms occur on large geographical scales and result in significant influxes of atmospheric carbon into the world's oceans. Viruses, particularly bacteriophages, are known drivers of dissolved organic matter (DOM) release back into the environment via a process known as the "viral shunt" [101,102]. With the large biomass of algae associated with these blooms, virus-mediated collapse by giant viruses may also be an important driver of dissolved and particulate organic matter release. Giant viruses that infect algae may be likened to bacteriophages in terms of participating in the viral shunt, and the release of nutrients back into the environment may be an important part of the ecological cycle in aquatic systems [102].

A recent estimate suggested that giant viruses available in culture were infectious to at least 22 different algal species [103]. Globally, it has been proposed that there are more than 350,000 algal species [104]. Given the possibility that all algae may be infected with one or more viruses [105,106], the possibility of a collection of unknown giants remains very real, and indeed molecular data point to at least a broad diversity within the known groups [107,108]. Building on the above, it is clear from a survey of the literature that researchers identified candidate protist-giant virus systems well before *Mimivirus* was documented (Table 3). In the late 1960s and early 1970s, the expanded availability of transmission electron microscopes to researchers resulted in a series of observations concerning the presence of large virus-like particles inside algal cells [107,109]. In many cases, these virus–host systems have been largely ignored by the scientific community, creating a broad spectrum of opportunities for researchers to begin to cultivate these plankton in an effort to isolate and characterize new giant viruses. Given the expansive putative host-range that has been observed, it is likely that many of these viruses could fill in knowledge gaps concerning the diversity and potential function of these particles. Indeed, one example of how new hosts can be used to discover new viruses are the *Faustovirus*, recently discovered using *Vermamoeba* (a protist found in both humans and natural systems) as a screen [110]: unique to these viruses is a collection of genes three times larger than the other members of the *Asfarviridae* family.

And while it is obvious that there is a dearth of knowledge concerning giant viruses that infect algae in the environment, there is an even larger knowledge gap regarding giant viruses infecting heterotrophic eukaryotes. The most studied of these viruses is CroV, which infects the heterotrophic grazer *Cafeteria roenbergensis* [111]. Given this organism is a grazer of primary producers it is possible that infection of this organism by CroV could have effects on lower trophic level organisms. It has been shown that grazing can be an important driver of algal bloom decline [112], so it stands to reason that the effects giant viruses have on mixo/heterotrophic-plankton are critical to understanding bloom dynamics. Almost no information, at this time, is available to discuss the impacts of these infections, but they will most likely result in interesting discoveries and further our understanding of how giant viruses alter the microbial food web.

Table 3. A chronological list of organisms shown in the literature to contain viruses consistent with the giant virus size class.

Year	Organism	Particle Size	References
1970	*Aphelidium* sp. (fungal parasite of algae)	190–210 nm	[113]
1972	*Oedogonium* spp. "L" (Chlorophyceae)	240 nm	[114]
	Chorda tomentosa (Phaeophyceae)	170 nm	[115]
1973	*Ectocarpus* sp.; *Ectocarpus fasciculatus* (Phaeophyceae)	150 nm, 170 nm	[116,117]
	Aulacomonas submarina (Chlorophyceae)	200–230 nm	[118]
1974	*Pylaiella littoralis* (Phaeophyceae)	130–170 nm	[119]
	Pyramimonas orientalis (Prasinophyceae)	200 nm	[120]
1975 [†]	*Chara corallina* (Charophyceae)	18 nm × 532 nm	[121]
1978	*Sorocarpus uvaeformis* (Phaeophyceae)	170 nm	[122]
1979	*Gymnodinium uberrimum* (Dinophyceae)	385 nm	[123]
	Mallomonas sp. (Synurophyceae)	175 nm	[123]
1980	*Uronema gigas* (Chlorophyceae)	390 nm	[124]
1984	*Paraphysomonas corynephora* (Chrysophyceae)	150–180 nm, 270–300 nm	[125]
1993	Various Phaeodarian food vacuoles	300–750 nm	[126]

[†] Although in length this virus qualifies as a giant, its rod shaped morphology is more consistent with Tobacco mosaic virus than any member of the *Mimiviridae*.

7. Intimate Interactions with the Host: Eco-Evolutionary Consequences

Only recently have we come to appreciate the possibility of gene transfer between giant viruses and their hosts. A large proportion of giant virus genes comes from diverse sources, including from their eukaryotic hosts [127]. In EhV, seven genes involved in sphingolipid biosynthesis pathway were putatively transferred from the host algae [128]. Upon infection, the host sphingolipid biosynthesis pathway is downregulated concomitant with the upregulation of the corresponding viral genes, leading to increased production of viral glycosphingolipids (vGSLs) [129]. EhV particles are covered by vGSLs, and this unique lipid molecule ultimately induces programmed cell death (PCD) in infected hosts [130].

A genome wide phylogenetic study of AaV identified a number of genes having their highest phylogenetic affinity to host (*Aureococcus*) homologs, [71]. This agrees with observations made by earlier studies on several other giant viruses [127,131]. While gene acquisition may be one of the evolutionary strategies of giant viruses, how these genes confer ecological advantages remains largely unknown. As the vast majority of viruses harbor streamlined genomes with few genes, the enormous genetic resource of giant viruses poses a paradox in terms of energetic cost of replication. Closer inspection of a number of sequenced eukaryotic genomes revealed a large number of genes originated from giant viruses [132,133]. In a recent study, large genomic islands, putatively derived from both giant viruses and a virophage, were found in *Bigelowiella natans*, a Cryptomonad algae [134]. In another study, "core" genes from giant viruses were detected in eight protists and a metazoan (*Hydra magnipapillata*) genome [132]. Remarkably, a 400-kb region in the *H. magnipapillata* was putatively identified to be of viral origin [132]. Major capsid gene phylogeny indicated the genes were likely from a *Mimiviridae* family member. Giant virus particles and marker genes have also recently also been observed associated with zooxanthellae from the genus *Symbiodinium*, a dinoflagellate typically found closely associated with corals [47]. Giant virus-like genes were also found in several other protists [133,135] and some plant genomes, namely *Physcomitrella patens* and *Selaginella moellendorffii* [136]. The role of giant virus-derived genes in host remain an open question.

Host–virus interactions result in an evolutionary arms race—leading to the emergence of new diversity in the host and virus population [137]. Hosts of giant viruses have evolved a variety of defense mechanisms against giant viruses. An elegant example is the 'Cheshire cat' strategy adopted by *Emiliania huxleyi* [138]. The diploid calcified cells of *E. huxleyi* are susceptible to EhV infection, while the haploid stage is 'invisible' to infection. It has been suggested that during the decline of the Brown tide blooms, a virus-resistant population of the *Aureococcus* persists, maintaining a relatively high abundance of *Aureococcus* even after the demise of the bloom [73,97].

8. Virophage

Another interesting characteristic of some giant viruses is their susceptibility to infection by other bioactive particles, termed "virophage". The first virophage to be isolated was named Sputnik [59], which replicates within the viral factory used by *Mamavirus* within *Acanthamoeba castellanii*. Because of this, Sputnik only replicates within *A. castellanii* co-infected with Mamavirus. Infection by the virophage causes abnormal capsid structure of *Mamavirus*, increasing capsid size and causing abnormal fiber localization on its surface, suggesting a parasitic relationship between the two [59]. Co-incubation of Sputnik and *Mamavirus* decreased infective *Mamavirus* particle titers by approximately 70% and increased the survival rate of the *A. castellanii* [59]. Similar virophages have been found infecting other giant viruses as well [139–142]. The discovery of "viruses that infect viruses" has strengthened the argument that viruses are living entities [143]. Some classes of *Mimivirus* appear to have developed a CRISPR-CAS-like system suggested to combat these virophages, called the *Mimivirus* virophage resistant element (MIMIVIRE) [144]. Interestingly, a number of genes homologous to those in the MIMIVIRE system are present in other giant viruses, suggesting that the MIMIVIRE-like defense systems might not be exclusive to *Mimivirus* [134,145]. Other interpretations, however, are questioning these conclusions [146]. Much has yet to be learned in these systems, but virophages may act like both 'provirophage' and 'provirus', depending on the genomic context at multiple levels.

9. Conclusions

The discovery of *Mimivirus* has driven both the nascence and evolution of a new area of scientific inquiry. Giant viruses are now the topics of evolutionary, ecological and biotechnological inquiries. Moreover, broad-scale efforts to identify new virus–host systems, ranging from classic culture-based approaches to newer bioinformatics efforts to link viruses and their hosts [147] will soon provide a larger data base of information concerning the key features of these novel virus particles. Indeed, a survey of older literature (Table 3) clearly demonstrates that there are many virus–host systems that have been observed but are yet to be isolated and characterized. Moving forward, there is little doubt that the study of giant viruses will shed new light not only on virus–host relationships, but also on key evolutionary processes including the natural occurrence rates of transduction and horizontal gene transfer.

Acknowledgments: The authors wish to thank the Department of Microbiology at the University of Tennessee: this review was assembled as part of a graduate course (Virology 604) instructed by S.W.W. and G.R.L. Funds to support publication were received from the Kenneth & Blaire Mossman Endowment to the University of Tennessee (S.W.W.).

Author Contributions: S.W.W. and G.R.L. conceived this exercise for the Department of Microbiology graduate journal club in virology. All authors participated in research and crafting of the document.

Conflicts of Interest: The authors declare no conflict of interest.

References

1. Fischer, M.G.; Condit, R.C. Editorial introduction to "giant viruses" special issue of virology. *Virology* **2014**, *466–467*, 1–2. [CrossRef] [PubMed]
2. Durzyriska, J. Giant viruses: Enfants terribles in the microbal world. *Future Virol.* **2015**, *10*, 795–806. [CrossRef]

3. Claverie, J.M.; Ogata, H.; Audic, S.; Abergel, C.; Suhre, K.; Fournier, P.E. Mimivirus and the emerging concept of "giant" virus. *Virus Res.* **2006**, *117*, 133–144. [CrossRef] [PubMed]

4. Yamada, T. Giant viruses in the environment: Their origins and evolution. *Curr. Opin. Virol.* **2011**, *1*, 58–62. [CrossRef] [PubMed]

5. Abergel, C.; Legendre, M.; Claverie, J.M. The rapidly expanding universe of giant viruses: Mimivirus, Pandoravirus, Pithovirus and Mollivirus. *FEMS Microbiol. Rev.* **2015**, *39*, 779–796. [CrossRef] [PubMed]

6. Gastrich, M.D.; Leigh-Bell, J.A.; Gobler, C.; Anderson, O.R.; Wilhelm, S.W. Viruses as potential regulators of regional brown tide blooms caused by the alga, *Aureococcus anophageefferens*: A comparison of bloom years 1999–2000 and 2002. *Estuaries* **2004**, *27*, 112–119. [CrossRef]

7. Serwer, P.; Hayes, S.J.; Thomas, J.A.; Hardies, S.C. Propagating the missing bacteriophages: A large bacteriophage in a new class. *Virol. J.* **2007**, *4*, 21. [CrossRef] [PubMed]

8. Martínez, J.M.; Swan, B.K.; Wilson, W.H. Marine viruses, a genetic reservoir revealed by targeted viromics. *ISME J.* **2014**, *8*, 1079–1088. [CrossRef] [PubMed]

9. Zhou, Y.; Qin, T.; Xiao, Y.; Qin, F.; Lei, C.; Sun, X. Genomic and biological characterization of a new cypovirus isolated from *Dendrolimus punctatus*. *PLoS ONE* **2014**, *9*, e113201. [CrossRef] [PubMed]

10. Corradi, N.; Pombert, J.-F.; Farinelli, L.; Didier, E.S.; Keeling, P.J. The complete sequence of the smallest known nuclear genome from the microsporidian *Encephalitozoon intestinalis*. *Nat. Commun.* **2010**, *1*, 77. [CrossRef] [PubMed]

11. Martínez-Cano, D.J.; Reyes-Prieto, M.; Martínez-Romero, E.; Partida-Martínez, L.P.; Latorre, A.; Moya, A.; Delaye, L. Evolution of small prokaryotic genomes. *Front. Microbiol.* **2015**, *5*, 742. [CrossRef] [PubMed]

12. Mesyanzhinov, V.V.; Robben, J.; Grymonprez, B.; Kostyuchenko, V.A.; Bourkaltseva, M.V.; Sykilinda, N.N.; Krylov, V.N.; Volckaert, G. The genome of bacteriophage φKZ of *Pseudomonas aeruginosa*. *J. Mol. Biol.* **2002**, *317*, 1–19. [CrossRef] [PubMed]

13. Sullivan, M.B.; Coleman, M.L.; Weigele, P.; Rohwer, F.; Chisholm, S.W. Three *Prochlorococcus* cyanophage genomes: Signature features and ecological interpretations. *PLoS Biol.* **2005**, *3*, e144. [CrossRef] [PubMed]

14. Ong, H.C.; Wilhelm, S.W.; Gobler, C.J.; Bullerjahn, G.; Jacobs, M.A.; McKay, J.; Sims, E.H.; Gillett, W.G.; Zhou, Y.; Haugen, E.; et al. Analyses of the complete chloroplast genome of two members of the pelagophyceae: *Aureococcus anophageefferens* CCMP1984 and *Aureoumbra lagunesis* CCMP1507. *J. Phycol.* **2010**, *46*, 602–615. [CrossRef]

15. Orsini, M.; Costelli, C.; Malavasi, V.; Cusano, R.; Alessandro, C.; Angius, A.; Cao, G. Complete sequence and characterization of mitochondrial and chloroplast genome of *Chlorella variabilis* NC64A. *Mitochondrial DNA A* **2015**, *27*, 3128–3130.

16. Korn, E.D.; Weisman, R.A. Phagocytosis of latex beads by *Acanthomoeba*. *J. Cell Biol.* **1967**, *34*, 219–227. [CrossRef]

17. Murray, A.G.; Jackson, G.A. Viral dynamics: A model of the effects of size, shape, motion and abundance of single-celled planktonic organisms and other particles. *Mar. Ecol. Prog. Ser.* **1992**, *89*, 103–116. [CrossRef]

18. Klose, T.; Rossmann, M.G. Structure of large dsDNA viruses. *Biol. Chem.* **2014**, *395*, 711–719. [CrossRef] [PubMed]

19. Philippe, N.; Legendre, M.; Doutre, G.; Coute, Y.; Poirot, O.; Lescot, M.; Arslan, D.; Seltzer, V.; Bertaux, L.; Bruley, C.; et al. Pandoraviruses: Amoeba viruses with genomes up to 2.5 mb reaching that of parasitic eukaryotes. *Science* **2013**, *341*, 281–286. [CrossRef] [PubMed]

20. Legendre, M.; Bartoli, J.; Shmakova, L.; Jeudy, S.; Labadie, K.; Adrait, A.; Lescot, M.; Poirot, O.; Bertaux, L.; Bruley, C.; et al. Thirty-thousand-year-old distant relative of giant icosahedral DNA viruses with a Pandoravirus morphology. *Proc. Natl. Acad. Sci. USA* **2014**, *111*, 4274–4279. [CrossRef] [PubMed]

21. Legendre, M.; Lartigue, A.; Bertaux, L.; Jeudy, S.; Bartoli, J.; Lescot, M.; Alempic, J.M.; Ramus, C.; Bruley, C.; Labadie, K.; et al. In-depth study of *Mollivirus sibericum*, a new 30,000-y-old giant virus infecting *Acanthamoeba*. *Proc. Natl. Acad. Sci. USA* **2015**, *112*, E5327–E5335. [CrossRef] [PubMed]

22. Pearson, H. 'Virophage' suggests viruses are alive. *Nature* **2008**, *454*, 677. [CrossRef] [PubMed]

23. Wilson, W.H.; Van Etten, J.L.; Allen, M.J. The phycodnaviridae: The story of how tiny giants rule the world. In *Lesser Known Large dsDNA Viruses*; VanEtten, J.L., Ed.; Springer Science & Business Media: Berlin, Germany, 2009; Volume 328, pp. 1–42.

24. Xiao, C.; Rossmann, M.G. Structures of giant icosahedral eukaryotic dsDNA viruses. *Curr. Opin. Virol.* **2011**, *1*, 101–109. [CrossRef] [PubMed]

25. Xiao, C.A.; Chipman, P.R.; Battisti, A.J.; Bowman, V.D.; Renesto, P.; Raoult, D.; Rossmann, M.G. Cryo-electron microscopy of the giant mimivirus. *J. Mol. Biol.* **2005**, *353*, 493–496. [CrossRef] [PubMed]

26. Chothi, M.P.; Duncan, G.A.; Armirotti, A.; Abergel, C.; Gurnon, J.R.; Van Etten, J.L.; Bernardi, C.; Damonte, G.; Tonetti, M. Identification of an l-rhamnose synthetic pathway in two nucleocytoplasmic large DNA viruses. *J. Virol.* **2010**, *84*, 8829–8838. [CrossRef] [PubMed]

27. Tonetti, M.; Chothi, M.P.; Abergel, C.; Seltzer, V.; Gurnon, J.; Van Etten, J.L. Glycosylation in nucleo-cytoplasmic large DNA viruses (NCLDV). *FEBS J.* **2011**, *278*, 420.

28. Piacente, F.; Gaglianone, M.; Laugieri, M.E.; Tonetti, M.G. The autonomous glycosylation of large DNA viruses. *Int. J. Mol. Sci.* **2015**, *16*, 29315–29328. [CrossRef] [PubMed]

29. Raoult, D. Viruses reconsidered. *Scientist* **2014**, *28*, 41.

30. Rodrigues, R.A.L.; Silva, L.K.D.; Dornas, F.P.; de Oliveira, D.B.; Magalhaes, T.F.F.; Santos, D.A.; Costa, A.O.; Farias, L.D.; Magalhaes, P.P.; Bonjardim, C.A.; et al. Mimivirus fibrils are important for viral attachment to the microbial world by a diverse glycoside interaction repertoire. *J. Virol.* **2015**, *89*, 11812–11819. [CrossRef] [PubMed]

31. Zhang, X.Z.; Xiang, Y.; Dunigan, D.D.; Klose, T.; Chipman, P.R.; Van Etten, J.L.; Rossmann, M.G. Three-dimensional structure and function of the *Paramecium bursaria* chlorella virus capsid. *Proc. Natl. Acad. Sci. USA* **2011**, *108*, 14837–14842. [CrossRef] [PubMed]

32. Suzan-Monti, M.; La Scola, B.; Raoult, D. Genomic and evolutionary aspects of mimivirus. *Virus Res.* **2006**, *117*, 145–155. [CrossRef] [PubMed]

33. Maat, D.S.; Bale, N.J.; Hopmans, E.C.; Baudoux, A.C.; Damste, J.S.S.; Schouten, S.; Brussaard, C.P.D. Acquisition of intact polar lipids from the prymnesiophyte *Phaeocystis globosa* by its lytic virus PgV-07t. *Biogeosciences* **2014**, *11*, 185–194. [CrossRef]

34. Martinez-Martinez, J.; Boere, A.; Gilg, I.C.; van Lent, J.W.M.; Witte, H.J.; van Bleijswijk, J.D.L.; Brussaard, C.P.D. New lipid envelop-containing dsDNA virus isolates infecting *Micromonas pusilla* reveal a separate phylogenetic group. *Aquat. Microb. Ecol.* **2015**, *74*, 17–28. [CrossRef]

35. Mackinder, L.C.M.; Worthy, C.A.; Biggi, G.; Hall, M.; Ryan, K.P.; Varsani, A.; Harper, G.M.; Wilson, W.H.; Brownlee, C.; Schroeder, D.C. A unicellular algal virus, *Emiliania huxleyi* virus 86, exploits an animal-like infection strategy. *J. Gen. Virol.* **2009**, *90*, 2306–2316. [CrossRef] [PubMed]

36. Frada, M.J.; Schatz, D.; Farstey, V.; Ossolinski, J.E.; Sabanay, H.; Ben-Dor, S.; Koren, I.; Vardi, A. Zooplankton may serve as transmission vectors for viruses infecting algal blooms in the ocean. *Curr. Biol.* **2014**, *24*, 2592–2597. [CrossRef] [PubMed]

37. Schwarz, K.B. Oxidative stress during viral infection: A review. *Free Radic. Biol. Med.* **1996**, *21*, 641–649. [CrossRef]

38. Kang, M.; Duncan, G.A.; Kuszynski, C.; Oyler, G.; Zheng, J.Y.; Becker, D.F.; Van Etten, J.L. Chlorovirus PBCV-1 encodes an active copper-zinc superoxide dismutase. *J. Virol.* **2014**, *88*, 12541–12550. [CrossRef] [PubMed]

39. Lartigue, A.; Burlat, B.; Coutard, B.; Chaspoul, F.; Claverie, J.M.; Abergel, C. The *Megavirus chilensis* Cu,Zn-superoxide dismutase: The first viral structure of a typical cellular copper chaperone-independent hyperstable dimeric enzyme. *J. Virol.* **2015**, *89*, 824–832. [CrossRef] [PubMed]

40. Fischer, M.G.; Kelly, I.; Foster, L.J.; Suttle, C.A. The virion of *Cafeteria roenbergensis* virus (CroV) contains a complex suite of proteins for transcription and DNA repair. *Virology* **2014**, *466*, 82–94. [CrossRef] [PubMed]

41. Hakim, M.; Ezerina, D.; Alon, A.; Vonshak, O.; Fass, D. Exploring ORFan domains in giant viruses: Structure of mimivirus sulfhydryl oxidase R596. *PLoS ONE* **2012**, *7*, e50649. [CrossRef] [PubMed]

42. Apperizeller-Herzog, C.; Ellgaard, L. The human PDI family: Versatility packed into a single fold. *Biochim. Biophys. Acta* **2008**, *1783*, 535–548. [CrossRef] [PubMed]

43. Ryser, H.J.P.; Levy, E.M.; Mandel, R.; Disciullo, G.J. Inhibition of human-immunodeficiency-virus infection by agents that interfere with thiol-disulfide interchange upon virus-receptor interaction. *Proc. Natl. Acad. Sci. USA* **1994**, *91*, 4559–4563. [CrossRef] [PubMed]

44. Schelhaas, M.; Malmstrom, J.; Pelkmans, L.; Haugstetter, J.; Ellgaard, L.; Grunewald, K.; Helenius, A. Simian virus 40 depends on ER protein folding and quality control factors for entry into host cells. *Cell* **2007**, *131*, 516–529. [CrossRef] [PubMed]

45. Moreira, D.; Brochier-Armanet, C. Giant viruses, giant chimeras: The multiple evolutionary histories of mimivirus genes. *BMC Evol. Biol.* **2008**, *8*, 12. [CrossRef] [PubMed]

46. Khan, M.; La Scola, B.; Lepidi, H.; Raoult, D. Pneumonia in mice inoculated experimentally with *Acanthamoeba polyphaga* mimivirus. *Microb. Pathog.* **2007**, *42*, 56–61. [CrossRef] [PubMed]

47. Correa, A.M.S.; Ainsworth, T.D.; Rosales, S.M.; Thurber, A.R.; Butler, C.R.; Vega Thurber, R.L. Viral outbreak in corals associated with an in situ bleaching event: Atypical herpes-like viruses and a new megavirus infecting symbiodinium. *Front. Microbiol.* **2016**, *7*, 127. [CrossRef] [PubMed]

48. Popgeorgiev, N.; Boyer, M.; Fancello, L.; Monteil, S.; Robert, C.; Rivet, R.; Nappez, C.; Azza, S.; Chiaroni, J.; Raoult, D.; et al. Marseillevirus-like virus recovered from blood donated by asymptomatic humans. *J. Infect. Dis.* **2013**, *208*, 1042–1050. [CrossRef] [PubMed]

49. Yolken, R.H.; Jones-Brando, L.; Dunigan, D.D.; Kannan, G.; Dickerson, F.; Severance, E.; Sabunciyan, S.; Talbot, C.C.; Prandovszky, E.; Gurnon, J.R.; et al. Chlorovirus ATCV-1 is part of the human oropharyngeal virome and is associated with changes in cognitive functions in humans and mice. *Proc. Natl. Acad. Sci. USA* **2014**, *111*, 16106–16111. [CrossRef] [PubMed]

50. Iyer, L.M.; Aravind, L.; Koonin, E.V. Common origin of four diverse families of large eukaryotic DNA viruses. *J. Virol.* **2001**, *75*, 11720–11734. [CrossRef] [PubMed]

51. Van Etten, J.L.; Meints, R.H. Giant viruses infecting algae. *Annu. Rev. Microbiol.* **1999**, *53*, 447–494. [CrossRef] [PubMed]

52. Garcia-Beato, R.; Salas, M.L.; Vinuela, E.; Salas, J. Role of the host cell nucleus in the replication of african swine fever virus DNA. *Virology* **1992**, *188*, 637–649. [CrossRef]

53. Goorha, R. Frog virus-3 DNA-replication occurs in 2 stages. *J. Virol.* **1982**, *43*, 519–528. [PubMed]

54. Moss, B. *Poxviridae: The Viruses and Their Replication*; Lippincott-Raven Publishers: Philadelphia, PA, USA, 1996; pp. 1163–1197.

55. Koonin, E.V.; Yutin, N. Origin and evolution of eukaryotic large nucleo-cytoplasmic DNA viruses. *Intervirology* **2010**, *53*, 284–292. [CrossRef] [PubMed]

56. Elde, N.C.; Child, S.J.; Eickbush, M.T.; Kitzman, J.O.; Rogers, K.S.; Shendure, J.; Geballe, A.P.; Malik, H.S. Poxviruses deploy genomic accordions to adapt rapidly against host antiviral defenses. *Cell* **2012**, *150*, 831–841. [CrossRef] [PubMed]

57. Filee, J. Route of NCLDV evolution: The genoic accordion. *Curr. Opin. Virol.* **2013**, *3*, 595–599. [CrossRef] [PubMed]

58. Desnues, C.; La Scola, B.; Yutin, N.; Fournous, G.; Robert, C.; Azza, S.; Jardot, P.; Monteil, S.; Campocasso, A.; Koonin, E.V.; et al. Provirophages and transpovirons as the diverse mobilome of giant viruses. *Proc. Natl. Acad. Sci. USA* **2012**, *109*, 18078–18083. [CrossRef] [PubMed]

59. La Scola, B.; Desnues, C.; Pagnier, I.; Robert, C.; Barrassi, L.; Fournous, G.; Merchat, M.; Suzan-Monti, M.; Forterre, P.; Koonin, E.; et al. The virophage as a unique parasite of the giant mimivirus. *Nature* **2008**, *455*, 100–104. [CrossRef] [PubMed]

60. Yutin, N.; Wolf, Y.I.; Raoult, D.; Koonin, E.V. Eukaryotic large nucleo-cytoplasmic DNA viruses: Clusters of orthologous genes and reconstruction of viral genome evolution. *Virol. J.* **2009**, *6*, 13. [CrossRef] [PubMed]

61. Claverie, J.M.; Abergel, C. Mimivirus and its virophage. In *Annual Review of Genetics*; Annual Reviews: Palo Alto, CA, USA, 2009; Volume 43, pp. 49–66.

62. Aherfi, S.; La Scola, B.; Pagnier, I.; Raoult, D.; Colson, P. The expanding family *Marseilleviridae*. *Virology* **2014**, *466–467*, 27–37. [CrossRef] [PubMed]

63. Colson, P.; de Lamballerie, X.; Fournous, G.; Raoult, D. Reclassification of giant viruses composing a fourth domain of life in the new order megavirales. *Intervirology* **2012**, *55*, 321–332. [CrossRef] [PubMed]

64. Colson, P.; De Lamballerie, X.; Yutin, N.; Asgari, S.; Bigot, Y.; Bideshi, D.K.; Cheng, X.W.; Federici, B.A.; Van Etten, J.L.; Koonin, E.V.; et al. "Megavirales", a proposed new order for eukaryotic nucleocytoplasmic large DNA viruses. *Arch. Virol.* **2013**, *158*, 2517–2521. [CrossRef] [PubMed]

65. Krupovic, M.; Bamford, D.H. Virus evolution: How far does the double beta-barrel viral lineage extend? *Nat. Rev. Microbiol.* **2008**, *6*, 941–948. [CrossRef] [PubMed]

66. Condit, R.C. Vaccinia, Inc.—Probing the functional substructure of poxviral replication factories. *Cell Host Microbe* **2007**, *2*, 205–207. [CrossRef] [PubMed]

67. Netherton, C.; Moffat, K.; Brooks, E.; Wileman, T. A guide to viral inclusions, membrane rearrangements, factories, and viroplasm produced during virus replication. *Adv. Virus Res.* **2007**, *70*, 101–182. [PubMed]

68. Mutsafi, Y.; Zauberman, N.; Sabanay, I.; Minsky, A. Vaccinia-like cytoplasmic replication of the giant mimivirus. *Proc. Natl. Acad. Sci. USA* **2010**, *107*, 5978–5982. [CrossRef] [PubMed]

69. Boyer, M.; Yutin, N.; Pagnier, I.; Barrassi, L.; Fournous, G.; Espinosa, L.; Robert, C.; Azza, S.; Sun, S.Y.; Rossmann, M.G.; et al. Giant marseillevirus highlights the role of amoebae as a melting pot in emergence of chimeric microorganisms. *Proc. Natl. Acad. Sci. USA* **2009**, *106*, 21848–21853. [CrossRef] [PubMed]

70. Netherton, C.L.; Wileman, T. Virus factories, double membrane vesicles and viroplasm generated in animal cells. *Curr. Opin. Virol.* **2011**, *1*, 381–387. [CrossRef] [PubMed]

71. Moniruzzaman, M.; LeCleir, G.R.; Brown, C.M.; Gobler, C.J.; Bidle, K.D.; Wilson, W.H.; Wilhelm, S.W. Genome of the brown tide virus (AaV), the little giant of the megaviridae, elucidates NCLDV genome expansion and host-virus coevolution. *Virology* **2014**, *466–467*, 60–70. [CrossRef] [PubMed]

72. Yutin, N.; Wolf, Y.I.; Koonin, E.V. Origin of giant viruses from smaller DNA viruses not from a fourth domain of cellular life. *Virology* **2014**, *466–467*, 38–52. [CrossRef] [PubMed]

73. Moniruzzaman, M.; Gann, E.R.; LeCleir, G.R.; Kang, Y.; Gobler, C.J.; Wilhelm, S.W. Diversity and dynamics of algal megaviridae members during a harmful brown tide caused by the pelagophyte, *Aureococcus anophagefferens*. *FEMS Microbiol. Ecol.* **2016**, *92*, fiw058. [CrossRef] [PubMed]

74. Yutin, N.; Koonin, E.V. Pandoraviruses are highly derived phycodnaviruses. *Biol. Direct* **2013**, *8*, 8. [CrossRef] [PubMed]

75. Lwoff, A. The concept of a virus. *J. Gen. Microbiol.* **1957**, *17*, 239–253. [CrossRef] [PubMed]

76. Twort, F.W. An investigation on the nature of ultra-microscopic viruses. *Lancet* **1915**, *2*, 1241–1243. [CrossRef]

77. D'Herelle, F. Sur un microbe invisible antagonistic des bacilles dysenterique. *C. R. Acad. Sci. Paris* **1917**, *165*, 373–375.

78. Lwoff, A.; Tournier, P. Classification of viruses. *Annu. Rev. Microbiol.* **1966**, *20*, 45–74. [CrossRef] [PubMed]

79. Raoult, D.; La Scola, B.; Birtles, R. The discovery and characterization of mimivirus, the largest known virus and putative pneumonia agent. *Clin. Infect. Dis.* **2007**, *45*, 95–102. [CrossRef] [PubMed]

80. Raoult, D.; Audic, S.; Robert, C.; Abergel, C.; Renesto, P.; Ogata, H.; La Scola, B.; Suzan, M.; Claverie, J.M. The 1.2-megabase genome sequence of mimivirus. *Science* **2004**, *306*, 1344–1350. [CrossRef] [PubMed]

81. Boyer, M.; Madoui, M.A.; Gimenez, G.; La Scola, B.; Raoult, D. Phylogenetic and phyletic studies of informational genes in genomes highlight existence of a 4th domain of life including giant viruses. *PLoS ONE* **2010**, *5*, e15530. [CrossRef] [PubMed]

82. Felsenstein, J. *Inferring Phylogenies*; Sinauer Associates: Sunderland, MA, USA, 2004.

83. Williams, T.A.; Embley, T.M.; Heinz, E. Informational gene phylogenies do not support a fourth domain of life for nucleocytoplasmic large DNA viruses. *PLoS ONE* **2011**, *6*, e21080. [CrossRef] [PubMed]

84. Doutre, G.; Philippe, N.; Abergel, C.; Claverie, J.M. Genome analysis of the first *Marseilleviridae* representative from Australia indicates that most of its genes contribute to virus fitness. *J. Virol.* **2014**, *88*, 14340–14349. [CrossRef] [PubMed]

85. Chow, C.-E.T.; Winget, D.M.; White, R.A.; Hallam, S.J.; Suttle, C.A. Combining genomic sequencing methods to explore viral diversity and reveal potential virus-host interactions. *Front. Microbiol.* **2015**, *6*, 265. [CrossRef] [PubMed]

86. La Scola, B.; Audic, S.; Robert, C.; Jungang, L.; de Lamballerie, X.; Drancourt, M.; Birtles, R.; Claverie, J.M.; Raoult, D. A giant virus in amoebae. *Science* **2003**, *299*, 2033. [CrossRef] [PubMed]

87. Abrahão, J.S.; Dornas, F.P.; Silva, L.C.; Almeida, G.M.; Boratto, P.V.; Colson, P.; La Scola, B.; Kroon, E.G. *Acanthamoeba polyphaga* mimivirus and other giant viruses: An open field to outstanding discoveries. *Virol. J.* **2014**, *11*, 120. [CrossRef] [PubMed]

88. Kim, M.S.; Park, E.J.; Roh, S.W.; Bae, J.W. Diversity and abundance of single-stranded DNA viruses in human feces. *Appl. Environ. Microbiol.* **2011**, *77*, 8062–8070. [CrossRef] [PubMed]

89. DeLong, J.P.; Al-Ameeli, Z.; Duncan, G.; Van Etten, J.L.; Dunigan, D.D. Predators catalyze an increase in chloroviruses by foraging on the symbiotic hosts of zoochlorellae. *Proc. Natl. Acad. Sci. USA* **2016**, *113*, 13780–13784. [CrossRef] [PubMed]

90. Kristensen, D.M.; Mushegian, A.R.; Dolja, V.V.; Koonin, E.V. New dimensions of the virus world discovered through metagenomics. *Trends Microbiol.* **2010**, *18*, 11–19. [CrossRef] [PubMed]

91. Williamson, S.J.; Allen, L.Z.; Lorenzi, H.A.; Fadrosh, D.W.; Brami, D.; Thiagarajan, M.; McCrow, J.P.; Tovchigrechko, A.; Yooseph, S.; Venter, J.C. Metagenomic exploration of viruses throughout the Indian Ocean. *PLoS ONE* **2012**, *7*, e42047. [CrossRef] [PubMed]

92. Khalil, J.Y.B.; Langlois, T.; Andreani, J.; Sorraing, J.-M.; Raoult, D.; Carmoin, L.; La Scola, B. Flow cytometry sorting to separate viable giant viruses from amoeba co-culture supernatants. *Front. Cell. Infect. Microbiol.* **2016**, *6*, 202. [CrossRef] [PubMed]

93. Martinez Martinez, J.; Poulton, N.J.; Stepanauskas, R.; Sieracki, M.E.; Wilson, W.H. Targeted sorting of single virus-infected cells of the coccolithophore *Emiliania huxleyi*. *PLoS ONE* **2011**, *6*, e22520. [CrossRef] [PubMed]

94. Zablocki, O.; van Zyl, L.; Adriaenssens, E.M.; Rubagotti, E.; Tuffin, M.; Cary, S.C.; Cowana, D. High-level diversity of tailed phages, eukaryote-associated viruses, and virophage-like elements in the metaviromes of Antarctic soils. *Appl. Environ. Microbiol.* **2014**, *80*, 10. [CrossRef] [PubMed]

95. Correa, A.M.; Welsh, R.M.; Vega Thurber, R.L. Unique nucleocytoplasmic dsDNA and +ssRNA viruses are associated with the dinoflagellate endosymbionts of corals. *ISME J.* **2013**, *7*, 13–27. [CrossRef] [PubMed]

96. Lysholm, F.; Wetterbom, A.; Lindau, C.; Darban, H.; Bjerkner, A.; Fahlander, K.; Lindberg, A.M.; Persson, B.; Allander, T.; Andersson, B. Characterization of the viral microbiome in patients with severe lower respiratory tract infections, using metagenomic sequencing. *PLoS ONE* **2012**, *7*, e30875. [CrossRef] [PubMed]

97. Brussaard, C.P.D.; Kuipers, B.; Veldhuis, M.J.W. A mesocosm study of *Phaeocystis globosa* population dynamics. *Harmful Algae* **2005**, *4*, 859–874. [CrossRef]

98. Castberg, T.; Thyrhaug, R.; Larsen, A.; Sandaa, R.-A.; Heldal, M.; Van Etten, J.L.; Bratbak, G. Isolation and characterization of a virus that infects *Emiliania huxleyi* (Haptophyta). *J. Phycol.* **2002**, *38*, 767–774. [CrossRef]

99. Gobler, C.J.; Anderson, O.R.; Gastrich, M.D.; Wilhelm, S.W. Ecological aspects of viral infection and lysis in the harmful brown tide alga *Aureococcus anophagefferens*. *Aquat. Microb. Ecol.* **2007**, *47*, 25–36. [CrossRef]

100. Nagasaki, K.; Tarutani, K.; Yamaguchi, M. Growth characteristics of *Heterosigma akashiwo* virus and its possible use as a microbiological agent for red tide control. *Appl. Environ. Microbiol.* **1999**, *65*, 898–902. [PubMed]

101. Brussaard, C.P.D.; Gast, G.J.; van Duyl, F.C.; Riegmen, R. Impact of phytoplankton bloom magnitude on a pelagic microbial food web. *Mar. Ecol. Prog. Ser.* **1996**, *144*, 211–221. [CrossRef]

102. Weitz, J.S.; Wilhelm, S.W. Ocean viruses and their effects on microbial communities and biogeochemical cycles. *F1000 Biol. Rep.* **2012**, *4*, 17. [CrossRef] [PubMed]

103. Wilhelm, S.W.; Suttle, C.A. Viruses and nutrient cycles in the sea. *Bioscience* **1999**, *49*, 781–788. [CrossRef]

104. Nagasaki, K.; Bratbak, G. Isolation of viruses infecting photosynthetic and nonphotosynthetic protists. In *Manual of Aquatic Viral Ecology*; Wilhelm, S.W., Weinbauer, M.G., Suttle, C.A., Eds.; ASLO: Waco, TX, USA, 2010; pp. 92–101.

105. Brodie, J.; Zuccarello, G.C. Systematics of the species rich algae: Red algal classificiation, phylogeny and speciation. In *Reconstructing the Tree of Life. Taxonomy and Systematics of Species Rich Taxa*; Hodkinson, T.R., Parnell, J.A.N., Eds.; CRC Press: New York, NY, USA, 2006; pp. 323–336.

106. Short, S.M. The ecology of viruses that infect eukaryotic algae. *Environ. Microbiol.* **2012**, *14*, 2253–2271. [CrossRef] [PubMed]

107. Johannessen, T.V.; Bratbak, G.; Larsen, A.; Ogata, H.; Egge, E.S.; Edvardsen, B.; Eikrem, W.; Sandaa, R.A. Characterisation of three novel giant viruses reveals huge diversity among viruses infecting Prymnesiales (Haptophyta). *Virology* **2015**, *476*, 180–188. [CrossRef] [PubMed]

108. Wilhelm, S.W.; Coy, S.R.; Gann, E.R.; Moniruzzaman, M.; Stough, J.M.A. Standing on the shoulders of giant viruses: 5 lessons learned about large viruses infecting small eukaryotes and the opportunities they create. *PLoS Pathog.* **2016**, *12*, e1005752. [CrossRef] [PubMed]

109. Moniruzzaman, M.; Wurch, L.L.; Alexander, H.; Dyhrman, S.T.; Gobler, C.J.; Wilhelm, S.W. Virus-host infection dynamics of marine single-celled eukaryotes resolved from metatranscriptomics. *bioRxiv* **2016**. [CrossRef]

110. Van Etten, J.L.; Lane, L.C.; Meints, R.H. Viruses and viruslike particles of eukaryotic algae. *Microbiol. Rev.* **1991**, *55*, 586–620. [PubMed]

111. Reteno, D.G.; Benamar, S.; Kahlil, J.B.; Andreani, J.; Armstrong, N.; Klose, T.; Rossmann, M.G.; Colson, P.; Raoult, D.; La Scola, B. Faustovirus, an asfarvirus-related new lineage of giant viruses infecting amoebae. *J. Virol.* **2015**, *89*, 6585–6594. [CrossRef] [PubMed]

112. Fischer, M.G.; Allen, M.J.; Wilson, W.H.; Suttle, C.A. Giant virus with a remarkable complement of genes infects marine zooplankton. *Proc. Natl. Acad. Sci. USA* **2010**, *107*, 19508–19513. [CrossRef] [PubMed]

113. Schnepf, E.; Soeder, C.J.; Hegewald, E. Polyhedral virus-like particles lysing the aquatic phycomycete *Aphedlidium* sp., a parasite of the grean algae *Scenedesmus armatus*. *Virology* **1970**, *42*, 482–487. [CrossRef]

114. Pickett-Heaps, J.D. A possible virus infection in the green alga *Oedogonium*. *J. Phycol.* **1972**, *8*, 44–47. [CrossRef]
115. Toth, R.; Wilce, R.T. Viruslike particles in the marine alga *Chorda tomentosa* lyngye (Phaeophyceae). *J. Phycol.* **1972**, *8*, 126–130.
116. Baker, J.R.J.; Evans, L.V. The ship fouling alga *Ectocarpus*. *Protoplasma* **1973**, *77*, 1–13. [CrossRef]
117. Clitheroe, S.B.; Evans, L.V. Virus like particles in the brown algae *Ectocarpus*. *J. Ultrastruct. Res.* **1974**, *49*, 211–217. [CrossRef]
118. Swale, E.M.F.; Belcher, J.H. A light and electron microscope study of the colourless flagellate *Aulacomonas* skuja. *Arch. Microbiol.* **1973**, *92*, 91–103. [CrossRef]
119. Markey, D.R. A possivle virus infection in the brown alga *Phlaiella littoralis*. *Protoplasma* **1974**, *80*, 223–232. [CrossRef] [PubMed]
120. Moestrup, O.; Thomsen, H.A. An ultrastructural study of the flagellate *Phyramimonas orientalis* with particular emphasis on golgi apparatus activity and the flagellar apparatus. *Protoplasma* **1974**, *81*, 247–269. [CrossRef]
121. Gibbs, A.; Skotnicki, A.H.; Gardiner, J.E.; Walker, E.S.; Hollings, M. A tabamovirus of a green alga. *Virology* **1975**, *64*, 571–574. [CrossRef]
122. Oliveira, L.; Bisalputra, T. A virus infection in the brown alga *Sorocarpus uvaeformis* (Lyngbye) pringsheim (Phaeophyta, Ecocarpales). *Ann. Bot.* **1978**, *42*, 439–445. [CrossRef]
123. Sicko-Goad, L.; Walker, G. Viroplasm and large virus-like particles in the dinoflagellate *Gymnodiunium uberrimum*. *Protoplasma* **1979**, *99*, 203–210. [CrossRef]
124. Dodds, J.A.; Cole, A. Microscopy and biology of *Uronema gigas*, a filamentous eucaryotic green alga, and its associated tailed virus-like particle. *Virology* **1980**, *100*, 156–165. [CrossRef]
125. Preisig, H.R.; Hibberd, D.J. Virus-like particles and endophytic bacteria in *Paraphysomonas* and *Chromophysomonas* (Chrysophyceae). *Nord. J. Bot.* **1984**, *4*, 279–285. [CrossRef]
126. Gowing, M.M. Large virus-like particles from vacuoles of phaeodarian radiolarians and from other marine samples. *Mar. Ecol. Prog. Ser.* **1993**, *101*, 33–43. [CrossRef]
127. Smayda, T.J. Complexity in the eutrophication-harmful algal bloom relationship, with comment on the important of grazing. *Harmful Algae* **2008**, *8*, 140–151. [CrossRef]
128. Filee, J.; Pouget, N.; Chandler, M. Phylogenetic evidence for extensive lateral acquisition of cellular genes by nucleocytoplasmic large DNA viruses. *BMC Evol. Biol.* **2008**, *8*, 320. [CrossRef] [PubMed]
129. Monier, A.; Pagarete, A.; de Vargas, C.; Allen, M.J.; Read, B.; Claverie, J.-M.; Ogata, H. Horizontal gene transfer of an entire metabolic pathway between a eukaryotic alga and its DNA virus. *Genome Res.* **2009**, *19*, 1441–1449. [CrossRef] [PubMed]
130. Rosenwasser, S.; Mausz, M.A.; Schatz, D.; Sheyn, U.; Malitsky, S.; Aharoni, A.; Weinstock, E.; Tzfadia, O.; Ben-Dor, S.; Feldmesser, E.; et al. Rewiring host lipid metabolism by large viruses determines the fate of *Emiliania huxleyi*, a bloom-forming alga in the ocean. *Plant Cell Online* **2014**, *26*, 2689–2707. [CrossRef] [PubMed]
131. Vardi, A.; Van Mooy, B.A.S.; Fredricks, H.F.; Popendorf, K.J.; Ossolinski, J.E.; Haramaty, L.; Bidle, K.D. Viral glycosphingolipids induce lytic infection and cell death in marine phytoplankton. *Science* **2009**, *326*, 861–865. [CrossRef] [PubMed]
132. Filee, J.; Siguier, P.; Chandler, M. I am what I eat and I eat what I am: Acquisition of bacterial genes by giant viruses. *Trends Genet.* **2007**, *23*, 10–15. [CrossRef] [PubMed]
133. Filee, J. Multiple occurrences of giant virus core genes acquired by eukaryotic genomes: The visible part of the iceberg? *Virology* **2014**, *466–467*, 53–59. [CrossRef] [PubMed]
134. Blanc, G.; Duncan, G.; Agarkova, I.; Borodovsky, M.; Gurnon, J.; Kuo, A.; Lindquist, E.; Lucas, S.; Pangilinan, J.; Polle, J.; et al. The *Chlorella variabilis* NC64A genome reveals adaptation to photosymbiosis, coevolution with viruses, and cryptic sex. *Plant Cell* **2010**, *22*, 2943–2955. [CrossRef] [PubMed]
135. Blanc, G.; Gallot-Lavallee, L.; Maumus, F. Provirophages in the bigelowiella genome bear testimony to past encounters with giant viruses. *Proc. Natl. Acad. Sci. USA* **2015**, *112*, E5318–5326. [CrossRef] [PubMed]
136. Read, B.A.; Kegel, J.; Klute, M.J.; Kuo, A.; Lefebvre, S.C.; Maumus, F.; Mayer, C.; Miller, J.; Monier, A.; Salamov, A.; et al. Pan genome of the phytoplankton *Emiliania* underpins its global distribution. *Nature* **2013**, *499*, 209–213. [CrossRef] [PubMed]
137. Maumus, F.; Epert, A.; Nogué, F.; Blanc, G. Plant genomes enclose footprints of past infections by giant virus relatives. *Nat. Commun.* **2014**, *5*, 4268. [CrossRef] [PubMed]

138. Breitbart, M. Marine viruses: Truth or dare. *Ann Rev Mar Sci* **2012**, *4*, 425–448. [CrossRef] [PubMed]

139. Frada, M.; Probert, I.; Allen, M.J.; Wilson, W.H.; de Vargas, C. The "cheshire cat" escape strategy of the coccolithophore *Emiliania huxleyi* in response to viral infection. *Proc. Natl. Acad. Sci. USA* **2008**, *105*, 15944–15949. [CrossRef] [PubMed]

140. Yau, S.; Lauro, F.M.; DeMaere, M.Z.; Brown, M.V.; Thomas, T.; Raftery, M.J.; Andrews-Pfannkoch, C.; Lewis, M.; Hoffman, J.M.; Gibson, J.A.; et al. Virophage control of antarctic algal host-virus dynamics. *Proc. Natl. Acad. Sci. USA* **2011**, *108*, 6163–6168. [CrossRef] [PubMed]

141. Campos, R.K.; Boratto, P.V.; Assis, F.L.; Aguiar, E.R.; Silva, L.C.; Albarnaz, J.D.; Dornas, F.P.; Trindade, G.S.; Ferreira, P.P.; Marques, J.T.; et al. Samba virus: A novel mimivirus from a giant rain forest, the Brazilian amazon. *Virol. J.* **2014**, *11*, 95. [CrossRef] [PubMed]

142. Fischer, M.G.; Suttle, C.A. A virophage at the origin of large DNA transposons. *Science* **2011**, *332*, 231–234. [CrossRef] [PubMed]

143. Gaia, M.; Benamar, S.; Boughalmi, M.; Pagnier, I.; Croce, O.; Colson, P.; Raoult, D.; La Scola, B. Zamilon, a novel virophage with *Mimiviridae* host specificity. *PLoS ONE* **2014**, *9*, e94923. [CrossRef] [PubMed]

144. Levasseur, A.; Bekliz, M.; Chabrière, E.; Pontarotti, P.; La Scola, B.; Raoult, D. Mimivire is a defence system in mimivirus that confers resistance to virophage. *Nature* **2016**, *531*, 249–252. [CrossRef] [PubMed]

145. Santini, S.; Jeudy, S.; Bartoli, J.; Poirot, O.; Lescot, M.; Abergel, C.; Barbe, V.; Wommack, K.E.; Noordeloos, A.A.M.; Brussaard, C.P.D.; et al. Genome of *Phaeocystis globosa* virus PgV-16t highlights the common ancestry of the largest known DNA viruses infecting eukaryotes. *Proc. Natl. Acad. Sci. USA* **2013**, *110*, 10800–10805. [CrossRef] [PubMed]

146. Claverie, J.M.; Abergel, C. CRISPR-CAS-like system in giant viruses: Why mimivire is not likely to be an adaptive immune system. *Virol. Sin.* **2016**, *31*, 193–196. [CrossRef] [PubMed]

147. Sullivan, M.B.; Weitz, J.S.; Wilhelm, S.W. Viral ecology comes of age. *Environ. Microbiol. Repo.* **2017**, *9*, 33–35. [CrossRef] [PubMed]

MDPI AG

St. Alban-Anlage 66

4052 Basel, Switzerland

Tel. +41 61 683 77 34

Fax +41 61 302 89 18

http://www.mdpi.com

Viruses Editorial Office

E-mail: viruses@mdpi.com

http://www.mdpi.com/journal/viruses

www.ingramcontent.com/pod-product-compliance
Lightning Source LLC
Chambersburg PA
CBHW051715210326
41597CB00032B/5484